Introduction to Biosystems Engineering

Introduction to Biosystems Engineering

Edited by

Nicholas M. Holden, Mary Leigh Wolfe, Jactone A. Ogejo, and Enda J. Cummins

A free version is available online at: https://doi.org/10.21061/IntroBiosystemsEngineering

American Society of Agricultural and Biological Engineers and Virginia Tech Publishing
St Joseph, MI and Blacksburg, VA, USA

Copyright © 2020 American Society of Agricultural and Biological Engineers (collection)

Authors retain copyright of their individual contributions.

This work is licensed with a Creative Commons Attribution 4.0 license: https://creativecommons.org/licenses/by/4.0. For more information, see p. page xv.

The work is published jointly by the American Society of Agricultural and Biological Engineers (ASABE) and Virginia Tech Publishing.

ASABE, 2950 Niles Rd., St. Joseph, MI 49085-9659 USA.
phone: 269-429-0300 fax: 269-429-3852 e-mail: hq@asabe.org
www.asabe.org

Virginia Tech Publishing, University Libraries at Virginia Tech, 560 Drillfield Drive,
Blacksburg, VA 24061, USA. publishing@vt.edu
publishing.vt.edu

Virginia Tech is a member of the Open Education Network.

Suggested citation: Holden, N. M., Wolfe, M. L., Ogejo, J. A. & Commins, E. J. (Eds.) (2020) *Introduction to Biosystems Engineering*. ASABE in association with Virginia Tech Publishing, Blacksburg, Virginia.
Available from https://doi.org/10.21061/IntroBiosystemsEngineering

Use of ASABE and Virginia Tech Publishing logos is authorized for this edition only.

ASABE is an educational and scientific organization dedicated to the advancement of engineering applicable to agricultural, food, and biological systems.

ASABE is not responsible for statements and opinions advanced in its meetings or printed in its publications. They represent the views of the individual to whom they are credited and are not binding on the Society as a whole.

Peer Review: This book has undergone editorial peer review.

Accessibility Statement: Virginia Tech Publishing is committed to making its publications accessible in accordance with the Americans with Disabilities Act of 1990. The screen reader–friendly ePub version of this book is tagged structurally and includes alternative text which allows for machine-readability.

Publication Cataloging Information
International Standard Book Number (ISBN) (PDF): 978-1-949373-97-4
International Standard Book Number (ISBN) (Print): 978-1-949373-93-6
Digital Object Identifier (DOI): https://doi.org/10.21061/IntroBiosystemsEngineering

Cover images:
Harvest [Public Domain] https://unsplash.com/photos/l_5MJnbrmrs
Crossing [Public Domain] Ricardo Gomez Angel https://unsplash.com/photos/TbwnUGV4kok
Sorting raspberries Frozen red raspberries in sorting and processing machines © Vladimir Nenezic, used under license from Shutterstock.com
Aerial view over Biogas plant and farm in green fields. Renewable energy from biomass. Modern agriculture in Czech Republic and European Union © Kletr, used under license from Shutterstock.com.

Copy editing by Peg McCann

Cover design: Robert Browder

Contents

Preface
Editorial Team
Features of This Open Textbook

Energy Systems

Bioenergy Conversion Systems
Biogas Energy from Organic Wastes
Biodiesel from Oils and Fats
Baling Biomass: Densification and Energy Requirements

Information Technology, Sensors, and Control Systems

Basic Microcontroller Use for Measurement and Control
Visible and Near Infrared Optical Spectroscopic Sensors for Biosystems Engineering
Data Processing in Biosystems Engineering

Machinery Systems

Traction
Crop Establishment and Protection
Grain Harvest and Handling
Mechatronics and Intelligent Systems in Agricultural Machinery

Natural Resources and Environmental Systems

Water Budgets for Sustainable Water Management
Water Quality as a Driver of Ecological System Health
Quantifying and Managing Soil Erosion on Cropland
Anaerobic Digestion of Agri-Food By-Products
Measurement of Gaseous Emissions from Animal Housing

Plant, Animal, and Facility Systems

Plant Production in Controlled Environments
Building Design for Energy Efficient Livestock Housing

Processing Systems

Freezing of Food
Principles of Thermal Processing of Packaged Foods
Deep Fat Frying of Food
Irradiation of Food
Packaging

Preface

The discipline of Biosystems Engineering emerged in the 1990s from the traditional strongholds of Agricultural Engineering and Food Engineering. Biosystems Engineering integrates engineering science and design with applied biological, environmental, and agricultural sciences. This book is targeted at 1st- and 2nd-year university-level students interested in Biosystems Engineering but not yet familiar with the breadth and depth of the subject. It is designed as a coherent educational resource, available as hard copy and for download as individual digital chapters.

The origins of the book date back to 2012, when a group of universities, led by University College Dublin (Ireland) in the European Union (EU) and Virginia Tech in the United States (US), began working on an EU-US Atlantis Programme project called *Trans-Atlantic Biosystems Engineering Curriculum and Mobility* (known as TABE.NET). One of the project activities was to explore the meaning of Biosystems Engineering. An important output of the work was the framework for an introductory course focused on Biosystems Engineering. This detailed how students might best be introduced to the subject and formed the basis for this textbook. The preparation of the textbook was supported by the American Society of Agricultural and Biological Engineers (ASABE) *Harold Pinches and Glenn Schwab Teaching Materials Fund*, intended for projects that "facilitate development or effective distribution of ASABE teaching materials, including textbooks," and the Virginia Tech University Libraries' Open Education Initiative (https://guides.lib.vt.edu/oer/grants). The writing and editing were provided as voluntary service by the authors and editors, and the ASABE funding supported the publication process so the online version of the book can be made freely available around the world (ASABE.org/BE).

The chapters are intended to stimulate interest and curiosity across the breadth of Biosystems Engineering and provide an international perspective. The goal of each chapter is to introduce the fundamental concepts needed to understand a specific topic within the discipline of Biosystems Engineering; it is not intended to provide complete coverage of the topic. The scope of each chapter is narrow enough to be addressed in one week.

All chapters follow the same structure: Introduction, Outcomes, Concepts, Applications, and Examples. Following a brief introduction, the learning outcomes state what the reader should be able to do after studying the chapter. The outcomes are consistent with expectations for a student engaging with the topic for the first time, and include being able to describe basic principles, complete fundamental calculations, and explain how the concepts are used in industry or by researchers. The concepts section covers basic principles and explains terminology, referencing commonly used sources and using internationally accepted units. Once the concepts have been described they are put into the context of applications in industry and research to help bring them to life. All chapters conclude with worked examples drawing on the concepts so that the reader can see how the information provided can be used to solve basic problems in Biosystems Engineering. The chapter structure used by this textbook could also be used to develop advanced textbooks delving deep into any topic relevant to Biosystems Engineering.

The inaugural chapters are aligned with six ASABE technical communities: Energy Systems; Information Technology, Sensors, and Control Systems; Machinery Systems; Natural Resources and Environmental Systems; Plant, Animal, and Facility Systems; and Processing Systems.

The *Energy Systems* section focuses on energy from biomass. The section places emphasis on feedstock and anaerobic digestion, but considers a wider range of technologies from a systems perspective. Specific topics addressed in this edition are densification of biomass; energy from organic wastes, fats and oils; and bioenergy systems analysis. There is scope for future chapters on other topics relevant to this community such as electrotechnology, feedstocks, and renewable power.

The *Information Technology, Sensors, and Control Systems* section focuses on processing optical sensor data and basic control. The section places emphasis on statistical methods as well as practical applications. Specific topics include use of optical sensors, multivariate data processing, and microcontrollers. Future chapters in this area could focus on topics such as automation, biosensors, robotics, sensors, and wireless technology.

The *Machinery Systems* section features agricultural technologies used for field crops, placing emphasis on the fundamental principles of equipment design. Current chapters focus on traction and mechatronics, which are topics that are relevant to multiple machinery systems, and on machinery systems for crop establishment and grain harvesting. Topics for future chapters could include automation, hydraulics, ISO bus, precision agriculture, and other machinery systems.

The *Natural Resources and Environmental Systems* section highlights the management, protection, and improvement of environmental resources including soil, water, and air. This section places emphasis on field, laboratory, and modeling studies related to environmental systems. Specific topics addressed are the measurement of gaseous emissions, water budgeting, water quality, soil erosion, and management of agri-food by-products using anaerobic digestion. Examples of technical areas for future chapters relevant to this community are drainage, irrigation, soil and water remediation, wetlands restoration, and sustainable land management.

The *Plant, Animal, and Facility Systems* section focuses on indoor plant and animal production. The current chapters emphasize mass and heat transfer for design and operation of agricultural buildings. Specific topics include plant production in controlled environments and energy efficiency for livestock housing. Future chapters relevant to this community could include topics such as grain handling, design of animal production structures, milk handling, manure management, feed storage and management, and aquaculture.

The *Processing Systems* section focuses on the safe processing and distribution of foods. This section places emphasis on heat and mass transfer and delivery of safe food to consumers. Specific topics addressed are food packaging, frying, and preservation by freezing, thermal processing, and irradiation. There is scope for future chapters on other topics relevant to this community such as food engineering, bioprocessing, bioconversion, drying, and unit operations.

While each chapter is placed in a single section, many of the topics are relevant to more than one section, or technical community. The "Anaerobic Digestion of Agri-Food By-Products" chapter in the *Natural Resources and Environmental Systems* section nicely illustrates the overlap between technical communities. Anaerobic digestion is both an environmental technology for the management of wastes and an energy technology for the provision of renewable energy. The chapters in the *Plant, Animal, and Facility Systems* section also overlap with the *Energy Systems* technical community because of the role of energy, or heat transfer, management in indoor plant and livestock production. The mechatronics chapter in *Machinery Systems* overlaps with the *Information Technology, Sensors, and Control Systems* community.

There is scope for contribution of introductory level chapters from the other ASABE technical communities, *Applied Science and Engineering* (e.g., forest engineering, fermentation, engineering and biological fundamentals), *Education, Outreach, and Professional Development* (e.g., ethics and professional conduct), and *Ergonomics, Safety, and Health* (e.g., vibration, farm safety, ergonomic design, health and safety training) for the ongoing development of the ASABE digital education resources.

Our ambition is for this textbook to continually evolve, with the addition of new online chapters every year and periodic publication of hard copy volumes. Each new chapter will follow the standard structure described above and focus on a specific topic. We believe that in time the textbook will provide a foundational resource used all over the world by students learning about Biosystems Engineering for the first time.

In parallel with the preparation of this book, the editors, with the support of the ASABE *Initiative Fund*, have been developing the *Biosystems Engineering Digital Library (BEDL)*. This resource will support instructors by providing additional teaching and learning materials for use in the classroom and for assignments. While the library will not be limited to the scope of this book, from the outset it will be used to support both instructors and students who use this book. We believe this textbook combined with the BEDL will provide a global digital teaching resource for Biosystems Engineering for many years (ASABE.org/BE).

Editorial Team

Editors-in-Chief

The editors-in-chief have overall responsibility for the book, including approval of content and editing. They established the team of technical community editors (listed below).

Nicholas M. Holden is Professor of Biosystems Engineering and Head of Teaching and Learning in the School of Biosystems and Food Engineering at University College Dublin, where he has worked for the last 25 years. His research is focused on the environmental impact and sustainability of agriculture, and food systems. He teaches life cycle assessment, precision agriculture, and green technology project modules and is the Programme Director of the BAgrSc Agricultural Systems Technology programme. He has been an ASABE member for over 20 years.

Mary Leigh Wolfe is Professor in the Department of Biological Systems Engineering (BSE) at Virginia Tech. After serving on the faculty at Texas A&M University for over six years, she moved to Virginia Tech in 1992. Recently, she served as head of the BSE department for over eight years. Her research and teaching has focused on hydrologic modeling, nonpoint source (NPS) pollution control strategies, and decision support tools for NPS pollution control and watershed management. She has also conducted research related to engineering education. She is a Fellow, past president, and life member of ASABE.

Jactone A. Ogejo is an Associate Professor in the Department of Biological Systems Engineering (BSE) at Virginia Tech. His research and extension programs focus on improving the management and use of bioresidues from production agriculture and food processing. His work encompasses recovering value-added products from bioresidues, agricultural air quality, and, more importantly, advancing knowledge to increase the acceptance and adoption of technology for manure management on animal production farms. He has been an ASABE member since 1992.

Enda J. Cummins is a Professor and Head of Research, Innovation and Impact in the School of Biosystems and Food Engineering at University College Dublin. His main research area is food safety, risk assessment, and predictive modelling, with a particular focus on implications for human health and environmental contamination. He teaches quantitative risk assessment, food physics, and research and teaching methods. He is Programme Director for the Masters of Engineering Science in Food Engineering at UCD. He has been an ASABE member since 2002.

Chapter Editors

The chapters were developed with the support of:

Amy Kaleita, Department of Agricultural and Biosystems Engineering, Iowa State University, USA

Aoife Gowen, UCD School of Biosystems and Food Engineering, University College Dublin, Ireland

Colette Fagan, Department of Food and Nutritional Sciences, University of Reading, UK

Elena Castell-Perez, Department of Agricultural and Biological Engineering, Texas A&M University, USA

Fionnuala Murphy, UCD School of Biosystems and Food Engineering, University College Dublin, Ireland

John Schueller, Department of Mechanical & Aerospace Engineering, University of Florida, USA

Ning Wang, Department of Biosystems and Agricultural Engineering, Oklahoma State University, USA

Panagiotis Panagakis, Department of Natural Resources Management and Agricultural Engineering, Agricultural University of Athens, Greece

Qin Zhang, Department of Biological Systems Engineering, Washington State University, Prosser Campus, USA

Ronaldo Maghirang, Department of Agricultural and Biological Engineering, University of Illinois at Urbana-Champaign, USA

Ruihong Zhang, Department of Biological and Agricultural Engineering, University of California—Davis, USA

Tom Curran, UCD School of Biosystems and Food Engineering, University College Dublin, Ireland

Contributors

Bedoić, Robert, University of Zagreb, Croatia

Both, A.J, Rutgers University, USA

Calvet, Salvador, Universitat Politècnica de València, Spain

Capareda, Sergio, Texas A&M University, USA

Castell-Perez, M. Elena, Texas A&M University, USA

Ćosić, Boris, University of Zagreb, Croatia

Costantino, Andrea, Politecnico di Torino, Italy

Duić, Neven, University of Zagreb, Croatia

El Mashad, Hamed, University of California—Davis, USA, Mansoura University, Egypt

Fabrizio, Enrico, Politecnico di Torino, Italy
Feng, Yao-Ze, Huazhong Agricultural University, China
Gates, Richard, Iowa State University, USA
Gorretta, Nathalie, University of Montpellier, France
Gowen, Aoife A, University College Dublin, Ireland
Hassouna, Mélynda, French National Institute for Agricultural Research, France
Hayes, Enda, University of the West of England, UK
He, Brian, University of Idaho, USA
Hutchinson, Stacy L, Kansas State University, USA
Krometis, Leigh-Anne H, Virginia Tech, USA
Moreira, Rosana G, Texas A&M University, USA
Morris, Scott A, University of Illinois at Urbana-Champaign, USA
Nuñez, Helena, Universidad Técnica Federico Santa María, Chile
Oberti, Roberto, University of Milano, Italy
Pryor, Scott, North Dakota State University, USA
Pukšec, Tomislav, University of Zagreb, Croatia
Qiu, Guangjun, South China Agricultural University, China
Queiroz, Daniel M, Universidade Federal de Vicosa, Brazil
Ramírez, Cristian, Universidad Técnica Federico Santa María, Chile
Rovira-Más, Francisco, Universitat Politècnica de València, Spain
Saibandith, Bandhita, Kasetsart University, Thailand
Saiz-Rubio, Verónica, Universitat Politècnica de València, Spain
Schrade, Sabine, Agroscope, Switzerland
Schueller, John K, University of Florida, USA
Schulze Lammers, Peter, University of Bonn, Germany
Shelford, Timothy J, Cornell University, USA
Shi, Yeyin, University of Nebraska-Lincoln, USA
Simpson, Ricardo, Universidad Técnica Federico Santa María, Chile
Sokhansanj, Shahab, University of Saskatchewan, Canada
Stombaugh, Tim, University of Kentucky, USA
Uusi-Kämppä, Jaana, Natural Resources Institute, Finland
Wang, Ning, Oklahoma State University, USA
Yamsaengsung, Ram, Prince of Songkla University, Thailand
Zhang, Ruihong, University of California–Davis, USA
Zhang, Qin, Washington State University, USA

Publishing Team

Managing Editor: Anita Walz
Production Manager: Corinne Guimont
Production Assistant: Sarah Mease
Cover Design: Robert Browder
Copyeditor: Peg McCann

Features of This Open Textbook

Additional Resources

The following resources are available at:
http://hdl.handle.net/10919/93254

Free downloadable PDF of the book
Free downloadable chapter-level PDFs
Print edition ordering details
Link to the *Biosystems Engineering Digital Library (BEDL)*
Links to additional resources
Errata

Tell Us About Your Use—Review / Adopt / Adapt / Build upon / Share

Are you an instructor reviewing, adopting, or adapting this textbook? Please help us understand your use by completing this form: https://bit.ly/IBSE_feedback

This work is licensed with a Creative Commons Attribution 4.0 International License. You are free to copy, share, adapt, remix, transform, and build upon the material for any purpose, even commercially, as long as you follow the terms of the license: https://creativecommons.org/licenses/by/4.0/legalcode

You must:

Attribute—You must give appropriate credit, provide a link to the license, and indicate if changes were made. You may do so in any reasonable manner, but not in any way that suggests the licensor endorses you or your use.

Suggested citation: Adapted by _[your name]_ from (c) Holden, N. M., Wolfe, M. L., Ogejo, J. A., and E. J. Cummins. (2020) Introduction to Biosystems Engineering, ASABE in association with Virginia Tech Publishing. https://doi.org/

10.21061/IntroBiosystemsEngineering, CC BY 4.0, https://creativecommons.org/licenses/by/4.0

You may not:

Add any additional restrictions—You may not apply legal terms or technological measures that legally restrict others from doing anything the license permits. If adapting or building upon, you are encouraged to:

Incorporate only your own work, works with a CC BY license, or works free from copyright.

Attribute all added content. If incorporating text or figures under an informed fair use analysis, mark them as such, including the country of the fair use exemption you're claiming, and cite them.

Include a transformation summary that describes changes, additions, accessibility features, and any subsequent peer review.

Have your work peer reviewed.

Suggestions for creating and adapting:

Create and share learning tools and study aids.

Add or modify problems and examples.

Translate, transform, or build upon in other formats.

Annotate using Hypothes.is http://web.hypothes.is

Modify an Open Textbook: https://press.rebus.community/otnmodify

For further ideas, contact a member of the Open Textbook Network https://open.umn.edu/otn

Feedback

Submit suggestions and comments https://bit.ly/IBSE_feedback

Bioenergy Conversion Systems

Sergio Capareda
Biological and Agricultural Engineering Department
Texas A&M University
College Station, Texas, USA

KEY TERMS		
Biodiesel	Pyrolysis	Energy balance
Bioethanol	Gasification	Economic evaluation
Biogas	Combustion	Sustainability issues

Variables

A = annuity or payment amount per period
η_e = overall conversion efficiency
n = total number of payments or periods
P = initial principal or amount of loan
R = rate of return or discount rate
r = interest rate per period

Introduction

This chapter introduces the importance of analyzing the energy balance and the economic viability of biomass conversion systems. In principle, the energy used for biomass production, conversion, and utilization should be less than the energy content of the final product. For example, one of the largest energy components for growing biomass is fertilizer (Pimentel, 2003), so this component must be included in the energy systems analyses. This chapter also introduces some biomass conversion pathways and describes the various products and co-products of conversions, with a focus on the techno-economic indicators for assessing the feasibility of a particular conversion system. Sustainability evaluation of biomass-derived fuels, materials, and co-products includes, among others, three key components: energy balance, environmental impact, and economic benefit. This chapter focuses primarily on energy balance and economic issues influencing bioenergy systems.

> **Outcomes**
>
> After reading this chapter, you should be able to:
>
> - Differentiate several common bioenergy conversion systems and pathways and identify some issues associated with their energy balance and economic viability
> - Calculate basic mass and energy balance of a particular bioenergy conversion system
> - Calculate basic economic parameters to get an idea of overall economic viability

Concepts

The major commercial fuels used in the world today are natural gas, gasoline (petrol), aviation fuel, diesel, fuel oils, and solid fuels such as coal. These commercial fossil fuels could be replaced with biofuels and solid fuels derived from biomass by using conversion technologies. There are specific biomass resources that are well-suited to each conversion technology. For example, sugar crops (sugarcane and sweet sorghum) are good feedstock materials for the conversion of bioethanol; oil crops (soybean and canola oil) are ideal feedstock for biodiesel production; and lignocellulosic biomass (e.g., wood wastes, animal manure or grasses) is the prime substrate for making biogas. Thermal conversion systems convert all other biomass resources into valuable products.

Replacement of these primary fuels with bio-based alternatives is one way to address energy sustainability. Heat and electrical power, needed worldwide, can also be produced through the conversion of biomass through thermo-chemical conversion processes such as pyrolysis and gasification to produce synthesis gas (or also called syngas, a shorter version). Syngas can be combusted to generate heat and can be thoroughly cleaned of tar and used in an internal combustion engine to generate mechanical or electrical power. Future world requirements for other basic energy and power needs can be met using a wide range of biomass resources, including oil and sugar crops, animal manure, crop residues, municipal solid wastes (MSW), fuel wood, aquatic plants like micro-algae, and dedicated energy farming for energy production. The three primary products of thermal conversion are solid bio-char, liquid, and synthesis gas.

A biomass energy conversion system can produce one or more of four major products: heat, electricity, fuel, and raw materials. The goal of any conversion process is to achieve the highest conversion efficiency possible by minimizing losses. The energy conversion efficiency for any type of product can be calculated as:

$$\text{Energy Conversion Efficiency}(\%) = \frac{\text{Energy Output(MJ)}}{\text{Energy Input(MJ)}} \times 100\% \quad (1)$$

There are three fundamental biomass conversion pathways (figure 1): physicochemical, biological, and thermal. Physicochemical conversion is the use of chemicals or catalysts for conversion at ambient or slightly elevated temperatures. Biological is the use of specific microbes or enzymes to generate valuable

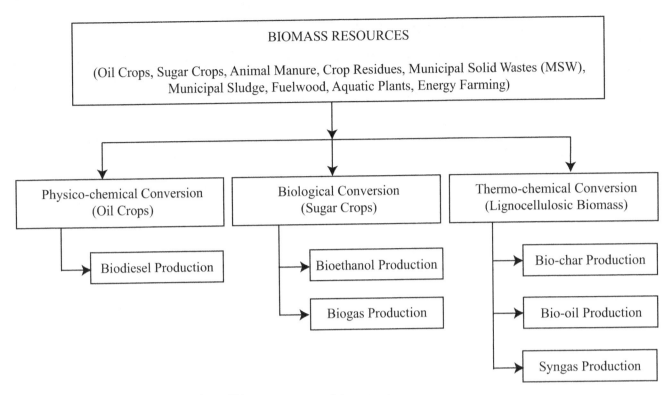

Figure 1. Pathways for the conversion of biomass resources into energy.

products. Thermo-chemical conversion occurs at elevated temperature (and sometimes pressure) for conversion. The products from biomass conversions can replace common fossil-resource-derived chemicals (e.g., lactic acid), fuel (e.g., diesel), and material (e.g., gypsum). This chapter focuses on energy derived by bioconversion.

Biodiesel Production

Refined vegetable oils and fats are converted into *biodiesel*, which is compatible with diesel fuel, by physicochemical conversion using a simple catalytic process using methanol (CH_3OH) and sodium hydroxide (NaOH) at a slightly elevated temperature. The process is called transesterification. Vegetable oils are also called triglycerides because their chemical structure is composed of a glycerol attached to three fatty acid molecules by ester bonds. When the ester bonds are broken by a catalyst, glycerin is produced and the fatty acid compound is converted into its methyl ester form, which is the technical term for biodiesel. The combination of methanol and sodium hydroxide results in a compound called sodium methoxide (CH_3ONa), which is the most common commercial catalyst for biodiesel production. The basic mass balance for the process is:

100 kg vegetable oil + 10 kg catalysts → 100 kg biodiesel + 10 kg glycerin

The energy balance depends on the specific facility design. For the biodiesel product to be considered viable, the energy in the biodiesel must exceed the

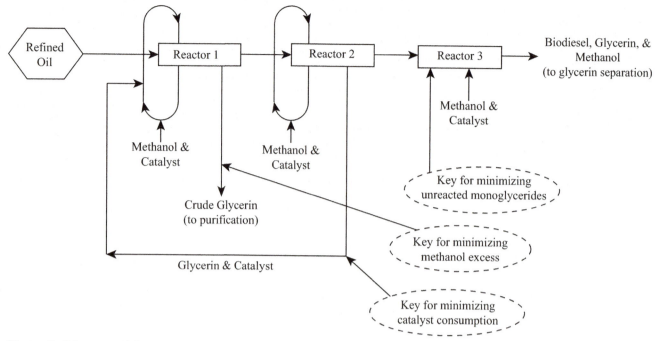

Figure 2. Schematic of the commercial process of making biodiesel fuel.

energy used to produce the vegetable oil used for the process. In a commercial system, the transesterification process is split into several stages (figure 2). Methanol and catalysts are recovered after all the stages to minimize catalyst consumption. Crude glycerin is also recovered at each stage to minimize the use of excess methanol. The remaining catalyst—the amount must be calculated accurately—is then introduced at the last stage of the process. This last stage reaction minimizes unreacted mono-glycerides (or remaining glycerol that still has a fatty acid chained to it via an ester bond). If soybean oil is used, the resulting biodiesel product is called soybean methyl ester, the most common biodiesel product in the United States. In Europe, canola (rapeseed) oil is the most common feedstock, which produces rapeseed methyl ester. The glycerin co-product is further purified to improve its commercial value.

Bioethanol Production

Bioethanol, which is compatible with gasoline or petrol, is produced from sugar, starchy, or lignocellulosic crops using microbes or enzymes. Sugars from crops are easily converted into ethanol using yeast (e.g., *Saccharomyces cerevisiae*) or other similar microbes, while starchy crops need enzymes (e.g., amylases) that convert starch to sugar, with the yeasts then acting on the sugars to produce bioethanol. Lignocellulosic crops need similar enzymes (e.g., enzymes produced by *Trichoderma reesei*) to break down cellulose into simple sugars. The basic mass balance for the conversion of plant sugars from biomass into ethanol (C_2H_6O) also yields heat:

$$C_6H_{12}O_6 + \text{yeast} \rightarrow 2C_2H_6O + 2CO_2\ (+\ \text{heat})$$

The most common feedstock for making bioethanol in the United States is dry milled corn (maize; *Zea mays*). In the process (figure 3), dry corn kernels are milled, then water is added to the powdered material while being heated (or gelatinized) in order to cook the starch and break it down using the amylase enzyme (saccharification). This process converts starch into sugars. The resulting product (mainly glucose) is then converted into bioethanol using yeasts fermentation for 3-5 days with a mass balance of:

$$2C_6H_{10}O_5 + H_2O + \text{amylase} \rightarrow C_{12}H_{22}O_{11}$$

or

$$C_{12}H_{22}O_{11} + H_2O + \text{invertase} \rightarrow 2C_6H_{12}O_6$$

In this representation, complex starch molecules are represented by repeating units of polymers of glucose [$(C_6H_{10}O_5)_n$] with n being any number of chains. The enzyme amylase reduces this polymer into simple compounds, such as sucrose ($C_{12}H_{22}O_{11}$), a disaccharide having just two molecules of glucose. Alternatively, the enzyme invertase is used to break down sucrose into glucose sugar. A yeast, such as the commercial yeast Ethanol Red (distributed by Fermentis of Lesaffre, France and sold worldwide) acts on the sugar product to convert the sugar into bioethanol. The resulting product (a broth) is called *beer* because its alcohol content is very close to 10%. The solid portion is called *distillers grain*, which is usually dried and fed to animals. The beer is distilled to yield solids (known as bottoms or still bottoms) and to recover 90-95% of the bioethanol (usually 180-190 proof), which is then purified using molecular sieves. (A molecular sieve is a crystalline substance with pores of carefully selected molecular dimensions that permit the passage of, in this case, only ethanol molecules.) The final separated and purified product may then be blended with gasoline or used alone.

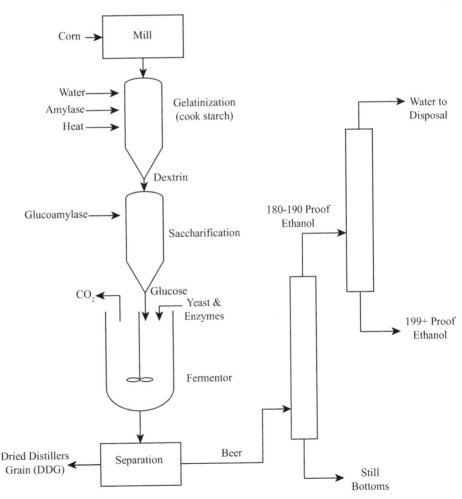

Figure 3. Schematic of the commercial process for making bioethanol via dry milling.

Biogas Production

Biogas, which is composed primarily of methane (CH_4; also called natural gas) and carbon dioxide (CO_2), is produced from lignocellulosic biomass by microbes under anaerobic conditions. Suitable microbes are commonly found in the stomachs of ruminant animals (e.g., cows). These microbes convert complex cellulosic materials into organic acids via hydrolysis or fermentation; these large organic acids are further converted into simpler organic acids (e.g., acetic acids) and hydrogen gas. Hydrogen gas and some organic acids that include CO_2 are further converted into CH_4 and CO_2 as the respiratory gases of these microbes. Biogas ($CO_2 + CH_4$) is the same as natural gas (CH_4) if the CO_2 component is removed. Natural gas is a common fuel derived by refining crude oil.

There are various designs of high-rate anaerobic digesters for biogas production (figure 4), which are commonly used in wastewater treatment plants worldwide. Simpler digesters use upflow and downflow anaerobic filters, basic fluidized beds, expanded beds, and anaerobic contact processes. One popular design from the Netherlands is the upflow anaerobic sludge blanket (or UASB) (Letingga et al., 1980). Improvements to the UASB include the anaerobic fluidized bed and expanded bed granular sludge blanket reactor designs. High-rate systems are commonly found in Europe, but there are few in the US. Most biogas plants in the US are simply covered lagoons.

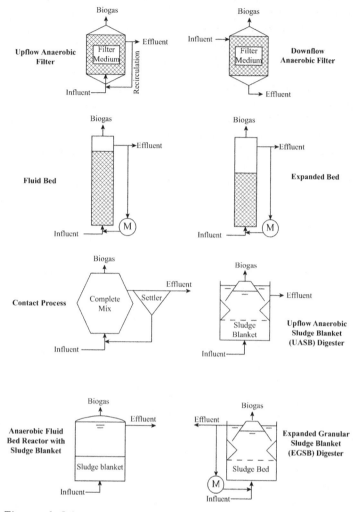

Figure 4. Schematic representation of various designs of high-rate biogas digesters.

Biomass Pyrolysis

Pyrolysis is a thermal conversion process at elevated temperatures in complete absence of oxygen or an oxidant. Figure 5 shows outputs and applications of pyrolysis. The primary products are solid bio-char, liquid, and gaseous synthesis gas. The ratios of these co-products depend on temperature, retention time, and type of biomass used. The quality and magnitude of products are also dependent on the reactor used. The simple rules of biomass pyrolysis processes are:

1. Solid bio-char (or charcoal) yield is maximized at the lowest pyrolysis temperature and the longest residence time.
2. Liquid yield is usually maximized at temperatures between 400°C and 600°C.

3. Synthesis gas, or syngas, is maximized at the highest operating temperature. The main components of syngas are carbon monoxide (CO) and hydrogen (H_2). Other component gases include lower molecular weight hydrocarbons such as CH_4, ethylene (C_2H_4), and ethane (C_2H_6).

Bio-char may be used as a soil amendment to provide carbon and nutrients when applied to agricultural land. A high-carbon bio-char may also be upgraded into activated carbon, a very high-value adsorbent material for water and wastewater treatment processes. The highest value for the bio-char is achieved when the carbon is purified of all inorganics to generate graphene products, which are among the hardest materials made from carbon.

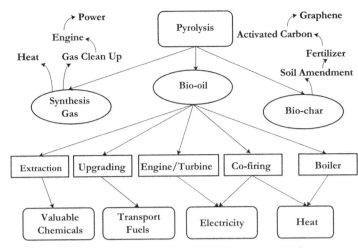

Figure 5. The outputs and applications of biomass pyrolysis.

The quality of liquid product (bio-oil) is enhanced or improved with short residence times such as those in fluidized bed pyrolysis systems but not with auger pyrolyzers. Auger pyrolyzers usually have long residence times. Short residence times give rise to less viscous bio-oil that is easy to upgrade into biofuel (gasoline or diesel) using catalysts. Bio-oil from pyrolysis process has a wide range of applications (figure 5). Valuable chemicals can be extracted; the unaltered bio-oil can be upgraded via catalytic processes to generate transport fuels; and may be co-fired in an engine to generate electricity or in a boiler to generate heat.

Syngas may simply be combusted as it is produced to generate heat. However, syngas may need to be cleaned of tar before use in an internal combustion engine. To generate electrical power, this internal combustion engine is coupled with a generator.

Biomass Gasification

Gasification is a partial thermal conversion of biomass to produce syngas. In older textbooks, this gas is also synonymously called "producer gas." Syngas can be combusted to generate heat or cleaned of tar and used in an internal combustion engine to generate electricity. The synthesis gas may also be used as feedstock to produce bio-butanol using microbes that also produce biofuel co-products. There are numerous types and designs of gasifiers, including fixed bed systems (updraft, downdraft, or cross-draft gasifiers) and moving bed systems (fluidized bed gasification systems).

A fluidized bed gasification system is shown in figure 6. Biomass is continuously fed to a large biomass bin. The fluidized bed reactor contains a bed material, usually refractory sand, to carry the heat needed for the reaction. The air-to-fuel ratio is controlled so the amount of air is below the stoichiometric requirement for combustion (i.e., combustion is incomplete) to ensure production of synthesis gas instead of heat and water vapor. The solid remaining after partial thermal conversion is high carbon bio-char that is removed via a series of cyclones. The simplest application of this system is the production of heat by combusting the synthesis gas. If electrical power is needed, then the synthesis gas must be cleaned of tar to be used in an internal combustion engine to generate electricity. The conversion efficiencies of gasification systems are typically less than 20%. An average value to use for a quick estimate of output is around 15% overall conversion efficiency.

Biomass Combustion

Direct combustion of biomass has been a traditional practice for centuries; burning wood to produce heat for cooking is an example. Combustion is the most efficient thermal conversion process for heat and power generation purposes. However, not many biomass products can be combusted because of the high ash and water content of most agricultural biomass products. The ash component can melt at higher combustion temperatures, resulting in phenomena called *slagging* and *fouling*. Melted ash forms slag that accumulates on conveying surfaces (fouls) as it cools.

Economic Evaluation of Bioenergy Systems

Commercial bioenergy facilities depreciate every year. There is no accurate estimate of depreciation values but a potential investor may use this parameter to save on capital costs each year from the proceeds of the commercial facility such that at the end of the life of the facility, the investor is prepared to invest in higher-yielding projects.

There are a number of simple methods that engineers may use for economic depreciation analyses of bioenergy facilities. A basic economic evaluation is required early in the design of the system to ascertain feasibility prior to significant capital investment. Evaluation of the economic feasibility begins with the analysis of the *fixed* (or capital) expenditures and *variable* (or operating) costs (Watts and Hertvik, 2018). Fixed expenditures include the capital cost of assets such as biomass conversion facilities, land, equipment, and vehicles, as well as depreciation of facilities and equipment, taxes, shelter, insurance, and interest on borrowed money. Variable costs are the daily or monthly operating costs for the production of a biomass product. Variable costs are associated with feedstock and chemicals, repair and maintenance, water, fuel, utilities, energy, labor, management, and waste disposal. Figure 7 shows the relationship between these two basic economic parameters. Fixed costs do not vary with time and output while variable costs increase with time and output of product. The total project cost is the sum of fixed and variable costs. Variable costs per unit of output decrease with increased amount of output, so the profitability of a product may depend on the amount produced.

In order to evaluate the economic benefits of a bioenergy project, some other economic parameters are commonly used (Stout, 1984), including net present value; benefit cost ratio, payback period, breakeven point analysis, and internal rate of return. The analyses must take into account the relationship between

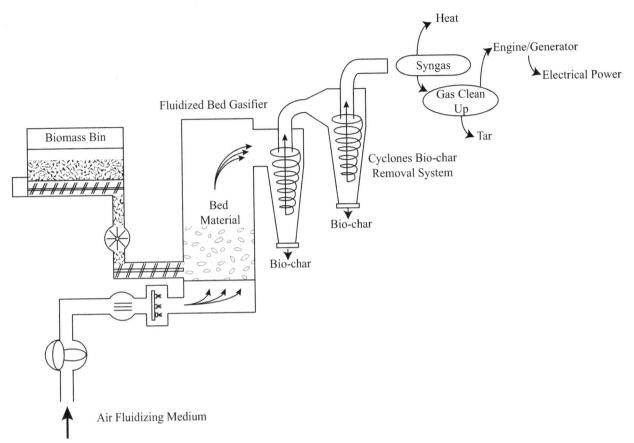

Figure 6. Schematic diagram of a fluidized bed gasifier.

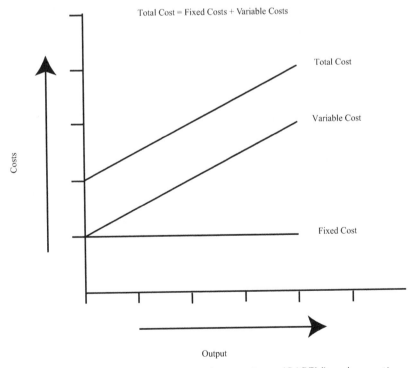

Figure 7. The relationships between capital expenditure (CAPEX) and operating expenditure (OPEX) for a bioenergy project.

> PV = present value of cash flow
> FV = future value of cash flow
> NPV = net present value
> BCR = benefit cost ratio
> PBP = payback period
> BEP = breakeven point
> IRR = internal rate of return
> DCFROR = discounted cash flow rate of return, another name for IRR
> SYD = sum-of-years' digit, a depreciation method

time and the value of money. The basic equations for estimating the present and future value of investments are:

$$\text{Present Value} = PV = FV \times \frac{1}{(1+R)^n} \qquad (2)$$

$$\text{Future Value} = FV = PV \times (1+R)^n \qquad (3)$$

where R = rate of return or discount rate (decimal)
n = number of periods (unitless)

The internal rate of return is a discounted rate that makes the net present value of all cash flows from a particular project equal to zero. The higher the internal rate of return, the more economically desirable the project. The net present value (equation 4) is the difference between the present value of cash inflows and the present value of cash outflows. A positive net present value means that the project earnings exceed anticipated costs. The benefit cost ratio (equation 5) is the ratio between the project benefits and costs. Values greater than 1 are desirable. The payback period (equation 6) is the length of time required to recover the cost of investment.

$$\text{Net Present Value} = NPV = \sum_{n=1}^{N} \frac{\text{cash inflow}}{(1+i)^n} - \text{cash outflow} \qquad (4)$$

$$\text{Benefit Cost Ratio} = BCR = \frac{\text{project benefits}}{\text{project costs}} \qquad (5)$$

$$\text{Payback Period} = PBP\,(\text{years}) = \frac{\text{project costs}}{\text{annual cash inflows}} \qquad (6)$$

When estimating the fixed cost of a project, the major cost components are the depreciation and the interest on borrowed money. There are many ways to estimate the depreciation of a facility. The two most common and simple methods are straight-line depreciation (equation 7), and the sum-of-years' digit depreciation method (SYD) (equation 8).

$$\text{Straight Line Depreciation}(\$) = \frac{\text{Principal} - \text{Salvage Value}}{\text{Life of Unit}} \qquad (7)$$

$$\text{SYD Depreciation}(\$) = \text{Depreciation Base} \times \frac{\text{Remaining Useful Life}}{\text{Sum of Years' Digits}} \qquad (8)$$

In equation 8, the depreciation base is the difference between the initial capital cost ($) and the salvage value of the asset ($). The sum of the years' digits is the sum series: 1, 2, 3, up to n, where n is the useful life of the asset in years, as shown in equation 9:

$$\text{Sum of Years' Digits} = \text{SYD} = \frac{n(n+1)}{2} \qquad (9)$$

The other large portion of capital cost is interest on borrowed money. This is usually the percentage (interest rate) charged by the bank based on the amount of the loan. The governing equation without including the salvage value (equation 10) is similar to the amortization calculation for a loan amount:

$$\text{Annuity} = A = P \times \left(\frac{r \times (1+r)^n}{(1+r)^n - 1} \right) \qquad (10)$$

where
A = annuity or payment amount per period ($)
P = initial principal or amount of loan ($)
r = interest rate per period (%)
n = total number of payments or period (unitless)

There are many tools used for economic evaluation of energy system, but one of the most popular is the HOMER Pro (or hybrid optimization model for energy renewal) developed by Peter Lilienthal of the US Department of Energy (USDOE) since 1993 (Lilienthal and Lambert, 2011). The model includes systems analysis and optimization for off grid connected power systems for remote, stand-alone, and distributed generation application of renewables. It has three powerful tools for energy systems simulation, optimization, and economic sensitivity analyses (Capareda, 2014). The software combines engineering and economics aspects of energy systems. This type of tool is used for planning and design of commercial systems, but its simple equations can be used first to assess the fundamental viability of a biomass conversion project.

Sustainability Issues in Biomass Energy Conversion Systems

The US Department of Energy (USDOE) and US Department of Agriculture (USDA) define sustainable biofuels as those that are "economically competitive, conserve the natural resource base, and ensure social well-being." The conservation of resource base points to the conservation of energy as well, that is, the fuel produced must have more energy than the total energy used to produce the fuel. One of the most common indicators of sustainability for biomass utilization is energy use throughout the life cycle of production. There are two measures used for this evaluation: the net energy ratio (NER) (equation 11) and net energy balance (NEB) (equation 12). NER must be greater than

1 and NEB must be positive for the system to be considered sustainable from an energy perspective.

$$\text{NER} = \frac{\text{Energy Content of Fuel(MJ)}}{\text{Energy Required to Produce the Biofuel(MJ)}} \quad (11)$$

$$\text{NEB} = \text{Biofuel Heating Value(MJ)} - \text{Energy Required to Produce the Biofuel(MJ)} \quad (12)$$

The heating value of the biofuel is defined as the amount of heat produced by the complete combustion of the fuel measured as a unit of energy per unit of mass.

Applications

Engineers assigned to design, operate, and manage a commercial biodiesel plant must decide what working system to adopt. The cheapest and most common is the use of gravity for separating the biodiesel (usually the top layer) and glycerin (the bottom layer). An example of this commercial operational facility is the 3 million gallon per year (MGY) (11.36 ML/yr) biodiesel plant in Dayton, Texas, operated by AgriBiofuels, LLC. This facility began operation in 2006 and is still in operation. The biodiesel recovery for this facility is slightly lower than those with computer-controlled advanced separation systems using centrifuges. This facility is also not following the ideal process flow (shown in figure 2) used by many other commercial facilities. Thus, one would expect their conversion efficiency and biodiesel recovery to be lower.

Biodiesel production is an efficient biomass conversion process. The ideal mass balance equation, presented earlier, is:

$$100 \text{ kg vegetable oil} + 10 \text{ kg catalysts} \rightarrow 100 \text{ kg biodiesel} + 10 \text{ kg glycerin}$$

The relationship shows that an equivalent mass of biodiesel is produced for every unit mass of vegetable oil used, but there are losses along the way and engineers must consider these losses when designing commercial facilities. In a commercial biodiesel facility, the transesterification process is split into several reactors (e.g., figure 2). However, to save on capital costs, some plant managers simply divide the process into two stages. Separating glycerin and biodiesel fuel is also an issue that the engineer will be faced with. Efficient separation systems that use centrifuges are expensive compared with physical separation, and this affects the overall economy of the facility. If the initial capital available is limited, investors will typically opt for cheaper, physical gravity separation instead of using centrifuges. Crown Iron Works (in Blaine, MN) sells low cost biodiesel facilities that employ gravity separation while GEA Wesfalia (Oelde, Germany) sell more expensive biodiesel facilities that use separation by centrifuge. The latter, expensive, system is more efficient at separating glycerin and biodiesel fuel and may be beneficial in the long term, allowing the facility to sell glycerin products with minimal contamination. The engineer may compare these systems in terms of costs and efficiencies. Ultimately equation 2 is used for designing and sizing a commercial plant to determine the daily, monthly, or yearly vegetable oil requirement. This means the engineer must determine the agricultural land area required both for the facility and the supply of biomass. There are standard tables of oil yields from crops that are used. For example, the highest oil yield comes from palm oils, with more than 7,018 kg oil production per hectare compared with 2,245 kg/ha for soybean oil (Capareda, 2014).

Designing, building, and operating a commercial bioethanol facility also requires knowledge primarily on the type of feedstock to use. Unlike a biodiesel plant, where the manager may have various options for using numerous vegetable oil types without changing the design, a bioethanol plant is quite limited to the use of a specific feedstock. The main choices are sugar crops, starchy crops, or lignocellulosic biomass. Designs for these three different types of feedstock are not the same; using lignocellulosic biomass as feedstock is the most complex. The simplest are sugar crops but sugary juice degrades very quickly and so the majority of commercially operating bioethanol plants in the US use starchy crops like corn. Corn grains may be dried, ground, and stored in sacks for future conversion without losing its potency. Examples of commercial bioethanol plants using lignocellulosic feedstocks are those being built by POET (Sioux Falls, South Dakota) in Emmetsburg, Iowa, using corncobs (25 MGY or 94.6 ML/yr), and another by Dupont (Wilmington, Delaware) in Nevada, Iowa, using corn stover (30 MGY or 113.6 ML/yr).

Bioethanol is an efficient biofuel product. Engineers must be aware of energy and mass balances required for biofuels production even though other waste materials are also used for the processes. As the potential bioethanol yields from crops varies, the design is for a specific feedstock. The greatest potential bioethanol yield comes from the Jerusalem artichoke (*Helianthus tuberosus*) (11,219 L/ha). Compare this to corn (maize, *Zea mays*) at a reported yield of only 2,001 L/ha (Capareda, 2014) and sorghum (*sorghum spp.*) cane (4,674 L/ha) or grain (1,169 L/ha).

While yields are important, the location of a project is also a significant factor in selecting the resource input for a bioethanol or biodiesel production facility. For example, the Jerusalem artichoke has the highest bioethanol yield but only grows in temperate conditions. When the bioethanol business started to boom in the US around 2013, there was an issue with the disposal of a by-product of the process, the distillers grain. During those initial periods, these co-products were simply disposed of with very minimal secondary processing (e.g., animal feed) or to a landfill. Options for secondary valorization (i.e., to enhance the price or value of a primary product) have now emerged such as further energy recovery and as a raw material for products such a films and membranes. Key issues for engineers include sizing of plants and determining the daily, weekly, and monthly resource requirements for the feedstock, which can be calculated using equations 3, 4, and 5, modified for inefficiency in practice.

A growing number of animal facilities have taken advantage of the additional energy recovered from anaerobic digestion of manure by converting their lagoons into biogas production facilities. In the US, the covered lagoon is still the predominant biogas digester design. The operation is very simple since the microorganisms needed for biogas production already exist in the stomachs of ruminants. Key issues for engineers are sizing (based on animal numbers), energy recovery rates, sizing of power production (engine) facilities, sludge production and energy remaining in the sludge and economic feasibility. There is increasing interest in designing systems that use sludge for pyrolysis to recover as much energy as possible from the feedstock. When these additional processes are adopted, the energy recovery from the waste biomass is improved and there is less overall waste. While the sludge is an excellent source of nutrients for crops, its energy value must be judged against its fertilizer value. Financially the energy case probably wins out, but a holistic analysis would be needed to judge the most desirable option from a sustainability perspective.

The economics of a biofuel facility are dependent on the price of the initial feedstock used. For example, 85% of the cost of producing biodiesel fuel comes from the cost of the initial feedstock. As a potential candidate for biodiesel production, if the price of refined vegetable oil is the same as the price of diesel fuel, it is not economical to turn the vegetable oil into a biofuel. The remaining 15% is usually the cost of catalysts used for the conversion process (Capareda, 2014). If biodiesel is made from any refined vegetable oil, the processing cost is the greatest component of the conversion process. The cost of chemicals and catalysts together usually amounts to approximately $0.06/L ($0.22/gal). Chemicals or catalysts are not the limiting factors in making biodiesel. This statement applies to biofuels in general. These production expenses are not a large part of the biofuels production expense. Biodiesel catalysts are rather cheap and abundant. They will not usually run out nor gets too expensive as production is increased.

In the bioethanol production process, the cost of the bioethanol fuel is also mainly affected by the price of the initial feedstock used, such as corn, as well as the enzymes used for the process. The process also uses significant volumes of water, but only minimal electricity. For example, for every 3.785 L (1 gallon) of bioethanol produced, 1.98 m^3 (70 ft^3) of natural gas and 155.5 L (41 gal) of water is required (Capareda, 2014). The electricity usage is around 0.185 kWh/L (0.7 kWh/gal). Hence, if the electricity cost is $0.10/kWh, then one would only spend around $0.0158/L (0.07/gal). Natural gas is used to heat up the beer and recover pure bioethanol. Because of the abundant use of water, this water input must be recycled for the process to be effective and efficient. The current industry standard for bioethanol production from grains is around 416.4 to 431.2 L/tonne (2.8–2.9 gal/bushel). Newer feedstocks for bioethanol production must exceed this value.

The economics of power production via thermal conversion such as pyrolysis or gasification is dependent upon the sale of electrical power. If a MW power plant using biomass is operated continuously for a year the electrical power should sell for $0.12/kWh to achieve a gross revenue of $1M. A preliminary economic evaluation of the economic return of a gasification for power facility can be completed by adjusting selling cost. Finally, the economics of biofuels production from biomass resources are also dependent on the price of crude oil from commercial distributors and importers. Biodiesel and bioethanol are mixed with commercial diesel and gasoline and are priced similarly. With a crude oil price below $100/barrel, the production cost for biodiesel and bioethanol must also be under $100/barrel.

The question of the sustainability of fuel production and usage must be addressed. Many biofuels produced from biomass resources in the USA are now being categorized according to their potential greenhouse gas reductions and are standardized under the renewable fuels standard (RFS) categories (figure 8). As shown, cellulosic biofuels—mainly bioethanol and biodiesel (also coded as D3/D7, respectively) coming from lignocellulosic biomass—have a reported 60% greenhouse gas reduction compared with biomass-based diesel, which only has a 50% GHG reduction potential (also coded as D4). Biodiesel from vegetable oils and ethanol from corn have lower GHG reductions potential than cellulosic biofuels and biomass-based diesel. The code D6 is for renewable fuels in general, produced from renewable biomass and is used to replace quantity of fossil fuel present in transportation fuel, heating fuel, or jet fuel (e.g., corn ethanol) and also not falling under any of the other categories. The code D5 is for advanced biofuels other than ethanol derived from corn starch (sugarcane ethanol), and biogas from other waste digesters.

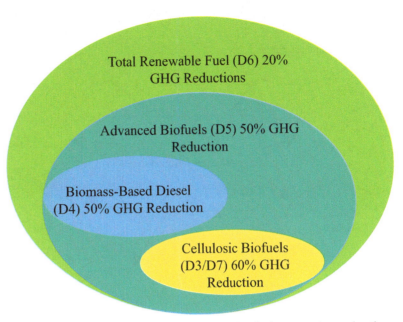

Figure 8. Schematic of USA nested renewable fuels categories under the renewable fuels standards.

While net energy ratio (NER) and net energy balance (NEB) are important, they have to be combined with estimates of CO_2 emissions and perhaps with land use to understand the foundations of sustainability of the use of biomass resources. A simple life cycle assessment (LCA) of coal and biomass for power generation (Mann and Spath, 1999) reported 1,022 g CO_2 emissions per kWh of electrical power produced by coal compared to only 46 g CO_2/kWh of electrical power by biomass. Contrary to the perception that using biomass would have zero net CO_2 emissions, there is actually some CO_2 produced for every kWh of electricity generated. It is also important to recognize the competing uses of land and biomass by society (figure 9). On one hand, biomass is used for food and feed (the food chain), and on the other for materials and energy (the bioeconomy). All uses have to consider climate change, food security, resource depletion, and energy security. Countries around the world need to create a balance of the use of biomass resources toward a better environment. Future engineers must be able to evaluate the use of biomass resources for materials and biofuels production

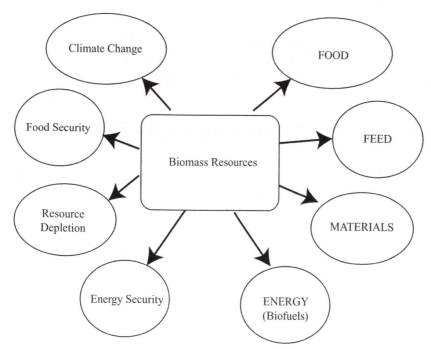

Figure 9. The role of biomass resources for a sustainable low-carbon future.

as well as relate this to climate change and energy security without depleting already limited resources.

The US Department of Energy created a hierarchy of materials and products from biomass resources (figure 10). On top of the pyramid are high-value fine chemicals, such as vanillin and phenol derivatives, worth more than $6,500 per tonne. Phenol derivatives have the potential to be further converted into expensive lubricants (Maglinao et al., 2019). Next are high-value carbon fibers such as graphene, followed by phenolic substances. There are also new products such as 100% biomass-based printed integrated circuit boards developed by IBM (International Business Machines Corporation, Armonk, NY). Biofuels are in the middle of the pyramid, valued around $650 per tonne, with simple energy recovery by combustion at the bottom.

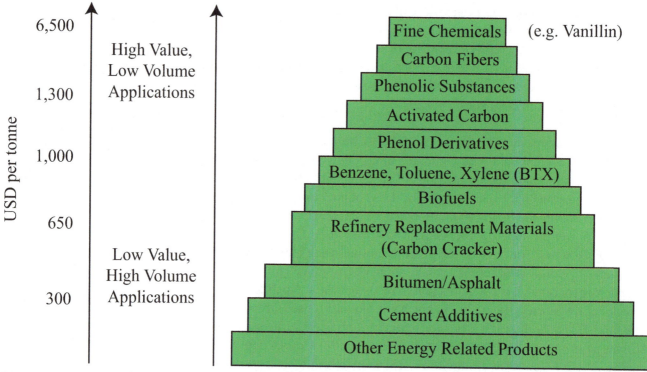

Figure 10. Hierarchy of biomass utilization from high-value, low-volume applications (top) to low-value, high-volume applications (bottom).

Examples

Engineers who manage biorefineries need to be aware of energy and mass balances to determine resource allocations, as well as conversion efficiencies to improve plant operations. The process of estimation includes simple conversion efficiency calculations and determining the economic feasibility of the biorefinery.

Example 1: Conversion efficiency calculations

Problem:
The ideal mass and energy balance is difficult to achieve. Plant managers must be able to estimate how close their operations are compared to the ideal conditions. The most common problem faced by a plant manager is to determine the conversion efficiency of refined vegetable oil into biodiesel. This example shows how the actual plant is operated and how close it is to the ideal mass balance. The energy content of refined canola oil is 39.46 MJ/kg and that of canola oil biodiesel was measured in the laboratory to be 40.45 MJ/kg. During an actual run, only about 95% biodiesel is produced from this refined canola oil input instead of the ideal 100% mass yield.

Determine the energy conversion efficiency of this facility from turning refined canola oil energy into fuel energy in biodiesel.

Solution:
1. The energy of the output biodiesel product unit of weight is calculated using the 95% mass yield of biodiesel as follows:

$$\text{Biodiesel Output (MJ)} = \frac{40.45 \text{ MJ}}{\text{kg}} \times 0.95 \text{ kg Biodiesel} = 38.43 \text{ MJ} = 36,424 \text{ Btu}$$

2. Using equation 1, the conversion efficiency per unit of weight is:

$$\text{Conversion Efficiency (\%)} = \frac{38.43 \text{ MJ}}{39.46 \text{ MJ}} \times 100\% = 97.4\%$$

Biodiesel production is perhaps one of the most efficient pathways for the conversion of vegetable oil into biofuel, having very close to 100% energy conversion efficiency.

Example 2: Sizing commercial biodiesel plants

Problem:
Planning to build a commercial biodiesel facility requires taking inventory of input resources needed. In this example, the engineer must determine the amount of soybean oil needed (L/year) to build and operate a 3.785-million liter (1-million gallon per year, MGY) biodiesel plant. The densities of soybean

oil and its equivalent biodiesel (also called soybean methyl ester) are as follows:

Soybean biodiesel density = 0.88 kg/L
Soybean oil density = 0.917 kg/L

Calculate the soybean oil requirements for a daily basis and a monthly basis.

Solution:

1. 3.785 million liters of biodiesel product is converted into its mass units as:

$$\text{Biodiesel Mass Requirement}\left(\frac{\text{tonnes}}{\text{year}}\right) = \frac{3.785 \times 10^6 \text{ L}}{\text{year}} \times 0.88 \frac{\text{kg}}{\text{L}} \times \frac{\text{tonne}}{1000 \text{ kg}} = 330.8 \frac{\text{tonnes}}{\text{year}} = 3{,}671.6 \frac{\text{tonnes}}{\text{year}}$$

2. This biodiesel mass of 3,330.8 tonnes per year is then equivalent to the mass of soybean oil required for the plant. This unit must be converted into volumetric units for trading vegetable oils, as:

$$\text{Soybean Oil Volume Requirement}\left(\frac{\text{L}}{\text{year}}\right) = \frac{3{,}330.8 \text{ tonnes}}{\text{year}} \times \frac{1{,}000 \text{ kg}}{1 \text{ tonne}} \times \frac{\text{L}}{0.917 \text{ kg}}$$

$$= 3{,}632{,}279 \frac{\text{L}}{\text{year}} = 959{,}651 \frac{\text{gallons}}{\text{year}}$$

3. Thus, the yearly soybean oil requirement for this biodiesel facility is more than 3.6 million liters (0.96 million gallons). The monthly and daily requirements are calculated as:

$$\text{Soybean Oil Mass Requirement}\left(\frac{\text{L}}{\text{month}}\right) = \frac{3{,}632{,}279 \text{ L}}{\text{year}} \times \frac{1 \text{ year}}{12 \text{ months}} = 302{,}689 \frac{\text{L}}{\text{month}} = 79{,}971 \frac{\text{gallons}}{\text{month}}$$

$$\text{Soybean Oil Mass Requirement}\left(\frac{\text{L}}{\text{day}}\right) = \frac{3{,}632{,}279 \text{ L}}{\text{year}} \times \frac{1 \text{ year}}{365 \text{ days}} = 9{,}951 \frac{\text{L}}{\text{day}} = 2{,}629 \frac{\text{gallons}}{\text{day}}$$

Further, this soybean oil requirement value may also be used to estimate the required acreage for soybean oil if one has data on soybean oil crop yield per acre. For example, a reported soybean oil yield of around 2,245 kg/ha (2000 lb/acre) (Capareda, 2014) will result in an estimated 1,483.6 ha (3,664 ac) needed for dedicated soybean land for this plant use year-round.

Example 3: Energy balance in the recovery of bioethanol

Problem:
Bio-ethanol may be produced from sweet sorghum via fermentation of its sugars. The sweet sorghum juice is fermented using yeasts (*Saccharomyces cerevisiae*). The resulting fermented product, called beer, has about 10% bio-ethanol. Higher percentage ethanol is required for engine use and may be separated from this fermented product through a simple distillation process. The liquid fermented product is simply heated until the bio-ethanol vapor is evaporated (around 80°C, the evaporation temperature of pure ethanol) and this vapor is condensed or liquefied in a simple condenser. In village-level systems, fuel wood is used to heat up the boiler where the fermented material is placed.

A village-level ethanol production scheme based on sweet sorghum has the following data for a series of experiments. In the first experiment, the operator was not mindful of the amount of fuel wood used for the recovery of highly concentrated ethanol and used too much, about 20 kg of waste fuel wood for the boiler. In addition, the boiler was not insulated during this run. In the second experiment, the operator insulated the boiler and was very careful in the use of fuel wood to adjust the boiler temperature below the boiling point of pure ethanol. Only about 10 kg of fuel wood was used, about half of the initial experiment. Assume that the energy of fuel wood is 20 MJ/kg and the heating value of ethanol is around 18 MJ/L. In both experiments, 120 liters of liquid fermented material (beer) was used and 13 liters of highly concentrated ethanol was recovered. Discuss the energy balance for each experiment.

Solution:
1. In the first experiment, the operator used about 400 MJ of input energy and produced 13 liters of ethanol with an energy content of 234 MJ:

$$\text{Energy from the Fuel Wood} = 20 \text{ kg fuel wood} \times \frac{20 \text{ MJ}}{\text{kg}} = 400 \text{ MJ}$$

$$\text{Energy from the Ethanol} = 13 \text{ L} \times \frac{18 \text{ MJ}}{\text{L}} = 234 \text{ MJ}$$

Clearly, the operator used more energy from the fuel wood than that of the recovered ethanol, demonstrating an unsustainable process.

2. The second experiment used only about 200 MJ of input wood energy, which is slightly less than the energy from the produced ethanol of 234 MJ.

$$\text{Energy from the Fuel Wood} = 10 \text{ kg fuel wood} \frac{20 \text{ MJ}}{\text{kg}} = 200 \text{ MJ}$$

By careful use of fuel wood, more energy from the bioethanol is recovered from a relatively efficient recovery process.

Note that there are other energy amounts expended from planting, harvesting, and transport of the sweet sorghum feedstock and this experiment is only one portion of the life cycle of bioethanol production, recovery, and use.

Example 4: Biogas production and use from animal manure

Problem:

Sizing a biogas facility is one task assigned to an engineer who operates a commercial biogas facility. One common calculation is to determine the electrical power produced from the manure collected from a 500-head dairy facility. Usually, one would need electrical power for 8 hours per day. The thermal conversion efficiency of an internal combustion engine is approximately 25% with a mechanical-to-electrical conversion efficiency of 80%. The specific methane yield was found to be 0.23 m³ biogas/kg volatile solids per day (Hamilton, 2012; ASABE Standard D384.2). Each mature dairy cow produces an average of 68 kg manure per head per day with a percentage of 7.5 volatile solids. The energy content of biogas was 24.2 MJ/m³ (650 Btu/ft³).

Size the generator to use for this facility.

Solution:

1. The amount of methane produced from a 500-head facility is calculated as follows:

$$\text{Biogas}\left(\frac{m^3}{day}\right) = 500 \text{ head} \times \frac{68 \text{ kg wet manure}}{\text{head per day}} \times \frac{0.075 \text{ kg VS}}{\text{kg wet manure}} \times \frac{0.23 \text{ m}^3 \text{ biogas}}{\text{kg VS}}$$

$$= 586.5 \frac{m^3}{day}$$

2. The theoretical power production is calculated as follows:

$$\text{Power}(kW) = \frac{586.5 \text{ m}^3}{day} \times \frac{1 \text{ day}}{8 \text{ hrs}} \times \frac{24,200 \text{ kJ}}{m^3} \times \frac{1 \text{ hr}}{3600 \text{ s}} \times \frac{kW}{kJ/s} = 492.8 \text{ kW}$$

3. The actual power produced based on 25% engine efficiency and 80% mechanical-to-electrical efficiency is calculated as follows:

$$\text{Actual Power}(kW) = 492.8 \text{ kW} \times 0.25 \times 0.80 = 98.6 \text{ kW}$$

A generator with a size close to 100 kW of power output will be required.

Example 5: Basic biomass pyrolysis energy and mass balances

Problem:

The thermal conversion of waste biomass into useful energy is a common calculation for an engineer. This simple example is the conversion of coconut shell (waste biomass) into bio-char (useable fuel). In the experiment, the engineer used 1 kg of coconut shell and pyrolyzed this at a temperature of 300°C. The measured energy content of this high-energy density biomass was 20.6 MJ/kg. The pyrolysis experiment produced about 0.80 kg of bio-char. The heating value of the bio-char was measured to be 22 MJ/kg. Minimal solids and gaseous products were produced in this low-temperature pyrolysis process. Determine the overall conversion efficiency (η_e) for the bio-char conversion process and also calculate the amount of energy retained in the bio-char and the energy lost through the process.

Solution:

1. Equation 1 is used directly to estimate the conversion efficiency for bio-char production.

$$\text{Energy Conversion Efficiency}(\%) = \frac{\text{Energy Output(MJ)}}{\text{Energy Input(MJ)}} \times 100\% \quad (1)$$

First, calculate the total energy of the bio-char per unit kg of material pyrolyzed as:

$$\text{Bio-char Energy (MJ)} = 0.80 \text{ kg} \times \frac{22 \text{ MJ}}{\text{kg}} = 17.6 \text{ MJ}$$

2. The overall conversion efficiency (η_e) is then calculated as follows:

$$\eta_e = \frac{17.6 \text{ MJ}}{20.6 \text{ MJ}} \times 100\% = 85.4\%$$

This value also indicates the percentage of energy retained in the bio-char.

3. The energy lost through the process is simply the difference between the original energy of the biomass and the energy retained in the bio-char as follows:

$$\text{Energy Loss (MJ)} = 20.6 \text{ MJ} - 17.6 \text{ MJ} = 3 \text{ MJ}$$

4. This energy loss is equivalent to 14.6%, the difference between 100% and the process conversion efficiency of 85.4%.

Notice the high yield of solid bio-char at this pyrolysis temperature, with a minimal yield of liquid and gaseous synthesis gas, which are considered losses at this point. However, at much higher pyrolysis temperatures, more liquid and gaseous synthesis gases are produced. Example 6 shows the uniqueness of the pyrolysis process in generating a wider range of co-products. Complete energy and mass balances of the process may also be estimated to evaluate overall conversion efficiencies.

Example 6: Basic biomass pyrolysis energy and mass balances

Problem:

An engineer conducted an experiment to pyrolyze 1.23 kg of sorghum biomass (heating value = 18.1 MJ/kg) at a temperature of 600°C in an auger pyrolyzer. The primary purpose of the experiment was to determine the energy contained in various co-products of the process. The input energy includes that from the auger motor (5 amps, 220 V) and tube furnace (2,400 Watts). The time of testing was 12 minutes. Data gathered during the experiments and other associated parameters needed to perform complete energy and mass balances are as follows:

Amount of bio-char produced = 0.468 kg
Volume of bio-oil produced = 225 mL
Density of = 1.3 g/mL
Volume of syngas produced = 120 L
Heating value of bio-char = 23.99 MJ/kg
Heating value of = 26.23 MJ/kg

The heating values of syngas produced as well as their composition is in the table below.

	Primary Gases		
	H_2	CH_4	CO
% Yield	20%	10%	15%
Density (kg/m³)	0.0899	0.656	1.146
HV (MJ/kg)	142	55.5	10.112

Determine an energy and mass balances for this process and report how much energy was contained in each of the co-products as well as the overall conversion efficiency.

Solution:

1. Draw a schematic of the complete mass and energy balance process as in figure 11.
2. Calculate the energy contained in the original biomass as:

$$\text{Biomass Energy}(MJ) = 1.23 \text{ kg} \times \frac{18.1 \text{ MJ}}{\text{kg}} = 22.26 \text{ MJ}$$

3. Calculate the input energy from the furnace as:

$$\text{Thermal Energy}(MJ) = 2.4 \text{ kW} \times \frac{12 \text{ hr}}{60} \times \frac{3.6 \text{ MJ}}{1 \text{ kWh}} = 1.728 \text{ MJ}$$

4. Calculate the input energy from the auger as:

$$\text{Auger Energy}(MJ) = 220 \text{ V} \times 5 \text{ A} \times \frac{\text{kW}}{1,000 \text{ VA}} \times \frac{12 \text{hr}}{60} \times \frac{3.6 \text{MJ}}{1 \text{kWh}} = 0.792 \text{ MJ}$$

5. Calculate the energy contained in the bio-char as:

$$\text{Bio-char Energy (MJ)} = 0.468 \text{ kg} \times \frac{23.99 \text{ MJ}}{\text{kg}} = 11.23 \text{ MJ}$$

6. Calculate the energy contained in the bio-oil as:

$$\text{Energy (MJ)} = 225 \text{ mL} \times \frac{26.23 \text{ MJ}}{\text{kg}} \times \frac{1.3 \text{ g}}{\text{mL}} \times \frac{\text{kg}}{1000 \text{ g}} = 7.67 \text{ MJ}$$

7. The total energy content of syngas is the sum of energy in the component gases. As given, about 120 L of syngas was produced, with 20% H_2 (24 L), 10% CH_4 (12 L) and 15% CO (18 L). The resulting energy content of the bio-oil is calculated as:

$$H_2 \text{ (MJ)} = 24 \text{ L} \times \frac{0.0899 \text{ kg}}{\text{m}^3} \times \frac{1 \text{ m}^3}{1{,}000 \text{ L}} \times \frac{142 \text{ MJ}}{\text{kg}} = 0.306 \text{ MJ}$$

$$CH_4 \text{ (MJ)} = 12 \text{ L} \times \frac{0.0656 \text{ kg}}{\text{m}^3} \times \frac{1 \text{ m}^3}{1{,}000 \text{ L}} \times \frac{55.5 \text{ MJ}}{\text{kg}} = 0.437 \text{ MJ}$$

$$CO \text{ (MJ)} = 18 \text{ L} \times \frac{1.145 \text{ kg}}{\text{m}^3} \times \frac{1 \text{ m}^3}{1{,}000 \text{ L}} \times \frac{10.112 \text{ MJ}}{\text{kg}} = 0.208 \text{ MJ}$$

The total energy content of syngas is:

$$\text{Syngas (MJ)} = 0.306 \text{ MJ} + 0.437 \text{ MJ} + 0.208 \text{ MJ} = 0.951 \text{ MJ}$$

Most of the energy is still retained in the bio-char (11.23 MJ), followed by the bio-oil (7.67 MJ), and the syngas (0.951 MJ).

8. The energy balance is:

$$\text{Input Energy (MJ)} = 22.2 \text{ MJ} + 1.73 \text{ MJ} + 0.792 \text{ MJ} = 24.722 \text{ MJ}$$

$$\text{Output Energy (MJ)} = 11.23 \text{ MJ} + 7.67 \text{ MJ} + 0.951 \text{ MJ} = 19.851 \text{ MJ}$$

9. Calculate the conversion efficiency as:

$$\text{Conversion Efficiency (\%)} = \frac{\text{Output}}{\text{Input}} \times 100\% = \frac{19.851}{24.722} \times 100\% = 80.3\%$$

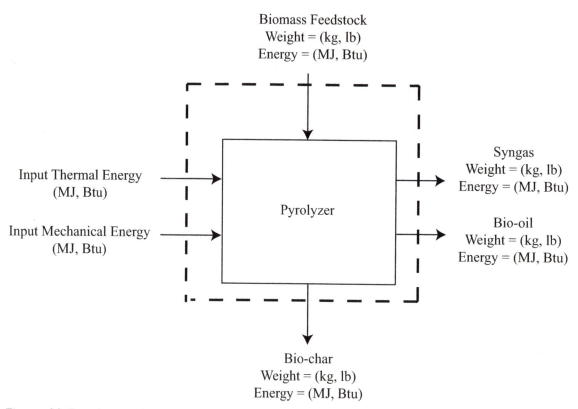

Figure 11. Distribution of mass and energy from all products of the pyrolysis process.

Example 7: Present and future value of investment in a bioenergy systems facility

Problem:

An investor deposited $1,100,000 in a bank in 2007 rather than investing it in a commercial biodiesel facility. Determine its estimated future value in 2018 using the present value equation, assuming a bank rate of return of 2.36%. Compare this to investing the money in operating a biodiesel facility with year 11 return of $2M.

Solution:

1. This is a simple future value calculation using equation 3:

$$FV = PV \times (1 + r)^n \quad (3)$$

where FV = future value of cash flow ($)
PV = present value of cash flow ($)
r = rate of return or discount rate (decimal)
n = number of periods (unitless)

$$\text{Future Value} = \$1,100,000 \times (1 + 0.0236)^{11} = \$1,421,758$$

2. If invested in a bank, the future value after 11 years would be about $1.4M compared to a return of $2M from investing in the biodiesel facility. In this example, investing $1.1M in a biodiesel facility generated more value than putting the money in a bank.

Example 8: Depreciation of a biodiesel plant

Problem:
An engineer was asked to report the yearly depreciation for a biodiesel facility whose initial asset value is $1,100,000. The lifespan of the facility is 20 years and the salvage value of all equipment and assets at the end of this life is 10% of the initial capital value of the facility. Use the straight-line method and sum-of-digits method for depreciation calculations. Describe the yearly variations in depreciation for each method.

Solution:
The straight-line method uses equation 7:

$$\text{Straight Line Depreciation}(\$) = \frac{\text{Principal} - \text{Salvage Value}}{\text{Life of Unit}} \quad (7)$$

$$= \frac{\$1,100,000 - \$110,000}{20} = \$50,000 / \text{year}$$

The yearly depreciation after year 1 is $50,000 per year.
The sum of digits method first calculates the sum of digits as follows:

$$(1 + 2 + 3 + 4 + 5 + 6 + 7 + 8 + 9 + 10 + 11 + 12 + 13 + 14 + 15 + 16 + 17 + 18 + 19 + 20) = 210$$

The factor to estimate the depreciation for year 1 uses the reverse order—the life of the facility in the numerator and the sum of digits in the denominator with year 1 having a factor of 20/210 and so on.

$$\text{Year 1 Depreciation}(\$) = \frac{20}{210} \times (\$1,100,000 - \$110,000) = \$94,285$$

$$\text{Year 2 Depreciation}(\$) = \frac{19}{210} \times (\$990,000) = \$89,571$$

$$\text{Year 3 Depreciation}(\$) = \frac{18}{210} \times (\$990,000) = \$84,857$$

Continue calculations for years 4 through 19 using years 17 through 2 in the numerator. . . .

$$\text{Year 20 Depreciation}(\$) = \frac{1}{210} \times (\$990,000) = \$4,714$$

Note that in both methods, the end of project asset cost is approximately equal to the salvage value given. If the data were plotted, a rapid decline in value over the first few years using the sum of digits method usually reflects the actual depreciation of many facilities.

Example 9: Calculation of net present value, benefit cost ratio, payback period, and internal rate of return for a biodiesel facility

Problem:

An engineer can be asked to evaluate a number of projects to estimate funding requirements. Comparative economic indicators can be used to compare one project proposal to another. Common indicators are net present value (NPV), benefit cost ratio (BCR), payback period (PBP), and internal rate of return (IRR). A 1,892,500 L/yr (half a million gallon per year) biodiesel facility with an initial capital cost of $1,100,000 and 10% salvage value has the following baseline data:

Capital Cost (CC)	$1,100,000	Repair & Maintenance Cost	3% of Capital Cost	Biodiesel Plant Cost per million L	$581,241/ML ($2,200,000/MG)
Interest	7.5%	Tax & Ins.	2% of CC	Conv. Eff.	99%
Life	20 years	Labor	8 hrs/day	Labor Cost	$15/hr
Vegetable Oil Price	$0.13/L ($0.50/gal)	Operation	365 days/yr	One Manager	$60,000/yr
Processing Cost	$0.13/L ($0.50/gal)	Personnel	6 full-time	Selling	$0.53/L ($2.00/gal)
Glycerin	10% Yield	Glycerin	0.18/L ($0.7/gal)	Biodiesel	1,873,575 L (495,000 gals)
Tax Credit	28%	Discount Rate	2.36%	Depreciation	Straight Line
Salvage Value	10% of initial capital cost				

The data can be used to calculate some economic performance data:

Average yearly gross income for the project with tax credit = $1,314,506
Average yearly net income for the project with tax credit = $279,305
Average discounted net benefits per year = $220,614
Average discounted costs per year = $817,670
Average discounted gross benefits = $1,038,284

Use these data to calculate NPV, BCR, PBP, and IRR.

Solution:

1. Calculate the NPV from equation 4,

$$\text{NPV} = \sum_{n=1}^{N} \frac{\text{cash inflow}}{(1+i)^n} - \text{cash outflow} \qquad (4)$$

or simply get the difference between the average yearly discounted benefits and the yearly average discounted costs:

$$\text{NPV} = \$1,038,284 - \$817,670 = \$220,614$$

The NPV value is positive. Hence, the project is economically feasible.
2. The BCR is the ratio between the discounted benefits and discounted costs as shown:

$$\text{BCR} = \frac{\text{project benefits}}{\text{project costs}} \quad (5)$$

$$= \frac{\$1,0.38,284}{\$817,670} = 1.27$$

The BCR is greater than 1, also showing the project is feasible.
3. The PBP is the ratio of the initial capital costs and the yearly average discounted net revenue:

$$\text{PBP (years)} = \frac{\text{project costs}}{\text{annual cash inflows}} \quad (6)$$

$$= \frac{\$1,100,000}{\$220,614} = 5 \text{ years}$$

4. To calculate the internal rate of return, compare the discounted net benefits throughout the life of the project with an assumed discount factor (discount rate). Manually, this is a trial-and-error method whereby the assumed discount rate results to net benefits greater than zero (positive) and less than zero (negative). The discount rate where the net benefit is exactly equal to zero is the internal rate of return for the project.

For example, when the above data is encoded in a spreadsheet and the assumed discount rate is 30%, the discounted net benefit is estimated at −$173,882, a negative value. However, when the discount rate of 20% is used, the discounted net benefit is calculated as $260,098, a positive value. Hence, the internal rate of return must be between these assumed values (that is, between 20% and 30%). By ratio and proportion (plotting these values in X-Y Cartesian coordinates, like cash flow, and comparing the smaller triangle with the larger triangle, the X being the discounted factor above 20% and Y the net benefits in $), the internal rate of return is then calculated as follows:

$$\frac{X}{\$260,098} = \frac{(30\% - 20\%)}{(\$260,097 + \$173,882)}$$

$$X = 6\%$$

$$\text{IRR} = 20\% + 6\% = 26\%$$

Thus the IRR must be around 26%, a positive value and higher than the bank interest rate of 7.5%. The project is then declared economically feasible using this parameter. (Note: When calculated by spreadsheet, the IRR values will be slightly different from this manual method).

Example 10: Determining net energy ratio and net energy balance for corn ethanol with and without co-products recycling

Problem:

To assess the merit of converting biomass into fuel as recommended by USDA it is possible to use of the net energy ratio (NER) and net energy balance (NEB) for a facility. Numerous studies conducted by USDA for corn ethanol production from wet milling and dry milling have established baseline data.

The total energy used for each process, without considering the use of co-products as sources of additional energy:

Total energy used for the dry milling process = 19.404 MJ/L
Total energy used for the wet milling process = 20.726 MJ/L
Heating value of ethanol produced = 21.28 MJ/L

The total energy used for each process when all by the products of the system are used to supply energy requirements for the facility:

Total energy used for the dry milling process = 15.572 MJ/L
Total energy used for the wet milling process = 16.482 MJ/L
Heating value of ethanol produced = 21.28 MJ/L

Determine whether it is better to use wet or dry milling, and whether it is better to use co-products as a source of energy within the facility.

Solution:

1. The NER is calculated using equation 11:

$$\text{NER} = \frac{\text{Energy Content of Fuel(MJ)}}{\text{Energy Required to Produce the Biofuel(MJ)}} \qquad (11)$$

$$\text{For the dry mill process, NER} = \frac{21.28 \text{ MJ/L}}{19.404 \text{ M/L}} = 1.10$$

$$\text{For the wet mill process, NER} = \frac{21.28 \text{ MJ/L}}{20.726 \text{ M/L}} = 1.03$$

The dry mill process has a higher NER than the wet mill process.

2. The NEB is calculated using equation 12:

NEB = Biofuel Heating Value(MJ) − Energy Required to Produce the Biofuel(MJ)   (12)

$$\text{For the dry mill process, NEB} = 21.28\frac{MJ}{L} - 19.404\frac{J}{L} = 1.876$$

$$\text{For the wet mill process, NEB} = 21.28\frac{MJ}{L} - 20.726\frac{J}{L} = 0.554$$

The dry milling process is better than the wet milling process according to both the NER and NEB when co-products are not used to supply energy.

3. The NER for the dry mill process with co-products allocation is:

$$NER = \frac{21.28 \text{ MJ/L}}{15.572 \text{ M/L}} = 1.37$$

The NER for the wet mill process when co-products are reused is:

$$NER = \frac{21.28 \text{ MJ/L}}{16.482 \text{ M/L}} = 1.29$$

4. The NEB for the dry mill process when co-products are reused for the process is:

$$NEB = 21.28\frac{MJ}{L} - 15.572\frac{J}{L} = 5.708$$

The NEB for the wet mill process when co-product are reused for the process is:

$$NEB = 21.28\frac{MJ}{L} - 16.482\frac{J}{L} = 4.798$$

The dry milling process remains the better option and both NER and NEB indicate that the co-products should be used as part of the system design.

Image Credits

Figure 1. Capareda, S. (CC By 4.0). (2020). Pathways for the conversion of biomass resources into energy.

Figure 2. Capareda, S. (CC By 4.0). (2020). Schematic of the commercial process of making biodiesel fuel.

Figure 3. Capareda, S. (CC By 4.0). (2020). Schematic of the commercial process for making bioethanol via dry milling.

Figure 4. Capareda, S. (CC By 4.0). (2020). Schematic representation of various designs of high-rate biogas digesters.

Figure 5. Capareda, S. (CC By 4.0). (2020). The outputs and applications of biomass pyrolysis.
Figure 6. Capareda, S. (CC By 4.0). (2020). Schematic diagram of a fluidized bed gasifier.
Figure 7. Capareda, S. (CC By 4.0). (2020). The relationships between capital expenditure (CAPEX) and operating expenditure (OPEX) for a bioenergy project.
Figure 8. Capareda, S. (CC By 4.0). (2020). Schematic of US nested renewable fuels categories under the renewable fuels standards.
Figure 9. Capareda, S. (CC By 4.0). (2020). The role of biomass resources for a sustainable low-carbon future.
Figure 10. Capareda, S. (CC By 4.0). (2020). Hierarchy of biomass utilization from high-value, low-volume applications (top) to low-value, high-volume applications (bottom).
Figure 11. Capareda, S. (CC By 4.0). (2020). Distribution of mass and energy from all products of the pyrolysis process.

References

Capareda, S. (2014). *Introduction to biomass energy conversions.* Boca Raton, FL: CRC Press. https://doi.org/10.1201/b15089.

Hamilton, D. W. (2012). Organic matter content of wastewater and manure, BAE 1760. Stillwater: Oklahoma Cooperative Extension Service.

Lettinga, G., van Velsen, A. F., Hobma, S. W., de Zeeuw, W., & Klapwijk, A. (1980). Use of the upflow sludge blanket (USB) reactor concept for biological wastewater treatment, especially for anaerobic treatment. *Biotech. Bioeng., 22*(4), 699-734. https://doi.org/10.1002/bit.260220402.

Lilienthal, P., & Lambert, T. W. (2011). HOMER. The micropower optimization model. Getting started guide for HOMER legacy. Ver. 2.68. Boulder, CO and Golden, CO: Homer Energy and NREL USDOE.

Maglinao, R. L., Resurreccion, E. P., Kumar, S., Maglinao, A. L., Capareda, S., & Moser, B. R. (2019). Hydrodeoxygenation-alkylation pathway for the synthesis of a sustainable lubricant improver from plant oils and lignin-derived phenols. *Ind. Eng. Chem. Res.,* 1-50. https://doi.org/10.1021/acs.iecr.8b05188.

Mann, M. K., & Spath, P. L. (1999). The net CO_2 emissions and energy balance of biomass and coal-fired power systems. Project Report NREL/TP-430-23076 and NREL/TP-57025119. Golden, CO: NREL USDOE.

Pimentel, D. (2003). Ethanol fuels: Energy balance, economics, and environmental impacts are negative. *Natural Resour. Res., 12*(2), 127-134. https://doi.org/10.1023/A:1024214812527.

Stout, B. A. 1984. *Energy use and management in agriculture.* Belmont, CA: Breton Publ.

Watts, S., & Hertvik, J. (2018). CapEx vs OpEx for IT & Cloud: What's the difference? The Business of IT Blog, January 2 issue. Retrieved from https://www.bmc.com/blogs/capex-vs-opex/.

Biogas Energy from Organic Wastes

Hamed El Mashad
Biological and Agricultural Engineering Department
University of California, Davis
Davis, CA, USA
and
Agricultural Engineering Department
Mansoura University
El Mansoura, Egypt

Ruihong Zhang
Biological and Agricultural Engineering Department
University of California, Davis
Davis, CA, USA

KEY TERMS		
Anaerobic digesters	Suspended growth	Sizing
Biochemical and physical processes	Biogas cleaning	Yield estimation
Fixed growth	Biogas upgrading	Commercial uses

Variables

ϕ_{om} = amount of organic matter to be treated per day
$a, b, c,$ and d = number of atoms of carbon, hydrogen, oxygen, and nitrogen, respectively
B_{dp} = daily biogas production
B_y = biogas yield production
C_{vb} = calorific value of biogas
C_{vm} = calorific value of methane
E_{dp} = daily energy production
k = first-order degradation kinetic rate constant
M_{dp} = daily methane production
M_c = methane content in the biogas
M_p = amount of methane produced
M_y = methane yield

Q = amount of feedstock to be treated day
S = concentration of the biodegradable organic matter (VS) in the digester
S_{deg} = degraded organic matter in the digester
t = digestion time
TS_c = total solids contents
V_{df} = volumetric feed to the digester
VS_c = volatile solids contents
V_h = head space volume of digester
V_t = total volume of digester
V_w = working volume of digester

Introduction

Fossil fuel is currently the main energy source in the world. With its limited supplies and the environmental pollution caused by its use, there is a need to increase the use of renewable energy. Sources of renewable energy include the sun, winds, tides, waves, rain, geothermal heat, and biomass. Biomass is plant or animal material that can be used to produce bioenergy as heat or fuel. The technologies for converting biomass into bioenergy can be classified as biochemical, physicochemical, and thermal-chemical technologies. The main biochemical technologies include anaerobic digestion to produce biogas and fermentation to produce alcohols such as ethanol and butanol. The main physicochemical technology is transesterification to produce biodiesel, and the main thermal-chemical technologies are combustion to produce heat, torrefaction to produce solid fuels, pyrolysis to produce oil, and gasification to produce syngas. The selection of a specific technology depends on the composition of the available biomass as well as the desired bioenergy considering economics, social implications, and environmental impact.

Biogas energy is produced by anaerobic digestion of organic matter, which is carried out by a consortium of microorganisms in the absence of oxygen. Airtight vessels called digesters or reactors are used for the process. Biogas is a mixture of methane (CH_4), carbon dioxide (CO_2), and traces of other gases, such as ammonia (NH_3) and hydrogen sulfide (H_2S). Anaerobic digestion technology can be used to treat organic materials, such as food residues and wastewater, thus reducing the amount of material to be disposed of, while generating bioenergy.

This chapter introduces biogas production using anaerobic digestion of organic waste (e.g., food scraps, animal manure, grass clippings and straws). It introduces the processes involved in anaerobic digestion, the major factors that influence these processes, the biogas produced, and common types of digesters. It also presents methods for determining biogas and methane yields.

> **Outcomes**
>
> After reading this chapter, you should be able to:
>
> - Explain the microbiological, chemical, and physical processes in anaerobic digestion
> - Describe the types of anaerobic digester used for biogas production and factors influencing their performance
> - Describe some methods of cleaning biogas for energy generation
> - Estimate the quantity of biogas, methane, and energy that can be produced from an organic material
> - Calculate the volume of a digester to treat a certain amount of a substrate

Concepts

Anaerobic digestion is a bioconversion process that is carried out by anaerobic microorganisms including anaerobic bacteria and methanogenic archaea to break down and convert organic matter into biogas, which is mainly a mixture of CH_4 and CO_2.

Biochemical Processes

Anaerobic digestion involves four major biochemical processes: hydrolysis, acidogenesis, acetogenesis, and methanogenesis. Figure 1 shows these processes for the conversion of organic substrates (such as proteins, carbohydrates, and lipids) into biogas.

Hydrolysis converts complex organic matter using extracellular and intracellular enzymes from the microorganisms to monomer or dimeric components, such as amino acids, single sugars, and long chain fatty acids (LCFA). During acidogenesis, the hydrolysis products are converted by acidogenic bacteria into smaller molecules such as volatile fatty acids (VFA), alcohols, hydrogen, and NH_3. In acetogenesis, alcohols and VFA (other than acetate) are converted to acetic acid or hydrogen and CO_2. The acidogenic and acetogenic bacteria are a diverse group of both facultative and

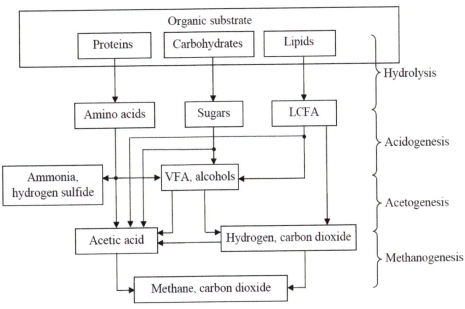

Figure 1. The steps of anaerobic digestion of complex organic matter into biogas. LCFA = long chain fatty acids; VFA = volatile fatty acids (derived from Pavlostathis and Giraldo-Gomez, 1991, and El Mashad, 2003).

obligate anaerobic microbes including *Clostridium, Peptococcus, Bifidobacterium, Corynebacterium, Lactobacillus, Actinomyces, Staphylococcus, Streptococcus, Desulfomonas, Pseudomonas, Selemonas, Micrococcus,* and *Escherichia coli* (Kosaric and Blaszczyk, 1992). During methanogenesis, acetic acid and methanol (an alcohol) are converted to CH_4 and CO_2. In addition, CO_2 and hydrogen are converted into CH_4. Methanogenic archaea include a diverse group of obligate anaerobes such as *Methanobacterium formicicum, Methanobrevibacter ruminantium, Methanococcus vannielli, Methanomicrobium mobile, Methanogenium cariaci, Methanospirilum hungatei,* and *Methanosarcina barkei* (Kosaric and Blaszczyk, 1992). Examples of the conversion of selected compounds during anaerobic digestion are shown in table 1.

Table 1. Examples of conversion of selected compounds during anaerobic digestion.

Sub-processes	Examples
Hydrolysis	Conversion of carbohydrates and proteins: $Cellulose + H_2O \rightarrow sugars$ $Proteins + H_2O \rightarrow amino\ acids$
Acidogenesis	Conversion of glucose into acetic and propionic acids: $C_6H_{12}O_6 \rightarrow 3CH_3COOH$ $C_6H_{12}O_6 + 2H_2 \rightarrow 2CH_3CH_2COOH + 2H_2O$
Acetogenesis	Conversion of propionate and butyrate into acetate and hydrogen as follows: $CH_3CH_2COO^- + 3H_2O \rightarrow CH_3COO^- + HCO_3^- + H^+ + 3H_2$ $CH_3CH_2CH_2COO^- + 2H_2O \rightarrow 2CH_3COO^- + H^+ + 2H_2$ $4H_2 + 2HCO_3^- + H^+ \rightarrow CH_3COO^- + 4H_2O$
Methanogenesis	Conversion of acetic acid, carbon dioxide and hydrogen, and methanol to methane: $4CH_3COOH \rightarrow 4\ CO_2 + 4\ CH_4$ $CO_2 + 4\ H_2 \rightarrow CH_4 + 2H_2O$ $4CH_3OH + 6\ H_2 \rightarrow 3CH_4 + 2H_2O$

Types of Anaerobic Digesters

Anaerobic digesters can be categorized based on how the microorganisms inside the digester interact with the substrate. There are three attributes used: (1) how the microorganisms are grown: suspended growth or fixed growth, (2) the feeding of substrate into the vessel as a batch, a plug, or continuously and (3) the number of stages, single or multistage. Further design considerations are whether the contents are actively mixed, whether the orientation is predominantly vertical or horizontal, and whether the flow through the vessel is downwards or upwards.

Suspended Growth Anaerobic Digesters

Suspended growth digesters are usually used for substrates with a high content of suspended solids, such as municipal wastewater and diluted solid waste. They can be operated as a batch process (figure 2) or as plug flow (figure 3), where a batch of substrate moves through the vessel as a block of material, called a plug. The microorganisms are dispersed throughout the reactor when the digester contents are mixed, such as continuous stirred tank reactors (CSTR), or anaerobic contact reactor (ACR) which is a CSTR with effluent solids recycled from a settling tank for solids. In a CSTR, solid retention time (SRT) equals the hydraulic retention time (HRT), which is the average time the solids and liquid remain in the bioreactor vessel. CSTR systems are operated at HRT and SRT ranging from 10 to 30 days. The ACR has a longer SRT (>50 days) than the HRT (0.5–5 days) because part of effluent solids is recycled back into the digester.

> HRT = hydraulic retention time = the average time the solids and liquid remain in the bioreactor vessel
>
> SRT = solid retention time
>
> ACR = anaerobic contact reactor
>
> ASBR = anaerobic sludge bed reactor
>
> CSTR = continuous stirred tank reactors
>
> UASB = upflow sludge blanket reactors

Figure 2 shows a schematic of a suspended growth batch anaerobic digester. These are simple to design and operate. They are usually an air-tight vessel with inflow and outflow ports to supply fresh substrate and remove spent substrate, a biogas outlet port, and a port for removing solids. These systems are commonly deployed at small scale and for testing the anaerobic biodegradability of different materials. Operation starts with mixing a fixed amount of substrate with inoculum, which is active bacterial culture taken directly from a running reactor. Afterwards, anaerobic conditions are maintained for the digestion time (i.e., the retention time), which should ensure the depletion of all the available substrate.

The CSTR is typically used to treat agricultural and municipal wastes with total solid (TS) contents of 3% to 12%. They are usually operated at controlled temperatures, so the vessel, constructed either below or above ground, is equipped with a heating system and thermal insulation to maintain a constant internal temperature. Plug flow digesters (figure 3) are constructed as long pipes or channels, above or below ground, with a gas tight cover. The digester contents travel through the vessel where they are converted into biogas until reaching the outlet. The residence time is determined by the time elapsed between the feed of fresh substrate and discharge of the digested materials. They are used to treat relatively high TS of 12% to 16%.

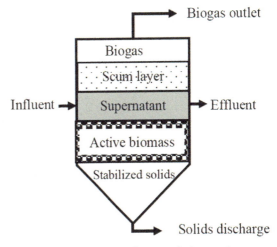

Figure 2. A schematic of suspended growth anaerobic digester.

Covered lagoons are commonly used to treat wastewater with low solids content (<3%), such as flushed animal manure. Manure lagoons on livestock farms can be upgraded to be anaerobic covered lagoons using a non-permeable covering to collect the biogas and double synthetic liners to prevent ground water contamination by seepage of the digester content. Covered lagoon digesters can be mixed or non-mixed (i.e., have mechanical agitation or not) and can be operated as plug flow or CSTR systems. They usually operate at ambient temperatures dictated by the local climate.

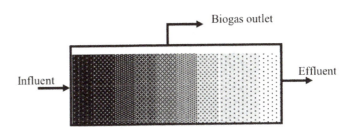

Figure 3. A schematic of a plug flow digester. The shading pattern represents the progressive decrease in substrate concentration as it moves through the digester.

There is also a class of suspended growth systems called high rate systems, which are characterized by using longer SRT than HRT. These systems are usually used for diluted wastewater with an SRT of >20 days, which is achieved by retaining the microorganisms in the digester. The long SRT enables treatment at high organic loading rates (amount of organic material processed per unit time). HRT can range from hours to days, depending on the characteristics of the wastewater. Designs include anaerobic sludge bed reactors (ASBR) and upflow sludge blanket reactors (UASB). In the ASBR, the retention of microorganisms is achieved by solids settling in the reactor prior to effluent removal. In the UASB, microorganisms form granules and are retained in the reactor.

Fixed Growth Anaerobic Digesters

In fixed growth digesters, microorganisms are grown on solid media allowing SRT longer than HRT. These systems are also high rate systems. Fixed growth anaerobic digesters are used to treat soluble organic wastes (i.e., low suspended solids content) that do not require hydrolysis. Media, such as plastic or rocks, are usually used to support the attachment and growth of microorganisms, which form biofilms. As wastewater passes over the growth media, contaminants are absorbed and adsorbed by the biofilms and degraded. Therefore, these digesters can be operated at higher organic loading rates than the suspended growth digesters.

Anaerobic filters are a type of fixed-growth anaerobic digester (figure 4). In these systems, much of the sludge containing active microorganisms is retained inside the digester by being attached as a biofilm to a solid (inert) carrier material. Anaerobic filters are operated in up-flow mode, meaning the inflow is below the outlet in the digestion chamber.

Factors Affecting Anaerobic Digestion and Biogas Production

Anaerobic digestion processes are affected by many factors, including substrate composition, temperature, pH, organic loading, retention time, and mixing, which in turn affect the yield and rate of biogas production. Process stability (i.e., the consistency of the biogas production rate) depends on maintenance of the biochemical balance between the acidogenic and methanogenic microorganisms. Process stability also depends on the chemical composition and physical properties of the substrate, digester configuration, and process parameters such as temperature, pH, and NH_3 concentration.

Substrate Composition and Characteristics

Substrate composition, particularly physical and chemical characteristics, is an important factor affecting design of biomass handling and digestion systems, performance of anaerobic digestion, biogas yield, and downstream processing of the digested materials. Materials with large particle sizes (e.g., crop residues and energy crops) may need to be ground before being fed into the anaerobic digester. The grinding process can aid in the conversion process because small particles can be degraded faster than large ones. Moreover, grinding

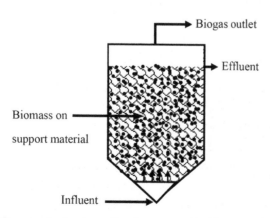

Figure. 4. A schematic of an anaerobic filter.

can help when handling the substrate and mixing the digester contents. Mixed wastes, such as municipal solid waste, usually contain inorganic materials (e.g., metals and construction debris) and need a separation process to remove these inorganic materials. Organic matter is mainly composed of carbon (C), hydrogen (H), and oxygen (O). It also contains many nutrient elements including macronutrients (e.g., nitrogen (N), potassium, magnesium, and phosphorus) and micronutrients (zinc, manganese, cobalt, nickel, and copper). Example compositions are given in table 2. All these nutrients are needed by microorganisms in order to break down and convert organic matter into biogas. An appropriate C: N ratio in the substrate is in the range of 20–25 C to 1 N. Most organic wastes, such as animal manure and food waste, contain enough nutrients to support the growth of microorganisms.

> TS = total solids
>
> VS = volatile solids, the organic fraction of TS or dry matter
>
> BOD = biochemical oxygen demand
>
> COD = chemical oxygen demand

The organic matter content of a substrate is described in terms of volatile solids (VS), chemical oxygen demand (COD), or biochemical oxygen demand (BOD). VS are used to characterize substrates with a high solids content, while COD and BOD are used to characterize substrates that have a low solids content, such as wastewater. VS is the organic fraction of total solids (TS) or dry matter. BOD is used to describe the biodegradability of a substrate, while COD is the amount of oxygen needed to chemically oxidize the organic matter in a substrate. If the chemical composition of a substrate is known, the COD can be calculated using the chemical reaction:

$$C_aH_bO_cN_d + \left(a + \frac{b}{4} - \frac{c}{2} - \frac{3d}{4}\right)O_2 \rightarrow aCO_2 + \left(\frac{b}{2} - \frac{3d}{2}\right)H_2O + dNH_3$$

where a, b, c, and d are number of atoms of carbon, hydrogen, oxygen, and nitrogen, respectively, and allow calculation of the amount of oxygen required for the reaction, i.e.,

$$\left[\left(a + \frac{b}{4} - \frac{c}{2} - \frac{3d}{4}\right)O_2\right] = COD$$

Table 2. Composition of selected organic wastes (dry weight basis), (Zhang, 2017)

Sample	C/N	C (%)	N (%)	P (%)	K (%)	S (%)	Ca (%)	Mg (%)	B (ppm)	Zn (ppm)	Mn (ppm)	Fe (ppm)	Cu (ppm)	Na (ppm)	Co (ppm)	Ni (ppm)
Tomato waste	13.0	40.3	3.1	0.3	1.1	0.3	2.4	0.7	72.9	40.1	183.6	4482.8	23.6	1528.5	2.5	14.0
Tomato pomace	17.0	57.8	3.5	0.5	1.0	0.2	0.3	0.3	17.6	40.1	53.8	510.3	14.3	477.0	0.4	3.0
Rice straw	77.0	38.6	0.5	0.1	2.8	0.1	0.2	0.2	6.6	33.5	492.2	432.2	4.9	2054.0	1.3	2.0
Egg liquid waste	8.0	61.8	7.8	0.6	0.7	0.7	0.4	0.1	1.3	18.1	1.5	68.0	15.9	7165.0	<0.1	5.0
Commercial food waste	16.0	43.7	2.7	0.5	2.4	0.3	3.5	0.2	18.7	170.8	34.1	443.7	9.1	3443.0	0.4	2.0
Supermarket vegetable waste	22.0	45.6	2.1	0.4	2.9	0.2	0.3	0.2	38.6	126.6	22.0	187.1	10.4	1669.5	0.2	15.0
Cardboard	231.0	46.2	0.2	0.0	0.0	0.2	0.4	0.0	42.4	18.6	26.3	255.8	10.3	1950.5	0.3	3.0
Dairy manure	18.0	34.0	1.9	0.8	2.6	0.5	1.5	1.5	70.0	280.0	210.0	2100	110.0	7790		<20
Chicken manure	9.0	31.9	3.7	1.8	2.8	0.6	10.3	0.6	34.6	325.3	312.2	739.4	36.1	4162.0	0.5	12.0

Temperature

Temperature is an important factor affecting the performance of anaerobic digestion because it affects the kinetics of the processes. Microorganisms are usually classified by the optimum temperature and the temperature range at which they grow. The normal classification is psychrophilic (<25°C), mesophilic (25 to 45°C), and thermophilic (45 to 65°C), but in theory there is the extreme of hyperthermophilic anaerobic archaea and bacteria that can grow in geothermal environments with optimal growth temperatures of 80°C to 110°C (Stetter, 1996). Thermophilic digestion may produce biogas with a higher CO_2 content than mesophilic digestion due to the low solubility of CO_2 in water at high temperatures.

The growth rate of microorganisms increases with increasing temperature up to an optimum. Above the optimum temperature, growth declines due to the thermal denaturation of the cell protein. The growth will cease when the essential protein of the cell is destroyed. Figure 5 shows the relative growth rate of methanogens at different temperature ranges. Within the temperature range of one species, the growth rate exponentially increases with temperature. Thermodynamically, most biochemical reactions require less energy to proceed at high temperatures. The rate of most chemical reactions approximately doubles with a temperature increase of 10°C (Stanier et al., 1972). The energy required to heat up the substrate and to keep the digester at the desired temperature is greater at higher temperatures.

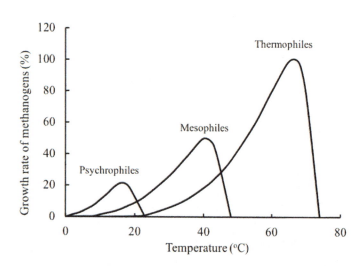

Figure 5. Relative growth rate of methanogens under psychrophilic, mesophilic, and thermophilic conditions (Lettinga et al., 2001).

pH

The pH of a digester is affected by the interaction between the composition of the substrate, its buffering capacity, and the balance between the rates of acidification and methanogenesis. If the rate of methanogenesis is lower than acidogenesis, the pH might reach values below 6, which can cause inhibition to methanogenic archaea. The relationship between pH and methanogenic activity is a bell shaped curve (figure 6) with a maximum methanogenic activity at pH values between about 6.8 and 8 (Speece, 1996; Khanal, 2008). An optimum pH near neutrality should be maintained in the anaerobic digester for biogas production.

Organic Loading

Organic loading (or initial loading) is a measure of the amount of organic matter, expressed in terms

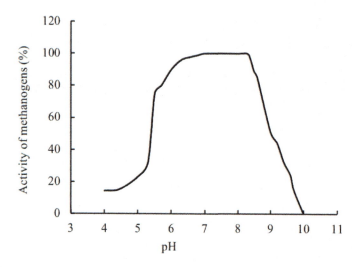

Figure 6. Relative activity of methanogenic archaea at different pH (Speece, 1996; Khanal, 2008).

of the amount of VS or COD that enters a batch digester at the beginning of a process cycle. It is an important parameter that affects digester sizing because it determines the concentration of functional microbial biomass per unit mass of substrate. For continuously fed digesters, organic loading rate (OLR), usually defined as the amount of organic matter fed per unit volume of the digester per day, depends on the biodegradation kinetics of the substrate, digester design and operating conditions. For example, a CSTR treating animal manure with a TS content of 1–6% is usually operated at an OLR of 1.6 to 4.8 kg m^3 day^{-1} and an HRT of 15 to 30 days.

Retention Time

Retention time is the time for the substrate to remain in the digester to be processed by the microorganisms. The appropriate retention time depends on the chemical and physical characteristics of the substrate and the rate of microbial metabolism. Complex substrates, such as agricultural wastes (e.g., animal manure), usually have low biodegradation rates and so need longer retention times (20–30 days), while highly biodegradable materials, such as food waste, may need shorter retention times (<15 days) to convert the biodegradable organic matter into biogas.

Mixing

Mixing affects the performance of anaerobic digesters by ensuring homogenization of the reactor contents by breaking of the substrate particles and exposing large surface areas of the substrate to microorganisms. Adequate mixing prevents development of stratification inside digesters, which could result in unfavorable micro-environments for the methanogens, such as regions rich in toxic compounds or with low pH. Mixing also helps to maintain a uniform temperature in the digester and prevents the formation of a scum layer. The requirement to achieve proper mixing depends on the digester shape, type of mixing systems, and solids content inside the digester. For example, a rectangular tank poses difficulty for mixing compared to cylindrical and egg-shaped reactors because it is difficult to mix into the corners. Digesters can be mixed with mechanical mixers, recirculation of biogas, or recirculation of reactor contents. The selection of the mixing system depends on the density of substrate (i.e., solid concentration), required mixing intensity, homogeneity, availability and cost of mixing equipment, and maintenance and energy consumption costs.

Process Configuration

Anaerobic digestion processes can be carried out in single stage digesters or in multistage digesters. Single stage digesters are usually used for materials that have balanced degradation rates of hydrolysis, acidogenesis, and methanogenesis and have enough buffer capacity to maintain the pH of the digester around neutral. However, for highly biodegradable materials, such as food waste, multiple stage (mostly two stages) digestion systems are usually used. In these systems, hydrolysis and acidogenesis are the predominant processes in the first stage, with low pH (4–6) due to the high concentrations of VFA.

The biogas produced from the first stage contains high contents of CO_2 and hydrogen and low content of CH_4. In the second stage, methanogenesis predominates when VFA are consumed by the methanogenic archaea and the pH is in the range of 6.8–8. The biogas produced from the second stage has high CH_4 content (50–70%).

Ammonia Concentration

The anaerobic digestion of protein-rich substrates may produce high NH_3 concentrations that can cause inhibition or even toxicity to anaerobic microorganisms. Microorganisms need N for their cell synthesis. Approximately 6.5% of the influent N is used for cell production. Fermentative bacteria can usually utilize both amino acids and NH_3, but methanogenic bacteria only use NH_3 for the synthesis of bacteria cells (Hobson and Richardson, 1983). High NH_3 concentrations can cause inhibition, or even toxicity, to methanogenic microorganisms. Inhibition is indicated by a decrease in NH_3 production and increasing VFA concentrations. When there is a total cessation of the methanogenic activity, free NH_3 is usually the main cause. This is because microorganism cells are more permeable to free NH_3 than to ammonium ions. The concentration of free NH_3 depends on the total NH_3, temperature, and pH.

Estimation of Biogas and Methane Yields

Theoretical Estimation of Yield

Biogas and CH_4 yields can be estimated theoretically from the chemical composition of the substrate or measured using batch digestion experiments. Biogas and CH_4 yield from a completely biodegradable organic substrate with the composition ($C_aH_bO_cN_d$) can be determined using Buswell's equation (Buswell and Mueller, 1952):

$$C_aH_bO_cN_d + \left(\frac{4a-b-2c+3d}{4}\right)H_2O \rightarrow \left(\frac{4a+b-2c-3d}{8}\right)CH_4 + \left(\frac{4a-b+2c+3d}{8}\right)CO_2 + dNH_3$$

This equation does not consider the needs of organic matter for cell maintenance and anabolism. From Buswell's equation, the total amount of biogas produced from one mole of the biodegradable organic substrate can be calculated as a summation of CH_4 and CO_2, i.e.:

$$\left[\left(\frac{4a+b-2c-3d}{8}\right)+\left(\frac{4a-b+2c+3d}{8}\right)\right]$$

and the amount of methane produced as $\left[\left(\frac{4a+b-2c-3d}{8}\right)\right]$, calculated (in moles) at a standard temperature (0°C) and pressure (101.325 kPa, or 1 atm).

The volume of the biogas or methane yield per each gram of the substrate L g^{-1} [VS] can be calculated using the molar volume of an ideal gas as 22.4 L at the standard temperature and pressure as:

$$M_y = \frac{\left(\frac{4a+b-2c-3d}{8}\right) \times 22.4}{12a+b+16c+14d} \quad (1)$$

where M_y = methane content in the biogas, % (mole/mole or v/v).

Assuming that biogas is composed mainly of methane and carbon dioxide and ammonia production is insignificant, methane content in the biogas can be calculated as follows:

$$M_c = \frac{\left(\frac{4a+b-2c-3d}{8}\right) \times 100}{\left(\frac{4a+b-2c-3d}{8}\right) + \left(\frac{4a-b+2c+3d}{8}\right)} \quad (2)$$

where M_c = methane content in the biogas, % (mole/mole or v/v).

The CH_4 production after a long degradation time is called the methane potential. Methane yield can be expressed as the volume of gas produced per unit mass of the substrate (L [CH_4]/kg [substrate]), VS (L [CH_4]/kg [VS]) or COD (L [CH_4]/kg [COD]). Theoretical CH_4 yield and content of selected substrates computed using the Buswell equation (table 3) are usually underestimated because CO_2 is more soluble in water than CH_4. In anaerobic digesters, CH_4 content of the biogas ranges from 55% to 70%, depending on the substrate and operation conditions of digesters (table 3). Substrates rich in lipids should produce biogas rich in methane.

Table 3. Composition and theoretical methane yield and methane content of selected substrates (e.g., Angelidaki & Sanders, 2004).

Substrate Type	Formula	Gas Yield[a] (L g^{-1} [VS])			Methane Content[b] (%)
		CH_4	CO_2	NH_3	
Carbohydrate	$(C_6H_{10}O_5)_n$	0.415	0.415	0.000	50.0
Protein	$C_5H_7NO_2$	0.496	0.496	0.198	50.0
Lipid	$C_{57}H_{104}O_6$	1.014	0.431	0.000	70.2
Acetate	$C_2H_4O_2$	0.374	0.374	0.000	50.0
Ethanol	C_2H_6O	0.731	0.244	0.000	75.0
Propionate	$C_3H_6O_2$	0.530	0.379	0.000	58.3

[a] Yields are at standard temperature and pressure (see text).
[b] Assuming the biogas is composed of methane and carbon dioxide.

Modeling the Anaerobic Digestion Process to Estimate Yield

There are mechanistic models that describe the anaerobic digestion process, which can be used to predict the performance of anaerobic digesters. One of the most used is the Anaerobic Digestion Model No. 1 (ADM1), developed by the International Water Association Task Group for Mathematical Modelling of Anaerobic Digestion Process (Batstone et al., 2002). ADM1 is structured around biochemical sub-processes, including hydrolysis, acidogenesis, acetogenesis, and methanogenesis. While a mechanistic modelling approach is necessary for advanced design, a simple first-order kinetic model can be used to calculate the methane yield from different substrates, such as food waste, animal manure, and crop residues, and be used for preliminary design. The first-order kinetics for a batch digester can be written as:

$$\frac{dS}{dt} = -k\,S \qquad (3)$$

where t = digestion time (days)
k = first-order degradation kinetic rate constant (day^{-1})
S = concentration of the biodegradable organic matter (expressed as VS, COD, or BOD) in the digester (kg m^{-3})

With the concentration of the biodegradable substrate at the beginning of the digestion time designated as S_0 (kg m^{-3}), the equation can be expressed as:

$$S = S_0 e^{-kt} \qquad (4)$$

Equation 4 can be used to predict the remaining substrate concentration (S) in the digester after a period of digestion time (t) if the initial substrate concentration (S_0) and degradation kinetic rate constant are known. The amount of degraded organic matter that is converted into methane, and the amount of methane produced can be calculated as:

$$S_{deg} = V_w (S_0 - S) \qquad (5)$$

$$M_p = M_y\, S_{deg} \qquad (6)$$

where S_{deg} = degraded organic matter in the digester (kg)
V_w = working volume of digester (i.e., volume of liquid inside the digester) (m^3)
M_p = amount of methane produced (m^3)
M_y = methane yield (m^3 kg^{-1})

Equations 4, 5 and 6 can be used to fit experimental data describing the substrate concentration at time steps throughout the process to determine the first-order degradation kinetic rate constant. They can also be used to predict degraded organic matter in the digester and methane yield at different digestion times if the first-order degradation kinetic rate constant is known from the literature or from experiments.

Estimation of Energy Production from a Substrate

The amount of energy contained in a fuel (e.g., biogas) is expressed using the higher heating value (HHV) or lower heating value (LHV). The HHV is the total heat produced from a complete combustion of a unit (usually 1 m^3) of the gas under a constant pressure and all the water formed by the combustion reaction condensed to the liquid state. The LHV is the net caloric value produced from the combustion of a unit amount of the fuel and all the water formed during the combustion reaction remains in the vapor state. Methane is used to calculate the amount of energy contained in the biogas because it is the main combustible gas. At standard temperature and pressure, methane has a LHV of approximately 36 MJ m^{-3}. Therefore, the LHV of biogas containing 65% methane is approximately 23.4 MJ m^{-3}, which is calculated by multiplying the LHV of methane with the methane content of the biogas.

The amount of energy that is produced from an anaerobic digester can be estimated using the amount of organic matter that is treated in a certain period of time (e.g., day), biogas yield of the substrate, and methane content of the biogas. Based on the TS and VS content of the substrate, the amount of organic matter to be treated can be calculated as:

$$\phi_{om} = Q \times T_{sc} \times V_{sc} \tag{7}$$

where ϕ_{om} = amount of organic matter to be treated per day, kg [VS] day^{-1}
Q = amount of feedstock to be treated (kg day^{-1})
T_{sc} = total solids contents, %, wet basis
V_{sc} = volatile solids contents, % of T_{sc}

The daily biogas and methane production can be calculated as:

$$B_{dp} = \phi_{om} \, B_y \tag{8}$$

$$M_{dp} = B_{dp} \, M_C \tag{9}$$

where B_{dp} = daily biogas production, m^3 day^{-1}
B_y = biogas yield production, m^3 kg^{-1} [VS]
M_{dp} = daily methane production, m^3 day^{-1}
M_c = methane content in the biogas, % vol vol^{-1}

The daily energy production from biogas can be calculated as:

$$E_{dp} = B_{dp} \times C_{vb} \tag{10}$$

or

$$E_{dp} = M_{dp} \times C_{vm} \tag{11}$$

where E_{dp} = daily energy production, MJ day^{-1}
C_{vb} = calorific value of biogas, MJ m^{-3}
C_{vm} = calorific value of methane, MJ m^{-3}

Sizing Anaerobic Digesters

Anaerobic digester performance is controlled by the number of active microorganisms that are in contact with the substrate. Therefore, increasing the number of active bacteria can increase the conversion rate and, consequently, higher organic loading rates can be used. The total volume (V_t) of a digester is calculated from the working volume (V_w) and head space volume (V_h) as:

$$V_t = V_w + V_h \tag{12}$$

The head space volume is the gas volume above the liquid that is sometimes used for gas storage. The head space volume is usually about 10% of the working volume. The required working volume of a continuously fed anaerobic digester can be determined from the amount of organic matter (expressed as VS or COD) to be treated per day and the OLR:

$$V_w = \frac{\phi_{om}}{\text{OLR}} \tag{13}$$

where V_w = working volume of digester, m³
OLR = organic loading rate

The working volume can also be determined from the volume of waste to be treated per day and the hydraulic retention time of the digester:

$$V_w = V_{df} \times \text{HRT} \tag{14}$$

where V_{df} = volumetric feed to the digester, m³ day⁻¹
HRT = hydraulic retention time

Biogas Cleaning and Upgrading

Biogas cleaning and upgrading processes are important to remove harmful and undesired compounds and increase the quality of the biogas as a fuel. Biogas cleaning is the removal of impurities such as hydrogen sulfide and organic compounds, and upgrading is the removal of CO_2 and water vapor, resulting in a relatively pure methane (biomethane) that can be used as automobile fuel or injected in a natural gas pipeline.

Table 4 shows a typical composition of biogas from agricultural waste digestion and municipal solid waste landfills.

Biogas Cleaning

Removing hydrogen sulfide is important prior to using biogas because it is corrosive and toxic. In the presence of water vapor, hydrogen sulfide forms sulfuric acid, which can cause serious corrosion of the metallic components of the digester and biogas handling equipment. The removal of hydrogen sulfide can be carried out using chemical precipitation by the addition of metal ions (usually ferric ions) to the digester vessel or chemical absorption by passing the biogas through a ferric solution (e.g., ferric choloride (known as an iron sponge)) as:

$$3H_2S + 2FeCl_3 \rightarrow Fe_2S_3 \downarrow + 6H^+ + 6Cl^-$$

In addition, hydrogen sulfide can be removed using biological oxidation by chemotrophic bacteria such as *Thiobacillus thioparus*. However, commercial application of biological oxidation is limited.

Siloxanes are volatile organic compounds that are usually found in the biogas produced from landfills. During combustion reactions, they are converted into silicon dioxide (SiO_2) and microcrystalline quartz that deposit on engine parts, causing problems such as wearing. Activated carbon or silica gel are commonly used as adsorbents to remove these organic compounds from the biogas.

Table 4. Typical composition of biogas from different materials (Coombs, 1990).

Component	Agricultural Wastes	Municipal Solid Waste Landfills
Methane	50–80%	45–65%
Carbon dioxide	30–50%	34–55%
Water vapor	Saturated	Saturated
Hydrogen sulfide	100–7,000 ppm	0.5–100 ppm
Hydrogen	0–2%	0–1%
Ammonia	50–100 ppm	Trace
Carbon monoxide	0–1%	Trace
Nitrogen	0–1%	0–20%
Oxygen	0–1%	0–5%
Organic volatile compounds	Trace	5–100 ppm

Biogas Upgrading

Removal of CO_2 is important to increase the energy content of biogas, to reduce the required volumes for biogas storage, and to achieve the quality needed for compliance with the specifications of natural gas for distribution with fossil gas and the specifications for compressed natural gas engines. Moreover, the presence of CO_2 can cause corrosion to equipment and pipelines if it mixes with water to form carbonic acid. Carbon dioxide can be removed from biogas with water or chemical scrubbing systems, in which water or chemical solvents (e.g., sodium hydroxide and amine) react with CO_2:

$$CO_2 + H_2O \leftrightarrow H_2CO_3$$

$$CO_2 + NaOH \rightarrow NaHCO_3$$

Carbon dioxide can also be removed from biogas by using membranes and pressure swing adsorption (PSA) systems. Membranes have selective permeability. They allow different compounds (e.g., gases) to move across the membrane at different rates. When biogas is pumped under pressure (up to 4000 kPa) through a membrane made of polymers, carbon dioxide is separated from methane. In the pressure swing adsorption system, biogas flows under pressure (up to 1000 kPa) through a porous material that allows methane to pass through while absorbing and removing carbon dioxide. The adsorbent materials in commercial systems include carbon molecular sieves, activated carbon, silica gel, and zeolites. Before the adsorbent material is completely saturated with carbon dioxide, it needs to be regenerated and then reused. The regeneration process is carried out by reducing the pressure in the vessel to pressures close to ambient and then to a vacuum.

Some adsorbent materials used for carbon dioxide can also adsorb hydrogen sulfide, oxygen, and nitrogen. However, the absorbance of hydrogen sulfide on these materials is not reversible.

Biogas collected from digesters is saturated with water vapor. The water content of biogas depends on the operating temperature of the digester. At lower temperatures there will be less water vapor in the biogas. Water vapor is removed to protect pipelines and equipment from corrosion through the formation of acids (e.g., sulfuric and carbonic acids). Water vapor can be removed by condensation or chemical drying (e.g., absorption). Water vapor condensation can be forced by reducing the dew point using a cooling system such as a chiller and heat exchanger. A fluid is cooled in the chiller and pumped through one side of the heat exchanger to reduce the temperature of the biogas that flows in the other side of the heat exchanger. In chemical drying, agents such as silica gel, magnesium oxide, aluminum oxide, or activated carbon are used to absorb the water vapor. After saturation, the drying agents are regenerated by heating to around 200°C. To maintain continuous operations, two columns filled with the drying agents are used to make sure that unsaturated drying agent is used while the saturated one is regenerated.

Applications

Experimentation to Determine Digestion Properties

The biogas and methane yields can be determined by using batch anaerobic digestion experimental set-ups ranging from the very simple (figure 7) to a sophisticated automated methane potential test system (AMPTS) (figure 8). Anaerobic batch digestion tests can be carried out at small scale (0.1–1 liter) to determine biogas and biomethane yields and biodegradability of a substrate. The simple batch method can be conducted using affordable laboratory equipment; an AMPTS is more expensive but can be automated and is more accurate. The AMPTS allows measurement of biogas production through time.

A simple anaerobic batch digestion system (figure 7) is composed of a vessel, which is normally a bottle sealed with a cap and an opening to let the biogas out. Based on the composition (TS and VS) of the substrate, an amount of the substrate that gives 3 g VS is used to start the digestion. The substrate is put in the vessel and inoculum added. The inoculum is a seed material taken from an active anaerobic digester. The pH of the digester should be approximately 7. The digester is flushed with an inert gas, such as helium or argon, for approximately two minutes to ensure anaerobic condition

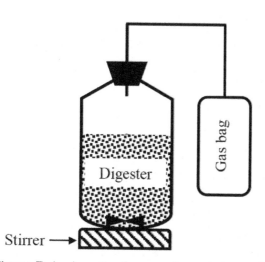

Figure 7. A schematic of an experimental set-up of a batch digester system.

by removing oxygen from both liquid and head space. The digester is sealed with a rubber stopper and connected to a gas bag (called a Tedlar bag) to collect the biogas. The digester is incubated at a constant temperature (35°–50°C) for up to 25 days. During the incubation time, the contents are mixed intermittently using a stirrer or by manual shaking for about one minute, but without breaking the seal of the bottle. Each treatment should be replicated and a control using just inoculum is used to estimate the biogas produced by the inoculum alone. The collected biogas can be measured using liquid displacement or gas tight syringe. The pH is measured at the end of the digestion time. Biogas yield (L g^{-1} VS) is determined by dividing the cumulative biogas by the initial amount of the VS in the digester at the beginning of the digestion. The methane yield is calculated by multiplying biogas yield by the methane content of biogas that can be measured using a gas chromatograph.

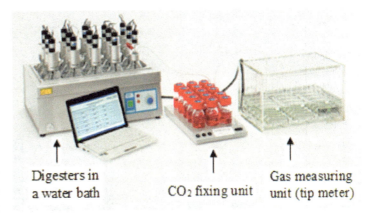

Figure 8. Experimental set-up of an automated methane potential test system (AMPTS).

An AMPTS (figure 8) is composed of three parts: a water bath with a temperature control, a CO_2 fixation unit, and a gas tip meter. The vessels are incubated in the water bath at a constant temperature. All the vessels are continuously mixed using mechanical mixers. The CO_2 fixation unit is used to remove CO_2 from the biogas. The gas measuring unit (tip meter) can determine the amount of methane production from each individual digester. The tip meter is connected to a data logger that continuously records the methane production. All procedures for preparing simple anaerobic batch digesters are also applied in the AMPTS.

Figure 9 shows daily biogas production and cumulative biogas yield determined from a batch anaerobic digester, with a capacity of 1 L, treating cafeteria food waste at an initial VS loading of 4 g L^{-1} and a temperature of 50°C. The biogas production rates are high at the beginning of the batch digestion and then decline until reaching almost zero. This is due to the reduction of the organic matter contained in the substrate over the digestion time until all the available organic matter is consumed by microorganisms.

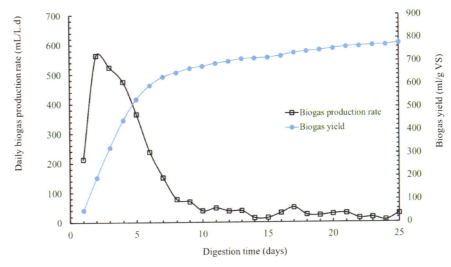

Figure 9. Daily biogas production and cumulative biogas yield of cafeteria food waste.

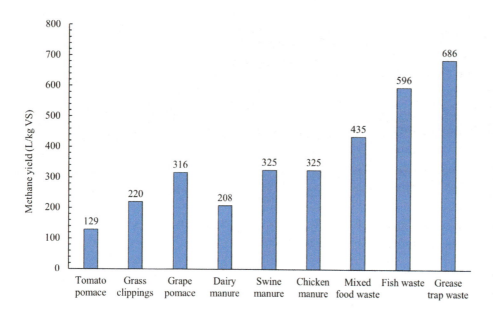

Figure 10. Methane yield of selected organic wastes.

The data of methane production and remaining substrate concentration in batch digestion tests can be used to determine the first-order degradation kinetic rate constant using equations 4, 5, and 6. Methane production is calculated by multiplying biogas production by methane content of the biogas (which is usually measured using gas chromatography). Figure 10 shows methane yields of various organic wastes after a digestion time of 25 days. As can be seen, the substrate composition affects the methane yield.

The experimental data from batch digestion tests could be used to determine the proper HRT and vessel size for pilot and full-scale systems to treat a specific amount of substrate. For example, the digestion time required to convert all or part of the biodegradable organic matter in a certain substrate to biogas could be used as a basis for determining the proper HRT to convert the substrate into biogas in a continuously fed digester at the same temperature. Once the HRT is determined, the effective volume can be determined using equation 14.

Commercial Uses of Biogas

In addition to utilizing biogas for electricity generation using generators and fuel cells, and for heating purposes, biogas can be upgraded to biomethane (also known as renewable natural gas, RNG). Biomethane is very similar to natural gas, therefore, most equipment used for natural gas can be operated with biomethane. Biomethane can be used as a transportation fuel in the form of renewable compressed natural gas (CNG) or liquefied natural gas (LNG). The U.S. Environmental Protection Agency defined renewable CNG and LNG as biogas or biogas-derived "pipeline quality gas" that is liquefied or compressed for transportation purposes. For these uses, biogas must be cleaned and upgraded, either onsite adjacent to the digester or pumped to a central facility that processes biogas from multiple digesters in the vicinity. Biomethane could also be sold to utility companies by injection into natural gas pipelines. Biomethane must meet high quality standards for injection in the pipelines (table 5).

Table 5. Quality specification of biomethane to be injected into the California gas pipeline (Coke, 2018).

Quality Parameter	Value
Water content (kg per 1000 m³ at 55.15 bar)	0.11
Hydrogen sulfide (ppm)	4
Total sulfur (ppm)	17
Carbon dioxide (%)	1
Hydrogen (%)	0.1

Examples

Example 1: Theoretical methane production

Problem:
A cafeteria wants to manage waste food by using it as the feedstock for an anaerobic digestor. What is the theoretical methane production at standard temperature and pressure from 1,000 kg of organic food waste with the chemical formula $C_{3.7}H_{6.4}O_{1.8}N_{0.2}$? What is the expected methane content of the biogas assuming it consists of only methane and carbon dioxide?

Solution:
Applying Buswell's equation:

$$C_{3.7}H_{6.4}O_{1.8}N_{0.2} + \left(\frac{4(3.7)-6.4-2(1.8)+3(0.2)}{4}\right) H_2O \rightarrow$$

$$\left(\frac{4(3.7)+6.4-2(1.8)-3(0.2)}{8}\right)CH_4 + \left(\frac{4(3.7)-6.4+2(1.8)+3(0.2)}{8}\right)CO_2 + 0.2\,NH_3$$

$$C_{3.7}H_{6.4}O_{1.8}N_{0.2} + 1.35\,H_2O \rightarrow 2.125\,CH_4 + 1.575\,CO_2 + 0.2\,NH_3$$

This means that 1 mole (82.4 g) of the organic food waste produces 2.125 mole of CH_4 and 1.575 mole of CO_2.

Calculate methane yield using equation 1:

$$M_y = \frac{\left(\frac{4a+b-2c-3d}{8}\right) \times 22.4}{12a+b+16c+14d} \tag{1}$$

$$M_y = \frac{2.125 \times 22.4}{82.4} = 0.577\ \text{L g}^{-1}\,[VS]$$

Amount of methane production from 1,000 kg $= 0.577 \times 1,000 \times 1,000 = 575,000$ L
$$= 577\ m^3$$

Calculate the methane content using equation 2:

$$M_C = \frac{\left(\frac{4a+b-2c-3d}{8}\right) \times 100}{\left(\frac{4a+b-2c-3d}{8}\right)+\left(\frac{4a-b+2c+3d}{8}\right)} \tag{2}$$

$$M_C = \frac{2.125 \times 100}{2.125+1.575} = 57.4\%$$

Example 2: Design of an anaerobic digester for dairy manure

Problem:

A dairy farmer wants to build an anaerobic digester to treat the manure produced from 1,000 cows. Each cow produces 68 kg of manure per day. The volatile solid (VS) of the manure is 11% (wet basis). The digester is to be operated at an organic loading rate of 2 kg [VS] m^{-3} day^{-1} and a temperature of 35°C. The gas headspace volume is 10% of the working volume. Biogas yield from manure is 288 L kg^{-1} [VS] and the methane content is 65%. Assume all the manure produced on the dairy will be treated in the digester. Calculate:

(a) the volume of the digester required,
(b) the daily biogas and methane production, and
(c) the daily energy production from biogas if the biogas has a calorific value of 23 MJ m^{-3}.

Solution:

The amount of organic matter to be treated per day (ϕ_{om}) can be calculated using the number of cows, the amount of manure produced from each cow per day, and the volatile solids contents of manure as follows:

The amount of organic matter to be treated per day (ϕ_{om}) =

number of cows × amount of manure produced from each cow

× volatile solids content of manure $= 1,000 \times 68 \times \left(\dfrac{11}{100}\right) = 7,480$ kg [VS] day^{-1}

Calculate the working volume of the digester using equation 13:

$$V_w = \dfrac{\phi_{om}}{OLR} = \dfrac{7,480}{2} = 3,740 \text{ m}^3$$

Calculate the total volume (V_t) of the digester using equation 12:

$$V_t = V_w + V_h$$

$$V_t = 3,740 + \left(\dfrac{10}{100}\right)(3,740) = 4,114 \text{ m}^3$$

Calculate the daily biogas production using equation 8:

$$B_{dp} = \phi_{om} B_y$$

$$B_{dp} = 7,480 \times 288 / 1,000 = 2,154.2 \text{ m}^3 \text{ day}^{-1}$$

Calculate the methane production using equation 9:

$$M_{dp} = B_{dp} \times M_C$$

$$M_{dp} = 2{,}154.2 \times 65/100 = 1{,}400.2 \text{ m}^3 \text{ day}^{-1}$$

Calculate the energy production using equation 10:

$$E_{dp} = B_{dp} \times CV_B$$

$$E_{dp} = 2{,}154.2 \times 23 = 49{,}546.6 \text{ MJ day}^{-1}$$

Example 3: Modeling and kinetics

Problem:
A batch digester with a volume of 5 L treats an organic substrate at an initial loading of 5 g [VS] L^{-1} for 25 days. The substrate has an ultimate methane yield of 350 mL g^{-1} [VS] degraded. Determine the concentration of the biodegradable substrate in the effluent and total amount of methane produced over 25 days if the first-order degradation kinetic rate constant is 0.12 day^{-1}.

Solution:
The concentration of the biodegradable VS in the digester effluent can be calculated using equation 4:

$$S = S_0 e^{-kt} \qquad (4)$$

After one day of digestion, the VS concentration is:

$$S = 5\left[e^{-0.12(1)}\right] = 4.434 \text{ g L}^{-1}$$

This calculation can be repeated for every day over the digestion time (25 days). The results of these calculations are plotted in figure 11.

The amounts of the degraded organic matter and methane produced can be predicted using equations 5 and 6. After one day of the digestion, these amounts can be calculated as:

$$S_{deg} = V_w(S_0 - S) \qquad (5)$$

$$S_{deg} = 5 \times (5 - 4.434) = 2.83 \text{ g}$$

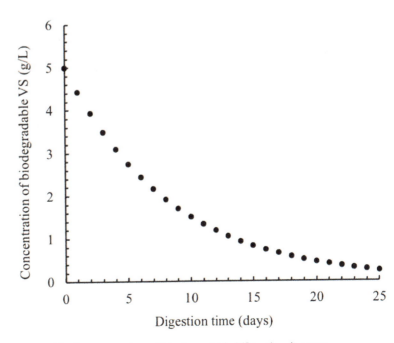

Figure 11. Concentration of biodegradable VS in the digester.

$$M_p = M_y S_{deg} \quad (6)$$

$$M_p = 350 \times 2.83 = 989.45 \text{ mL} = 0.9894 \text{ L}$$

These calculations can be repeated for every day over the digestion time (25 days). The results of these calculations are plotted in figure 12.

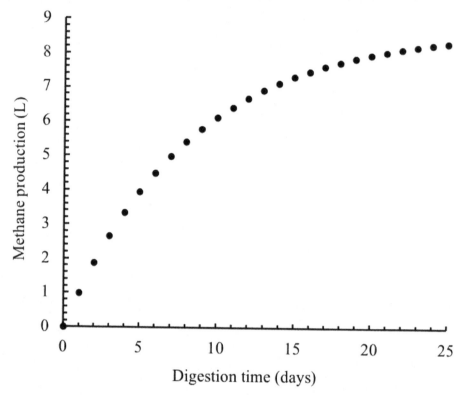

Figure 12. Total cumulative methane production.

Image Credits

Figure 1. El Mashad, H. & Zhang, R. (CC By 4.0). (2020). The steps of anaerobic digestion of complex organic matter into biogas (derived from Pavlostathis and Giraldo-Gomez, 1991 and El Mashad, 2003).

Figure 2. El Mashad, H. & Zhang, R. (CC By 4.0). (2020). A schematic of suspended growth anaerobic digester.

Figure 3. El Mashad, H. & Zhang, R. (CC By 4.0). (2020). A schematic of a plug flow digester.

Figure 4. El Mashad, H. & Zhang, R. (CC By 4.0). (2020). A schematic of an anaerobic filter.

Figure 5. El Mashad and Zhang. (CC By 4.0). (2020). Relative growth rate of methanogens under psychrophilic, mesophilic and thermophilic conditions (derived from Lettinga, et al., 2001. https://www.sciencedirect.com/science/article/pii/S0167779901017012#FIG1)

Figure 6. El Mashad, H. and Zhang, R. (CC By 4.0). (2020). Relative activity of methanogenic archaea at different pH (derived from Speece, 1996; Khanal, 2008).

Figure 7. El Mashad, H. & Zhang, R. (CC By 4.0). (2020). A schematic of an experimental set-up of a batch digester system.

Figure 8. Bioprocess Control. (2020). Experimental set-up of Automated Methane Potential Test System (AMPTS). Adapted from https://www.bioprocesscontrol.com/. [Fair Use].

Figure 9. El Mashad, H. & Zhang, R. (CC By 4.0). (2020). Daily biogas production and cumulative biogas yield of cafeteria food waste.
Figure 10. El Mashad, H. & Zhang, R. (CC By 4.0). (2020). Methane yield of selected organic wastes.
Figure 11. El Mashad, H. & Zhang, R. (CC By 4.0). (2020). Concentration of biodegradable VS in the digester.
Figure 12. El Mashad, H. & Zhang, R. (CC By 4.0). (2020). Total cumulative methane production.

References

Angelidaki, I., & Sanders, W. (2004). Assessment of the anaerobic biodegradability of macropollutants. *Rev. Environ. Sci. Biotechnol., 3*(2), 117-129.

Batstone, D. J., Keller, J., Angelidaki, I., Kalyuzhnui, S. V., Pavlostanthis, S. G., Rozzi, A., . . . Vavilin, V. A. (2002). *Anaerobic Digestion Model No.1. IWA task group for mathematical modelling of anaerobic digestion processes.* London: IWA Publishing.

Buswell, A. M., & Mueller, H. F. (1952). Mechanism of methane fermentation. *Ind. Eng. Chem., 44*(3), 550-552. https://doi.org/10.1021/ie50507a033.

Coke, C. (2018). Pipeline injection of biomethane in California. *BioCycle, 59*(3), 32. Retrieved from https://www.biocycle.net/2018/03/12/pipeline-injection-biomethane-california/.

Coombs, J. (1990). The present and future of anaerobic digestion. In A. Wheatley (Ed.), *Anaerobic digestion: A waste treatment technology. Critical Rep. Appl. Chem., 31,* 93-138.

El Mashad, H. M. (2003). Solar thermophilic anaerobic reactor (STAR) for renewable energy production. PhD thesis. The Netherlands: Wageningen University. Retrieved from https://edepot.wur.nl/121501.

Hobson, P. N., & Richardson, A. J. (1983). The microbiology of anaerobic digestion. In B. F. Pain & R. Q. Hepherd (Eds.), *Anaerobic digestion of farm waste.* Technical Bulletin 7. Reading, UK: The National Institute for Research in Dairying.

Khanal, S. K. (2008). *Anaerobic biotechnology for bioenergy production: Principles and applications.* Ames, IA: Wiley-Blackwell.

Kosaric, N., & Blaszczyk, R. (1992). Industrial effluent processing. In: J. Lederberg (Ed.), *Encyclopedia of microbiology* (Vol. 2, pp. 473–491). New York, NY: Academic Press.

Lettinga, G., Rebac, S., & Zeeman, G. (2001). Challenge of psychrophilic anaerobic wastewater treatment. *Trends in Biotechnol. 19*(9), 363-370. https://doi.org/10.1016/S0167-7799(01)01701-2.

Pavlostathis, S. G,. & Giraldo-Gomez, E. (1991). Kinetics of anaerobic treatment: A critical review. *Critical. Rev. Environ. Control, 21* (5/6), 411-490. https://doi.org/10.1080/10643389109388424.

Speece, R. E. (1996). *Anaerobic biotechnology for industrial wastewater treatments.* Nashville, TN: Archae Press.

Stanier, R. Y., Doudoroff, M., & Adelberg, E. A. (1972). *General microbiology.* Englewood Cliffs, NJ: Prentice-Hall.

Stetter, K. O. (1996). Hyperthermophilic procaryotes. *FEMS Microbiol. Rev., 18,* 149-158. https://doi.org/10.1111/j.1574-6976.1996.tb00233.x.

Zhang, M. (2017). Energy and nutrient recovery from organic wastes through anaerobic digestion and digestate treatment. PhD diss. Davis, CA: University of California, Davis.

Biodiesel from Oils and Fats

B. Brian He
Biological Engineering
University of Idaho
Moscow, ID, USA

Scott W. Pryor
Agricultural and Biosystems Engineering and College of Engineering
North Dakota State University, Fargo, ND, USA

KEY TERMS		
Feedstocks	Conversion process	Properties
Chemistry	Process configuration	Storage and handling

Variables

$C_{i,FA}$ = mass fraction of a particular fatty acid

$MW_{ave,FA}$ = average molecular weight of fatty acids in the oil

$MW_{ave,FAME}$ = average molecular weight of fatty acid methyl esters or biodiesel

MW_{gly} = molecular weight of glycerol (92.09 kg/kmol)

$MW_{i,FA}$ = molecular weight of a particular fatty acid

MW_{water} = molecular weight of water (18.02 kg/kmol)

Introduction

Biodiesel is the term given to a diesel-like fuel made from biologically derived lipid feedstocks, such as vegetable oils, animal fats, and their used derivatives such as waste cooking oils. Biodiesel is a renewable fuel that can be made from a diverse array of domestic feedstocks, has low safety concerns for use and handling, and can have relatively low environmental impact from production and use.

Biodiesel has several properties that make it a safer fuel than conventional petroleum-based diesel. While conventional diesel is categorized as a flammable fuel, biodiesel is rated as combustible, which means it has a low vapor pressure, is resistant to static sparks, and is much less likely to self-ignite during storage. During transportation, tankers carrying pure biodiesel are not required to display warning signs in the United States.

Biodiesel is especially of interest to farmers because of the potential for on-farm production using harvested crops. Oil can be extracted from oilseeds relatively easily, and this oil can then be used to make biodiesel to run farm machinery. It provides farmers an additional resource for economic welfare and an additional choice for managing cropland. In addition, using biodiesel from domestically grown feedstocks can decrease a country's dependence on imported oil, thus enhancing national energy security. On the other hand, concerns are sometimes raised about converting oils and fats, which could serve as food resources, into fuels (Prasad and Ingle, 2019).

Biodiesel is typically considered an environmentally friendly fuel. Production and combustion of biodiesel results in less air pollution than using conventional diesel. According to a study sponsored by the U.S. Department of Agriculture and the Department of Energy, using biodiesel in urban buses can reduce total particulate matter (PM), carbon monoxide (CO) and sulfur oxides (SO_x) by 32%, 35% and 8%, respectively (Sheehan et al., 1998).

The diesel engine is named for Rudolf Diesel, who invented it in the 1890s. Diesel's engines could run on various fuels including vegetable oils. At the Paris Exposition in 1900, Diesel demonstrated his engines running on peanut oil and made this famous statement:

> The use of vegetable oils for engine fuels may seem insignificant today. But such oils may become in course of time as important as petroleum and the coal tar products of the present time.

Diesel's vision was valid in that vegetable oils can still be used directly as a fuel for diesel engines. However, raw vegetable oils without pre-processing are not an ideal fuel for modern diesel engines due to their high viscosity and other chemical properties. Burning raw vegetable oils in today's diesel engines results in heavy carbon deposits in the cylinders, which can stall the engine in a short period of time.

To overcome this problem, research was conducted starting in the late 1930s to chemically process vegetable oils into a mixture of short-chained alkyl fatty acid esters. This fuel has a much lower viscosity and is thus better suited for use in diesel engines. During the petroleum crisis in the 1970s, the use of alkyl fatty acid esters as a fuel for diesel engines became more popular. Two decades later, in the 1990s, the name "biodiesel" was coined and gained popularity.

In the early 1980s, Mittelbach and his team at the Technical University of Graz in Austria were the first to research biodiesel as a diesel fuel. The commercialization of biodiesel started with a pilot biodiesel production facility by an Austrian company, Gaskoks, in 1987. The European Biodiesel Board (EBB), a non-profit organization promoting the use of biodiesel in Europe, was founded in 1997.

Biodiesel research and utilization in the U.S. started around the same time as in Europe. Dr. Charles Peterson and his research team at the University of Idaho conducted a series of research projects on the use of vegetable oil as tractor fuel. The team worked on biodiesel production, engine testing, emission assessment, and field utilization. The National Biodiesel Board (NBB) was founded in the U.S. in 1992 and has conducted health and environmental assessments on biodiesel utilization. The NBB also registered biodiesel with the U.S. Environmental Protection Agency (USEPA) as a substitute fuel for diesel engines. Supported by the NBB and the biodiesel research community, biodiesel was established as an industry sector. Total biodiesel production reached approximately 7.2 billion L in the USA in 2018 with an additional 39.4 billion L produced globally.

Although biodiesel can be used as a pure diesel-replacement fuel called B100, it is typically available as a diesel/ biodiesel blend at retail pumps. Biodiesel blends are designated to indicate a volumetric mixture such as B5 or B20 for 5% or 20% biodiesel, respectively, in conventional diesel.

Outcomes

After reading this chapter, you should be able to:

- Describe the advantages and limitations of using biodiesel in diesel-powered engines
- Describe biodiesel production processes
- Explain how biodiesel is similar to and different from conventional petroleum-based diesel
- Describe how feedstock composition and properties affect biodiesel properties
- Explain the important unit operations commonly used for producing biodiesel
- Calculate proportions of vegetable oil, methanol, and catalyst needed to make a given quantity of biodiesel, and the size of the reactor required for conversion

Concepts

Biodiesel Chemistry

To qualify as biodiesel in the U.S., a fuel must strictly comply with the ASTM definition of a "fuel comprised of mono-alkyl esters of long chain fatty acids derived from vegetable oils or animal fats, designated B100" (ASTM, 2015). It must also meet all of the quality parameters identified in that standard. In Europe, the definition of biodiesel is covered by the European standard EN 14214 (CEN, 2013). The generic name for vegetable oils (more generally plant oils) or animal fats is simply fat or lipid. The primary distinguishing factor between a fat and an oil is that a fat is a solid at room temperature while an oil is a liquid. The primary compounds in both oils and fats are a group of chemicals called triglycerides (figure 1a).

Glycerol (figure 1b), also known as glycerin, is a poly-hydric alcohol with three alcoholic hydroxyl groups (-OH). Pure glycerol is colorless, odorless, and hygroscopic. Fatty acids (figure 1c) are a family of carboxylic acids with relatively long carbon chains.

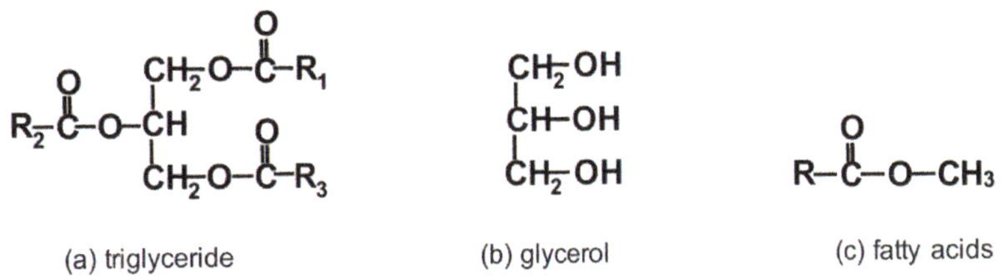

Figure 1. Chemical structure of triglycerides, glycerol, and fatty acids. R, R_1, R_2, and R_3 represent alkyl groups typically with carbon chain lengths of 15–17 atoms.

(a) triglyceride (b) glycerol (c) fatty acids

Triglycerides, also called triacylglycerols, are the glycerol esters of fatty acids, in which three fatty acids attach chemically to a glycerol carbon backbone where the hydroxyl (OH) groups are attached. Triglycerides in oils and fats may contain fatty acid chains of 10 to 24 carbons (C_{10}-C_{24}) but are most commonly 16 to 18 carbons (C_{16}-C_{18}) in length. The three fatty acids attached to the glycerol molecule can be the same or different. The alkyl chain length of fatty acids, the presence and number of double bonds contained in the fatty acid chains, and the position and orientation of the double bonds collectively determine the chemical and physical properties of the triglyceride. Some examples are provided in table 1.

Table 1. Fatty acids commonly seen in oils and fats.

Abbreviation	Common Name	Formula	Chemical Structure	MW[1]
$C_{12:0}$[2]	lauric acid	$C_{12}H_{24}O_2$	$CH_3(CH_2)_{10}COOH$	200.3
$C_{14:0}$	myristic acid	$C_{14}H_{28}O_2$	$CH_3(CH_2)_{12}COOH$	228.4
$C_{16:0}$	palmitic acid	$C_{16}H_{32}O_2$	$CH_3(CH_2)_{14}COOH$	256.5
$C_{18:0}$	stearic acid	$C_{18}H_{36}O_2$	$CH_3(CH_2)_{16}COOH$	284.5
$C_{18:1}$	oleic acid	$C_{18}H_{34}O_2$	$CH_3(CH_2)_7CH{:}CH(CH_2)_7COOH$	282.5
$C_{18:2}$	linoleic acid	$C_{18}H_{32}O_2$	$CH_3(CH_2)_3(CH_2CH{:}CH)_2(CH_2)_7COOH$	280.5
$C_{18:3}$	linolenic acid	$C_{18}H_{30}O_2$	$CH_3(CH_2CH{:}CH)_3(CH_2)_7COOH$	278.5
$C_{20:0}$	arachidic acid	$C_{20}H_{40}O_2$	$CH_3(CH_2)_{18}COOH$	312.6
$C_{20:1}$	eicosenoic acid	$C_{20}H_{38}O_2$	$CH_3(CH_2)_7CH{:}CH(CH_2)_9COOH$	310.5
$C_{20:5}$	eicosapentaenoic	$C_{20}H_{30}O_2$	$CH_3(CH_2CH{:}CH)_5(CH_2)_3COOH$	302.5
$C_{22:1}$	erucic acid	$C_{22}H_{42}O_2$	$CH_3(CH_2)_7CH{:}CH(CH_2)_{12}COOH$	338.6

[1] *MW* = molecular weight, g/mol
[2] $C_{x:y}$ stands for a chain of *x* carbon atoms with *y* double bonds in that chain.

Biodiesel Properties

Biodiesel is a commercialized biofuel used by consumers around the globe. Several international standards have been developed and approved to assure engine manufacturers and diesel engine customers that biodiesel meets specified fuel quality requirements. As a commercial product, biodiesel must comply with the specifications defined by the ASTM Standard D6751 (ASTM, 2015) in North America or EN14214 (CEN, 2013) in Europe. Several other countries have also developed their own standards; in many cases, they are based on the ASTM and EN standards. Table 2 summarizes the specifications for biodiesel fuel according to these two standards.

Biodiesel properties are affected by both the feedstock and the conversion process. Meeting specification for all parameters in the relevant standards must be documented before a fuel can be marketed. However, some fuel properties are more critical than others in terms of use. In the USA, biodiesel *sulfur content* must be no more than 15 ppm for Grade S15, and 500 ppm for Grade S500, to qualify as an ultra-low sulfur fuel. If virgin vegetable oils are used as the feedstock, sulfur content in the biodiesel is typically very low. However, if

Table 2. Major specifications for biodiesel (B100).

Property	Units	ASTM D6751[a] Grade 1B (S15)	ASTM D6751[a] Grade 2B (S15)	EN14214
Sulfur (15 ppm or lower level) (maximum)	ppm	15	15	[b]
Cold soak filterability (maximum)	Sec.	200	360	[b]
Mono-glyceride (maximum)	% mass	0.40	[b]	0.8
Calcium & magnesium combined (maximum)	ppm (µg/g)	5	5	
Flash point (closed cup) (minimum)	°C	93		101
Alcohol control (one of the following shall be met)				
a) Methanol content (maximum)	mass %	0.2		0.2
b) Flash point (minimum temperature)	°C	130		[b]
Water and sediment (maximum)	% volume	0.050		0.005
Kinematic viscosity (40°C)	mm²/s	1.9–6.0		3.5–5.0
Sulfated ash (maximum)	% mass	0.02		0.02
Copper strip corrosion		No. 3		No. 1
Cetane number (minimum)		47		51
Cloud point	°C	Must be reported		[b]
Carbon residue (maximum)	% mass	0.05		0.03
Acid number (maximum)	mg KOH/g	0.50		0.5
Free glycerol (maximum)	% mass	0.02		0.02
Total glycerol (maximum)	% mass	0.24		0.25
Phosphorus content (maximum)	% mass	0.001		0.001
Distillation temperature (90%) (maximum)	°C	360		[b]
Sodium and potassium combined (maximum)	ppm (µg/g)	5		5
Oxidation stability (minimum)	hours	3		6

[a] Grade refers to specification for monoglycerides and cold soak filterability. S15 indicates maximum sulphur content of 15 ppm.
[b] Not specified in the standard

used cooking oils or animal fats are used, the sulfur content in biodiesel must be carefully monitored to meet the required specification.

A liquid fuel's *flash point* refers to the lowest temperature at which its vapor will be combustible. Biodiesel has a high flash point, making it safe for handling and storage. The flash point, however, may drop if the residual alcohol from the biodiesel production process is inadequately removed. To maintain a high flash point, biodiesel *alcohol content* cannot be more than 0.2%. *Cloud point* and *cold soak filterability* are both properties relating to flowability at cold temperatures and are important for biodiesel use in relatively low temperature environments. Cloud point refers to the temperature at which dissolved solids begin to precipitate and reduce clarity. Cold soak filterability refers to how well biodiesel flows through a filter at a specified temperature (4.4°C). Biodiesel is limited in its use in colder climates because it typically has a much higher cloud point (−6°C to 0°C for rapeseed and soybean based biodiesel and up to 14°C for palm oil based biodiesel) than conventional No. 2 diesel (−28°C to −7°C).

Generally, methyl esters of long-chain, saturated fatty acids have high cloud points, especially in comparison to conventional diesel fuel. Although there are commercial additives available for improving biodiesel cold flow properties, their effectiveness is limited. Cold flow properties can be a limiting factor related to the biodiesel blend used (e.g., B2 vs. B10 or B20) in colder climates or at colder times of the year.

The presence of *monoglycerides* in biodiesel is an indicator of incomplete feedstock conversion and can adversely affect fuel combustion in an engine. Monoglycerides also contribute to measurements of both *total glycerine* and *free glycerol*. Total glycerol should be 0.24% or lower to avoid injector deposits and fuel filter clogging problems in engine systems.

Biodiesel *viscosity* is significantly lower than that of vegetable oil but is higher than conventional diesel in most cases. Biodiesel viscosity will vary based primarily on the fatty acid carbon chain length and level of saturation in the feedstock. Although specified biodiesel viscosity levels range from 2.8 to 6.1 mm^2/s at 40°C, typical values are greater than 4 mm^2/s at that temperature (Canackci and Sanli, 2008), while No. 2 conventional diesel has a specified viscosity range of 1.9-4.1 mm^2/s at 40°C with typical values less than 3.0 mm^2/s (ASTM, 2019).

Most biodiesel fuels have a higher *cetane number* than conventional diesel. *Cetane number* measures the ability of a fuel to ignite under pressure and a high cetane number is generally advantageous for combustion in diesel engines. Typical values are approximately 45–55 for soybean-based biodiesel and 49–62 for rapeseed-based biodiesel. The higher cetane number of biodiesel is largely attributed to the long carbon chain and high degree of unsaturation in fatty acid esters. *Acid number* of biodiesel fuel is an indication of free fatty acid content in biodiesel, which affects the oxidative and thermal stabilities of the fuel. To ensure biodiesel meets the specification of acid number, feedstocks with high free fatty acid content must be thoroughly treated and the finished product adequately washed.

Mineral ash contents of combined *calcium and magnesium*, combined *sodium and potassium*, and *carbon residue* have a harmful effect on biodiesel quality by leading to abrasive engine deposits. *Phosphorus content* is also regulated closely because of its adverse impact on the catalytic converter. Good quality control practices are vital in controlling residual mineral content in biodiesel. Biodiesel instability can also be affected negatively by excess *water and sediment* because of inadequate refining, or from contamination during transport or storage. Biodiesel tends to absorb moisture from the air, making it susceptible to such contamination. It can absorb 15–25 times more moisture than conventional petroleum-based diesel (He et al., 2007). Excess water can be controlled by adequately drying the moisture from biodiesel after water washing, and through proper handling and storage of the fuel.

Biodiesel Feedstocks

The primary feedstocks for making biodiesel are vegetable oils and animal fats. Typical properties are given in table 3. The feedstocks for biodiesel production can be any form of triglycerides. The most commonly used feedstocks include

Table 3. Typical fatty acid composition of common oils and fats.[1]

Oils and Fats	$C_{12:0}$	$C_{14:0}$	$C_{16:0}$	$C_{18:0}$	$C_{18:1}$	$C_{18:2}$	$C_{18:3}$	$C_{20:1}$
Plant Oils								
Algae oil		12–15	10–20		4–19	1–2	5–8	35–48[2]
Camelina					12–15	15–20	30–40	12–15
Canola, general			1–3	2–3	50–60	15–25	8–12	
Canola, high oleic			1–3	2–3	70–80	12–15	1–3	
Coconut oil	45–53	16–21	7–10	2–4	5–10	1–2.5		
Corn		1–2	8–16	1–3	20–45	34–65	1–2	
Cottonseed		0–2	20–25	1–2	23–35	40–50		
Grape seed oil			5–11	3–6	12–28	58–78		
Jatropha			11–16	6–15	34–45	30–50	3–5[4]	
Flax (linseed) oil			4–7	2–4	25–40	35–40	25–60	
Mustard seed oil				1–2	8–23	10–24	6–18	5–13 & 20–50[3]
Olive			9–10	2–3	72–85	10–12	0–1	
Palm oil		0.5–2	39–48	3–6	36–44	9–12		
Palm kernel oil	45–55	14–18	6–10	1–3	12–19			
Peanut			8–9	2–3	50–65	20–30		
Rapeseed (high erucic/oriental)			1–3	0–1	10–15	12–15	8–12	45–60[3] & 7–10[4]
Rapeseed (high oleic /canola)			1–5	1–2	60–80	16–23	10–15	
Safflower (high linoleic)			3–6	1–3	7–10	80–85		
Safflower (high oleic)			1–5	1–2	70–75	12–18	0–1	
Sesame oil			8–12	4–7	35–45	37–48		
Soybean oil			6–10	2–5	20–30	50–60	5–11	
Soybean (high oleic)			2–3	2–3	80–85	3–4	3–5	
Sunflower			5–8	2–6	15–40	30–70		
Sunflower (high oleic)			0–3	1–3	80–85	8–10	0–1	
Tung oil			3–4	0–1	4–15		75–90	
Animal Fats								
Butter		7–10	24–26	10–13	28–31	1–3	0–1	
Chicken fat								
Lard		1–2	25–30	10–20	40–50	6–12	0–1	
Tallow		3–6	22–32	10–25	35–45	1–3		

[1] Compiled from various sources: Peterson et al., 1983; Peterson, 1986; Goodrum and Geller 2005; Dubois et al., 2007; Kostik et al., 2013; Knothe et al., 2015.
[2] $C_{20:5}$
[3] $C_{22:1}$
[4] $C_{20:0}$

soybean oil, rapeseed/canola oil, and animal fats. Used cooking oils and/or yellow/trap greases can also be used but may be better as supplements to a feedstock supply with more consistent quality and quantity. Feedstock choice for biodiesel production is generally based on local availability and price. Vegetable oils and/or animal fats all have existing uses and markets. The availability of each type of feedstock varies widely depending on current market conditions, and changes almost on a yearly basis. Before a biodiesel production facility is constructed, securing adequate feedstock supply is always the number one priority. Based on their availability, soybean oil and corn oil are the major feedstocks in the U.S., while rapeseed/canola oil is the most common feedstock used in Europe. Other major producing countries include Brazil and Indonesia which rely on soybean oil and palm oil, respectively.

Compared to other oilseeds, soybeans have a relatively low oil content, typically 10–20% of the seed mass. However, soybean yields are relatively high, typically 2,500–4,000 kg/ha (2,200–3,600 lb/acre), and the U.S. and Brazil are the two largest soybean producers in the world. Due to the large production and trade of soybeans, approximately 11 million metric tons (24.6 billion lbs) of soybean oil were on the market in the 2016–2017 season; of that, 2.8 million metric tons (6.2 billion lbs) were used for biodiesel production (USDA ERS, 2018a).

In recent years, corn oil has been used increasingly and has become the second largest feedstock for making biodiesel in the U.S. Corn planted in the U.S. is mainly used for animal feed, corn starch or sweeteners, and for ethanol production. Corn oil can be extracted in a facility producing corn starch or sweeteners and is also increasingly being extracted from different byproducts of the ethanol industry. The total supply of corn oil in the U.S. was approximately 2.63 million metric tons (5.795 billion lbs) in 2017 (USDA ERS, 2018b). The quantity of corn oil used for biodiesel production was approximately 717,000 metric tons (1.579 billion lb), or approximately 10% of the total biodiesel market. Canola oil is the third largest feedstock with a use of approximately 659,000 metric tons (1.452 billion lbs) in 2017 (USDA EIA, 2018).

Rapeseed belongs to the *Brassica* family of oilseed crops. Original rapeseed, including the cultivars planted in China and India, contains very high contents of erucic acid and glucosinolates, chemicals undesirable in animal feed. Canola is a cultivar of rapeseed developed in Canada with very low erucic acid and glucosinolates contents. While the oilseed crop planted in Europe is still called rapeseed there, it is essentially the same plant called canola in North America. The yield of rapeseed in Europe is high, in the range of 2,000–3,500 kg/ha (1,800–3,100 lb/acre) and is planted almost exclusively for biodiesel production.

Other plant oils, including palm and coconut oil, can also be used for producing biodiesel and are especially popular in tropical nations due to very high oil yields per acre. Plant species with high oil yields, requiring low agricultural inputs and with the ability to grow on marginal lands, such as camelina and jatropha, are of particular interest and have been researched for biodiesel production. Oils from safflower, sunflower, and flaxseed can be used for making biodiesel, but their high value in the food industry makes them uneconomical for biodiesel production.

Some strains of microalgae have a high lipid content and are also widely researched and used to produce algal oil as a biodiesel feedstock. They are considered a promising feedstock because of their potential to be industrialized or produced in an industrial facility rather than on agricultural land. Microalgae can be cultivated in open ponds, but high-oil strains may be better suited to production in closed photo-bioreactors. The potential yield of microalgal oil per unit land can be as high as 6,000 L/ha/y (1600 gal/ac/y), more than 10 times that of canola or soybeans. Currently, however, microalgal lipids are not used for industrial biodiesel production because of their high production cost.

Like plant oils, animal fats contain similar chemical components and can be used directly for biodiesel production. In 2017, approximately 1.2 million metric tons (2.6 billion lbs) of used cooking oils and animal fats were used for biodiesel production in the U.S., accounting for 23% of the total used cooking oils and animal fats in the U.S. market (Swisher, 2018) and less than 20% of U.S. biodiesel production.

Conversion Process

Biodiesel is made by reacting triglycerides (the chemicals in oils and fats) with an alcohol. The chemical reaction is known as *transesterification*. In transesterification of oils and/or fats, which are the glycerol esters of fatty acids (figure 2), the glycerol needs to be transesterified by another alcohol, most commonly methanol. The three fatty acids (R_1, R_2, and R_3) react with the alkyl groups of the alcohol to produce fatty acid esters, or biodiesel. Those fatty acids from the triglyceride are replaced by the hydroxyl groups from the alcohol to produce glycerol, a by-product. The glycerol can be separated from the biodiesel by gravity, but the process is typically accelerated through a centrifugation step. If methanol (CH_3–OH) is used as the alcohol for the transesterification reaction, methyl groups attach to the liberated triglyceride fatty acids (R_x–CH_3), as illustrated in figure 2. The resulting mixture after glycerol separation is referred to as fatty acid methyl esters (or FAME as commonly called in Europe), and biodiesel after further refining. Without the glycerol skeleton, the mixture of FAME is much less viscous than the original vegetable oil or animal fat, and its fuel properties are suitable for powering diesel engines.

The transesterification of oils and fats involves a series of three consecutive reactions. Each fatty acid group is separated from the glycerol skeleton and transesterified individually. The intermediate products are diglycerides (when two fatty acid groups remain on the glycerol backbone) and monoglycerides (when one fatty acid group remains on the glycerol backbone). Transesterification reactions are also reversible. The diglyceride and monoglyceride intermediate products can react with a free fatty acid and reform triglycerides and diglycerides, respectively, under certain conditions. The degree of reverse reaction depends on the chemical kinetics of transesterification

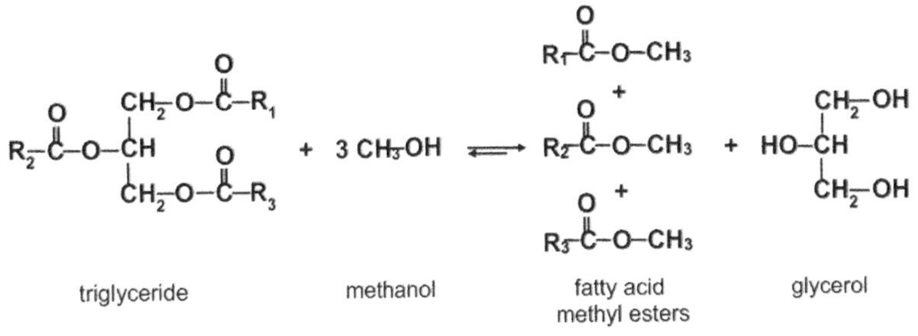

Figure 2. Transesterification of triglycerides with methanol. R_1, R_2, and R_3 are alkyl groups in chain lengths of, most commonly, 15–17 carbons.

and the reaction conditions. In practical application, approximately twice the stoichiometric methanol requirement is added in order to drive the forward reactions and to ensure more complete conversion of oils and fats into biodiesel. The excess methanol can be recovered and purified for reuse in the system.

The density of vegetable oil at 25°C is in the range of 903–918 kg/m³ (7.53–7.65 lb/gal) depending on the specific feedstock (Forma et al., 1979). The density of biodiesel is approximately 870–880 kg/m³ (7.25–7.34 lb/gal) (Pratas et al., 2011). Comparison reveals that vegetable oil is approximately 4% heavier than biodiesel. While planning for biodiesel production, it is an acceptable assumption that each volume of biodiesel produced requires an equal volume of vegetable oil.

To calculate the exact volume of chemicals (i.e., reactant methanol and catalyst) needed for the transesterification, the molecular weight of the vegetable oil is needed. However, as seen from table 3, vegetable oils vary in fatty acid composition depending on oil source and even on the specific plant cultivar. There is no defined molecular weight for all vegetable oil, but an average molecular weight is used for calculations. Based on the hydrolysis of fatty acid esters of glycerol, the molecular weight of vegetable oil (a mixture of fatty acid glycerol esters), MW_{ave}, can be calculated as:

$$MW_{ave} = MW_{gly} - 3\, MW_{water} + 3\, MW_{ave,FA} \tag{1}$$

where MW_{gly} = molecular weight of glycerol = 92.09 kg/kmol
MW_{water} = molecular weight of water = 18.02 kg/kmol
$MW_{ave,FA}$ = average molecular weight of fatty acids in the oil

The water is subtracted in the equation because three individual fatty acids are joined to the single glycerol molecule in a condensation reaction that produces three water molecules in the process. The opposite reaction, hydrolysis, would split the fatty acid from the glycerol through incorporation of the water molecule ions into the products. The overall average molecular weight of vegetable oil fatty acids is calculated as:

$$\frac{1}{MW_{ave,FA}} = \sum \frac{C_{i,FA}}{MW_{i,FA}} \tag{2}$$

where $C_{i,FA}$ = mass fraction of a particular fatty acid
$MW_{i,FA}$ = molecular weight of that particular fatty acid

The difference between the weight of the methyl group (–CH_3; 15 kg/kmol) and that of the hydrogen atom (–H; 1 kg/kmol) on the carboxyl group of fatty acids is 14 atomic mass units. To find the average molecular weight of fatty acid methyl esters (FAME) or biodiesel, $MW_{ave,FAME}$, the following formula can be used:

$$MW_{ave,FAME} = MW_{ave,FA} + 14 \tag{3}$$

Use of a Catalyst

The transesterification reaction will occur even at room temperature if a vegetable oil is mixed with methanol, but would take an extraordinarily long time to approach equilibrium conditions. A catalyst and elevated temperatures are typically used to help the reaction move forward and dramatically reduce the reaction time. The catalysts suitable for transesterification of oils and fats are either strong acids or strong bases; the latter are most commonly used, especially for virgin vegetable oils. Sodium hydroxide (NaOH) and potassium hydroxide (KOH) are inexpensive choices for use as base catalysts; they are typically available commercially as solid flakes or pellets. Before being used as a catalyst for transesterification, the solid form of NaOH or KOH needs to be prepared by reacting with methanol to form a homogenous solution. This dissolving process is a chemical reaction to form soluble methoxide (–OCH$_3$), as shown in figure 3.

Figure 3. Chemical reaction between methanol and potassium hydroxide to form potassium methoxide.

The methoxide is the active species for catalysis in the system. Therefore, the solution of sodium methoxide (NaOCH$_3$) or potassium methoxide (KOCH$_3$) in methanol are the preferred form of the catalysts for large continuous-flow biodiesel production. Solutions of NaOCH$_3$ or KOCH$_3$ in methanol are commercially available in 25–30% concentrations.

Other Factors Affecting Conversion

Note in figure 3 that one mole of water is formed per mole of KOH reacted. Water in the transesterification of oils and/or fats is undesirable because it potentially leads to the hydrolysis of triglycerides to free fatty acids, which in turn react with the base catalyst, either KOH or KOCH$_3$, to form soap. This soap-making process is called saponification (figure 4). Soap in the system will cause the reaction mixture to form a uniform emulsion, making the separation of biodiesel from its by-product glycerol impossible. Therefore, special attention is needed to avoid significant soap formation. Thus, prepared methoxide is preferred to hydroxide as the catalyst for use in biodiesel production, so water can be minimized in the system.

Transesterification of oils and/or fats requires a catalyst for realistic conversion rates, but the reaction will still take up to eight hours to complete if it is carried out at room temperature. Therefore, the process temperature also plays a very important role in the reaction rate, and higher reaction temperatures reduce the required reaction time. When the reaction temperature is maintained at 40°C (104°F), the time for complete transesterification can be shortened to 2–4 hours. If the reaction temperature is at 60°C (140°F), the time can be reduced even further to 1–2 hours for a batch reactor. The highest reaction temperature that can be applied under atmospheric pressure is limited by the boiling temperature of methanol, 64.5°C (148°F). Typical reaction temperatures for transesterification of oils and fats in large batch operations are in the range of 55–60°C (130–140°F). Higher temperatures can be used but require a closed system under pressure.

There are situations in which high amounts of free fatty acids (higher than 3% on a mass basis) exist naturally in feedstocks, such as used

Figure 4. Saponification between potassium hydroxide and a fatty acid.

$$R-\underset{\underset{\text{fatty acid}}{}}{\overset{\overset{O}{\|}}{C}}-O-H + \underset{\text{methanol}}{CH_3OH} \rightleftharpoons R-\underset{\underset{\text{fatty acid ester}}{}}{\overset{\overset{O}{\|}}{C}}-O-CH_3 + \underset{\text{water}}{H_2O}$$

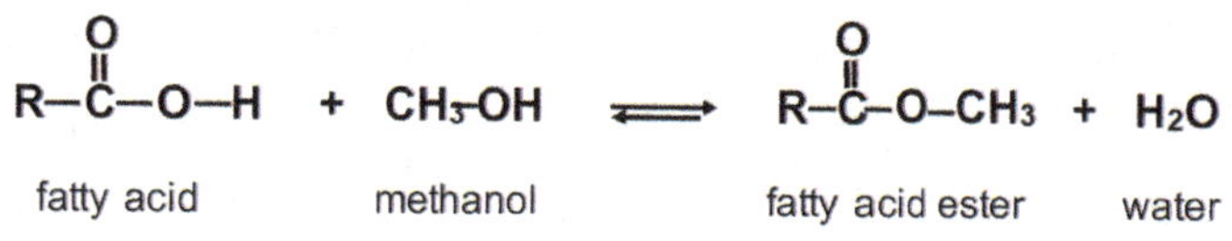

Figure 5. Esterification of a fatty acid reacting with methanol (in the presence of an acid catalyst) to yield a methyl ester and water.

vegetable oils and microalgal lipids. To transesterify feedstocks with high free fatty acid content, direct application of base catalysts, either as hydroxide (–OH) or methoxide (–OCH$_3$), is not recommended because of the increased likelihood of soap formation. Instead, a more complicated two-step transesterification process is used. In the first step, a strong acid, such as sulfuric acid (H$_2$SO$_4$), is used as a catalyst to convert most of the free fatty acids to biodiesel via a chemical process called esterification (figure 5). In the second step, a base catalyst is used to convert the remaining feedstock (mainly triglycerides) to biodiesel.

Safe Handling of Chemicals in Biodiesel Production

Conversion of oils and/or fats to biodiesel is a chemical reaction so a good understanding of the process chemistry, safe chemical processing practices, and all regulations is necessary to ensure safe and efficient biodiesel production. First aid stations must be in place in biodiesel laboratories and production facilities. Although biodiesel itself is a safe product to handle, some of the components involved in production can be hazardous. The chemicals in biodiesel production can include methanol, sodium or potassium hydroxide, and sulfuric acid, all of which have safety concerns related to storage and use. Extreme caution must be practiced in handling these chemicals during the whole process of biodiesel production. The appropriate Material Safety and Data Sheets for all chemicals used should be reviewed and followed to maintain personal and environmental safety.

Applications

Biodiesel Production Systems

The fundamental unit operations for transesterification of a feedstock with low free fatty acid content, such as virgin soybean or canola oil, using KOH as catalyst are illustrated in figure 6. The catalyst solution is prepared by reacting it with methanol, in the case of hydroxide flakes, or by mixing it with a measured amount of methanol, in the case of methoxide solution, in a mixer. The prepared catalyst/methanol solution is added to the vegetable oil/fat in the reactor under gentle agitation. The reactor may be an individual or a series of stirred tanks, or some other reactor type. As discussed above, the transesterification reaction typically takes place in 1–2 hours at 55–60°C (130–140°F).

Crude glycerol is the term used for the glycerol fraction after initial separation. It contains some residual methanol, catalyst and a variety of other chemical impurities in the triglyceride feedstock. Crude glycerol is either refined on site or sold to a market for further processing. Although there are many uses of glycerol in industries from food to cosmetics to pharmaceuticals, the economics of refining severely limits its use. The grey water from biodiesel washing is a waste product containing small quantities of methanol, glycerol, and catalyst. It needs adequate treatment before it can be discharged to a municipal wastewater system.

Process Configuration

Biodiesel can be produced in a batch, semi-continuous, or continuous process. The economics of process configuration are largely dependent on production capacity. Batch processes require less capital investment and are easier to build. A major advantage of batch processing is the flexibility to accommodate variations in types and quantities of feedstock. Challenges of batch processing include lower productivity, higher labor needs, and inconsistent fuel quality. Continuous-flow biodiesel production processes can be scaled more easily and are preferred by larger producers. In continuous-flow processes, fuel quality is typically very consistent. The higher initial capital costs, including costs for complicated process control and process monitoring, are mitigated in large operations by greater throughput and higher quality product. As a result, the net capital and operating costs per unit product is less than that of batch processes. The types of reactors for transesterification can be simple stirred tanks for batch processes and continuously stirred tank reactors (CSTR) for continuous-flow processes.

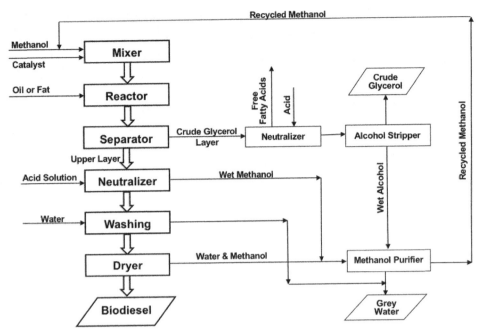

Figure 6. Schematic illustration of a biodiesel production system.

Upon completion of the reaction, the product mixture passes to a separator, which can be a decanter for a batch process or a centrifuge for the continuous-low system. The crude glycerol, which is denser than the biodiesel layer, is removed. Any residual catalyst in the biodiesel layer is then neutralized by a controlled addition of an acid solution. In the same unit, most of the excess methanol and some residual glycerol is concentrated in the aqueous acid solution layer and withdrawn to a methanol recovery unit, where the methanol is concentrated, purified, and recirculated for reuse.

The neutralized biodiesel layer is washed by gentle contact with softened water to further remove residual methanol and glycerol. The washed biodiesel layer is dried by heating to approximately 105°C (220°F) until all moisture is volatized. The finished biodiesel after drying is tested for quality before being transferred to storage tanks for end use or distribution.

Biodiesel Storage and Utilization

Biodiesel has relatively low thermal and oxidative stabilities. This is due to the unsaturated double bonds contained in the oil and fat feedstocks. Therefore, biodiesel should be stored in cool, light-proof containers, preferably in underground storage facilities. The storage containers should be semi-sealed to minimize air exchange with the environment, reducing the possibility of oxidation and moisture absorption of biodiesel. Where permitted, the headspace of the storage containers can be filled with nitrogen to prevent biodiesel from coming into contact with oxygen. If biodiesel will be stored for longer than six months before use, adding a biocide and a stability additive is necessary to avoid microbial activity in the biodiesel. Biodiesel storage and transportation containers should not be made of aluminum, bronze, copper, lead, tin, or zinc because contact with these types of metals will accelerate degradation. Containers made of steel, fiberglass, fluorinated polyethylene, or Teflon can be used.

Biodiesel is a much stronger solvent than conventional diesel. Storage tanks for conventional diesel may have organic sludge build-up in them. If using such tanks for biodiesel storage, they should be thoroughly cleaned and dried to prevent the sludge from being dissolved by biodiesel and potentially causing problems to fuel lines and fuel filters. Similar problems can occur when using biodiesel in older engines with petroleum residues in fuel tanks or transfer lines. For more information on handling and storing biodiesel, readers are recommended to consult "Biodiesel Handling and Use Guide" (5th ed.) prepared by the National Renewable Energy Laboratory of the U.S. Department of Energy (Alleman et al., 2016).

Examples

Example 1: Volumes of soybean oil for biodiesel production

Problem:

Last year, a farmer used a total of 13,250 L of diesel fuel to run the farm's machinery and trucks. After attending a workshop on using biodiesel on farms for both economic and environmental benefits, the farmer has decided to use a B20 blend of biodiesel in all the farm's vehicles. The average annual yield of soybeans on the farm is 2,800 kg/ha. The soybeans contain 18.5% oil on a mass basis, and the efficiency of soybean oil extraction through mechanical pressing is approximately 80%. The density of soybean oil is 916 kg/m^3.

Answer the following questions to help the farmer develop the details needed:

(a) How much pure biodiesel (B100) is needed to run the farm's vehicles using a B20 blend (i.e., a mixture of 20% biodiesel and 80% of conventional diesel on a volume basis)?
(b) How much soybean oil is needed to produce sufficient B100 to blend with conventional diesel?
(c) What field area will yield enough soybeans for the needed quantity of oil?

Solution:

(a) Given that the farmer uses 13,250 L of diesel fuel yearly, if 20% of the quantity is replaced by biodiesel, the quantity of pure biodiesel must be:

$$13{,}250 \text{ L} \times 0.20 = 2{,}650 \text{ L}$$

The farmer will still need to purchase conventional diesel fuel, which is 80% of the total consumption:

$$13{,}250\ L \times 0.80 = 10{,}600\ L$$

Therefore, 2,650 L of pure biodiesel (B100) is needed to blend with 10,600 L of conventional diesel to make a total of 13,250 L of a B20 blend for the farm's vehicles.

(b) As an estimate of how much soybean oil (in kg) is needed, each volume of biodiesel requires approximately one volume of soybean oil (or other oil) to produce it, as noted in the Conversion Process section. Therefore, the initial estimate for the quantity of soybean oil is the same as the required quantity of pure biodiesel, i.e., 2,650 L of soybean oil.

Calculate the mass quantity of soybean oil by multiplying the volume of soybean oil by the density of soybean oil (916 kg/m³ or 0.916 kg/L):

$$2{,}650\ L \times 0.916\ kg/L = 2{,}427\ kg$$

(c) The given soybean yield is 2,800 kg/ha, the oil content of soybean is 18.5%, and the oil extraction efficiency is 80%. Therefore, each ha planted in soybean will yield:

$$(2800\ kg)(0.185)(0.80) = 414.4\ kg\ of\ soybean\ oil$$

The area of soybean field for produce the needed 2427 kg of soybean oil is:

$$2{,}427\ kg\ /\ 414.4\ kg/ha = 5.86\ ha$$

In summary, the farmer needs to plant at least 5.86 ha of soybeans to have enough soybean oil to produce the biodiesel needed to run the farm's vehicles.

Example 2: Average molecular weight of soybean oil

Problem:

The farmer had the farm's soybean oil analyzed by a commercial laboratory via a gas chromatographic analysis and obtained the following fatty acid profile on a mass basis:

	Palmitic (C16:0)	Stearic (C18:0)	Oleic (C18:1)	Linoleic (C18:2)	Linolenic (C18:3)
Profile	9%	4%	22%	59%	6%
$MW_{i,FA}$ (kg/kmol)	256.5	284.5	282.5	280.5	278.5

(a) What is the average molecular weight of the soybean oil?
(b) What is the average molecular weight of biodiesel from this soybean oil?

Solution:

(a) First, calculate the average molecular weight of fatty acids ($MW_{ave,FA}$) in the soybean oil using equation 2:

$$\frac{1}{MW_{ave,FA}} = \sum \frac{C_{i,FA}}{MW_{i,FA}} \qquad (2)$$

$$\frac{1}{MW_{ave,FA}} = \frac{9\%}{256.5} + \frac{4\%}{284.5} + \frac{22\%}{282.5} + \frac{59\%}{280.5} + \frac{6\%}{278.5}$$

$$= \frac{0.09}{256.5} + \frac{0.04}{284.5} + \frac{0.22}{282.5} + \frac{0.59}{280.5} + \frac{0.06}{278.5}$$

$$= 0.003589 \text{ kmol/kg}$$

Therefore, $MW_{ave,FA} = 1 / 0.003589 = 278.6$ kg/kmol.

Next, calculate the average molecular weight of the soybean oil using equation 1:

$$MW_{ave} = MW_{gly} - 3MW_{water} + 3MW_{ave,FA} \qquad (1)$$

$$= 92.09 - (3 \times 18.02) + (3 \times 278.6) \text{ kg/kmol}$$

Therefore, $MW_{ave} = 873.7$ kg/kmol.

(2) Calculate the average molecular weight of biodiesel using equation 3:

$$MW_{ave,FAME} = MW_{ave,FA} + 14 \qquad (3)$$

$$= 278.6 + 14 \text{ kg/kmol}$$

Therefore, $MW_{ave,FAME} = 292.6$ kg/kmol.

In summary, the average molecular weights of the soybean oil and biodiesel are 873.7 and 292.6 kg/kmol, respectively.

Example 3: Chemicals in converting soybean oil to biodiesel

Problem:

As determined in example 1, the farmer needs to produce 2,650 L of pure biodiesel (B100; molecular weight = 880 kg/kmol) to run the farm's vehicles and machinery with B20 blends. In converting soybean oil into biodiesel, the methanol (CH_3OH, molecular weight = 32.04 kg/kmol) application rate needs to be 100% more than the stoichiometrically required rate to ensure a complete reaction. The application rate of the potassium hydroxide catalyst (KOH, molecular weight = 56.11 kg/kmol) is 1% of the soybean oil on a mass basis. How much methanol and potassium hydroxide, in kg, are needed to produce the required biodiesel? The average molecular weights of the soybean oil and biodiesel are 873.7 and 292.6 kg/kmol, respectively.

Solution:

First, write out the transesterification of soybean oil to biodiesel with known molecular weights (MW) (similar to figure 2):

Next, convert the quantity of biodiesel from volume to mass by the biodiesel density, 880 kg/m³ = 0.880 kg/L:

$$2{,}650 \text{ L} \times 0.880 \text{ kg/L} = 2{,}332 \text{ kg}$$

Next, calculate the amount of methanol from the stoichiometric ratio of the transesterification reaction.
methanol : biodiesel

The stoichiometric mass ratio $3 \times 32.04 : 3 \times 292.6$
The unknown mass ratio (kg) $M : 2{,}332$
Or $(3 \times 292.6) \times M = (3 \times 32.04) \times 2{,}332$

Therefore, the quantity of methanol is

$$M = (3 \times 32.04) \times 2{,}332 \text{ kg} / (3 \times 292.6) = 255.5 \text{ kg}$$

Next, calculate the total amount of methanol with 100% excess, as required:

$$M' = 2M = 2 \times 255.5 = 511 \text{ kg}$$

Finally, calculate the quantity of catalyst KOH needed. Since the application rate of the catalyst KOH is 1% of the soybean oil, before the quantity of KOH can be calculated, the quantity of soybean oil must be obtained from the stoichiometric ratio of the transesterification reaction.
soybean oil : biodiesel

The stoichiometric mass ratio $873.7 : 3 \times 292.6$
The unknown mass ratio (kg) $S : 2{,}332$
Or $(3 \times 292.6) \times S = 873.7 \times 2{,}332$ kg

The quantity of soybean oil is, then:

$$S = 873.7 \times 2{,}332 \text{ kg} / (3 \times 292.6) = 2{,}321 \text{ kg}$$

Therefore, the quantity of catalyst KOH is calculated as 1% of the oil:

$$2{,}321 \text{ kg} \times 0.01 = 23.2 \text{ kg}$$

In summary, the quantities of methanol and potassium hydroxide are 511 kg and 23.2 kg, respectively.

Image Credits

Figure 1. He, B. (CC By 4.0). (2020). Chemical structure of triglycerides, glycerol, and fatty acids. R, R1, R2, and R3 represent alkyl groups typically with carbon chain lengths of 15–17 atoms.

Figure 2. He, B. (CC By 4.0). (2020). Transesterification of triglycerides with methanol. R1, R2, and R3 are alkyl groups in chain lengths of, most commonly, 15–17 carbons.

Figure 3. He, B. (CC By 4.0). (2020). Chemical reaction between methanol and potassium hydroxide to form potassium methoxide.

Figure 4. He, B. (CC By 4.0). (2020). Saponification between potassium hydroxide and a fatty acid.

Figure 5. He, B. (CC By 4.0). (2020). Esterification of a fatty acid reacting with methanol (in the presence of an acid catalyst) to yield a methyl ester and water.

Figure 6. He, B. (CC By 4.0). (2020). Schematic illustration of a biodiesel production system.

The chemical formula in Example 3. He, B. (CC By 4.0). (2020).

References

Alleman, T. L., McCormick, R. L., Christensen, E. D., Fioroni, G., Moriarty, K., & Yanowitz, J. (2016). *Biodiesel handling and use guide* (5th ed.). Washington, DC: National Renewable Energy Laboratory, U.S. Department of Energy. DOE/GO-102016-4875. https://doi.org/10.2172/1332064.

ASTM. (2015). D6751-15ce1: Standard specification for biodiesel fuel blend stock (B100) for middle distillate fuels. West Conshohocken, PA: ASTM Int. https://doi.org/10.1520/D6751-15CE01.

ASTM, D975. (2019). D975-19c: Standard specification for diesel fuel. West Conshohocken, PA: ASTM Int. https://doi.org/10.1520/D0975-19C.

Canakci, M., & Sanli, H. (2008). Biodiesel production from various feedstocks and their effects on the fuel properties. *J. Ind. Microbiol. Biotechnol.*, *35*(5), 431-441. https://doi.org/10.1007/s10295-008-0337-6.

CEN. (2013). EN14214+A1 Liquid petroleum products—Fatty acid methyl esters (FAME) for use in diesel engines and heating applications—Requirements and test methods. Brussels, Belgium: European Committee for Standardization.

Dubois, V., Breton, S., Linder, M., Fanni, J., & Parmentier, M. (2007). Fatty acid profiles of 80 vegetable oils with regard to their nutritional potential. *European J. Lipid Sci. Technol.*, *109*(7), 710-732. https://doi.org/10.1002/ejlt.200700040.

Forma, M. W., Jungemann, E., Norris, F. A., & Sonntag, N. O. (1979). Bailey's industrial oil and fat products. D. Swern (Ed.), (4th ed., Vol. 1, pp 186-189). New York, NY: John Wiley & Sons.

Goodrum, J. W., & Geller, D. P. (2005). Influence of fatty acid methyl esters from hydroxylated vegetable oils on diesel fuel lubricity. *Bioresour. Technol.*, *96*(7), 851-855. https://doi.org/10.1016/j.biortech.2004.07.006.

He, B. B., Thompson, J. C., Routt, D. W., & Van Gerpen, J. H. (2007). Moisture absorption in biodiesel and its petro-diesel blends. *Appl. Eng. Agric.*, *23*(1), 71-76. https://doi.org/10.13031/2013.22320.

Knothe, G., Krabl, J., & Van Gerpen, J. (2015). *The biodiesel handbook* (2nd ed.). AOCS Publ.

Kostik, V., Memeti, S., & Bauer, B. (2013). Fatty acid composition of edible oils and fats. *J. Hygienic Eng. Design, 4*, 112-116. Retrieved from http://eprints.ugd.edu.mk/11460/1/06.%20Full%20paper%20-%20Vesna%20Kostik%202.pdf.

Peterson, C. L. (1986). Vegetable oil as a diesel fuel: Status and research priorities. *Trans. ASAE, 29*(5), 1413-1422. https://doi.org/10.13031/2013.30330.

Peterson, C. L., Wagner, G. L., & Auld, D. L. (1983). Vegetable oil substitutes for diesel fuel. *Trans. ASAE, 26*(2), 322-327. https://doi.org/10.13031/2013.33929.

Prasad, S., & Ingle, A. P. (2019). Chapter 12. Impacts of sustainable biofuels production from biomass. In M. Rai, & A. P. Ingle (Eds.), *Sustainable bioenergy—Advances and impacts* (pp. 327-346). Cambridge, MA: Elsevier. https://doi.org/10.1016/B978-0-12-817654-2.00012-5.

Pratas, M. J., Freitas, S. V., Oliveira, M. B., Monteiro, S. l. C., Lima, A. l., & Coutinho, J. A. (2011). Biodiesel density: Experimental measurements and prediction models. *Energy Fuels, 25*(5), 2333-2340. https://doi.org/10.1021/ef2002124.

Sheehan, J., Camobreco, V., Duffield, J., Graboski, M., & Shapouri, H. (1998). Life cycle inventory of biodiesel and petroleum diesel for use in an urban bus. NREL/SR-580-24089. Golden, CO: National Renewable Energy Laboratory. https://doi.org/10.2172/658310.

Swisher, K. (2018). U.S. market report: Fat usage up but protein demand down. *Render Magazine*. Retrieved from http://www.rendermagazine.com/articles/.

USDA EIA. (2018). Monthly biodiesel production report. Table 3, U.S. inputs to biodiesel production. Washington, DC: USDA EIA. Retrieved from https://www.eia.gov/biofuels/biodiesel/production/.

USDA ERS. (2018a). U.S. bioenergy statistics. Table 6, Soybean oil supply, disappearance and share of biodiesel use. Washington, DC: USDA ERS. Retrieved from https://www.ers.usda.gov/data-products/us-bioenergy-statistics/.

USDA ERS. (2018b). U.S. bioenergy statistics. Table 7, Oils and fats supply and prices, marketing year. Washington, DC: USDA ERS. Retrieved from https://www.ers.usda.gov/data-products/us-bioenergy-statistics/.

Baling Biomass
Densification and Energy Requirements

Shahab Sokhansanj, Ph.D., P.Eng.
University of Saskatchewan
Saskatoon, Saskatchewan, Canada

KEY TERMS		
Field capacity	Bale density	Round baling
Material capacity	Bale storage	Square baling

Variables

- ρ = density
- a = machine specific parameter
- A = area on which the force is exerted
- b = machine specific parameter
- B_n = factor that represents the type of surface on which the equipment is towed
- c = machine specific parameter
- C_a = field area covered per unit of time
- C_m = material capacity
- D = draft force
- E = specific energy
- F = force
- F_w = material feed rate, wet mass
- E_f = field efficiency
- g = gravitational acceleration constant (9.81 m/s^2)
- k = positive constant
- m = mass of pulled equipment and its load
- M = mass of bale

n = positive constant
P = pressure
P_d = tractor draft (pull) power
P_{max} = maximum pressure
P_{op} = power-takeoff required to operate the baler
P_{opt} = theoretical power to operate the baler
r = ratio, resistance to travel
sl = decimal value representing tractor wheel slippage
S = average forward speed
V = volume of bale
W = effective width over which the machine works
Y = average yield of the crop

Introduction

An important issue in a biomass-based bioenergy system is the transportation of feedstock from the field to the processing facility. Baling, which is the dense packing of biomass into a manageable form, is of importance because it is an energy-consuming process that determines the efficiency of the bioenergy system. Bale density is the most important factor influencing the logistics (number of vehicles, storage volume, duration of use) and cost (labor and energy) of harvesting and delivering biomass to a biorefinery. Unless the biomass is packed to sufficient density, the energy required for transport may exceed the energy release by the bioconversion processes. This chapter discusses two types of balers, round and square; the relationship between biomass density and energy required to make bales; and the pros and cons of the different bale types. The chapter discusses proper methods for handling bales in order to minimize dry matter losses during storage.

Outcomes

After reading this chapter, you should be able to:

- Describe the attributes, advantages, and disadvantages of square and round bale systems
- Calculate square and round bale density
- Calculate the energy use required to make a square bale and a round bale
- Compare performance factors and energy requirements to make square and round bales
- Describe a proper method of storing bales to minimize dry matter losses

Concepts

The contribution of energy and power to the quality of life is indispensable. Countries that enjoy a high quality of life are the ones that consume the most energy per capita. Fuels generate power to run factories, to mobilize motorized transport, and to heat and cool buildings. Until the sixteenth century, most energy came either directly from the sun or indirectly by burning biomass, mostly wood and other plant material. The introduction of coal brought a new era in industrial development. By the nineteenth century, oil and gas revolutionized industrial development.

The development of agriculture happened in parallel with the exploitation of new sources of energy. Farmers abandoned back-breaking farming practices and adopted powered equipment like tractors and cultivators. Farmers who had used hand tools to cut and stack a crop in the field started to use machines that were able to do these tasks more efficiently. Large land preparation equipment, fertilizer applicators, and crop protection equipment, along with new harvest, handling, and transport equipment, were developed. This was possible because of fossil fuel products like gasoline and diesel.

Fossil fuels powered mechanization tools to produce ample food and clothing for humankind to this day. Unfortunately, the use of fossil fuels resulted in some unexpected consequences. The additional carbon dioxide (CO_2) and other gases released from fossil fuel combustion increased the concentration of greenhouse gases in the atmosphere impacting or contributing to climate change effects. With time, fossil fuels will become expensive for farmers because of limited availability and policy-based penalties for causing unwanted air polluting emissions. Focus on renewable energy sources as an alternative to displace fossil fuels in society has increased. Biomass, for example, can be used much more efficiently beyond conventional burning as a feedstock for producing biofuels.

The farm equipment manufacturing industry has developed a number of machines for harvesting and post-harvest handling of grains, fruits, and vegetables. Residues such as straws and leaves have traditionally had little financial value, so the industry had not developed many machines to exploit whole crops or residues, instead focusing on extracting only the valuable part of crops, like grain and fruit. The remaining part of the crop such as straw, leaves, and branches were left on the field mostly unused.

Since the late twentieth century, there has been a demand for equipment to collect and package straws, grasses, and whole plants, which coincided with other developments such as restrictions on burning residues (because of air quality) and the operation of conservation tillage systems. The farm equipment industry has developed equipment, such as balers, to gather whole plants, straws, and grasses into round or square packages of much higher density than can be achieved by passive stacking of the material. The dense bales take less space for storing and transporting biomass.

Densification

Baling is the most used method for on-farm densifying and packaging of biomass (figure 1). Density is the mass of biomass in the bale divided by the volume of the bale:

$$\rho = \frac{M}{V} \quad (1)$$

where ρ = density (kg/m³)
M = mass of bale (kg)
V = volume of bale (m³)

The density of bales typically varies from 140 to 240 kg/m³ depending on the type of biomass and the pressure used on the biomass when forming the bale. Bale density influences the cost of baling and delivering biomass to the point of its use. Harvesting, storage, transportation, and processing can contribute up to 50% of bioenergy feedstock cost (Shinners and Friede, 2018), so this is an important consideration when operating the system. Transport equipment has a maximum volume and mass (payload) per trailer, so optimizing bale density minimizes transport costs. Creating a dense bale requires power to form the bale and power to transport it during the operation. Bales can be stacked at a location on the farm prior to transport to a bioenergy facility. For energy applications, the dense bales are typically transported to a pelleting facility where the bales are broken and re-compacted into denser pellets.

A range of biomass crops are baled, such as alfalfa (*Medicago sativa*), timothy (*Phleum pretense*), grasses in general (Poaceae), wheat (*Triticum spp.*), maize/corn (*Zea mays*) and soybean (*Glycine max*). Biomass crops may be harvested as whole plants (mowed) or separated in the field using a combine harvester that splits the grain from the other plant material. Regardless of the crop, when cut in the field, the material that will be baled is left as a windrow (a low-density linear pile of material parallel to the direction of machine travel). Materials are usually left in the windrow to dry. The ideal moisture content for safe baling and storage depends on the crop but is typically less than 30%. There can be losses due to shattering if the field moisture content is too low, or the equipment must use energy needlessly if too wet. Wet biomass may spoil due to fungal and bacterial growth during storage, which can interfere with refining processes to make biofuels. Depending on the weather, the length of time the plant remains in the field to dry ranges from a few hours to a few days. When ready, a baler picks up the material from the windrow to form bales. Modern balers are mobile, i.e., the equipment moves around the field.

A number of factors determine the choice between round or square bales. Round bales are preferred in wetter regions as they can shed the rain. Square bales are preferred in dry regions as they stack better. In North America, smaller farms tend to use round balers and

Figure 1. Illustration of a square baler processing straw. (Photo Courtesy of Krone.)

larger farms tend to use large square balers. Table 1 lists some characteristics of the balers operating on North American farms. In some countries, such as Ireland, small farmers tend to favor small square bales as they are easier to handle once made.

Table 1. Typical values for bales.

Bale Categories	Dimensions (width × depth × length for square; diameter × depth for round)	Mass (kg)	Farm Size	Productivity	Typical Cost ($/h)
Small square	356 × 457 × 914 mm	24	Small farms	Low	120
Large square	914 × 1219 × 2438 mm	435	Large farms	High	250
Small round	1219 × 1219 mm	228	Small farms	Medium	130
Large round	1829 × 1829 mm	769	Large farms	High	150

Square Baling

A square baler (figure 2) consists of a pick-up mechanism (1) to lift the biomass from the windrow and delivers it into feed rolls (2). A set of knives on a rotating wheel (3) cuts the biomass to a set length. A pitman arm (5) connects a flywheel (4) eccentrically (off-center) to a plunger (6). This arrangement converts the rotation of the flywheel into a reciprocating motion, to move the plunger back and forth in the bale chamber.

The power needed to form the bale comes from the tractor power takeoff (PTO). Each rotation plunges biomass as it enters the baling chamber. The reciprocating plunger compresses the loose material to form a bale. The process of feeding hay into the bale chamber and compressing it is repeated until the bale is formed. The density of the bale is determined by adjusting spring-loaded or hydraulic upper and lower tension bars on the bale chamber. A bale-measuring wheel (8) rotates as the bale moves through the bale chamber.

The bale length is controlled by adjusting the number of rotations of the measuring wheel. The tying mechanism (9) is synchronized with the plunger movement. When the plunger is at its rear position and the biomass is fully compressed, a set of needles delivers the twine to the tying mechanism. As the twine is grasped by the tying mechanism, the needles retract, and the bale is tied. Once compressed and tied, the bale is ejected from the bale chamber. Square bales are usually produced in several sizes (table 1) and

Figure 2. Inline square baler operation: ①pickup mechanism; ②feed rolls; ③cutting wheel; ④flywheel; ⑤pitman arm; ⑥plunger; ⑦compressed biomass; ⑧measuring wheel. (Adapted from Krone.)

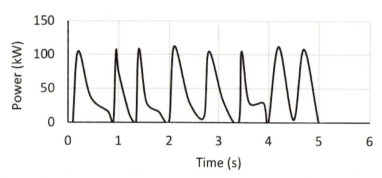

Figure 3. An experimental plot of power in a square baler using a plunger to compact the material. Plotted data are extracted from PAMI Evaluation Report 628 (PAMI, 1990).

the weight depends upon the baler design, type of biomass, and moisture content, but typically ranges from 24 to 908 kg.

Figure 3 shows a plot of instantaneous power requirements for a square baler (PAMI, 1990). The peak power requirements are a result of the plunger action. In a typical alfalfa crop, average power input varied from 23 to 30 kW while the instantaneous peak power input was 110 kW. Average drawbar power requirements for towing the baler in the field varied from 5 to 8 kW and peaked at 20 kW in soft or hilly fields. To fully utilize baler capacity, PAMI (1990) recommends a tractor with a minimum PTO rating of 68 kW (90 hp). A tractor with an 83 kW (110 hp) PTO rating would be required in hilly conditions.

Round Baling

A round baler (figure 4) forms the biomass into a cylindrical bale. The round baler collects the biomass from the windrow using finger-like teeth, made from spring steel or a strong polyethylene material, and rolls the biomass inside the bale chamber using wide rollers or belts.

A round baler comes in two types. Those with a fixed-size chamber use fixed position rollers (figure 5a), and those with a variable chamber use flexible belts (figure 5b). A fixed chamber makes bales with a soft core. A variable chamber makes bales with a hard core. A soft-core bale is "breathable," meaning the porosity is sufficient for the bale to continue drying when left in the field. The bale size remains fixed by the size of the chamber. In a variable chamber, a series of springs and retractable levers ensures a tight bale is formed from core to circumference. The operator sets the diameter of the bale and a target mass to achieve a required density. Following the formation of the bale, the forward motion of the machine and the inflow of biomass are stopped. Twine or a net is wrapped around the circumference of the bale using a moveable arm. Once the net has encircled the bale enough times to maintain shape and sufficiently contain the material, the arms return to the start position and the twine or net strands are cut. The net wrap covers more of the surface area of the bale, preventing material loss and easily maintaining the shape of the bale.

Once the bale is formed and wrapped, it is ejected from the bale chamber. Some round balers have hydraulic "kickers," while others are spring loaded or rely on the spinning of the bale to roll the bale out of the chamber. Once the bale has been ejected from the baler, the back door to the chamber is closed, and the machine starts moving forward, taking in biomass until the next bale is ready. Variable-chamber, large round balers typically produce bales from 1.2 to 1.8 m diameter and up to 1.5 m width, weighing

Figure 4. The round baler makes a cylindrical bale, wrapping the biomass in a net before it is ejected behind the baler. (Photo Courtesy of Krone.)

from 500 to 1000 kg, depending upon size, material, and moisture content. A typical round bale density ranges from 140 kg/m³ to 180 kg/m³.

Figure 6 shows typical PTO and drawbar power requirements for the John Deere 535 round baler at a material capacity of 16.1 t/h (PAMI, 1992). The instantaneous power recorded by the tractor is plotted against the bale weight to show the increase in power input while each bale is formed. The curves are an average of the highly fluctuating measured PTO data, which varied from 5 to 8 kW at no load to a maximum of 32 kW in alfalfa for full sized bales. PTO power input is highly dependent on material capacity (t/h). Drawbar power requirements at 11.5 km/h were about 8 kW when the bale reached to a full size. Although maximum horsepower requirements did not exceed 38 kW, additional power was required to suit other field conditions such as soft and hilly fields. The manufacturer suggested a 56 kW (75 hp) tractor to fully utilize baler capacity.

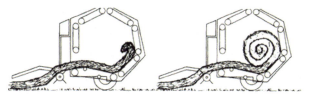

(a) Fixed chamber configuration

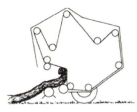 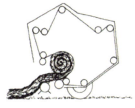 

(b) Variable chamber configuration

Figure 5. The two types of round balers, (a) fixed chamber and (b) variable chamber (Freeland and Bledsoe, 1988).

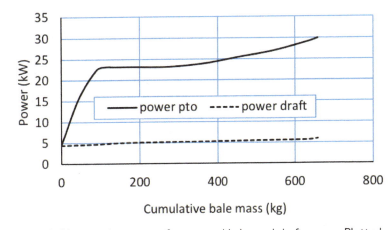

Figure 6. Measured power to form round bales and draft power. Plotted data are extracted from PAMI Evaluation report 677 (PAMI, 1992).

Assessing Baling Performance

ASABE Standards EP496 and D497 (ASABE Standards, 2015a,b) define the performance of field equipment in terms of field capacity and material capacity.

Field Capacity

Field capacity quantifies the rate of land processed (area per unit time) as:

$$C_a = \frac{SWE_f}{10} \qquad (2)$$

where C_a = field area covered per unit of time (ha/h)
  S = average field speed of the equipment while harvesting (km/h)
  W = effective width (m)
  E_f = field efficiency (decimal) (table 2)

Field speed, S, can range from 4 to 13 km/h (table 2). This range represents the variability in field conditions that affects the travelling speed of the equipment.

Effective width, W, is the width over which the machine works. It may be wider or narrower than the measured width of the machine depending on design, how the machine is used in the field with other equipment, and operator experience and skill. The effective width might be determined by the cut width of a mower ahead of the baler, when a wheel rake gathers the mowed crop into a swath for the baler to pick up.

Field efficiency, E_f (table 2), is the ratio of the productivity of a machine under field conditions and the theoretical maximum productivity. Field efficiency accounts for failure to utilize the theoretical operating width of the machine, time lost because of operator's lack of skill, frequent stoppages, and field characteristics that cause interruptions in regular operation. Travel to and from a field, major repairs, preventive maintenance, and daily service activities are not included in field time or field efficiency calculations.

Field efficiency is not a constant for a particular machine but varies with the size and shape of the field, pattern of field operation, crop yield, crop moisture, and other conditions. The majority of time lost in the field is due to turning and idle travel, material handling time, cleaning of clogged equipment, machine adjustment, lubrication, and refueling. Round balers have a lower efficiency than square balers because the shape of the round bale makes handling, transportation, and storage of the bale inefficient compared to handling a square bale (Kemmerer and Liu, 2011).

Table 2. Range and typical values for biomass harvest equipment including balers (ASABE Standards, 2015a).

Biomass Harvest Equipment	Field Efficiency Range (%)	Field Efficiency Typical (%)	Field speed Range (km/h)	Field speed Typical (km/h)	Remarks
Small square baler	60–85	75	4.0–10.0	6.5	Small to mid-size bales
Large square baler	70–90	80	6.5–13.0	8.0	Mid-size to large bales
Large round baler	55–75	65	5.0–13.0	8.0	Commercial round bales

Material Capacity

Material capacity is the mass of crop baled per hour, and is calculated using the field capacity (C_a) and the field yield:

$$C_m = C_a Y \tag{3}$$

where C_m = material capacity (t/h)

Y = average yield of the field (t/ha); it is the amount of biomass that is cut and placed in the swath ready for baling, not the total above-ground biomass in the field.

For crops grown for energy supply purposes, typically no more than 50% of the above ground biomass is cut and baled. In practice, yield (Y) may be as low as 25–30% of the total above ground biomass. The remaining 70–75% of the biomass is left in the field for soil conservation purposes. Removal of a higher percentage may also pick up undesired dirt and foreign material along with the biomass.

Energy Requirements

The bale density that can be achieved is dependent on the specifications of the machine (its dimensions and efficiency) and the mechanical energy that can be supplied to the baler.

Energy Requirements for Square Bales

We start from defining pressure and density in order to calculate energy and power input to make a square bale.

Pressure, P, is calculated using force over area,

$$P = F/A \quad (4)$$

where A = area on which the force is exerted (m²)
F = force (kN)

Force (kN) is derived from mass (M, kg),

$$F = M\,(\text{kg}/1000) \times g\,(\text{m/s}^2) \quad (5)$$

where g = acceleration due to gravity (9.81 m/s²).

The power requirement is related to bale density. The relationship is determined by first relating pressure to density, then calculating energy from force vs. displacement, and finally estimating power from the time rate of energy.

For the first step, a commonly used equation to relate pressure and density is (Van Pelt, 2003; Afzalinia and Roberge, 2013):

$$P = \left(\frac{1}{k}\rho\right)^{1/n} \quad k, n > 0 \quad (6)$$

where P = the pressure exerted by the plunger (kPa)
k and n = positive constants
ρ = density (kg/m³)

Hofstetter and Liu (2011) suggested values for k and n for several crops (table 3).

Table 3. Coefficients *k* and *n* of pressure density (Hofstetter and Liu, 2011).

Biomass crop	k	n
Stover	29.48	0.330
Wheat straw	38.79	0.293
Switchgrass	100.99	0.137

During bale formation, the initial density is zero (empty chamber), and steadily increases to the maximum density possible given the plunger pressure.

The next step is to calculate energy from force and displacement. The total energy input required to make a bale is calculated by integrating the area under the pressure-displacement curve from 0 to P_{max}. This integration yields energy input per unit mass (E) for a single stroke of the plunger to form what is known as a wafer. Equation 7 represents integration of force vs. displacement:

$$E = \int_0^{P_{max}} \left(\frac{1}{\rho}\right) dP \tag{7}$$

where P = pressure (kN/m²)
E = energy input per unit mass (kJ/kg)

Substituting ρ from equation 6 and integrating yields:

$$E = \frac{1}{(1-n)k} P_{max}^{(-n+1)} \tag{8}$$

Replacing P_{max} with ρ_{max} allows an estimate of specific energy, E (kJ/kg):

$$E = \frac{1}{(1-n)k}\left(\frac{1}{k}\rho_{max}^{\frac{1-n}{n}}\right) \tag{9}$$

When making a square bale, each stroke of the plunger makes a wafer of around 51 mm thickness. It would require around 19 strokes to make a 915 mm bale. For a complete bale the energy required, (E_{op}, kJ), can be calculated from E multiplied by the final mass of the bale,

$$E_{op} = E \times M \tag{10}$$

For the last step, the power (energy per unit time) required to make one bale is calculated by multiplying the specific energy (E) by the material capacity (C_m)

$$P_{opt} = \frac{C_m E_{op}}{3.6e} \tag{11}$$

where P_{opt} = theoretical power to operate the square baler (kW)
e = efficiency factor that accounts for inefficiency of transmission of power from the PTO to the baler

In practice, ASABE Standard D497 (ASABE Standards, 2015a) suggests that about 4 kW is needed for a baler to run empty so this power overhead must be added to P_{opt}.

Energy Requirements for Round Bales

For a round baler, ASABE Engineering Practice EP 496 (ASABE Standards, 2015b) recommends estimating the operating power for balers and other rotating machines using:

$$P_{op} = a + b\,W + c\,F_w \qquad (12)$$

where P_{op} = power-takeoff required to operate the round baler (kW)
W = working width of the baler (m)
F_w = material feed rate, wet mass (t/h)
a, b, and c = machine-specific parameters (table 4)

Comparing the power requirements, Tremblay et al. (1997) found that the variable chamber baler required an average PTO power of 10.2 kW compared to a fixed chamber baler that required an average PTO power of 13.3 kW. Also, the peak PTO power required was considerably less for the variable chamber (14.5 kW) compared to fixed chamber (37.5 kW). This means a much larger tractor would normally be required to operate a fixed chamber baler. For flexible operation in terms of tractor required and size and density of bales, a flexible chamber round baler is perhaps the best option.

Table 4. Typical parameter values for equation 12 for balers from ASAE D497 (ASABE Standards, 2015a).

Baler Type	a (kW)	b (kW/m)*	c (kWh/t)
Small square	2.0	0	1.0
Large square	4.0	0	1.3
Large round, variable chamber	4.0	0	1.1
Large round, fixed chamber	2.5	0	1.8

* Non-zero values are reported for machinery such as mowers and rakes.

Energy Requirements for Pulling a Baler

The power required to drive the tractor and tow the baler is determined from the draft force (D, kN):

$$D = r\,m\,g\,/\,1000 \qquad (13)$$

where r = ratio, resistance to travel
m = mass of pulled equipment and its load (kg)
g = gravitational acceleration constant = 9.81 m/s^2

Resistance to travel is an additional draft force that must be included in computing power requirements. Values of resistance to travel depend on transport wheel dimensions, tire pressure, soil type, and soil moisture. Motion resistance ratios are defined in ASAE S296 (ASABE Standards, 2018). The value of r can be estimated using (ASABE Standards, 2015a):

$$r = \frac{1}{B_n} + 0.04 + \frac{0.5\,sl}{\sqrt{B_n}} \qquad (14)$$

where B_n = soil index factor (table 5)
sl = decimal value representing tractor wheel slippage (table 5)

Table 5. Values of soil index factor B_n, slippage sl, and draft coefficient X_d for various surfaces on which equipment is towed (ASABE Standards, 2015a).

Surface Condition	B_n	sl	Drawbar X_d[a]
Hard—concrete	80	0.04–0.08	0.88
Firm soil	55	0.08–0.10	0.77
Tilled soil	40	0.11–0.13	0.75
Soft soil	20	0.14–0.16	0.70

[a] X_d represents the ratio of draft power to PTO power. The listed values are for 4-wheel drive tractors.

Given the speed and draft force (kN), draft power is calculated by:

$$P_d = \frac{DS}{3.6} \qquad (15)$$

where P_d = the tractor draft (pull) power (kW)
S = the average forward speed of the baler (km/h)

Applications

Handling and Storing Bales

Bale Stacking

Once the bales are made, they must be removed from the field before the land can be prepared for the next crop. Tractors and loaders equipped with grabbing devices pick up and load the bales onto a trailer for transport out of the field. The bales are then stacked either next to the field or in a central storage site by using a tractor or a loader. HSE (2012) recommends building stacks on firm, dry, level, freely draining ground, which should be open and well ventilated, away from overhead electric poles. Use of stones or crushed rock on the ground beneath a stack to make it level and to stop water rising into the stack is recommended. The site should be away from any potential fire hazards and sources of ignition with good road access so bales can be transported to and from the stack safely. There must be sufficient space to allow tractors, trailers and other vehicles adequate room to maneuver.

Figure 7 shows the correct configuration of stacking square bales and round bales, with a wide base that narrows as the stack gets higher. The maximum height of the stack should not be greater than 1.5 times the width of the base. Generally, a stack of no more than 10 bales on hard surfaces and 8 bales on soft surfaces is recommended. Square bales must be laid with each row is offset from the row below, such that there is no continuous gap between them. Round bales are stacked in a pyramid with fewer bales in each direction than in the layer below. The outside round bales need a chock at each of the bales in the lowest layer to prevent them from rolling out (figure 7). As with square bales, round bales should be laid to cover the gap between two bales underneath.

Once a stack is formed, the weight of each bale becomes an issue for the stability of the pile. The weight of a large bale may range from 300 kg to more than 500 kg. The bales at the top press onto the lower bales causing their slow deformation. The degree of deformation depends upon bale density and moisture content, and the length of time

Figure 7. Examples of stacking large square bales and round bales.

they remain in the stack. A lower density and a higher moisture bale tends to deform more than a higher density and a dryer bale.

Dry Matter Loss

Moisture content at the time of baling plays an important role in the amount of dry matter loss that may happen during baling and later during storage. For leafy biomass like alfalfa, the recommended moisture content for baling is less than 30% and for storage less than 15% to 20%; however, for longer storage, a lower moisture content of 10% to 12% is preferred. Square bales tend to lose less moisture than round bales, but regardless of shape, it is important to make bales as near to the target moisture as possible.

Losses can be mechanical and microbial. Mechanical losses mostly occur during bale handling, such as building the stack or removing the bales from the stack. Some physical removal of biomass (known as leaching) may also take place due to rain wash. Also, the carbohydrates in freshly cut green biomass can decay to CO_2, water, and heat.

The most prevalent dry matter loss is due to microbial activity, which causes the deterioration of the plant material and loss of dry matter. The growth of microbes on the biomass is directly related to the moisture content. Dry biomass adsorbs moisture from rain when exposed and becomes a host for mold to develop. Cover and duration of storage both influence dry matter loss (table 6). For example, the dry matter loss from an uncovered bale on the ground may range from 5% to 20% within 9 months of storage. If storage time increases to 12 to 18 months, dry matter loss can increase to 20% to 35% of the mass of the bales. Storing bales under a roof will limit losses to 2% to 5%. Research shows there is not much difference between dry matter loss for round bales vs. square bales when stored in similar conditions (Wilcke et al., 2018).

Table 6. Percent dry matter loss for different methods of storing biomass (Lemus, 2009).

Storage Method		Storage Period (months)	
		0 to 9	12 to 18
Ground	Covered with a tarp	5–9	10–15
	Exposed	5–20	20–35
Elevated	Covered with a tarp	2–4	5–10
	Exposed	3–15	12–35
Barn	Enclosed	~2	2–5
	Open sides	2–5	3–10

The range of dry matter loss (table 6) stems from differences in climate, crop type, and initial moisture content of the biomass. Nevertheless, these numbers are good for making a decision on the kind of storage system to be chosen for bales. In terms of capital expenditure, storing on the ground is the least expensive and storing in an enclosed barn is the most expensive.

Decision Factors for Square vs. Round Bales

The selection of round or square bales depends on several factors including crop species to be baled, regional climate conditions, volume of crop to be harvested, types of storage available, tractor power, and ancillary services available. Key advantages and disadvantages for round and square bales are listed in table 7.

Table 7. Advantages and disadvantages of square bales and round bales.

Square Bale	Round Bale
Advantages	**Advantages**
• More efficiently uses space in transport and storage • Better shape retention during storage • Easier to stack	• Greater availability of balers and handling equipment • Lower price for balers • Greater ability to shed water if bales are stored uncovered
Disadvantages	**Disadvantages**
• Greater moisture absorption by bales stored without cover	• Less efficient use of space in hauling and storing bales • A tendency for bales to lose their shape during storage

Examples

Example 1: Field and material capacity

Problem:

A field of hay is cut by using a disk mower cutting 5 m swaths. Following a few days of drying, a rotary rake is used to windrow the hay for baling. Calculate the field capacity and material capacity of three balers: small square, large square, and round for a yield of 7 t/ha of hay. Which machine would you choose?

Solution:

The effective width is 5 m as this is the swath width of the mower. Calculate field capacity using equation 2 and material capacity using equation 3:

$$C_a = \frac{SWE_f}{10} \tag{2}$$

$$C_m = C_a Y \tag{3}$$

where C_a = field area covered per unit of time (ha/h)
S = average field speed of the equipment while harvesting (km/h)
W = effective width (m)
E_f = field efficiency (decimal)
C_m = material capacity (t/h)
Y = average yield of the field (t/ha)

Use typical values from table 2 for speed and efficiency of each type of baler. Table 8 lists the input values and calculation results for field capacity and material capacity. The large square baler can process the largest area per hour, therefore it can also process the greatest mass per hour. Thus, with typical values for speed and efficiency, the large square baler would be selected if the only criteria were field and material capacity.

Table 8. Input values and calculation results for example 1.

Baler	Width of cut, W (m)	Field speed, S (km/h)	Field efficiency, E_f (%)	Field capacity, C_a (ha/h)	Yield, Y (t/ha)	Material capacity, C_m (t/h)
Small square baler	5	6.5	75	2.44	7	17.06
Large square baler	5	8.0	80	3.20	7	22.40
Round baler	5	8.0	65	2.60	7	18.20

Example 2: Maximum bale density and mass

Problem:
A farmer is making square bales of cornstalk at 35% moisture content (wet mass basis). The compressed bale dimensions are 914 mm × 1219 mm × 2438 mm. Determine the maximum density and mass of each bale given the mass equivalent of force exerted on the cross section (914 mm × 1219 mm) bale is 20 tonne (t).

Solution:
The maximum density is a function of the maximum pressure exerted on the pressure exerted on the bale cross section. First, calculate the force on the cross section of the bale (equation 5) using the given mass equivalent of force as 20 t, which is 20,000 kg, and acceleration due to gravity as 9.8 m/s²:

$$F = M\,(\text{kg}/1000) \times g\,(\text{m/s}^2) \qquad (5)$$

$$F = 20000 \times 9.8 / 1000 = 196 \text{ kN}$$

Calculate the pressure exerted on the bale cross section using equation 4:

$$P = F/A \qquad (4)$$

$$P = 196 \text{ kN}/(0.914 \times 1.219 \text{ m}^2) = 175.92 \text{ kPa}$$

Calculate bale density by solving equation 6 for ρ, using values of k and n from table 3:

$$P_{max} = \left(\frac{1}{k}\rho\right)^{1/n} \quad k,n > 0 \qquad (6)$$

$$\rho = kP^n = 29.48(175.92)^{0.33} = 162.1 \text{ kg/m}^3$$

The mass of the bale can be calculated from density and the dimensions of the bale:

$$M = \rho V = 162.1 \text{ kg/m}^3 \times (0.914 \times 1.219 \times 2.438 \text{ m}^3) = 440.32 \text{ kg}$$

Example 3: Specific and operating energy

Problem:
For the baler specified in example 2, calculate specific energy of the baler and, from this, the operating energy required to make one bale.

Solution:
Calculate specific energy using equation 9:

$$E = \frac{1}{(1-n)k}\left(\frac{1}{k}\rho_{max}^{\frac{1-n}{n}}\right) \tag{9}$$

$$= \frac{1}{(1-0.33)29.48}\left(\frac{162.36}{29.48}\right)^{\frac{1-033}{0.33}} = 1.62 \text{ kJ/kg}$$

Now, calculate the operating energy using equation 10:

$$E_{op} = E \times M \tag{10}$$

$$E_{op} = 1.62 \text{ kJ/kg} \times 440.32 \text{ kg} = 713.32 \text{ kJ}$$

Example 4: Operating power

Problem:
For the baler in examples 2 and 3, power transmission from the tractor PTO to the baler will not be 100% efficient. Assuming 50% transmission efficiency of power from the tractor to the baler, estimate the operating power that must be supplied to the baler.

Solution:
Estimate the theoretical operating power, P_{opt}, using equation 11, with $e = 0.50$:

$$P_{opt} = \frac{C_m E_{op}}{3.6e} \tag{11}$$

$$P_{opt} = \frac{(22,400 \text{ kg/h}) \times 1.62 \text{ kJ/kg}}{(3600 \text{ s/h})(0.50)} = 20.16 \text{ kW}$$

Applying the ASABE D497 assertion that about 4 kW is needed for the machine to run when empty, the P_{opt} is:

$$P_{opt} = 20.16 + 4 = 24.16 \text{ kW}$$

Example 5: Power requirements of a round baler

Problem:

A farmer has the option of using a round baler with a fixed chamber, an operating width of 2 m, a feed intake of 18.2 t/h, and a mass of 15,800 kg, that produces bales of 1.83 m diameter, 1.83 m width or depth, and 180 kg/m³ density. Calculate (a) the power requirement for the fixed chamber round baler, (b) the draft force of the machine, and (c) the draft power of the tractor required to pull the machine through the field.

Solution:

(a) Equation 12 can be used to estimate the power requirement. A bale of almost 2 m wide would be regarded as a large bale (table 1), so the parameters for equation 12 can be taken from table 4 accordingly:

$$P_{opt} = a + b\,W + c\,F \tag{12}$$

$$P_{opt} = 2.5 + (0 \times 2) + (1.8 \times 18.2) = 35.26 \text{ kW}$$

(b) The draft force of the machine can be calculated using equation 13:

$$D = r\,m\,g\,/\,1000 \tag{13}$$

First, calculate the motion resistance, r, using equation 14 with values from table 5. Assume the machine is working on a soft soil surface and with average slippage. Thus, from table 5, $B_n = 20$, $sl = 0.15$ (average of 0.14 and 0.16):

$$r = \frac{1}{B_n} + 0.04 + \frac{0.5\,sl}{\sqrt{B_n}} \tag{14}$$

$$r = \frac{1}{20} + 0.04 + \frac{0.5(0.15)}{\sqrt{20}} = 0.16771$$

Next, calculate the mass of bale plus baler:

Bale volume: $V = \pi\,r^2\,L = 3.14\,(0.915\text{ m})^2\,(1.83\text{ m}) = 4.814\text{ m}^3$
Bale mass: $M = V\rho = 4.814\text{ m}^3 \times 180\text{ kg/m}^3 = 866.5\text{ kg}$
Mass of bale plus baler: $m = 866.5 + 15{,}800 = 16{,}666.5\text{ kg}$
Substituting values in equation 13 yields the draft force of the baler:

$$D = r\,m\,g\,/\,1000 = (0.16771 \times 16{,}666.5 \times 9.81)/1000 = 27.4\text{ kN}$$

(c) From the draft force, calculate the draft power, P_d, for the given speed, S, using equation 15:

$$P_d = \frac{D(kN)\,S\left(\frac{km}{h}\right)}{3.6} \quad (15)$$

$$P_d = \frac{27.4\,(kN)\,8\left(\frac{km}{h}\right)}{3.6} = 60.89 \text{ kW}$$

Example 6: Dry matter loss

Problem:
A stack of round bales from example 5 are to be stored with an average moisture content of 15% (wet mass basis). Estimate the dry matter loss from the bales when covered with tarp and stored on the ground for 9 months and 18 months.

Solution:
The bale wet mass is 866.5 kg (calculated in example 5). Calculate the bale dry mass using the given average moisture content of 15% (wet mass basis):

$$\text{Bale dry mass} = 866.5 \times (1 - 0.15) = 736.53 \text{ kg}$$

Assume a midrange dry matter loss from table 6, or percent dry mass of 7.5% for 9 months and 12.5% for 18 months. Use the values of percent dry mass loss to calculate the dry matter loss:

$$\text{Dry matter loss after 9 months} = 736.53 \times 0.075 = 55.2 \text{ kg}$$

$$\text{Dry matter loss after 18 months} = 736.53 \times 0.125 = 92.1 \text{ kg}$$

Image Credits

Figure 1. Krone. (CC by 4.0). (2020). Illustration of a square baler processing straw. Used with written permission. Retrieved from https://www.krone-northamerica.com/.
Figure 2. Krone. (CC by 4.0). (2020). Inline square baler operation. Used with written permission. Retrieved from https://www.krone-northamerica.com/.
Figure 3. Sokhansanj, S. (CC by 4.0). (2020). An experimental plot of power in a square baler.
Figure 4. Krone. (CC by 4.0). (2020). The round baler makes a cylindrical bale. Used with written permission. Retrieved from https://www.krone-northamerica.com/.
Figure 5. Freeland and Bledsoe. (CC By 1.0). (1988). The two types of round balers. Retrieved from ASABE publication Transactions.
Figure 6. Sokhansanj, S. (CC By 4.0). (2020). Measured power to form round bales.
Figure 7. Examples of stacking large square bales and round bales. Square bale photo adapted from background removed: Courtesy of Ryley Schmidt, Barr-Ag Inc. Alberta. Round bale picture credit: Evelyn Simak / *A stack of straw bales* / CC BY-SA 2.0. (details of licence

can be found here: https://commons.wikimedia.org/wiki/File:A_stack_of_straw_bales_-_geograph.org.uk_-_1501535.jpg)

References

Afzalinia, S., & Roberge, M. (2013). Modeling of the pressure-density relationship in a large cubic baler. *J. Agric. Sci. Technol., 15*(1), 35-44.

ASABE Standards. (2015a). ASAE D497.7 MAR2011 (R2015): Agricultural machinery management data. St. Joseph, MI: ASABE. http://elibrary.asabe.org

ASABE Standards. (2015b). ASAE EP496.3 FEB2006 (R2015) Cor.1: Agricultural Machinery Management St. Joseph, MI: ASABE. http://elibrary.asabe.org

ASABE Standards. (2018). ANSI/ASAE S296.5 DEC2003 (R2018): General terminology for traction of agricultural traction and transport devices and vehicles. St. Joseph, MI: ASABE. http://elibrary.asabe.org

Freeland, R. S., & Bledsoe, B. L. (1988). Energy required to form large round hay bales—Effect of operational procedure and baler chamber type. *Trans. ASAE, 31*(1), 63-67. http://dx.doi.org/10.13031/2013.30666.

Hofstetter, D. W., & Liu, J. (2011). Power requirement and energy consumption of bale compression. ASABE Paper No. 1111266, St. Joseph, MI: ASABE.

HSE (2012). Safe working with bales in agriculture. The Health and Safety Executive 05/12 INDG125(rev3). 10 pages. Retrieved from https://www.hse.gov.uk/pubns/indg125.pdf.

Kemmerer, B., & Liu, J. (2011). Large square baling and bale handling efficiency—A case study. *Agric. Sci., 3*(2), 178-183. http://dx.doi.org/10.4236/as.2012.32020.

Lemus, R. (2009). Hay storage: Dry matter losses and quality changes. Retrieved from http://pss.uvm.edu/pdpforage/Materials/CuttingMgt/Hay_Storage_DM_Losses_MissSt.pdf.

PAMI (1992). Evaluation report 677. John Deere 535 round baler. Retrieved from http://pami.ca/pdfs/reports_research_updates/(4a)%20Balers%20and%20Baler%20Attachments/677.PDF.

PAMI (1990). Evaluation report 628. Vicon MP800 square baler. Retrieved from http://pami.ca/pdfs/reports_research_updates/(4a)%20Balers%20and%20Baler%20Attachments/628.PDF.

Shinners, K., & Friede, J. (2018). Energy requirements for biomass harvest and densification. *Energies, 11*(4), 780. http://doi.org/10.3390/en11040780.

Tremblay, D., Savoie, P., & Lepha, Q. (1997). Power requirements and bale characteristics for a fixed and a variable chamber baler. *Canadian Agric. Eng., 39*(1), 73-75. Retrieved from https://pdfs.semanticscholar.org/cb81/3812beeb7dcee3ecd34e5dbf39617869b8a6.pdf.

Van Pelt, T. (2003). Maize, soybean, and alfalfa biomass densification. *Agric. Eng. Intl.* Manuscript EE 03 002. Retrieved from https://pdfs.semanticscholar.org/8d9f/46c0431869b9f2b8edbedb4fcc5e657b7ac2.pdf.

Wilcke, W., Cuomo, G., Martinson, K., & Fox, C. (2018). Preserving the value of dry stored hay. Retrieved from https://extension.umn.edu/forage-harvest-and-storage/preserving-value-dry-stored-hay.

Basic Microcontroller Use for Measurement and Control

Yeyin Shi
Department of Biological Systems Engineering, University of Nebraska-Lincoln, Lincoln, Nebraska, USA

Ning Wang
Department of Biosystems and Agricultural Engineering, Oklahoma State University, Stillwater, Oklahoma, USA

Guangjun Qiu
College of Engineering, South China Agricultural University, Guangzhou, Guangdong, China

---

KEY TERMS

Architecture and hardware   Operating principles   Greenhouse control
Programming

Introduction

Measurement and control systems are widely used in biosystems engineering. They are ubiquitous and indispensable in the digital age, being used to collect data (measure) and to automate actions (control). For example, weather stations measure temperature, precipitation, wind, and other environmental parameters. The data can be manually interpreted for better farm management decisions, such as flow rate and pressure regulation for field irrigation. Measurement and control systems are also part of the foundation of the latest internet of things (IoT) technology, in which devices can be remotely monitored and controlled over the internet.

A key component of a measurement and control system is the microcontroller. All biosystems engineers are required to have a basic understanding of what microcontrollers are, how they work, and how to use them for measurement and control. This chapter introduces the concepts and applications of microcontrollers illustrated with a simple project.

> **Outcomes**
>
> After reading this chapter, you should be able to:
>
> - Describe the architecture and operating principles of microcontrollers
> - Explain how to approach programming a microcontroller
> - Develop a simple program to operate a microcontroller for measurement and control systems

Concepts

Measurement and Control Systems

Let's talk about measurement and control systems first. As shown in figure 1, signals can be generated by mechanical actuators and measured by sensors, for example, the voltage signal from a flow rate sensor. The signal is then input to a central control unit, such as a microcontroller, for signal processing, analysis, and decision making. For example, to see if the flow rate is in the desired range or not. Finally, the microcontroller outputs a signal to control the actuator, e.g., adjust the valve opening, and/or at the same time display the system status to users. Then the actuator is measured again. This forms an endless loop that runs continuously until interrupted by the user or time out. If we view the system from the signal's point of view, the signal generated by the actuators and measured by the sensors are usually analog signals which are continuous and infinite. They are often pre-processed to be amplified, filtered, or converted to a discrete and finite digital format in order to be processed by the central control unit. If the actuator only accepts analog signals, the output signal to control the actuator from the central control unit needs to be converted back to the analog format. As you can tell, the central control unit plays a

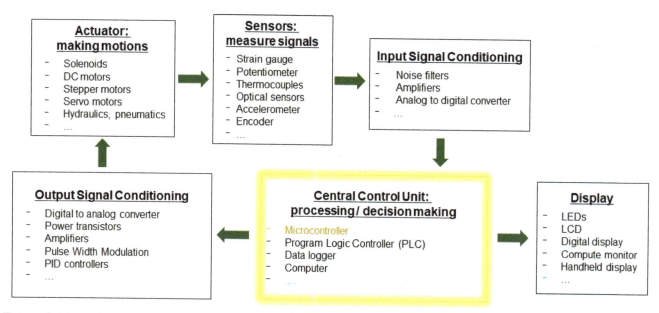

Figure 1. Main components in a measurement and control system (adapted from figure 1.1 in Alciatore and Histand, 2012).

critical role in the measurement and control loop. Microcontroller is one of the most commonly used central control units. We will focus on microcontrollers in the rest of the chapter.

Microcontrollers

A microcontroller is a type of computer. A computer is usually thought of as a general-purpose device configured as a desktop computer (personal computer; PC or workstation), laptop, or server. The "invisible" type of computer that is widely used in industry and our daily life is the microcontroller. A microcontroller is a miniature computer, usually built as a single integrated circuit (IC) with limited memory and processing capability. They can be embedded in larger systems to realize complex tasks. For example, an ordinary car can have 25 to 40 electronic control units (ECUs), which are built around microcontrollers. A modern tractor can have a similar number of ECUs with microcontrollers handling power, traction, and implement controls. Environmental control in greenhouses and animal houses, and process control in food plants all rely on microcontrollers. Each microcontroller for these applications has a specific task to measure and control, such as air flow (ventilation, temperature) or internal pressure, or to perform higher-level control of a series of microcontrollers. Understanding the basic components of a microcontroller and how it works will allow us to design a measurement and control system.

A microcontroller mainly consists of a central processing unit (CPU), memory units, and input/output (I/O) hardware (figure 2). Different components interact with each other and with external devices through signal paths called buses. Each of these parts will be discussed below.

The *CPU* is also called a microprocessor. It is the brain of the microcontroller, in charge of the primary computation and system internal control. There are three types of information that the CPU handles: (1) the data, which are the digital values to be computed or sent out; (2) the instructions, which indicate which data are required, what calculations to impose, and where the results are to be stored; and (3) the addresses, which indicate where a data or an instruction comes from or is sent to. An arithmetic logic unit (ALU) within the CPU executes mathematical functions on the data structured as groups of binary digits, or "bits." The value of a bit is either 0 or 1. The more bits a microcontroller CPU can handle at a time, the faster the CPU can compute. Microcontroller CPUs can often handle 8, 16, or 32 bits at a time.

A *memory unit* (often simply called memory) stores data, addresses, and instructions, which can be retrieved by

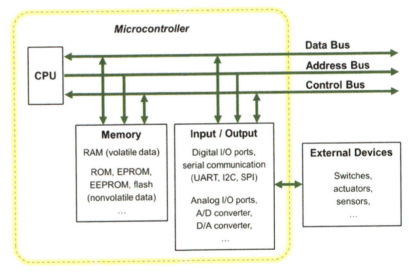

Figure 2. Microcontroller architecture.

the CPU during processing. There are generally three types of memory: (1) random-access memory (RAM), which is a volatile memory used to hold the data and programs being executed that can be read from or written to at any time as long as the power is maintained; (2) read-only memory (ROM), which is used for permanent storage of system instructions even when the microcontroller is powered down. Those instructions or data cannot be easily modified after manufacture and are rarely changed during the life of the microcontroller; and (3) erasable-programmable read only memory (EPROM), which is semi-permanent memory that can store instructions that need to be changed occasionally, such as the instructions that implement the specific use of the microcontroller. *Firmware* is a program usually permanently stored in the ROM or EPROM, which provides for control of the hardware and a standardized operating environment for more complex software programmed by users. The firmware remains unchanged until a system update is required to fix bugs or add features. Originally, EPROMS were erased using ultraviolet light, but more recently the flash memory (electrically erasable programmable read-only memory; EEPROM) has become the norm. The amount of RAM (described in bytes, kilobytes, megabytes, or gigabytes) determines the speed of operation, the amount of data that can be processed and the complexity of the programs that can be implemented.

Digital input and output (I/O) ports connect the microcontroller with external devices using digital signals only. The high and low voltage in the signal correspond to on and off states. Each digital port can be configured as an input port or an output port. The input port is used to read in the status of the external device and the output port is used to send a control instruction to an external device. Most microcontrollers operate over 0 to +5V with limited current because the voltage signal is not used directly, only the binary status. If the voltage and current are to be used to directly drive a device, a relay or voltage digital analog convertor is required between the port and device. Usually digital I/O ports communicate or "talk" with external devices through standard communication protocols, such as *serial communication protocols*. For example, a microcontroller can use digital I/O pins to form serial communication ports to talk to a general-purpose computer, external memory, or another microcontroller. Common protocols for serial communication are UART (universal asynchronous receiver-transmitter), USB (universal serial bus), I²C (inter-integrated circuit), and SPI (serial peripheral interface). *Analog input and output (analog I/O) ports* can be connected directly to the microcontroller. Many sensors (e.g., temperature, pressure, strain, rotation) output analog signals and many actuators require an analog signal. The analog ports integrate either an *analog to digital (A/D) converter* or *digital to analog (D/A) converter*.

The CPU, memory, and I/O ports are connected through electrical signal conductors known as *buses*. They serve as the central nervous system of the computer allowing data, addresses, and control signals to be shared among all system components. Each component has its own bus controller. There are three types of buses: the data bus, the address bus, and the control bus. The data bus transfers data to and from the data registers of various system components. The address bus carries the address of a system component that a CPU would like to communicate with or a specific data location in memory that a CPU would like to access. The control bus transmits the operational signal between the CPU and system components such as the read and write signals, system clock signal, and system interrupts.

Finally, *clock/counter/timer* signals are used in a microcontroller to synchronize operations among components. A clock signal is typically a pulse sequence with a known constant frequency generated by a quartz crystal oscillator. For example, a CPU clock is a high frequency pulse signal used to time and coordinate various activities in the CPU. A system clock can be used to synchronize many system operations such as the input and output data transfer, sampling, or A/D and D/A processes.

Microcontroller Software and Programming

The specific functions of a microcontroller depend on its software or how it is programed. The programs are stored in the memory. Recall that the CPU can only execute binary code, or *machine code*, and performs low-level operations such as adding a number to a register or moving a register's value to a memory location. However, it is very difficult to write a program in machine code. Hence, programming languages were developed over the years to make programming convenient. Low-level programming languages, such as assembly language, are the most similar to machine code. They are typically hardware-specific and not interchangeable among different types of microcontrollers. High-level programming languages, such as BASIC, C, or C++, tend to be more generic and can be deployed among different types of microcontrollers with minor modifications.

The programming languages for a specific microcontroller are determined by the microcontroller manufacturer. High-level programming languages are dominant in today's microcontrollers since they are much easier for learning, interpretation, implementation, and debugging. Programming a microcontroller often requires references to manuals, tutorials, and application notes from manufacturers. Online digital courses and online community-based learning are often good resources as well.

The example presented later in this chapter is a hands-on project using a microcontroller board called Arduino UNO. Arduino is a family of open-source hardware and software, single-board microcontrollers. They are popular and there are many online resources available to help new users develop applications. The microcontrollers are easy to understand and easy to use in real world applications with sensors and actuators (Arduino, 2019). The programming language of the Arduino microcontrollers is based on a language called Processing, which is similar to C or C++ but much simpler (https://processing.org/). The code can be adapted for other microcontrollers. In order to convert codes from a high-level language to the machine code to be executed by a specific CPU, or from one language to another language, a computer program called a *compiler* is necessary.

Programs can be developed by users in an *integrated development environment (IDE)*, which is a software that runs on a PC or laptop to allow the microcontroller code to be programmed and simulated on the PC or laptop. Most programming errors can be identified and corrected during the simulation. An IDE typically consists of the following components:

- An *editor* to program the microcontroller using a relevant high-level programming language such as C, C++, BASIC, or Python.
- A *compiler* to convert the high-level language program into low-level assembly language specific to a particular microcontroller.
- An *assembler* to convert the assembly language into machine code in binary bit (0 or 1) format.
- A *debugger* to error check (also called "debug") the code, and to test whether the code does what it was intended to do. The debugger typically finds syntax errors, which are statements that cannot be understood and cannot be compiled, and redundant code, which are lines of the program that do nothing. The line number or location of the error is shown by the debugger to help fix problems. The programmer can also add error testing components when writing the code to use the debugger to help confirm the program does what was originally intended.
- A *software emulator* to test the program on the PC or laptop before testing on hardware.

Not all components listed above are always presented to the user in an IDE, but they always exist. For the development of some systems, a hardware *emulator* might also be available. This will consist of a printed circuit board connected to the PC or laptop by ribbon cable joining I/O ports. The emulator can be used to load and run a program for testing before the microcontroller is embedded on a live measurement or control system.

Designing a Microcontroller-Based Measurement and Control System

The following workflow can help us design and build a microcontroller-based measurement and control system.

***Step 1.* Understand the problem and develop design objectives** of the measurement and control system with the end-users. Useful questions to ask include:

- What should be the functions of the system? For example, a system is needed to regulate the room temperature of a confined animal housing facility within an optimal range.
- Where or in what environment does the measurement or control occur? For example, is it an indoor or outdoor application? Is the operation in a very high or low temperature, a very dusty, muddy, or noisy environment? Is there anything special to be considered for that application?
- Are there already sensors or actuators existing as parts of the system or do appropriate ones need to be identified? For example, are there already thermistors installed to measure the room temperature, or are there fans or heaters installed?

- How frequently and how fast should things be measured or controlled? For example, it may be fine to check and regulate a room temperature every 10 seconds for a greenhouse; however, the flow rate and pressure of a variable-rate sprayer running at 5 meters per second (about 12 miles per hour) in the field need to be monitored and controlled at least every second.
- How much precision does the measurement and control need? For example, is a precision of a Celsius degree enough or does the application need sub-Celsius level precision?

Step 2. **Identify the appropriate sensors and/or actuators** if needed for the desired objectives developed in the previous step.

Step 3. **Understand the input and output signals** for the sensors and actuators by reading their specifications.

- How many inputs and outputs are necessary for the system functions?
- For each signal, is it a voltage or current signal? Is it a digital or analog signal?
- What is the range of each signal?
- What is the frequency of each signal?

Step 4. **Select a microcontroller** according to the desired system objective, the output signals from the sensors, and the input signals required by the actuators. Read the technical specifications of the microcontroller carefully. Be sure that:

- the number and types of I/O ports are compatible with the output and input signals of the sensors and actuators;
- the CPU speed and memory size are enough for the desired objectives;
- there are no missing components between the microcontroller, the sensors, and actuators such as converters or adapters, and if there are any, identify them; and
- the programming language(s) of the microcontroller is appropriate for the users.

Step 5. **Build a prototype** of the system with the selected sensors, actuators, and microcontroller. This step typically includes the physical wiring of the hardware components. If preferred, a virtual system can be built and tested in an emulator software to debug problems before building and testing with the physical hardware to avoid unnecessary hardware damage.

Step 6. **Program the microcontroller.** Develop a program with all required functions. Load it to the microcontroller and debug with the system. All code should be properly commented to make the program readable by other users later.

Step 7. **Deploy and debug** the system under the targeted working environment with permanent hardware connections until everything works as expected.

Step 8. **Document the system** including, for example, specifications, a wiring diagram, and a user's manual.

Applications

Microcontroller-based measurement and control systems are commonly used in agricultural and biological applications. For example, a field tractor has many microcontrollers, each working with different mechanical modules to realize specific functions such as monitoring and maintaining engine temperature and speed, receiving GPS signals for navigation and precise control of implements for planting, spraying, and tillage. A linear or center pivot irrigation system uses microcontrollers to ensure flow rate, nozzle pressure, and spray pattern are all correct to optimize water use efficiency. Animal logging systems use microcontrollers to manage the reading of ear tags when the animals pass a weighing station or need to be presented with feed. A food processing plant uses microcontroller systems to monitor and regulate processes requiring specific throughput, pressure, temperature, speed, and other environmental factors. A greenhouse control system for vegetable production will be used to illustrate a practical application of microcontrollers.

Modern greenhouse systems are designed to provide an optimal environment to efficiently grow plants with minimal human intervention. With advanced electronic, computer, automation, and networking technologies, modern greenhouse systems provide real-time monitoring as well as automatic and remote control by implementing a combination of PC communication, data handling, and storage, with microcontrollers each used to manage a specific task (figure 3). The specific tasks address the plants' need for correct air composition (oxygen and carbon dioxide), water (to ensure transpiration is optimized to drive nutrient uptake and heat dispersion), nutrients (to maximize yield), light (to drive photosynthesis), temperature (photosynthesis is maximized at a specific temperature for each type of plant, usually around 25°C) and, in some cases, humidity (to help regulate pests and diseases as well as photosynthesis). In a modern greenhouse, photosynthesis, nutrient and water supplies, and temperature are closely monitored and controlled using multiple sensors and microcontrollers.

As shown in figure 3, the overall control of the greenhouse environment is divided into two levels. The upper-level control system (figure 4) integrates an array of lower-level microcontrollers, each responsible for specific tasks in specific parts of the greenhouse, i.e., there may be multiple microcontrollers regulating light and shade in a very large greenhouse.

At the lower level, microcontrollers may work in

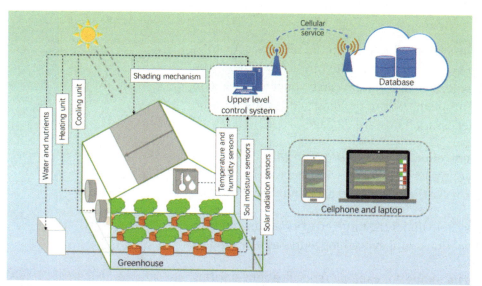

Figure 3. A diagram of a modern greenhouse system.

sub-systems or independently. Each microcontroller has its own suite of sensors providing inputs, actuators controlled by outputs, an SD (secure digital) card as a local data storage unit, and a CPU to run a program to deliver functionality. Each program implements its rules or decisions independently but communicates with the upper-level control system to receive time-specific commands and to transmit data and status updates. Some sub-systems may be examined in more detail and more frequently.

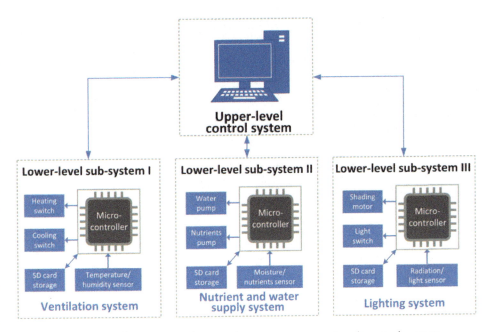

Figure 4. The overall structure of a greenhouse measurement and control system.

The ventilation sub-system is designed to maintain the temperature and humidity required for optimal plant growth inside the greenhouse. A schematic of a typical example (figure 5) shows the sub-system structure. Multiple temperature and humidity sensors are installed at various locations in the greenhouse and connected to the inputs of a microcontroller. Target temperature and humidity values can be input using a keypad connected to the microcontroller (figure 6) or set by the upper-level control system. Target values are also called "control set points" or simply "set points." They are the values the program is designed to maintain for the greenhouse. The microcontroller's function is to compare the measured temperature and humidity with the set point values to make a decision and adjust internal temperature. If a change is needed, the microcontroller controls actuators to turn on a heating device to raise the temperature (if temperature is below set point) or a cooling system fan (if temperature is above set point) to bring the greenhouse to the desired temperature and humidity.

The control panel in a typical ventilation system is shown in figure 6. Here a green light indicates that the heating unit is running, while the red lights indicate

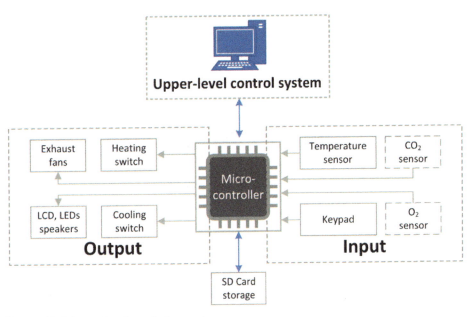

Figure 5. Schematic of ventilation system.

Basic Microcontroller Use for Measurement and Control • 9

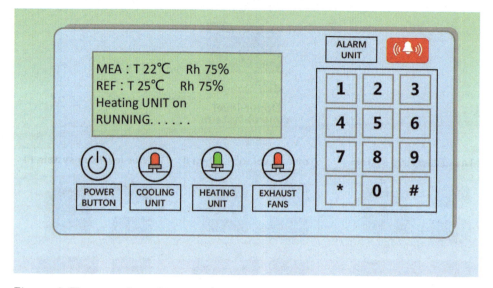

Figure 6. The control panel in a ventilation system.

that both the cooling unit and exhaust fans are off. The LCD displays the measured temperature and relative humidity inside the greenhouse (first line of text), the set point temperature and humidity values (second line of text), the active components (third line of text) and system status (fourth line of text). As the measured temperature is cooler than the set point, the heating unit has been turned on to increase the temperature from 22°C to 25°C. When the measured temperature reaches 25°C, the heating unit will be switched off. It is also possible to program alarms to alert an operator when any of the measured values exceed critical set points.

The nutrient and water supply sub-system (figure 7) provides plants with water and nutrients at the right time and the right amounts. It is possible to program a preset schedule and preset values or to respond to sensors in the growing medium (soil, peat, etc.). As in the temperature and humidity sub-system, the user can manually input set point values, or the values can be received from the upper-level system. Ideally, multiple sensors are used to measure soil moisture and nutrient levels in the root zone at various locations in the greenhouse. The readings of the sensors are interpreted by the microcontroller. When measured water or nutrient availability drops below a threshold, the microcontroller controls an actuator to release more water and/or nutrients.

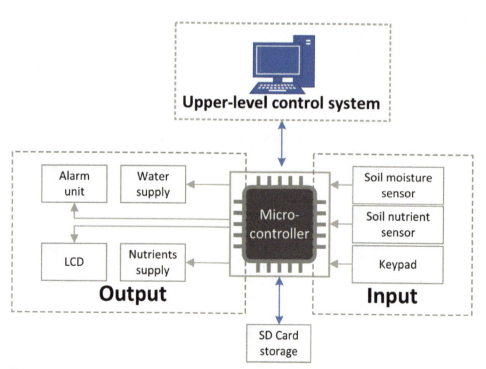

Figure 7. The nutrient and water supply system.

The lighting sub-system (figure 8) is designed to replace or supplement solar radiation provided to the plants for photosynthesis. Solar radiation and light sensors are installed in the greenhouse. The microcontroller reads data from these sensors and compares them with set points. If the measured value is too high, the microcontroller actuates a shading mechanism to cover the roof area. If the measured value is too low, the microcontroller activates the shading mechanism to remove all shading and, if necessary, turns on supplemental light units.

The upper-level control system is usually built on a PC or a server, which provides overall control through an integration of the subsystems. All of the sub-systems are connected to the central control computer through serial or wireless communication, such as an RS-232 port, Bluetooth, or Ethernet. The central control computer collects the data from all of the subsystems for processing analysis and record keeping. The upper-level control system can make optimal control decisions based on the data from all subsystems. It also provides an interface for the operator to manage the whole system, if needed. The central control computer also collects all data from all sensors and actuators to populate a database representing the control history of the greenhouse. This can be used to understand failure and, once sufficient data are collected, to implement machine learning algorithms, if required.

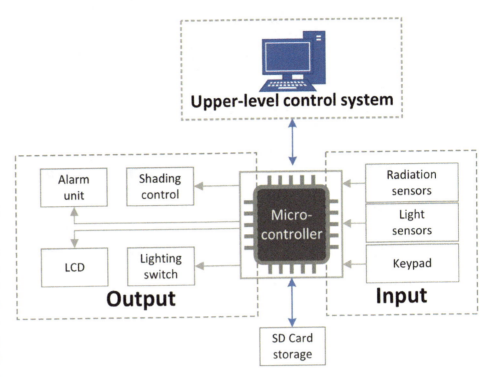

Figure 8. The schematic of lighting system.

This greenhouse application is a simplified example of a practical complex control system. Animal housing and other environmental control problems are of similar complexity. Modern agricultural machinery and food processing plants can be significantly more complex to understand and control. However, the principle of designing a hierarchical system with local automation managed by a central controller is very similar. Machine learning and artificial intelligence are now being used to achieve precise and accurate controls in many applications. Their control algorithms and strategies can be implemented on the upper-level control system, and the control decisions can be sent to the lower-level subsystems to implement the control functions.

Example

Example 1: Low-Cost Temperature Measurement and Control System

Problem:

A farmer wants to develop a low-cost measurement and control system to help address heat and cold stresses in confined livestock production. Specifically, the farmer wants to maintain the optimal indoor temperature of 18° to 20°C for a growing-finishing pig barn. A heating/cooling system needs to be activated if the temperature is lower or higher than the optimal range. The aim is to make a simple indicator to alert the stock handlers when the temperature is out of the target range, so that they can take action. (Automatic heating and cooling control is not required here.) Design and build a microcontroller-based measurement and control system to meet the specified requirements.

Solution:

Complete the recommended steps discussed above.

Step 1. Understand the problem.

- Functions—We need a system to monitor the ambient temperature and make alerts when the temperature is out of the 18° to 20°C range. The alert needs to indicate whether it is too cold or too hot, and the size of the deviation from that range.
- Environment—As a growing-finishing pig barn can be noisy, we will use a visual indicator as an alert rather than a sound alert.
- Existing sensors or actuators—For this example, assume that heating and cooling mechanisms have been installed in the barn. We just need to automate the temperature monitoring and decision-making process.
- Frequency—The temperature in a growing-finishing pig barn usually does not change rapidly. In this example, let's assume the caretakers require the temperature to be monitored every second.
- Precision—In this project, let's set the requirements for the precision at one degree Celsius for the temperature control.

Step 2. Identify the appropriate sensors and/or actuators.

The sensor that will be used in this example to measure temperature is the Texas Instruments LM35. It is one of the most widely used, low-cost temperature sensors in measurement and control systems in industry. Its output voltage is linearly proportional temperature, so the relationship between the sensor output and the temperature is straightforward.

We will use an RGB LED to light in different colors and blink at different rates to indicate the temperature and make alerts. This type of LED is a combination of a red LED, a green LED, and a blue LED in one package. By adjusting the intensity of each LED, a series of colors can be made. In this example, we will light the LED in blue when a temperature is lower than the optimal range, in green when the temperature is within the optimal range, and in red when the temperature is higher than the optimal range. In addition, the further the temperature has deviated from the optimal range, the faster the LED will blink. In this way, we alert the caretakers that a heating or cooling action needs to be taken and how urgent the situation is.

Step 3. Understand the input and output signals.

The LM35 series are precision integrated circuit temperature sensors with an output voltage linearly proportional to the Celsius (C) temperature (LM35 datasheet; http://www.ti.com/lit/ds/symlink/lm35.pdf). There are three pins in the LP package of the sensors as shown in figure 9. A package is a way that a block of semiconductors is encapsulated in a metal, plastic, glass, or ceramic casing.

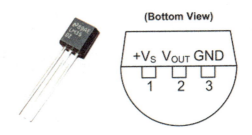

Figure 9. Texas Instruments LM35 precision centigrade temperature sensor in the LP package and its pin configuration and functions (from LM35 datasheet http://www.ti.com/lit/ds/symlink/lm35.pdf).

- The $+V_S$ pin is the positive power supply pin with voltage between 4V and 20V (in this project, we use +5V);
- The V_{OUT} pin is the temperature sensor analog output of no more than 6V (5V for this project);
- The GND pin is the device ground pin to be connected to the power supply negative terminal.

The accuracy specifications of the LM35 temperature sensor are given with respect to a simple linear transfer function:

$$V_{OUT} = 10 \text{ mV}/°C \times T \tag{1}$$

where V_{OUT} is the temperature sensor output voltage in millivolts (mV) and T is the temperature in °C.

In an RGB LED, each of the three single-color LEDs has two leads, the anode (or positive pin) where the current flows in and the cathode (or negative pin) where the current flows out. There are two types of RGB LEDs: common anode and common cathode. Assume we use the common cathode RGB LED as show in figure 10 but the other type would also work. The common cathode (−) pin 2 will connect to the ground. The anode (+) pins 1, 3, and 4 will connect to the digital output pins of the microcontroller.

> The sensor measurement needs to be calibrated. To do this, you can use an ice-water bath to create a 0°C environment, a cup of boiling water to create a 100°C environment, and an accurate thermometer to measure a room temperature. Derive a regression line. Its slope and intercept represent the relationship between the sensor measurements and the true values. For the example below, the slope is 1 and the intercept is 0.5°C.

Step 4. Select a microcontroller.

There are many general-purpose microcontrollers available commercially, such as the Microchip PIC, Parallax BASIC Stamp 2, ARM, and Arduino (Arduino, 2019). In this example, we will select an Arduino UNO microcontroller board based on the ATmega328P microcontroller (https://store.arduino.cc/usa/arduino-uno-rev3) (figure 11). The microcontroller has three types of memory: a 2KB RAM where the program creates and manipulates variables when it runs; a 1KB EEPROM where long-term information such as the firmware of the

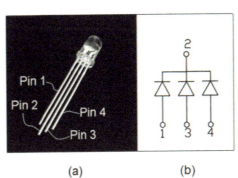

Figure 10. (a) a 5-mm common cathode RGB LED and (b) its pin configuration (https://www.sparkfun.com/products/105).

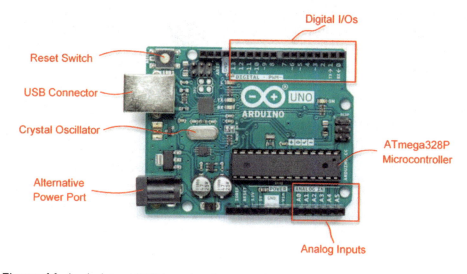

Figure 11. An Arduino UNO board and some major components (adapted from https://store.arduino.cc/usa/arduino-uno-rev3).

microcontroller is stored, and 32KB flash memory that can be used to store the programs you developed. The flash memory and EEPROM memory are non-volatile, which means the information persists after the power is turned off. The RAM is volatile, and the information will be lost when the power is removed. There are 14 digital I/O pins and 6 analog input pins on the Arduino UNO board. There is a 16 MHz quartz crystal oscillator. ATmega-based boards, including the Arduino UNO, take about 100 microseconds (0.0001 s) to read an analog input. So, the maximum reading rate is about 10,000 times a second, which is more than enough for our desired sampling frequency of every second. The board runs at 5 V. It can be powered by a USB cable, an AC-to-DC adapter, or a battery. If an USB cable is used, it also serves for loading, running, and debugging the program developed in the Arduino IDE. The Arduino UNO microcontroller is compatible with the LM35 temperature sensor and the desired control objectives of this project.

Step 5. Build a prototype.

The materials you need to build the system are:

- Arduino UNO board × 1
- Breadboard × 1
- Temperature sensor LM35 × 1
- RGB LED × 1
- 220 Ω resistor × 3
- Jumper wires

Figure 12 shows the hardware wiring.

- Pin 1 of the temperature sensor goes to the +5V power supply on the Arduino UNO board;
- Pin 2 of the temperature sensor goes to the analog pin A0 on the Arduino UNO board;
- Pin 3 of the temperature sensor goes to one of the ground pin GND on the Arduino UNO board;
- Digital I/O pin 2 on the Arduino UNO board connects with pin 4 (the blue LED) of the RGB LED through a 220 Ω resistor;

- Digital I/O pin 3 on the Arduino UNO board connects with pin 3 (the green LED) of the RGB LED through a 220 Ω resistor;
- Digital I/O pin 4 on the Arduino UNO board connects with pin 1 (the red LED) of the RGB LED through a 220 Ω resistor; and
- Pin 2 (cathode) of the RGB LED connects to the ground pin GND on the Arduino UNO board.

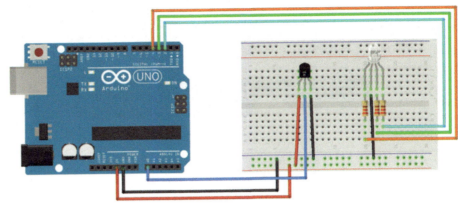

Figure 12. Wiring diagram for setting up the test platform. The Arduino is wired to a breadboard with three resistors and the LM35 temperature sensor.

An electronics breadboard (figure 13) is used to create a prototyping circuit without soldering. This is a great way to test a circuit. Each plastic hole on the breadboard has a metal clip where the bare end of a jumper wire can be secured. Columns of clips are marked as +, −, and a to j; and rows of clips are marked as 1 to 30. All clips on each one of the four power rails on the sides are connected. There are typically five connected clips on each terminal strip.

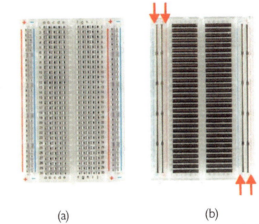

(a) (b)

Figure 13. A breadboard: (a) front view (b) back view with the adhesive back removed to expose the bottom of the four vertical power rails on the sides (indicated with arrows) and the terminal strips in the middle. (Picture from Sparkfun, https://learn.sparkfun.com/tutorials/how-to-use-a-breadboard/all).

Step 6. **Program the microcontroller.** The next step is to develop a program that runs on the microcontroller. As we mentioned earlier, programs are developed in IDE that runs either on a PC, a laptop, or a cloud-based online platform. Arduino has its own IDE. There are two ways to access it. The Arduino Web Editor (https://create.arduino.cc/editor/) is the online version that enables developers to write code, access tutorials, configure boards, and share projects. It works within a web browser so there is no need to install the IDE locally; however, a reliable internet connection is required. The more conventional way is to download and install the Arduino IDE locally on a computer (https://www.arduino.cc/en/main/software). It has different versions that can run on Windows, Mac OS X, and Linux operating systems. For this project, we will use the conventional IDE installed on a PC running Windows. The way the IDE is set up and operates is similar between the conventional one and the web-based one. You are encouraged to try both and find the one that works best for you.

Follow the steps on the link https://www.arduino.cc/en/Main/Software#download to download and install the Arduino IDE with the right version

> Resistors are passive components that can reduce current and divide voltage. The resistors used in this project all have a resistance of 220 Ω. If you are interested in learning how to recognize the resistance of a resistor by the color codes, check here: https://www.allaboutcircuits.com/textbook/reference/chpt-2/resistor-color-codes/.

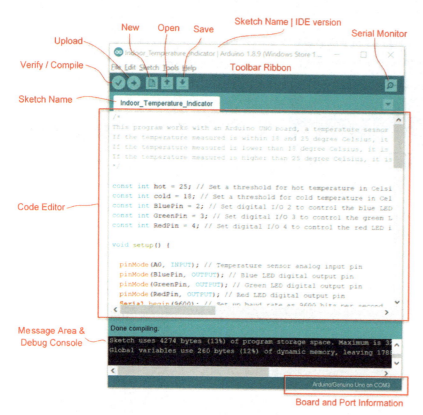

Figure 14. The interface and anatomy of the Arduino IDE.

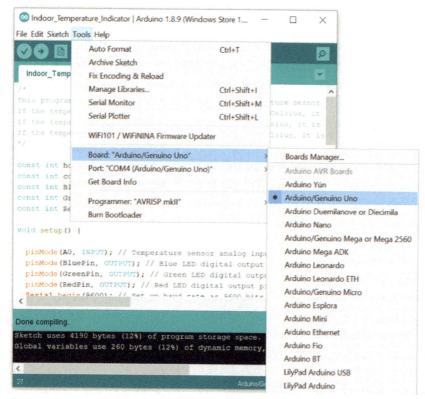

Figure 15. Select the right board and COM port in the Arduino IDE.

for your operating system. Open the IDE. It contains a few major components as shown in figure 14: a Code Editor to write text code, a Message Area and Debug Console to show compile information and error messages, a Toolbar Ribbon with buttons for common functions, and a series of menus.

Make sure that you disconnect the plug-ins of all the wires and pins the first time you power on the Arduino board either with a USB cable or a DC power port. It is a good habit to never connect or disconnect any wires or pins when the board is power on. Connect the Arduino UNO board and your PC or laptop using the USB cable. Under "Tools" in the main menu (figure 15) of the Arduino IDE, select the right board from the drop-down menu of "Board:" and the right COM port from the drop-down menu of "Port:" (which is the communication port the USB is using). Then disconnect the USB cable from the Arduino UNO board.

Now let's start coding in the Code Editor of the IDE. An Arduino board runs with a programming language called Processing, which is similar to C or C++ but much simpler (https://processing.org/). We will not cover the details about the programming syntax here; however, we will explain some of them along with the programming structure and logic. At the same time, you are encouraged to go to the websites of Arduino and the Processing language to learn more details about the syntax of Arduino programming.

Arduino programs have a minimum of 2 blocks—a setup block and an execution loop block. Each block has a set of statements enclosed in a pair of curly braces:

```
/*
Setup Block
*/

void setup() {              // Opening brace here

  Statements 1;             //Semicolon after every statement
  Statements 2;
  ...
  Statements n;

}                           // Closing brace here
```

```
/*
Execution Loop Block
*/

void loop() {               // Opening brace here

  Statements 1;             // Semicolon after every statement
  Statements 2;
  ...
  Statements n;

}                           // Closing brace here
```

There must be a semicolon (;) after every statement to indicate the finish of a statement; otherwise, the IDE will return an error during compiling. Statements after "//" in a line or multiple lines of statements between the pair of "/*" and "*/" are comments. Comments will not be compiled and executed, but they are important to help the readers understand the code.

The program logic flowchart is shown in figure 16. To better understand the code, we will separate the code into a few parts according to the logic flowchart. Each part will have its associated code shown in a grey box with explanations. You can copy and paste them into the Code Editor in the Arduino IDE. When writing the codes, be sure to save them frequently.

Program Part 1—Introductive Comments

Here we use multiple lines of statements to summarize the general purpose and function of the code.

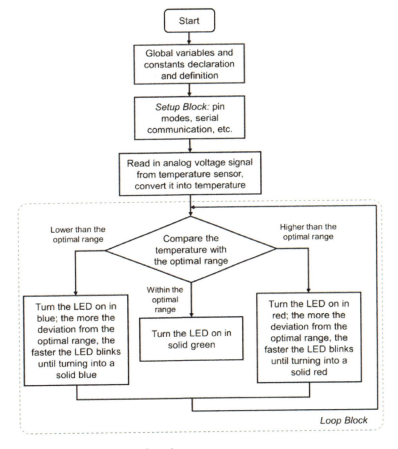

Figure 16. Program logic flowchart.

```
/*
This program works with an Arduino UNO board, a temperature sensor and an RGB
LED to measure and indicate the ambient temperature.
If the temperature measured is within 18 and 20 degree Celsius, it is considered
as optimal temperature and the LED is lit in green color.
If the temperature measured is lower than 18 degree Celsius, it is considered
as cold and the LED is lit in blue color and blinks. The colder the temperature,
the faster the LED blinks.
If the temperature measured is higher than 20 degree Celsius, it is considered
as hot and the LED is lit in red color and blinks. The hotter the temperature,
the faster the LED blinks.
*/
```

Program Part 2—Declarations of Global Variables and Constants

In this part of the program, we define a few variables and constants that will be used later for the entire program, including the upper and lower thresholds of the optimal temperature range and the numbers of the digital pins for the red, green, and blue LEDs inside the RGB LED, respectively. For example, the first statement here, "*const int hot = 20*" means that a constant ("*const*") integer ("*int*") called "*hot*" is created and assigned to the value of "*20*" which is the upper limit of the optimal temperature range. The third statement here, "*const int BluePin = 2*," means that a constant ("*const*") integer ("*int*") called "*BluePin*" is created and assigned to the value of "*2*" which will be used later in the setup block of the program to set digital pin 2 as the output pin to control the blue LED.

```
const int hot = 20;
// Set a threshold for hot temperature in Celsius
const int cold = 18;
// Set a threshold for cold temperature in Celsius
const int BluePin = 2;
// Set digital I/O 2 to control the blue LED in the RGB LED
const int GreenPin = 3;
// Set digital I/O 3 to control the green LED in the RGB LED
const int RedPin = 4;
// Set digital I/O 4 to control the red LED in the RGB LED
```

Program Part 3—Setup Block

As mentioned earlier, the setup block must exist even if there are no statements to execute. It is executed only once before the microcontroller executes the loop block repeatedly. Usually the setup block includes the initialization of the pin modes and the setup and start of serial communication between the microcontroller and the PC or laptop where the IDE runs. In this example, we set the analog pin A0 as the input of the temperature sensor measurements, digital pins defined earlier in Part 2 of the code as output pins to control the RGB LED, and start the serial communication with a typical communication speed (9600 bits per second) so that everything is ready for the microcontroller to execute the loop block.

```
void setup() {
    pinMode(A0, INPUT);
    // Temperature sensor analog input pin
    pinMode(BluePin, OUTPUT);
    // Blue LED digital output pin
    pinMode(GreenPin, OUTPUT);
    // Green LED digital output pin
    pinMode(RedPin, OUTPUT);
    // Red LED digital output pin
    Serial.begin(9600);
    // Set up baud rate as 9600 bits per second
}
```

Program Part 4—Execution Loop Block

The loop part of the program is what the microcontroller runs repeatedly unless the power of the microcontroller is turned off.

Program Part 4.1—Start the loop and read in the analog input from the temperature sensor:

```
void loop() {
    int sensor = analogRead(A0);
    // Read in the value from the analog pin connected to
    // the temperature sensor
    float voltage = (sensor / 1023.0) * 5.0;
    // Convert the value to voltage
    float tempC = (voltage - 0.5) * 100;
    // Convert the voltage to temperature using the
    /* scale factor; 0.5 is the deviation of the output voltage versus
    temperature from the best-fit straight line derived from sensor
    calibration */
    Serial.print("Temperature: ");
    Serial.print(tempC);
    // Print the temperature on the Arduino IDE output console
```

Here you see two types of variables, the integer ("int") and the float ("float"). For an Arduino UNO, an "int" is 16 bit long and can represent a number ranging from −32,768 to 32,767 (−2^15 to (2^15) − 1). A "float" in Arduino UNO is 32 bit long and can represent a number that has a decimal point, ranging from −3.4028235E+38 to 3.4028235E+38. Here, we define the variable of the temperature measured from the LM35 sensor as a float type so that it can represent a decimal number and is more accurate.

Program Part 4.2—Check if the temperature is lower than the optimal temperature range. If yes, turn on the LED in blue and blink it according to how much the temperature deviated from the optimal range:

```
    if (tempC < cold) {
// If the temperature is colder than the optimal temperature range
    Serial.println("It's cold.");
    float temp_dif = cold - tempC;
// Calculate how much the temperature deviated from
// the optimal range
      if (temp_dif <= 10) {
        int LED_blink_interval = (1.0 - (temp_dif / 10.0)) * 1000;
        /* Calculate LED blink interval in milliseconds based on the
        temperature deviation from the optimal range; the further the
        deviation, the faster the LED blinks until turning into a solid blue */
        // Blink the LED in blue:
        digitalWrite(BluePin, HIGH);
      // Turn on the blue LED
        digitalWrite(GreenPin, LOW);
      // Turn off the green LED
        digitalWrite(RedPin, LOW);
      // Turn off the red LED
        delay(LED_blink_interval);
      // Keep this status for a certain amount of time in milliseconds
        digitalWrite(BluePin, LOW);
      // Turn off the blue LED
        delay(LED_blink_interval);
      // Keep this status for a certain amount of time in milliseconds
    }
    else {
        digitalWrite(BluePin, HIGH);
      // Turn off the blue LED
        digitalWrite(GreenPin, LOW);
      // Turn off the green LED
        digitalWrite(RedPin, LOW);
      // Turn on the red LED
    }
  }
```

Here, we define an integer variable called "*LED_blink_interval*" which is inversely proportional to the deviation of the temperature from the optimal range "*temp_dif.*" A coefficient 4000 is used here to convert the number to something close to 1000. Arduino always measures the time duration in millisecond, so delay(1000) means delay for 1000 millisecond, or 1 second.

Program Part 4.3—Check if the temperature is higher than the optimal temperature range. If yes, turn on the LED in red and blink it according to how much the temperature deviated from the optimal range:

```
  else if (tempC > hot) {
  // If the temperature is hotter than the optimal temperature range
    Serial.println("It's hot.");
    // Calculate how much the temperature deviated from the optimal range
    float temp_dif = tempC - hot;
  if (temp_dif <= 10) {
    int LED_blink_interval = (1.0 - (temp_dif / 10.0)) * 1000;
   /* Calculate LED blink interval in milliseconds based on the temperature
       deviation from the optimal range; the further the deviation, the faster
    the LED blinks until turning into a solid red */
    // Blink the LED in red:
    digitalWrite(BluePin, LOW); // Turn off the blue LED
    digitalWrite(GreenPin, LOW); // Turn off the green LED
    digitalWrite(RedPin, HIGH); // Turn on the red LED
    delay(LED_blink_interval); // Keep this status for certain time in ms
    digitalWrite(RedPin, LOW); // Turn off the red LED
    delay(LED_blink_interval); // Keep this status for certain time in ms
  }
  else {
    digitalWrite(BluePin, LOW); // Turn off the blue LED
    digitalWrite(GreenPin, LOW); // Turn off the green LED
    digitalWrite(RedPin, HIGH); // Turn on the red LED
  }
```

Program Part 4.4—If the temperature is within the optimal range, turn on the LED in green:

```
    else {
  // Otherwise the temperature should be fine; turn the LED on in solid green
      Serial.println("The temperature is fine.");
      digitalWrite(BluePin, LOW);
    // Turn off the blue LED
      digitalWrite(GreenPin, HIGH);
    // Turn on the green LED
      digitalWrite(RedPin, LOW);
    // Turn off the red LED
    }
  delay(10);
}
```

After the program is written, use the "verify" button in the IDE to compile the code and debug errors if there are any. If the code has been transcribed accurately, there should be no syntax errors or bugs. If the IDE indicates errors, it is necessary to work through each line of code to make sure the program is correct. Be aware that sometimes the real error indicated by the debugger is in the lines before or after the location indicated. Some common errors include missing variable definition, missing braces, wrong spelling for a function, and letter capitalization error. Some other errors, such as the wrong selection of variable type, often cannot be caught during the compile stage, but we can use the "*Serial.print*" function to print the results or intermediate results on the serial monitor to see if they look reasonable.

Once the program code has no errors, connect the PC or laptop with the Arduino UNO board without any wire or pin plug-ins using the USB cable. Check if the selections for the type of board and port options under "Tools" in the main menu are still right. Use the "upload" button in the IDE to upload the program code to the Arduino board. Disconnect the USB cable from the board, and now plug in all the wires and pins. Re-connect the board and open the "Serial Monitor" from the IDE. The current ambient temperature should display in the serial monitor, and the LED lights color and blink accordingly. If any further errors occur, they will show in the message area at the bottom part of the IDE window. Go back to debugging if this happens. If there are no errors and everything runs correctly, test how the measurement system works by changing the temperature around the sensor to see the corresponding response of the LED color and blinking frequency. This can be done by breathing over the sensor or placing it close to a cup of iced water or in a fridge for a short time. When the room temperature is in the set point range (about 18°C to 20°C) the green LED should be lit. Once the temperature is too high, only the red LED should be lit. When the temperature is too low, only the blue LED should be lit. If this does not work, check that you have created different temperatures by using a laboratory thermometer and then check the program code.

Step 7. **Deploy and debug.**
Deploy and debug the system under the targeted working environment with permanent hardware connections until everything works as expected.

We leave this step of making the permanent hardware connections for you to complete if interested. In practice, the packaging of the overall system will be designed to accommodate the working environment. The completed final product will be tested extensively for durability and reliability.

Step 8. **Document the system.**
Write documentation such as system specifications, wiring diagram, and user's manual for the end users. At this stage, an instruction and safety manual would be written, and, if necessary, the product can be sent for local certification. Now the system you developed is ready to be signed off and handed over to the end users!

Image Credits

Figure 1. Alciatore, D.G., and Histand, M.B. (CC By 4.0). (2012). Main components in a measurement and control system. Introduction to mechatronics and measurement systems. Fourth edition. McGraw Hill.
Figure 2. Alciatore, D.G., and Histand, M.B. (2013). Microcontroller architecture. Adapted from Introduction to mechatronics and measurement systems. Fourth edition. McGraw Hill.
Figure 3. Qiu, G. (CC By 4.0). (2020). A diagram of a modern greenhouse system.
Figure 4. Qiu, G. (CC By 4.0). (2020). The core structure of a greenhouse control system.
Figure 5. Qiu, G. (CC By 4.0). (2020). The schematic of ventilation system.
Figure 6. Qiu, G. (CC By 4.0). (2020). The schematic of the control panel in ventilation system.
Figure 7. Qiu, G. (CC By 4.0). (2020). The nutrient and water supply system.
Figure 8. Qiu, G. (CC By 4.0). (2020). The schematic of lighting system.
Figure 9. Texas Instrument. (2020). Texas Instruments LM35 precision centigrade temperature sensor in LP package (a) and its pin configuration and functions (b). Retrieved from http://www.ti.com/lit/ds/symlink/lm35.pdf
Figure 10. Amazon. (2020). A 5 mm common cathode RGB LED and its pinout. Retrieved from https://www.amazon.com/Tricolor-Diffused-Multicolor-Electronics-Components/dp/B01C3ZZT8W/ref=sr_1_36?keywords=rgb+led&qid=1574202466&sr=8-36
Figure 11. Arduino. (2020). An Arduino UNO board and some major components. Retrieved from https://store.arduino.cc/usa/arduino-uno-rev3
Figure 12. Shi, Y. (CC By 4.0). (2020). Wiring diagram for setting up the test platform.
Figure 13. Sparkfun. (CC By 4.0). (2020). A breadboard: (a) front view (b) back view with the adhesive back removed to expose the bottom of the four vertical power rails on the sides (indicated with arrows) and the terminal strips in the middle Retrieved from https://learn.sparkfun.com/tutorials/how-to-use-a-breadboard/all
Figure 14. Shi, Y. (CC By 4.0). (2020). The interface and anatomy of Arduino IDE.
Figure 15. Shi, Y. (CC By 4.0). (2020). Select the right board and COM port in Arduino IDE.
Figure 16. Shi, Y. (CC By 4.0). (2020). Program logic flowchart.

References

Alciatore, D. G., and Histand, M. B. 2012. *Introduction to mechatronics and measurement systems.* 4th ed. McGraw Hill.
Arduino, 2019. https://www.arduino.cc/ Accessed on March 15, 2019.
Bolton, W. 2015. *Mechatronics, electronic control systems in mechanical and electrical engineering.* 6th ed. Pearson Education Limited.
Carryer, J. E., Ohline, R. M., and Kenny, T.W. 2011. *Introduction to mechatronic design.* Prentice Hall.
de Silva, C. W. 2010. *Mechatronics—A foundation course.* CRC Press.
University of Florida. 2019. What makes plants grow? http://edis.ifas.ufl.edu/pdffiles/4h/4H36000.pdf.

Visible and Near Infrared Optical Spectroscopic Sensors for Biosystems Engineering

Nathalie Gorretta
University of Montpellier, INRAe and SupAgro
Montpellier, France

Aoife A. Gowen
UCD School of Biosystems and Food Engineering
University College Dublin, Ireland

KEY TERMS

Electromagnetic spectrum	Spectroscopic measurements	Food authentication
Light and matter interaction	Spectral indices	Food quality control
Beer-Lambert Law	Contaminant detection	Vegetation monitoring in agriculture

Variables

- ε = molar absorptivity or molar extinction coefficient = Beer-Lambert proportionality constant
- λ = wavelength
- v = frequency
- $\bar{v}$ = wave number
- A = absorbance
- b = path length
- c = speed of light (3×10^8 m s^{-1})
- C = concentration
- E = energy of photons of light

h = Planck's constant (6.6260693 × 10^{-34} J·s)
I = transmitted light intensity
I_0 = incident light intensity
R_{NIR} = reflectance NIR
R_R = reflectance in the red part of spectral range
R_{SWIR} = reflectance SWIR
T = transmittance

Introduction

Optical sensors are a broad class of devices for detecting light intensity. This can be a simple component for notifying when ambient light intensity rises above or falls below a prescribed level, or a highly sensitive device with the capacity to detect and quantify various properties of light such as intensity, frequency, wavelength, or polarization. Among these sensors, optical spectroscopic sensors, where light interaction with a sample is measured at many different wavelengths, are popular tools for the characterization of biological resources, since they facilitate comprehensive, non-invasive, and non-destructive monitoring. Optical sensors are widely used in the control and characterization of various biological environments, including food processing, agriculture, organic waste sorting, and digestate control.

The theory of spectroscopy began in the 17th century. In 1666, Isaac Newton demonstrated that white light from the sun could be dispersed into a continuous series of colors (Thomas, 1991), coining the word *spectrum* to describe this phenomenon. Many other researchers then contributed to the development of this technique by showing, for example, that the sun's radiation was not limited to the visible portion of the electromagnetic spectrum. William Herschel (1800) and Johann Wilhelm Ritter (1801) showed that the sun's radiation extended into the infrared and ultraviolet, respectively. A major contribution by Joseph Fraunhofer in 1814 laid the foundations for quantitative spectrometry. He extended Newton's discovery by observing that the sun's spectrum was crossed by a large number of fine dark lines now known as Fraunhofer lines. He also developed an essential element of future spectrum measurement tools (spectrometers) known as the diffraction grating, an array of slits that disperses light. Despite these major advances, Fraunhofer could not give an explanation as to the origin of the spectral lines he had observed. It was only later, in the 1850s, that Gustav Kirchoff and Robert Bunsen showed that each atom and molecule has its own characteristic spectrum. Their achievements established spectroscopy as a scientific tool for probing atomic and molecular structure (Thomas, 1991; Bursey, 2017).

Many terms are used to describe the measurement of electromagnetic energy at different wavelengths, such as spectroscopy, spectrometry, and spectrophotometry. The word *spectroscopy* originates from the combination of *spectro* (from the Latin word *specere*, meaning "to look at") with *scopy* (from the Greek word *skopia*, meaning "to see"). Following the achievements of Newton, the term spectroscopy was first applied to describe the study of visible light dispersed by a prism as a function of its wavelength. The concept of spectroscopy was extended, during a lecture by Arthur Schuster in 1881 at the Royal Institution, to incorporate any interaction with radiative energy according to its wavelength or frequency (Schuster, 1911). Spectroscopy, then, can be summarized as the scientific study of the electromagnetic radiation emitted, absorbed, reflected, or scattered by atoms or molecules. *Spectrometry* or *spectrophotometry* is the quantitative measurement of the electromagnetic energy emitted, reflected, absorbed, or scattered by a material as a function of wavelength. The suffix "*-photo*" (originating from the Greek term *phôs*, meaning "light") refers to visual observation, for example, printing on photographic film, projection on a screen, or the use of an observation scope, while the suffix "*-metry*" (from the Greek term *metria*, meaning the process of measuring) refers to the recording of a signal by a device (plotter or electronic recording).

Spectroscopic data are typically represented by a spectrum, a plot of the response of interest (e.g. reflectance, transmittance) as a function of wavelength or frequency. The instrument used to obtain a spectrum is called a spectrometer or a spectrophotometer. The spectrum, representing the interaction of electromagnetic radiation with matter, can be analyzed to gain information on the identity, structure, and energy levels of atoms and molecules in a sample.

Two major types of spectroscopy have been defined, atomic and molecular. Atomic spectroscopy refers to the study of electromagnetic radiation absorbed or emitted by atoms, whereas molecular spectroscopy refers to the study of the light absorbed or emitted by molecules. Molecular spectroscopy provides information about chemical functions and structure of matter while atomic spectroscopy gives information about elemental composition of a sample. This chapter focuses on molecular spectroscopy, particularly in the visible-near infrared wavelength region due to its relevance in biosystems engineering.

Outcomes

After reading this chapter, you should be able to:

- Describe basic concepts of light and matter interaction, the electromagnetic spectrum, and the fundamental processes involved in absorption spectroscopy
- Use the Beer-Lambert law to predict the concentration of an unknown solution
- Calculate spectral indices from spectral imaging data

Concepts

Light and Matter Interaction

Spectroscopy is based on the way electromagnetic energy interacts with matter. All light is classified as electromagnetic radiation consisting of alternating electric and magnetic fields and is described classically by a continuous

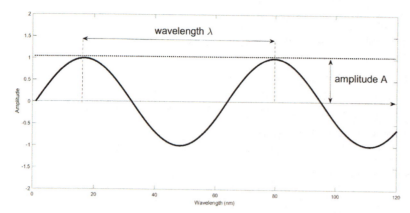

Figure 1. Schematic of a sinusoidal wave described by its wavelength λ and its amplitude A.

sinusoidal wave-like motion of the electric and magnetic fields propagating transversally in space and time. Wave motion can be described by its wavelength λ (nm), the distance between successive maxima or minima, or by its frequency ν (Hz), the number of oscillations of the field per second (figure 1). Wavelength is related to the frequency via the speed of light c (3×10^8 m s^{-1}) according to the relationship given in equation 1.

$$\lambda = \frac{c}{\nu} \tag{1}$$

Sometimes it is convenient to describe light in terms of units called "wavenumbers," where the wavenumber is the number of waves in one centimeter. Thus, wavenumbers are frequently used to characterize infrared radiation. The wavenumber, $\bar{\nu}$, is formally defined as the inverse of the wavelength, λ, expressed in centimeters:

$$\bar{\nu} = \frac{1}{\lambda} \tag{2}$$

The wavenumber is therefore directly proportional to frequency, ν:

$$\nu = c\bar{\nu} \tag{3}$$

leading to the following conversion relationships:

$$\bar{\nu}\,(\text{cm}^{-1}) = \frac{10^7}{\lambda\,(\text{nm})} \tag{4}$$

$$\lambda\,(\text{nm}) = \frac{10^7}{\bar{\nu}\,(\text{cm}^{-1})} \tag{5}$$

Reminder: 1 nanometer = 10^{-7} cm

The propagation of light is described by the theory of electromagnetic waves proposed by Christian Huygens in 1878 (Huygens, 1912). However, the interaction of light with matter (emission or absorption) also leads to the particle nature of light and electromagnetic waves as proposed by Planck and Einstein in the early 1900s. In this theory, light is considered to consist of particles called photons, moving at the speed c. Photons are "packets" of elementary energy, or quanta, that are exchanged during the absorption or emission of light by matter.

The energy of photons of light is directly proportional to its frequency, as described by the fundamental Planck relation (equation 6).

Table 1. Conversion relationships between λ and $\bar{\nu}$.

Unit	Wavelength λ	Wavenumber $\bar{\nu}$	Relation
	cm	cm^{-1}	$\bar{\nu} = \dfrac{1}{\lambda}$
	nm	cm^{-1}	$\bar{\nu} = \dfrac{10^7}{\lambda}$

Thus, high energy radiation (such as X-rays) has high frequencies and short wavelengths and, inversely, low energy radiation (such as radio waves) has low frequencies and long wavelengths.

$$E = h\nu = \frac{hc}{\lambda} = hc\bar{\nu} \tag{6}$$

where E = energy of photons of light (J)
h = Plank's constant = $6.62607004 \times 10^{-34}$ J·s
ν = frequency (Hz)
c = speed of light (3×10^8 m s^{-1})
λ = wavelength (m)

The *electromagnetic spectrum* is the division of electromagnetic radiation according to its different components in terms of frequency, photon energy or associated wavelengths, as shown in figure 2. The highest energy radiation corresponds to the γ-ray region of the spectrum. At the other end of the electromagnetic spectrum, radio frequencies have very low energy (Pavia et al., 2008). The visible region only makes up a small part of the electromagnetic spectrum and ranges from 400 to about 750 nm. The infrared (IR) spectral region is adjacent to the visible spectral region and extends from about 750 nm to about 5×10^6 nm. It can be further subdivided into the *near-infrared* region (NIR) from about 750 nm to 2,500 nm which contains the *short wave-infrared* (SWIR) from 1100–2500 nm, the *mid-infrared* (MIR) region from 2,500 nm to 5×10^4 nm, and the *far-infrared* (FIR) region from 5×10^4 nm to 5×10^6 nm (Osborne et al., 1993).

When electromagnetic radiation collides with a molecule, the molecule's electronic configuration is modified. This modification is related to the wavelength of the radiation and consequently to its energy. The interaction of a wave with matter, whatever its energy, is governed by the Bohr atomic model and derivative laws established by Bohr, Einstein, Planck, and De Broglie (Bohr, 1913; De Broguie, 1925). Atoms and molecules can only exist in certain quantified energy states. The energy exchanges between matter and radiation can, therefore, only be done by

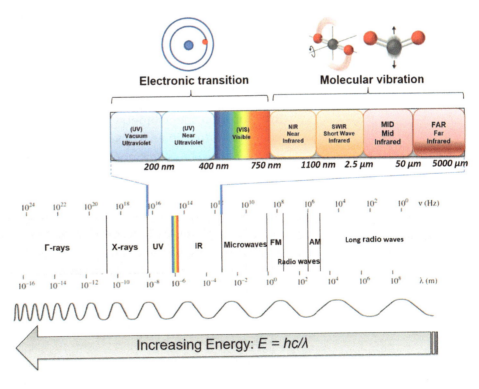

Figure 2. Electromagnetic spectrum.

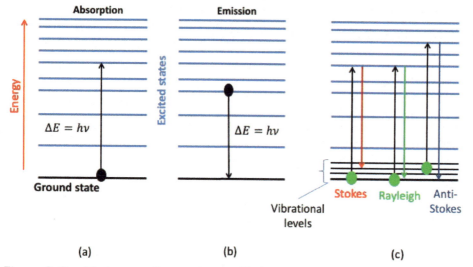

Figure 3. Simplified energy diagram showing (a) absorption, (b) emission of a photon by a molecule, (c) diffusion process.

specific amounts of energy or quanta $\Delta E = h\nu$. These energy exchanges can be carried out in three main ways (figure 3): absorption, emission, or diffusion.

In absorption spectroscopy, a photon is absorbed by a molecule, which undergoes a transition from a lower-energy state E_i to a higher energy or excited state E_j such that $E_j - E_i = h\nu$. In emission spectroscopy, a photon can be emitted by a molecule that undergoes a transition from a higher energy state E_j to a lower energy state E_i such that $E_j - E_i = h\nu$. In diffusion or scattering spectroscopy, a part of the radiation interacting with matter is scattered in many directions by the particles of the sample. If, after an interaction, the photon energy is not modified, the interaction is known as *elastic*. This corresponds to Rayleigh or elastic scattering, which maintains the frequency of the incident wave. When the photon takes or gives energy to the matter and undergoes a change in energy, the interaction is called *inelastic*, corresponding, respectively, to Stokes or anti-Stokes Raman scattering. Transitions between energy states are referred to as absorption or emission lines for absorption and emission spectroscopy, respectively.

Absorption Spectrometry

In absorption spectrometry, transitions between energy states are referred to as *absorption lines*. These absorption lines are typically classified by the nature of the electronic configuration change induced in the molecule (Sun, 2009):

- *Rotation lines* occur when the rotational state of a molecule is changed. They are typically found in the microwave spectral region ranging between 100 µm and 1 cm.
- *Vibrational lines* occur when the vibrational state of the molecule is changed. They are typically found in the IR, i.e., in the spectral range between 780 and 25,000 nm. Overtones and combinations of the fundamental vibrations in the IR are found in the NIR range (figure 2).
- *Electronic lines* correspond to a change in the electronic state of a molecule (transitions of the energetic levels of valence orbitals). They are typically found in the ultraviolet (approx. 200–400 nm) and visible

region (approx. 200–400 nm). In the visible region (350–800 nm), molecules such as carotenoids and chlorophylls absorb light due to their molecular structure. This visible spectral range is also used to evaluate color (for instance, of food or vegetation). In the ultraviolet spectral range, fluorescence and phosphorescence can be observed. While fluorescence and phosphorescence are both spontaneous emission of electromagnetic radiation, they differ in the way the excited molecule loses its energy after it has been irradiated. The glow of fluorescence stops right after the source of excitatory radiation is switched off, whereas for phosphorescence, an afterglow can last from fractions of a second to hours.

The spectral ranges selected for measurement and analysis depend on the application and the materials to be characterized. Absorption spectroscopy in the visible and NIR is commonly used for the characterization of biological systems due to the many advantages associated with this wavelength range, including rapidity, non-invasivity, non-destructive measurement, and significant incident wave penetration. Moreover, the NIR range enables probing of molecules containing C-H, N-H, S-H, and O-H bonds, which are of particular interest for characterization of biological samples (Pasquini, 2018; 2003). In addition to the chemical characterization of materials, it is possible to quantify the concentration of certain molecules using the Beer-Lambert law, described in detail below.

> **Fundamental vibrations, overtones, and combinations**
>
> Several vibrational modes could occur linked to a specific functional group of atoms: a characteristic frequency named the fundamental vibration, which usually occurs in the IR, as well as overtones and combinations of these fundamental frequencies. Overtone frequencies occur at integer multiples of the fundamental. For example, given a fundamental frequency at 1000 cm^{-1}, the first overtone would occur at 2000 cm^{-1} and the second overtone at 3000 cm^{-1}. Given two fundamental frequencies at 1500 cm^{-1} and 1000 cm^{-1}, their combination frequency would be 2500 cm^{-1}.

Beer-Lambert Law

Incident radiation passing through a medium undergoes several changes, the extent of which depends on the physical and chemical properties of the medium. Typically, part of the incident beam is reflected, another part is absorbed and transformed into heat by interaction with the material, and the rest passes through the medium. Transmittance is defined as the ratio of the transmitted light intensity to the incident light intensity (equation 7). Absorbance is defined as the logarithm of the inverse of the transmittance (equation 8). Absorbance is a positive value, without units. Due to the inverse relationship between them, absorbance is greater when the transmitted light is low.

$$T = \frac{I}{I_0} \qquad (7)$$

$$A = \log\left(\frac{1}{T}\right) = \log\left(\frac{I_0}{I}\right) \qquad (8)$$

where T = transmittance
I = transmitted light intensity
I_0 = incident light intensity
A = absorbance (unitless)

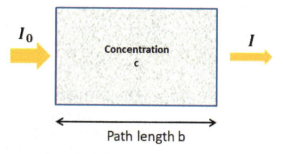

Figure 4. Absorption of light by a sample.

The Beer-Lambert law (equation 9) describes the linear relationship between absorbance and concentration of an absorbing species. At a given wavelength λ, absorbance A of a solution is directly proportional to its concentration (C) and to the length of the optical path (b), i.e., the distance over which light passes through the solution (figure 4, equation 9). When the concentration is expressed in moles per liter (mol L^{-1}), the length of the optical path in centimeters (cm), the molar absorptivity or the molar extinction coefficient ε is expressed in L mol^{-1} cm^{-1}.

Molar absorptivity is a measure of the probability of the electronic transition and depends on the wavelength but also on the solute responsible for absorption, the temperature and, to a lesser extent, the pressure.

$$A = \varepsilon b C \qquad (9)$$

where A = absorbance (unitless)
ε = molar absorptivity or molar extinction coefficient = Beer-Lambert proportionality constant (L mol^{-1} cm^{-1})
b = path length of the sample (cm)
C = concentration (mol L^{-1})

Beer-Lambert Law Limitations

Under certain circumstances, the linear relationship between the absorbance, the concentration, and the path length of light can break down due to chemical and instrumental factors. Causes of nonlinearity include the following:

- Deviation of absorptivity coefficient: The Beer-Lambert law is capable of describing the behavior of a solution containing a low concentration of an analyte. When analyte concentration is too high (typically >10 mM), electrostatic interactions between molecules close to each other result in deviations in absorptivity coefficients.
- High analyte concentrations can also alter the refractive index of the solution which in turn could affect the absorbance obtained.
- Scattering: Particulates in the sample can induce scattering of light.
- Fluorescence or phosphorescence of the sample.
- Non-monochromatic radiation due to instrumentation used.

Non-linearity can be detected as deviations from linearity when the absorbance is plotted as a function of concentration (see example 1). This is usually overcome by reducing analyte concentration through sample dilution.

Spectroscopic Measurements

Spectrometers are optical instruments that detect and measure the intensity of light at different wavelengths. Different measurement modes are available, including transmission, reflection, and diffuse reflection (figure 5). In transmission mode, the spectrometer captures the light transmitted through a sample, while in reflectance mode, the spectrometer captures the light reflected by the sample. In some situations, e.g., for light-diffusing samples such as powders, reflected light does not come solely from the front surface of the object; radiation that penetrates the material can reappear after scattering of reflection within the sample. These radiations are called diffuse reflection.

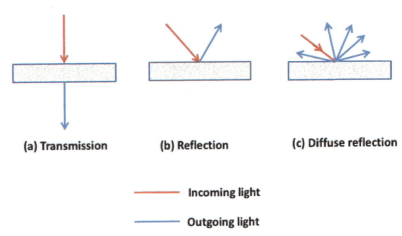

Figure 5. Schematic diagram showing the path of light for different modes of light measurement, i.e. (a) transmission, (b) reflection, and (c) diffuse reflection.

Spectrometers share several common basic components, including a source of light energy, a means for isolating a narrow range of wavelengths (typically a dispersive element), and a detector. The dispersive element must allow light of different wavelengths to be separated (figure 6).

The light source is arguably the most important component of any spectrophotometer. The ideal source is a continuous one that contains radiation of uniform intensity over a large range of wavelengths. Other desirable properties are stability over time, long service life, and low cost. Quartz-tungsten halogen lamps are commonly used as light sources for the visible (Vis) and NIR regions, and deuterium lamps or high-powered light emitting diodes may be used for the ultraviolet region.

The light produced by the light source is then focused and directed to the monochromater by an entrance slit. A grating diffraction element is then used to split the white light from the lamp into its components. The distance between the lines on gratings ("grating pitch") is of the same order of magnitude as the wavelength of the light to be analyzed. The separated wavelengths then propagate towards the sample compartment through the exit slit.

Depending on the technology used for the detector, the sample can be positioned before or after the monochromater. For simplicity, this chapter describes a positioning of the sample after the monochromater; the entire operation described above is valid regardless of the positioning of the sample.

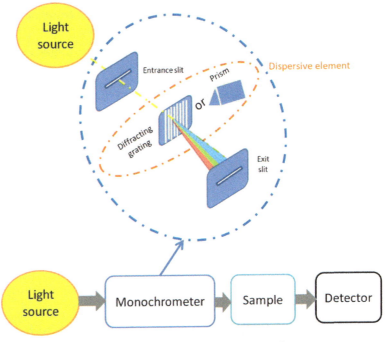

Figure 6. Spectrometer configuration: transmission diffraction grating.

In some spectrometers, an interferometer (e.g. Fabry-Pérot or Fourier-transform interferometer for UV and IR spectral range, respectively) is used instead of a diffraction grating to obtain spectral measurements. In this case, the initial beam light is split into two beams with different optical paths by using mirror arrangements. These two beams are then recombined before arriving at the detector. If the optical path lengths of the two beams do not differ by too much, an interference pattern is produced. A mathematical operation (Fourier transform) is then applied to the obtained interference pattern (interferogram) to produce a spectrum.

Once the light beams have passed through the samples, they will continue to the detector or photodetector. A photodetector absorbs the optical energy and converts it into electrical energy. A photodetector is a multichannel detector and can be a photodiode array, a charge coupled device (CCD), or a complementary metal oxide semiconductor (CMOS) sensor. While photodetectors can be characterized in many different ways, the most important differentiator is the detector material. The two most common semiconductor materials used in Vis-NIR spectrometers are silicon (Si) and indium gallium arsenide (InGaAs).

Spectral Imaging

Spectral imaging is a technique that integrates conventional imaging and spectroscopy to obtain both spatial and spectral information from an object. Multispectral imaging usually refers to spectral images in which <10 spectral bands are collected, while hyperspectral imaging is the term used when >100 contiguous spectral bands are collected. The term spectral imaging is more general. Spectral images can be represented as three-dimensional blocks of data, comprising two spatial and one wavelength dimension.

Two sensing modes are commonly used to acquire hyperspectral images, i.e., reflectance and transmission modes (figure 7). The use of these modes depends on the objects to be characterized (e.g., transparent or opaque) and the properties to be determined (e.g. size, shape, chemical composition, presence of defects). In reflectance mode, the hyperspectral sensor and light are located on the same side of the object and the imaging system acquires the light reflected by the object. In this mode, the lighting system should be designed to avoid any specular reflection. Specular reflection occurs when a light source can be seen as a direct reflection on the surface of an object. It is characterized by an angle of reflection being equal to the angle of incidence of the incoming light source on the sample. Specular reflection appears as bright saturated spots on acquired images impacting their quality. In transmittance mode, the detector is located in the opposite side of the light source and captures the transmitted light through the sample.

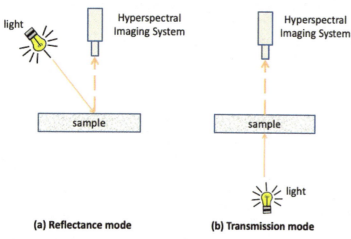

Figure 7. Hyperspectral imaging sensing mode: (a) reflectance mode, (b) transmission mode.

Applications

Vegetation Monitoring in Agriculture

The propagation of light through plant leaves is governed primarily by absorption and scattering interactions and is related to chemical and structural composition of the leaves. Spectral characteristics of radiation reflected, transmitted, or absorbed by leaves can thus provide a more thorough understanding of physiological responses to growth conditions and plant adaptations to the environment. Indeed, the biochemical components and physical structure of vegetation are related to its state of growth and health. For example, foliar pigments including chlorophyll a and b, carotenoids, and anthocyanins are strong absorbers in the Vis region and are abundant in healthy vegetation, causing plant reflectance spectra to be low in the Vis relative to NIR wavelength range (Asner, 1998; Ollinger, 2011) (figure 8). Chlorophyll pigments absorb violet-blue and red light for photosynthesis, the process by which plants use sunlight to synthesize organic matter. Green light is not absorbed by photosynthesis and reflectance spectra of green vegetation in the visible range are maximum around 550 nm. This is why healthy leaves appear to be green. The red edge refers to the area of the sudden increase in the reflectance of green vegetation between 670 and 780 nm. The reflectance in the NIR plateau (800–1100 nm) is a region where biochemical absorptions are limited and is affected by the scattering of light within the leaf, the extent of which is related to the leaf's internal structure. Reflectance in the short wave-IR (1100–2500 nm) is characterized by strong water absorption and minor absorptions of other foliar biochemical contents such as lignin, cellulose, starch, protein, and cellulose.

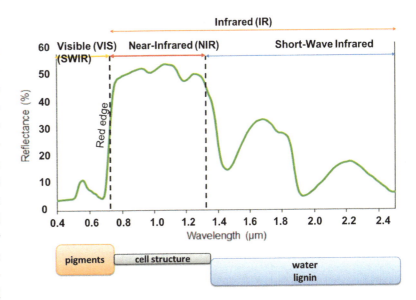

Figure 8. A green vegetation spectrum.

Stress conditions on plants, such as drought and pathogens, will induce changes in reflectance in the Vis and NIR spectral domain due to degradation of the leaf structure and the change of the chemical composition of certain tissues. Consequently, by measuring crop reflectance in the Vis and NIR regions of the spectrum, spectrometric sensors are able to monitor and estimate crop yield and crop water requirements and to detect biotic or abiotic stresses on vegetation. Vegetation indices (VI), which are combinations of reflectance images at two or more wavelengths designed to highlight a particular property of vegetation, can then be calculated over these images to monitor vegetation changes or properties at different spatial scales.

The normalized difference vegetation index (NDVI) (Rouse et al., 1974) is the ratio of the difference between NIR and red reflectance, divided by the sum of the two:

$$\text{NDVI} = \frac{R_{NIR} - R_R}{R_{NIR} + R_R} \qquad (10)$$

where R_{NIR} = reflectance in the NIR spectral region (one wavelength selected over the 750–870 nm spectral range) and R_R = reflectance in the red spectral region (one wavelength selected over 580–650 nm spectral range). Dividing by the sum of the two bands reduces variations in light over the field of view of the image. Thus, NDVI maintains a relatively constant value regardless of the overall illumination, unlike the simple difference which is very sensitive to changes in illumination. NDVI values can range between –1 and +1, with negative values corresponding to surfaces other than plant cover, such as snow or water, for which the red reflectance is higher than that in the NIR. Bare soils, which have red and NIR reflectance about the same order of magnitude, NDVI values are close to 0. Vegetation canopies have positive NDVI values, generally in the range of 0.1 to 0.7, with the highest values corresponding to the densest vegetation coverage.

NDVI can be correlated with many plant properties. It has been, and still is, used to characterize plant health status, identify phenological changes, estimate green biomass and yields, and in many other applications. However, NDVI also has some weaknesses. Atmospheric conditions and thin cloud layers can influence the calculation of NDVI from satellite data. When vegetation cover is low, everything under the canopy influences the reflectance signal that will be recorded. This can be bare soil, plant litter, or other vegetation. Each of these types of ground cover will have its own spectral signature, different from that of the vegetation being studied. Other indices to correct NDVI defects or to estimate other vegetation parameters have been proposed, such as the normalized difference water index or NDWI (Gao, 1996), which uses two wavelengths located respectively in the NIR and the SWIR regions (750–2500 nm) to track changes in plant moisture content and water stress (eq. 11). Both wavelengths are located in a high reflectance plateau (fig. 8) where the vegetation scattering properties are expected to be about the same. The SWIR reflectance is affected by the water content of the vegetation. The combination of the NIR and the SWIR wavelength is thus not sensitive to the internal structure of the leaf but is affected by vegetation water content. The normalized difference water index is:

$$\text{NDWI} = \frac{R_{NIR} - R_{SWIR}}{R_{NIR} + R_{SWIR}} \qquad (11)$$

where R_{NIR} is the reflectance in the NIR spectral region (one wavelength selected over the 750–870 nm spectral range) and R_{SWIR} is the reflectance in the SWIR spectral region around 1240 nm (water absorption band). Gao (1996) proposed using R_{NIR} equal to reflectance at 860 nm and R_{SWIR} at 1240 nm.

Absorption spectroscopy is widely used for monitoring and characterizing vegetation at different spatial, spectral, and temporal scales. Sensors are available mainly for broad-band multispectral or narrow-band hyperspectral data acquisition. Platforms are space-borne for satellite-based sensors, airborne for sensors on manned and unmanned airplanes, and ground-based for field and laboratory-based sensors.

Satellites have been used for remote sensing imagery in agriculture since the early 1970s (Bauer and Cipra, 1973; Doraiswamy et al., 2003) when Landsat 1 (originally known as Earth Resources Technology Satellite 1) was launched. Equipped with a multispectral scanner with four wavelength channels (one green, one red and two IR bands), this satellite was able to acquire multispectral images with 80 m spatial resolution and 18-day revisit time (Mulla 2013). Today, numerous multispectral satellite sensors are available and provide observations useful for assessing vegetation properties far better than Landsat 1. Landsat 8, for example, launched in 2013, offers nine spectral bands in the Vis to short-wave IR spectral range (i.e., 400–2500 nm) with a spatial resolution of 15–30 m and a 16-day revisit time. Sentinel-2A and Sentinel-2B sensors launched in 2015 and 2017, respectively, have 13 spectral bands (400–2500 nm) and offer 10–30 m multi-spectral global coverage and a revisit time of less than 10 days. Hyperspectral sensors, however, are still poorly available on satellites due to their cost and their relatively short operating life. Among them, Hyperion (EO-1 platform) has 220 spectral bands over the 400–2500 nm spectral range, a spatial resolution of 30 m, and a spectral resolution of 10 nm. The next generation, such as PRISMA (PRecursore IperSpettrale della Missione Applicativa) with a 30 m spatial resolution and a wavelength range of 400–2505 nm and the EnMAP (Environmental Mapping and Analysis Program) with a 30 m spatial resolution and a wavelength range of 400–2500 nm (Transon et al., 2018), indicate the future for this technology.

Some companies now use satellite images to provide a service to help farmers manage agricultural plots. Farmstar (http://www.myfarmstar.com/web/en) and Oenoview (https://www.icv.fr/en/viticulture-oenology-consulting/oenoview), for example, support management of inputs and husbandry in cereal and vine crops, respectively. However, satellite-based sensors often have an inadequate spatial resolution for precision agriculture applications. Some farm management decisions, such as weed detection and management, require images with a spatial resolution in the order of one centimeter and, for emergent situations (such as to monitor nutrient stress and disease), a temporal resolution of less than 24 hours (Zhang and Kovacs, 2012).

Airborne sensors are today able to produce data from multispectral to hyperspectral sensors with wavelengths ranging from Vis to MIR, with spatial resolutions ranging from sub-meter to kilometers and with temporal frequencies ranging from 30 min to weeks or months. Significant advancements in unmanned aerial vehicle (UAV) technology as well as in hyperspectral and multispectral sensors (in terms of both weight and image acquisition modes) allow for the combination of these tools to be used routinely for precision agricultural applications. The flexibility of these sensors, their availability and the high achievable spatial resolutions (cm) make them an alternative to satellite sensors. Multispectral sensors embedded on UAV platforms have been used in various agricultural studies, for example, to detect diseases in citrus trees (Garcia-Ruiz et al., 2013), grain yield in rice (Zhou et al., 2017) and for mapping vineyard vigor (Primicerio et al., 2012). UAV systems with multispectral imaging capability are used routinely by companies to estimate the nitrogen needs of plants. This information, given in near real-time to farmers, helps them to make decisions about management. Information extracted from airborne images are also used for precision farming to enhance planning of agricultural interventions or management of agricultural production at the scale of farm fields.

Ground-based spectroscopic sensors have also been developed for agricultural purposes. They collect reflectance data from short distances and can be mounted on tractors or held by hand. For example, the Dualex Force A handtool leaf clip (https://www.force-a.com/fr/produits/dualex) is adapted to determine the optical absorbance of the epidermis of a leaf in the ultraviolet (UV) optical range through the differential measurement of the fluorescence of chlorophyll as well as the chlorophyll content of the leaf using different wavelengths in the red and NIR ranges. Using internal model calibration, this tool calculates leaf chlorophyll content, epidermal UV-absorbance and a nitrogen balance index (NBI). This information could then be used to obtain valuable indicators of nitrogen fertilization, plant senescence, or pathogen susceptibility. Other examples are the nitrogen sensors developed by Yara (https://www.yara.fr/fertilisation/outils-et-services/n-sensor/) that enable adjustment of the nitrogen application rate in real time and at any point of the field, according to the crop's needs.

Food-Related Applications

Conventional, non-imaging, spectroscopic methods are widely used for routine analysis and process control in the agri-food industry. For example, NIR spectroscopy is commonly used in the prediction of protein, moisture, and fat content in a wide range of raw materials and processed products, such as liquids, gels, and powders (Porep et al., 2015). Ultraviolet-Vis (UV-Vis) spectroscopy is a valuable tool in monitoring bioprocesses, such as the development of colored phenolic compounds during fermentation of grapes in the process of winemaking (Aleixandre-Tudo et al., 2017). The Beer-Lambert law (equation 9) can be used to predict the concentration of a given compound given its absorbance at a specific wavelength.

While conventional spectroscopic methods are useful for characterizing homogeneous products, the lack of spatial resolution leads to an incomplete assessment of heterogeneous products, such as many foodstuffs. This is particularly problematic in the case of surface contamination, where information on the location, extent, and distribution of contaminants over a food sample is required. Applications of Vis-NIR spectral imaging for food quality and safety are widespread in the scientific literature and are emerging in the commercial food industry. The heightened interest in this technique is driven mainly by the non-destructive and rapid nature of spectral imaging, and the potential to replace current labor- and time-intensive analytical methods in the production process.

This section provides a brief overview of the range and scope of such applications. For a more comprehensive description of these and related applications, several informative reviews have been published describing advances in hyperspectral imaging for contaminant detection (Vejarano et al., 2017), food authentication (Roberts et al., 2018), and food quality control (Gowen et al. 2007; Baiano, 2017).

Contaminant Detection

The ability of spectral imaging to detect spatial variations over a field of view, combined with chemical sensitivity, makes it a promising tool for contaminant detection. The main contaminants that can be detected in the food chain using Vis-NIR include polymers, paper, insects, soil, bones, stones, and fecal matter. Diffuse reflectance is by far the most common mode of spectral imaging utilized for this purpose, meaning that primarily only surface or peripheral contamination can be detected. Of concern in the food industry is the growth of spoilage and pathogenic microorganisms at both pre-harvest and post-harvest processing stages, since these result in economic losses and potentially result in risks to human health. Vis-NIR spectral imaging methods have been demonstrated for pre-harvest detection of viral infection and fungal growth on plants, such as corn (maize) and wheat. For instance, decreases in the absorption of light in wavebands related to chlorophyll were found to be related to the destruction of chloroplasts in corn ears due to *Fusarium* infection (Bauriegel et al., 2011). Fecal contamination acts as a favorable environment for microbial growth, thus many studies have focused on the detection of such contamination over a wide variety of foods, including fresh produce, meat, and poultry surfaces. For example, both fluorescence and reflectance modalities have been shown to be capable of detecting fecal contamination on apples with high accuracy levels (Kim et al., 2007). Recent studies have utilized spectral imaging transmittance imaging for insect detection within fruits and vegetables, resulting in high detection levels (>80% correct classification) (Vejarano et al., 2017).

Food Authentication

Food ingredient authentication is necessary for the ever expanding global supply chain to ensure compliance with labeling, legislation, and consumer demand. Due to the sensitivity of vibrational spectroscopy to molecular structure and the development of advanced multivariate data analysis techniques such as chemometrics, NIR and MIR spectroscopy have been used successfully in authentication of the purity and geographical origin of many foodstuffs, including honey, wine, cheese, and olive oil. Spectral imaging, having the added spatial dimension, has been used to analyze non-homogeneous samples, where spatial variation could improve information on the authentication or prior processing of the food product, for example, in the detection of fresh and frozen-thawed meat or in adulteration of flours (Roberts et al., 2018).

Food Quality Control

Vis-NIR spectral imaging has been applied in a wide range of food quality control issues, such as bruise detection in mushrooms, apples, and strawberries, and in the prediction of the distribution of water, protein, or fat content in heterogeneous products such as meat, fish, cheese, and bread (Liu et al., 2017). The dominant feature in the NIR spectrum of high moisture foods is the oxygen-hydrogen (OH) bond-related peak centered around 1450 nm. The shape and intensity of this peak is sensitive to the local environment of the food matrix, and can provide information on changes in the water present in food products. This is useful since many deteriorative biochemical processes, such as microbial growth and non-enzymatic browning, rely on the availability of free water in foods. Vis-NIR spectral imaging has also been applied to quality assessment of semi-solid foods, as reviewed by Baiano (2017). For instance, transmittance spectral imaging has been used to non-destructively assess the interior quality of eggs (Zhang et al., 2015), while diffuse reflectance spectral imaging has been used to study the microstructure of yogurt (Skytte et al., 2015) and milk products (Abildgaard et al., 2015).

Examples

Example 1: Using the Beer-Lambert law to predict the concentration of an unknown solution

Problem:

Data were obtained from a UV-Vis optical absorption instrument, as shown in table 2. Light absorbance was measured at 520 nm for different concentrations of a compound that has a red color. The path length was 1 cm. The goal is to use the Beer-Lambert law to calculate the molar absorptivity coefficient and determine the concentration of an unknown solution that has an absorbance of 1.52.

Table 2. Concentration (mol L^{-1}) and corresponding absorbance at 520 nm for a red colored compound.

Concentration (mol L^{-1})	Absorbance at 520 nm
0.001	0.21
0.002	0.39
0.005	1.01
0.01	2.02

Solution:

The first step required in calculating the molar absorptivity coefficient is to plot a graph of absorbance as a function of concentration, as shown in figure 9. The data follow a linear trend, indicating that the assumptions of the Beer-Lambert law are satisfied.

To calculate the molar absorptivity coefficient, it is first necessary to calculate the line of best linear fit to the data. This is achieved here using the "add trendline" function in Excel. The resultant line of best fit is shown in figure 10. The equation of this line is y = 201.85x.

Compare this equation to the Beer-Lambert law (equation 9):

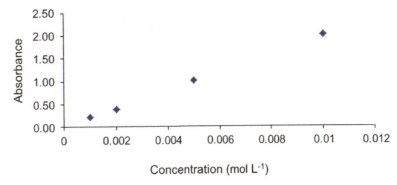

Figure 9. Plot of absorbance at 520 nm as a function of concentration.

$$A = \varepsilon b c \qquad (9)$$

where A = absorbance (unitless)
  ε = molar absorptivity or molar extinction coefficient = Beer-Lambert proportionality constant (L mol^{-1} cm^{-1})
  b = path length of the sample (cm)
  C = concentration (mol L^{-1})

In this example, $\varepsilon\, b$ = 201.85, where b is the path length, defined in the problem as 1 cm. Consequently, ε = 201.85 (L mol^{-1} cm^{-1}). To calculate the concentration of the unknown solution, substitute the absorbance of the unknown solution (1.52) into the equation of best linear fit, resulting in a concentration of 0.0075 mol L^{-1}.

This type of calculation can be used for process or quality control in the food industry or for environmental monitoring such as water quality assessment.

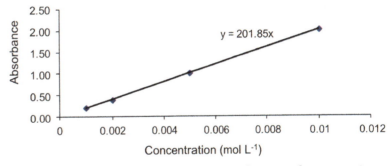

Figure 10. Plot of absorbance at 520 nm as a function of concentration showing line and equation of best linear fit to the data.

Example 2: Calculation of vegetation indices from a spectral image

Problem:
The Airborne Visible/Infrared Imaging Spectrometer (AVIRIS) developed by the National Aeronautics and Space Administration (NASA) is one of the foremost spectral imaging instruments for Earth remote sensing (NASA, n. d.). An agricultural scene was gathered by flying over the Indian Pines test site in northwestern Indiana (U.S.) and consists of 145 × 145 pixels and 224 spectral reflectance bands in the wavelength range 400–2500 nm. The Indian Pines scene (freely available at https://doi.org/10.4231/R7RX991C; Baumgardner et al., 2015) contains two-thirds agricultural land and one-third forest or other natural perennial vegetation. There are also two major dual lane highways and a rail line, as well as some low-density housing, other structures, and smaller roads present in the scene. The ground truth image shows the designation of various plots and regions in the scene, and is designated into sixteen classes, as shown in figure 11. The average radiance spectrum of four classes of land cover in the scene is plotted in

> **Radiance**
>
> Radiance is the flux of light that reaches a measurement system per unit of area and unit of solid angle perpendicular to the surface of the detector. It is expressed in $W\,sr^{-1}\,m^{-2}$.

figure 12. Table 3 shows the data corresponding to the plots shown in figure 11. Using the mean radiance values, calculate the NDVI and NDWI for each class of land cover. Please note: In this example, the mean radiance values are being used for illustration purposes. This simplification is based on the assumption that the radiation receipt is constant across all wavebands so radiance is assumed to be linearly proportional to reflectance (ratio of reflected to total incoming energy). Typically, vegetation indices are calculated from pixel-level reflectance spectra.

Solution:

The NDVI (calculated using NIR wavelength = 764 nm and red wavelength = 647 nm) and NDWI (calculated using NIR wavelengths 860 nm and 1244 nm) were calculated from the Indian Pines image by selecting the appropriate wavebands and calculating their normalized differences as described in equations 10 and 11:

$$\text{NDVI} = \frac{R_{NIR} - R_R}{R_{NIR} + R_R} = \frac{R_{764} - R_{647}}{R_{764} + R_{647}} \quad (10)$$

where $R_{NIR} = R_{764}$ = radiance in the NIR spectral region at 764 nm in this example.

$R_R = R_{647}$ = radiance in the red spectral region (one wavelength selected over the 580–650 nm spectral range) and at 647 nm in this example.

$$\text{NDWI} = \frac{R_{NIR} - R_{SWIR}}{R_{NIR} + R_{SWIR}} = \frac{R_{860} - R_{1244}}{R_{860} + R_{1244}} \quad (11)$$

where $R_{NIR} = R_{860}$ = radiance in the NIR spectral region (at 860 nm in this example)

$R_{SWIR} = R_{1244}$ = radiance at in the SWIR spectral region (at 1244 nm in this example)

Using the radiance values given in table 3 for the grass-pasture category, equation 11 becomes

$$\text{NDWI} = \frac{114 - 38}{114 + 38} = 0.5$$

The results are shown in table 4. The grassland classes have a positive NDVI value, with grass-pasture having the highest NDVI among the selected classes, while the stone-steel towers class has a negative NDVI.

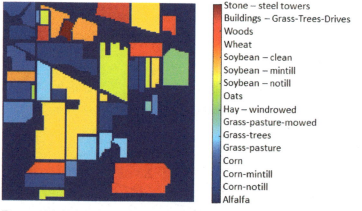

Figure 11. Indian Pines ground truth image showing various plots and regions in the scene, designated into sixteen classes (Baumgardner et al., 2015).

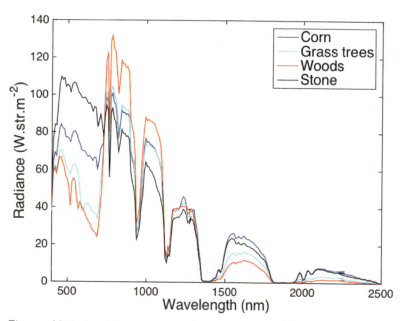

Figure 12. Indian Pines average radiance spectrum of four classes of land cover in the scene shown in figure 11.

Table 3. Mean radiance values for selected classes of land cover from the Indian Pines dataset (Baumgardner et al., 2015).

	Radiance (W sr^{-1} m^{-2})				
Wavelength (nm)	Grass-Pasture	Grass-Trees	Grass-Pasture-Mowed	Hay-Windrowed	Stone-Steel Towers
647	35	39	56	50	94
657	34	37	54	43	91
667	34	37	54	44	94
677	34	36	53	43	93
687	31	34	40	42	84
697	35	40	57	55	86
687	31	34	39	42	83
697	35	40	59	55	86
706	50	52	69	68	90
716	61	62	71	71	84
725	72	70	71	71	79
735	97	88	84	85	88
745	116	101	93	94	94
754	118	100	92	92	92
764	77	64	59	60	56
774	123	103	95	95	90
783	127	105	97	98	93
793	120	99	93	94	87
803	119	98	92	93	85
812	109	90	84	86	78
822	98	80	76	78	70
831	105	86	82	84	75
841	115	95	91	93	82
851	114	93	90	92	80
860	114	93	90	93	79
870	112	92	89	92	78
880	113	92	90	93	78
889	110	90	89	92	76
899	86	71	70	73	59
908	81	67	66	69	53
918	79	65	64	67	50
928	67	52	55	58	43
937	31	26	26	28	23
947	35	29	29	31	25
956	39	33	34	37	29
966	53	41	38	50	40
976	70	59	63	66	46

(continued)

Table 3. Mean radiance values for selected classes of land cover from the Indian Pines dataset (*continued*).

	Radiance (W sr⁻¹ m⁻²)				
Wavelength (nm)	Grass-Pasture	Grass-Trees	Grass-Pasture-Mowed	Hay-Windrowed	Stone-Steel Towers
985	81	69	73	77	59
995	88	74	78	83	64
1004	86	73	77	81	62
1014	87	73	77	81	61
1024	86	72	76	80	59
1033	86	72	75	80	59
1043	86	72	74	79	57
1052	85	71	73	78	55
1062	82	68	71	75	52
1072	80	67	69	74	49
1081	78	65	67	72	48
1091	74	61	64	68	44
1100	61	45	54	58	40
1110	39	39	40	42	32
1120	18	15	16	17	13
1129	14	12	12	14	10
1139	20	17	18	20	15
1148	17	15	17	18	14
1158	27	25	28	30	23
1168	39	35	40	42	33
1177	41	38	43	41	34
1187	41	37	42	41	34
1196	42	39	44	39	35
1206	41	39	43	40	35
1216	43	40	43	38	36
1225	42	41	34	40	37
1235	38	42	35	46	39
1244	38	42	34	45	38
1254	43	41	38	38	36

Table 4. NDVI and NDWI calculated from mean radiance of selected classes of land cover from the Indian Pines dataset.

	Grass-Pasture	Grass-Trees	Grass-Pasture-Mowed	Hay-Windrowed	Stone-Steel Towers
NDVI	0.38	0.24	0.03	0.09	−0.25
NDWI	0.5	0.38	0.45	0.35	0.35

By applying the calculation to each pixel spectrum in the image, it is possible to create images of the NDVI and NDWI, as shown in figure 13. The NDVI highlights regions of vegetation in red, regions of crop growth and soil in light green-blue, and regions of stone in darker blue. The NDWI, sensitive to changes in water content of vegetation canopies, shows regions of high water content in red, irregularly distributed in the wooded regions.

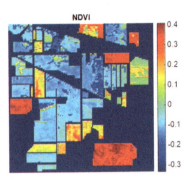 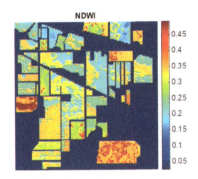

Figure 13. NDVI and NDWI calculation on Indian Pines images.

Image Credits

Figure 1. Gorretta, N. (CC By 4.0). (2020). Schematic of a sinusoidal wave described by its wavelength.
Figure 2. Gorretta, N. (CC By 4.0). (2020). Electromagnetic spectrum.
Figure 3. Gorretta, N. (CC By 4.0). (2020). Simplified energy diagram showing (a) absorption, (b) emission of a photon by a molecule, (c) diffusion process.
Figure 4. Gorretta, N. (CC By 4.0). (2020). Absorption of light by a sample.
Figure 5. Gorretta, N. (CC By 4.0). (2020). Schematic diagram showing the path of light for different modes of light measurement, i.e. (a) transmission, (b) reflection, and (c) diffuse reflection.
Figure 6. Gorretta, N. (CC By 4.0). (2020). Spectrometer configuration: transmission diffraction grating.
Figure 7. Gorretta, N. (CC By 4.0). (2020). Hyperspectral imaging sensing mode: (a) reflectance mode, (b) transmission mode.
Figure 8. Gorretta, N. (CC By 4.0). (2020). A green vegetation spectrum.
Figure 9. Gowen, A. A. (CC By 4.0). (2020). Plot of absorbance at 520 nm as a function of concentration.
Figure 10. Gowen, A. A. (CC By 4.0). (2020). Plot of absorbance at 520 nm as a function of concentration showing line and equation of best linear fit to the data.
Figure 11. Gowen, A. A. (CC By 3.0). (2015). Indian Pines ground truth image showing various plots and regions in the scene, designated into sixteen classes. Citation might be: Baumgardner, M. F., L. L. Biehl, and D. A. Landgrebe. 2015. "220 Band AVIRIS Hyperspectral Image Data Set: June 12, 1992 Indian Pine Test Site 3." Purdue University Research Repository. doi:10.4231/R7RX991C. This item is licensed CC BY 3.0.
Figure 12. Gowen, A. A. (CC By 4.0). (2020). Indian Pines average reflectance spectrum of four classes of land cover in the scene shown in figure 11.
Figure 13. Gowen, A. A. (CC BY 4.0). (2020). NDVI and NDWI calculation of Indian Pines images.

References

Abildgaard, O. H., Kamran, F., Dahl, A. B., Skytte, J. L., Nielsen, F. D., Thomsen, C. L., ... Frisvad, J. R. (2015). Non-invasive assessment of dairy products using spatially resolved diffuse reflectance spectroscopy. *Appl. Spectrosc.*, *69*(9), 1096–1105. https://doi.org/10.1366/14-07529.

Aleixandre-Tudo, J. L., Buica, A., Nieuwoudt, H., Aleixandre, J. L., & du Toit, W. (2017). Spectrophotometric analysis of phenolic compounds in grapes and wines. *J. Agric. Food Chem.*, *65*(20), 4009-4026. https://doi.org/10.1021/acs.jafc.7b01724.

Asner, G. P. (1998). Biophysical and biochemical sources of variability in canopy reflectance. *Remote Sensing Environ.*, *64*(3), 234-253. https://doi.org/10.1016/S0034-4257(98)00014-5.

Baiano, A. (2017). Applications of hyperspectral imaging for quality assessment of liquid based and semi-liquid food products: A review. *J. Food Eng.*, *214*, 10-15. https://doi.org/10.1016/j.jfoodeng.2017.06.012.

Bauer, M. E., & Cipra, J. E. (1973). Identification of agricultural crops by computer processing of ERTS MSS Data. *Proc. Symp. on Significant Results Obtained from the Earth Resources Technology Satellite*. Retrieved from http://agris.fao.org/agris-search/search.do?recordID=US201302721443.

Baumgardner, M. F., Biehl, L. L., & Landgrebe, D. A. (2015). 220 Band AVIRIS hyperspectral image data set: June 12, 1992 Indian Pine Test Site 3. Purdue University Research Repository. https://doi.org/10.4231/R7RX991C.

Bauriegel, E., Giebel, A., & Herppich, W. B. (2011). Hyperspectral and chlorophyll fluorescence imaging to analyse the impact of *Fusarium culmorum* on the photosynthetic integrity of infected wheat ears. *Sensors*, *11*(4), 3765-3779. https://doi.org/10.3390/s110403765.

Bohr, N. (1913). I. On the constitution of atoms and molecules. *London Edinburgh Dublin Philosophical Magazine J. Sci.*, *26*(151), 1-25. https://doi.org/10.1080/14786441308634955.

Bursey, M. M. (2017). A brief history of spectroscopy. *Access Science*. https://doi.org/10.1036/1097-8542.BR0213171.

De Broguie, L. V. 1925. On the theory of quanta. Paris, France.

Doraiswamy, P. C., Moulin, S., Cook, P. W., & Stern, A. (2003). Crop yield assessment from remote sensing. *Photogrammetric Eng. Remote Sensing*, *69*(6), 665-674. https://doi.org/10.14358/PERS.69.6.665.

Farmstar. (n. d.). Farmstar: Have everything you need to manage your crops! Retrieved from http://www.myfarmstar.com/web/en.

Force A. (n. d.). Dualex scientific. Retrieved from https://www.force-a.com/fr/produits/dualex.

Gao, B.-c. (1996). NDWI: A normalized difference water index for remote sensing of vegetation liquid water from space. *Remote Sensing Environ.*, *58*(3), 257-266. https://doi.org/10.1016/S0034-4257(96)00067-3.

Garcia-Ruiz, F., Sankaran, S., Maja, J. M., Lee, W. S., Rasmussen, J., & Ehsani, R. (2013). Comparison of two aerial imaging platforms for identification of Huanglongbing-infected citrus trees. *Comput. Electron. Agric.*, *91*, 106-115. https://doi.org/10.1016/j.compag.2012.12.002.

Gowen, A. A., O'Donnell, C. P., Cullen, P. J., Downey, G., & Frias, J. M. (2007). Hyperspectral imaging—An emerging process analytical tool for food quality and safety control. *Trends Food Sci. Technol.*, *18*(12), 590-598. https://doi.org/10.1016/j.tifs.2007.06.001.

Huygens, C. (1912). *Treatise on light*. Macmillan. Retrieved from http://archive.org/details/treatiseonlight031310mbp.

Kim, M. S., Chen, Y.-R., Cho, B.-K., Chao, K., Yang, C.-C., Lefcourt, A. M., & Chan, D. (2007). Hyperspectral reflectance and fluorescence line-scan imaging for online defect and fecal contamination inspection of apples. *Sensing Instrumentation Food Qual. Saf.*, *1*(3), 151. https://doi.org/10.1007/s11694-007-9017-x.

Liu, Y., Pu, H., & Sun, D.-W. (2017). Hyperspectral imaging technique for evaluating food quality and safety during various processes: A review of recent applications. *Trends Food Sci. Technol.*, *69*, 25-35. https://doi.org/10.1016/j.tifs.2017.08.013.

Mulla, D. J. (2013). Twenty five years of remote sensing in precision agriculture: Key advances and remaining knowledge gaps. *Biosyst. Eng., 114*(4), 358-371. https://doi.org/10.1016/j.biosystemseng.2012.08.009.

NASA (n. d.). Airborne visible/infrared imaging spectrometer: AVIRIS overview. NASA Jet Propulsion Laboratory, California Institute of Technology. https://www.jpl.nasa.gov/missions/airborne-visible-infrared-imaging-spectrometer-aviris/.

Ollinger, S. V. (2011). Sources of variability in canopy reflectance and the convergent properties of plants. *New Phytol., 189*(2), 375-394. https://doi.org/10.1111/j.1469-8137.2010.03536.x.

Osborne, B. G., Fearn, T., Hindle, P. H., & Osborne, B. G. (1993). *Practical NIR spectroscopy with applications in food and beverage analysis* (Vol. 2). Longman Scientific & Technical.

Pasquini, C. (2003). Near infrared spectroscopy: Fundamentals, practical aspects and analytical applications. *J. Brazilian Chem. Soc., 14*(2), 198-219. https://doi.org/10.1590/S0103-50532003000200006.

Pasquini, C. (2018). Near infrared spectroscopy: A mature analytical technique with new perspectives: A review. *Anal. Chim. Acta, 1026*, 8-36. https://doi.org/10.1016/j.aca.2018.04.004.

Pavia, D. L., Lampman, G. M., Kriz, G. S., & Vyvyan, J. A. (2008). *Introduction to spectroscopy*. Cengage Learning.

Porep, J. U., Kammerer, D. R., & Carle, R. (2015). On-line application of near infrared (NIR) spectroscopy in food production. *Trends Food Sci. Technol., 46*(2, Part A), 211-230. https://doi.org/10.1016/j.tifs.2015.10.002.

Primicerio, J., Di Gennaro, S. F., Fiorillo, E., Genesio, L., Lugato, E., Matese, A., & Vaccari, F. P. (2012). A flexible unmanned aerial vehicle for precision agriculture. *Precision Agric., 13*(4), 517-523. https://doi.org/10.1007/s11119-012-9257-6.

Roberts, J., Power, A., Chapman, J., Chandra, S., & Cozzolino, D. (2018). A short update on the advantages, applications and limitations of hyperspectral and chemical imaging in food authentication. *Appl. Sci., 8*(4), 505. https://doi.org/10.3390/app8040505.

Rouse Jr., J. W., Haas, R. H., Schell, J. A., & Deering, D. (1974). Monitoring vegetation systems in the Great Plains with ERTS. NASA Special Publ. 351.

Schuster, A. (1911). Encyclopedia Britannica, *2*:477.

Skytte, J., Moller, F., Abildgaard, O., Dahl, A., & Larsen, R. (n. d.). Discriminating yogurt microstructure using diffuse reflectance images. *Proc. Scandinavian Conf. on Image Analysis* (pp. 192-203). Springer. https://doi.org/10.1007/978-3-319-19665-7_16.

Sun, D.-W. (2009). Infrared spectroscopy for food quality analysis and control. Academic Press.

Thomas, N. C. (1991). The early history of spectroscopy. *J. Chem. Education, 68*(8), 631. https://doi.org/10.1021/ed068p631.

Transon, J., D'Andrimont, R., Maugnard, A., & Defourny, P. (2018). Survey of hyperspectral earth observation applications from space in the sentinel-2 context. *Remote Sensing, 10*(2). https://doi.org/10.3390/rs10020157.

Vejarano, R., Siche, R., & Tesfaye, W. (2017). Evaluation of biological contaminants in foods by hyperspectral imaging: A review. *Int. J. Food Properties, 20*(sup2), 1264-1297. https://doi.org/10.1080/10942912.2017.1338729.

Zhang, C., & Kovacs, J. M. (2012). The application of small unmanned aerial systems for precision agriculture: A review. *Precision Agric., 13*(6), 693-712. https://doi.org/10.1007/s11119-012-9274-5.

Zhang, W., Pan, L., Tu, S., Zhan, G., & Tu, K. (2015). Non-destructive internal quality assessment of eggs using a synthesis of hyperspectral imaging and multivariate analysis. *J. Food Eng., 157*, 41-48. https://doi.org/10.1016/j.jfoodeng.2015.02.013.

Zhou, X., Zheng, H. B., Xu, X. Q., He, J. Y., Ge, X. K., Yao, X., . . . Tian, Y. C. (2017). Predicting grain yield in rice using multi-temporal vegetation indices from UAV-based multispectral and digital imagery. *ISPRS J. Photogrammetry Remote Sensing, 130*, 246-255. https://doi.org/10.1016/j.isprsjprs.2017.05.003.

Data Processing in Biosystems Engineering

Yao Ze Feng
College of Engineering, Huazhong Agricultural University
and Key Laboratory of Agricultural Equipment in Mid-lower
Yangtze River, Ministry of Agriculture and Rural Affairs
Wuhan, Hubei, China

KEY TERMS		
Pretreatment	Normalization	Partial least square regression (PLSR)
Smoothing	Linear regression	Model performance
Derivatives	Principal component analysis (PCA)	Model evaluation

Variables

β = regression coefficient
E = residual vector
$\boldsymbol{E}$ = residual matrix
n = number of samples
$\boldsymbol{P}$ and $\boldsymbol{C}$ = loadings (PLS)
$\boldsymbol{P}^{\mathrm{T}}$ = loading matrix
$\boldsymbol{T}$ = score matrix
$\boldsymbol{W}_a$ = partial least squares weighting
W_i = weighting term for i^{th} data point
x = represents any variable
X = original signal/independent variable vector
$\boldsymbol{X}$ = independent variable matrix
X_{nor} = normalized value of X
XS = smoothed signal

Introduction

Novel sensing technologies and data processing play a very important role in most scenarios across the wide varieties of biosystems engineering applications, such as environmental control and monitoring, food processing and safety control, agricultural machinery design and its automation, and biomass and bioenergy production, particularly in the big data era. For instance, to achieve automatic, non-destructive grading of agricultural products according to their physical and chemical properties, raw data from different types of sensors should be acquired and carefully processed to accurately describe the samples so that the products can be classified into different categories correctly (Gowen et al., 2007; Feng et al., 2013; O'Donnell et al., 2014; Baietto and Wilson, 2015; Park and Lu, 2016). For the environmental control of greenhouses, temperature, humidity, and the concentration of particular gases should be determined by processing the raw data acquired from thermistors, hydrometers, and electronic noses or optical sensors (Bai et al., 2018). Successful use of measurements relies heavily on data processing that converts the raw data into meaningful information for easier interpretation and understanding the targets of interest.

The purpose of data processing is to turn raw data into useful information that can help understand the nature of objects or a process. To make this whole procedure successful, particular attention should be paid to ensure the quality of raw data. However, the raw data obtained from biological systems are always affected by environmental factors and the status of samples. For example, the optical profiles of meat are vulnerable to temperature variation, light conditions, breeds, age and sex of animals, type of feeds, and geographical origins, among other factors. To ensure the best quality of raw data, data pretreatment is essential.

In this chapter, data pretreatment methods, including smoothing, derivatives, and normalization, are introduced. With good quality data, a modeling process correlating the raw data with features of the object or process of interest can be developed. This can be realized by employing different modeling methods. After validation, the established model can then be used for real applications.

Outcomes

After reading this chapter, you should be able to:

- Describe the principles of various data processing methods

- Determine appropriate data processing methods for model development

- Evaluate the performance of established models

- List examples of the application of data processing

Concepts

Data Pretreatment

Data Smoothing

To understand the features of biological objects, different sensors or instruments can be employed to acquire signals representing their properties. For example, a near-infrared (NIR) spectrometer is used to collect the optical properties across different wavelengths, called the spectrum, of a food or agricultural product. However, during signal (i.e., spectrum) acquisition, random noise will inevitably be introduced, which can deteriorate signal quality. For example, short-term fluctuations may be present in signals, which may be due to environmental effects, such as the dark current response and readout noise of the instrument. Dark current is composed of electrons produced by thermal energy variations, and readout noise refers to information derived from imperfect operation of electronic devices. Neither of them contribute to the understanding of the objects under investigation. In order to decrease such effects, *data smoothing* is usually applied. Some popular data smoothing methods include moving average (MV) and S-G (Savitzky and Golay) smoothing.

The idea of moving average is to apply "sliding windows" to smooth out random noises at each segment of the signal by calculating the average value in the segment so that the random noise in the whole signal can be reduced. Given a window with an even number of data points at a certain position, the average value of the original data within the window is calculated and used as the smoothed new value for the central point position. This procedure is repeated until reaching the end of the original signal. For the data points at the two edges of the signal that cannot be covered by a complete window, one can still assume the window is applied but only calculate the average of the data available in the window. The width of window is a key factor that should be determined carefully. It is not always true that the signal-to-noise ratio increases with window width since a too-large window will tend to smooth out useful signal as well. Moreover, since the average value is calculated for each window, all data points in the window are considered as equal contributors for the signal; this will sometimes result in signal distortion. To avoid this problem, S-G smoothing can be introduced.

Instead of using a simple average in the moving average process, Savitzky and Golay (1964) proposed assigning weights to different data in the window. Given an original signal X, the smoothed signal XS can be obtained as:

$$XS_i = \frac{\sum_{j=-r}^{r} X_{i+j} W_j}{\sum_{j=-r}^{r} W_j} \qquad (1)$$

where $2r+1$ is window width and W_i is the weight for the ith data point in the window. W is obtained by fitting the data points in the window to a polynomial form following the least squares principle to minimize the errors between the original signal X and the smoothed signal XS and calculating the central points of the window from the polynomial. In applying S-G smoothing, the smoothing points and order of polynomials should be decided first. Once the two parameters are determined, the weight coefficients can then be applied to the data points in

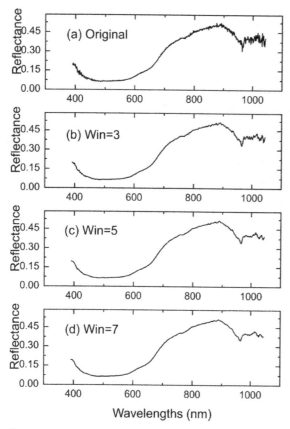

Figure 1. S-G smoothing of a spectral signal. (a) The original spectrum; (b),(c) and (d) are S-G smoothing results under window widths (Win) of 3, 5, and 7, respectively.

the window to calculate the value of the central point using equation 1.

Figure 1 shows the smoothing effect by applying S-G smoothing to a spectrum of beef sample (Figure 1b-d). It is clearly shown that after S-G smoothing, the random noise in the original signal (Figure 1a) is greatly suppressed when the window width is 3 (Figure 1b). An even better result is achieved when the window width increases to 5 and 7, where the curve becomes smoother (Figure 1d) and the short fluctuations are barely seen.

Derivatives

Derivatives are methods for recovering useful information from data while removing slow change of signals (or low frequency signals) that could be useless in determining the properties of biological samples. For example, for a spectrum defined as a function $y = f(x)$, the first and second derivatives can be calculated as:

$$\frac{dy}{dx} = \frac{f(x+\Delta x) - f(x)}{\Delta x} \quad (2)$$

$$\frac{d^2y}{dx^2} = \frac{f(x+\Delta x) - 2f(x) + f(x-\Delta x)}{\Delta x^2} \quad (3)$$

From equations 2 and 3, it can be understood that the offset (e.g., constant shift of signals) of the signal can be eliminated after first derivative processing, while both offset and slope in the original signal can be excluded after second derivative processing. Specifically, for the first derivative, the constant values (corresponding to the offset) can be eliminated due to the difference operation in the numerator of equation 2. After the first derivative, the spectral curve with the same slope can be converted to a new offset and this can be further eliminated by a second derivative. Since offset variations and slope information always indicate environmental effects on the signal and irrelevant factors that are closely correlated with independent variables, application of derivative methods will help reduce such noises. Moreover, processing signals with derivatives offer an efficient approach to enhance the resolution of signals by uncovering more peaks, particularly in spectral analysis.

For biological samples with complicated chemical components, the spectra are normally the combination of different absorbance peaks arising from these components. Such superimposed peaks, however, can be well separated in second derivative spectra. Nevertheless, it should be noted that the signal-to-noise ratio of the signal will deteriorate with the increase of derivative orders since the noise is also enhanced substantially, particularly for the higher order derivatives, though high order derivatives are sometimes found to be useful in understanding the detailed properties of the objects. To avoid noise enhancement, a S-G derivative can be introduced where signal derivatives are attained by computing the derivatives of the polynomial. Specifically, the data points in a sliding window are fitted to a polynomial of a certain order following the procedure of S-G smoothing. Within the window, derivatives of the fitted polynomial are then calculated to produce new weights for the central point. When the sliding window reaches the end of the signal, derivatives of the current signal are then attained.

Figure 2 shows absorbance and derivative spectra of bacterial suspensions (Feng et al., 2015). It is demonstrated that after S-G derivative operation with 5 smoothing points and polynomial order of 2, the constant offset and linear baseline shift in the original spectrum (Figure 2a) are effectively removed in the first (Figure 2b) and second (Figure 2c) derivative spectra, respectively. Particularly, the second derivative technique is also a useful tool to separate overlapped peaks where a peak at ~1450 nm is resolved into two peaks at 1412 and 1462 nm.

Normalization

The purpose of data *normalization* is to equalize the magnitude of sample signals so that all variables for a sample can be treated equally for further analysis. For example, the surface temperature of pigs and environmental factors (temperature, humidity, and air velocity) can be combined to detect the rectal temperature of sows. Since the values for pig surface temperature can be around 39°C while the air velocity is mostly below 2 m/s, if these values are used directly for further data analysis, the surface temperature will intrinsically play a more dominant role than air velocity does simply due to its larger values. This may lead to biased interpretation of the importance of variables. Data normalization is also helpful when signals from different sensors are combined as variables (i.e., data fusion) to characterize biological samples that are complex in composition and easily affected by environmental conditions. However, since data normalization removes the average as well as the standard deviation of the sample variables,

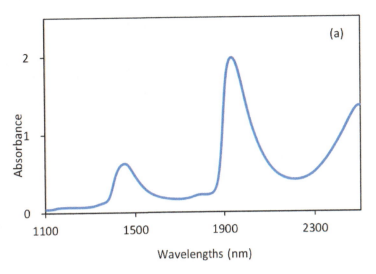
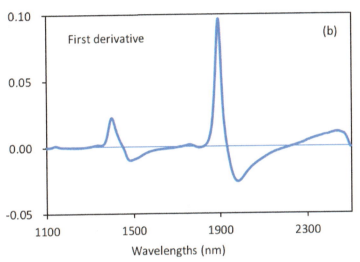
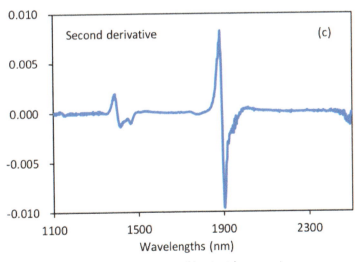

Figure 2. NIR derivative spectra of bacterial suspensions. (a): original spectrum; (b): First derivative spectrum; (c) second derivative spectrum.

it might give confusing information about the samples if variabilities of variables in different units are important in characterizing sample properties.

Standard normal variate (SNV), or standardization, is one of the most popular methods used to normalize sample data (Dhanoa et al., 1994). Given a sample data X, the normalized X_{nor} can be obtained as:

$$X_{nor} = \frac{X - \text{mean}(X)}{\text{SD}(X)} \qquad (4)$$

where mean(X) and SD(X) are the mean and standard deviation of X, respectively.

After SNV transformation, a new signal with a mean value of 0 and unit standard deviation is produced. Therefore, SNV is useful in eliminating dimensional variance among variables since all variables are compared at the same level. In addition, as shown in figure 3, SNV is capable of correcting the scattering effect of samples due to physical structure of samples during light-matter interactions (Feng and Sun, 2013). Specifically, the large variations in visible NIR (vis-NIR) spectra of beef samples (Figure 3a) are substantially suppressed as shown in Figure 3b.

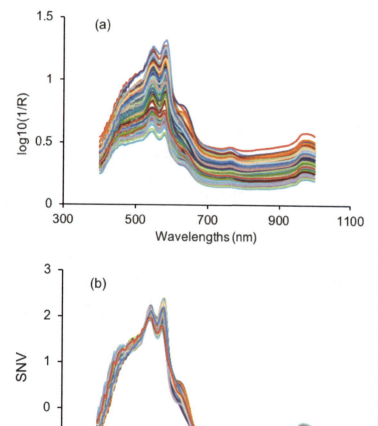

Figure 3. SNV processing of vis-NIR spectra of beef samples adulterated with chicken meat. (a) Original spectra; (b) SNV processed spectra.

Modeling Methods

The purpose of modeling in data processing is mainly to establish the relationship between independent variables and dependent variables. *Independent variables* are defined as stand-alone factors that can be used to determine the values of other variables. Since the values of other variables depend on the independent variables, they are called *dependent variables*. For example, if size, weight, and color are used to classify apples into different grades, the variables of size, weight, and color are the independent variables and the grade of apples is the dependent variable. The dependent variables are calculated based on measured independent variables. During model development, if only one independent variable is used, the resultant model is a *univariate model*, while two or more independent variables are involved in *multivariate models*. If dependent variables are used during model calibration or training, the methods applied in model development are termed *supervised*. Otherwise, an *unsupervised* method is employed. The dataset used for model development is called the *calibration set* (or training set) and a new dataset where the model is applied for validation is the *validation set* (or prediction set).

The developed models can be used for different purposes. Basically, if the model is used to predict a discrete class (categorical), it is a *classification* model; and if it aims to predict a continuous quantity, it is a *regression model*. For instance, if spectra of samples are used to identify the geographical origins of beef, the spectra (optical properties at different wavelengths) are the independent variables and the geographical origins are the dependent variables. The established multivariate model describing the relationship between spectra and geographical origins is a classification model. In a classification model, the dependent variables are dummy variables (or labels) where different arbitrary numbers are used to represent different classes but with no physical meaning. On the other hand, if spectra of samples are used to determine the water content of beef, the developed model is then a regression model. The dependent variables are meaningful numbers indicating the actual water content. Simply, a classification model tries to answer the question of "What is it?" and a regression model tries to determine "How much is there?" There is a wide range of methods for regression or classification models. Some are described below.

Linear Regression

Linear regression is an analytical method that explores the linear relationship between independent variables (**X**) and dependent variables (**Y**). Simple linear regression is used to establish the simplest model that can be used to illustrate the relationship between one independent variable **X** and one dependent variable **Y**. The model can be described as:

$$Y = \beta_0 + \beta_1 X + E \qquad (5)$$

where **X** is the independent variable; **Y** is the dependent variable; β_0, β_1, are the regression coefficients; and **E** is the residual vector.

Simple linear regression is used when only one independent variable is to be correlated with the dependent variable. In the model, the two important coefficients, β_0 and β_1, can be determined by finding the best fit line through the scatter curve between **X** and **Y** via the least squares method. The best fit line requires minimization of errors between the real **Y** and the predicted $\hat{Y}$. Since the errors could be either positive or negative, it is more appropriate to use the sum of squared errors. Based on this, β_0 and β_1 can be calculated as:

$$\beta_1 = \frac{\sum_{i=1}^{n}(X_i - \bar{X})(Y_i - \bar{Y})}{\sum_{i=1}^{n}(X_i - \bar{X})^2} \qquad (6)$$

$$\beta_0 = \bar{Y} - \beta_1 \bar{X} \qquad (7)$$

where $\bar{X}$ and $\bar{Y}$ are mean values of **X** and **Y**, respectively, and n is the number of samples.

Multiple linear regression (MLR) is a linear analysis method for regression in which the corresponding model is established between multiple independent variables and one dependent variable (Ganesh, 2010):

$$Y = \beta_0 + \sum_{j=1}^{n} \beta_j X_j + E \qquad (8)$$

where X_j is the jth independent variable; Y is the dependent variable; β_0 is the intercept; $\beta_1, \beta_2, \ldots, \beta_n$ are regression coefficients, and E is the residual matrix.

Although MLR tends to give better results compared with simple linear regression since more variables are utilized, MLR is only suitable for situations where the number of variables is less than the number of samples. If the number of variables exceeds the number of samples, equation 8 will be underdetermined and infinite solutions can be produced to minimize residuals. Therefore, multiple linear regression is generally employed based on important feature variables (such as important wavelengths in spectral analysis) instead of all variables, if the number of variables is larger than that of samples.

Similar to simple linear regression, the determination of regression coefficients also relies on the minimization of prediction residuals (i.e., the sum of squared residuals between true Y values and predicted $\hat{Y}$). Specific procedures can be found elsewhere (Friedman et al., 2001).

Principal Component Analysis (PCA)

Due to the complicated nature of biological samples, data acquired to characterize samples usually involve many variables. For example, spectral responses at hundreds to thousands of wavelengths may be used to characterize the physical and chemical components of samples. Such great dimensionality inevitably brings difficulties in data interpretation. With the original multivariate data, each independent variable or variable combinations can be used to draw one-, two-, or three-dimensional plots to understand the distribution of samples. However, this process requires a huge workload and is unrealistic if more than three variables are involved.

Principal component analysis (PCA) is a powerful tool to compress data and provides a much more efficient way for visualizing data structure. The idea of PCA is to find a set of new variables that are uncorrelated with each other and attach the most data information onto the first few variables (Hotelling, 1933). Initially, PCA tries to find the best coordinate that can represent the most data variations in the original data and record it as PC1. Other PCs are subsequently extracted to cover the greatest variations of the remaining data. The established PCA model can be expressed as:

$$X = TP^T + E \qquad (9)$$

where X is the independent variable matrix, T is the score matrix, P^T is the loading matrix, and E is the residual matrix. The score matrix can be used to visualize the relationship between samples and the loadings can be used to express the relations between variables.

After PCA, the data can be represented by a few PCs (usually less than 10). These PCs are sorted according to their contribution to the explanation of data variance. Specifically, an accumulated contribution rate, defined as explained variance from the first few PCs over the total variance of the data, is usually employed to evaluate how many new variables (PCs) should be used to represent the data. Nevertheless, by applying PCA, the number of variables required for characterizing data variance is substantially reduced. After projecting the original data into the new PC spaces, data structure can be easily seen, if it exists.

Partial Least Squares Regression (PLSR)

As illustrated above, MLR requires that the number of samples be more than the number of variables. However, biological data normally contain far more variables than samples, and some of these variables may be correlated with each other, providing redundant information. To cope with this dilemma, *partial least squares regression* (PLSR) can be used to reduce the number of variables in the original data while retaining the majority of its information and eliminating redundant variations (Mevik et al., 2011). In PLSR, both **X** and **Y** are projected to new spaces. In such spaces, the multidimensional direction of **X** is determined to best account for the most variance of multidimensional direction of **Y**. In other words, PLSR decomposes both predictors **X** and dependent variable **Y** into combinations of new variables (scores) by ensuring the maximum correlation between **X** and **Y** (Geladi and Kowalski, 1986). Specifically, the score **T** of **X** is correlated with **Y** by using the following formulas:

$$Y = XB + E = XW_a^* C + E = TC + E \qquad (10)$$

$$W_a^* = W_a (P^T W_a)^{-1} \qquad (11)$$

where **B** is the regression coefficients for the PLSR model established; **E** is the residual matrix; W_a represents the PLS weights; a is the desired number of new variables adopted; **P** and **C** are loadings for **X** and **Y**, respectively. The new variables adopted are usually termed as *latent variables* (LVs) since they are not the observed independent variables but inferred from them.

The most important parameter in PLS regression is the determination of the number of LVs. Based on the PLSR models established with different LVs, a method named *leave-one-out cross validation* is commonly utilized to validate the models. That is, for the model with a certain number of LVs, one sample from the data set is left out with the remaining samples used to build a new model. The new model is then applied to the sample that is left out for prediction. This procedure is repeated until every sample has been left out once. Finally, every sample would have two values, i.e., the true value and the predicted value. These two types of values can then be used to calculate root mean squared errors (RMSEs; equation 13 in the Model Evaluation section below) for different numbers of LVs. Usually, the optimal number of LVs is determined either at the minimum value of RMSEs or the one after which the RMSEs are not significantly different from the minimum RMSE. In Figure 4 for

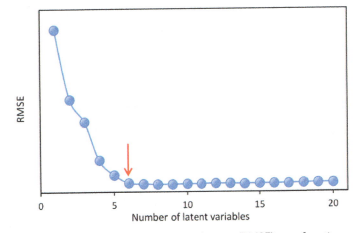

Figure 4. Plot of root mean squared error (RMSE) as a function of number of latent variables (LVs) for a PLSR model. The minimum RMSE is attained when 11 latent variables are used. However, using 6 LVs, as indicated by the red arrow, is better in terms of model simplicity.

instance, using 6 latent variables would produce a very similar RMSE value to the minimum RMSE that is attained with 11 LVs; therefore, 6 latent variables would be more suitable for simpler model development.

In addition to the methods introduced above, many more algorithms are available for model development. With the fast growth of computer science and information technologies, modern machine learning methods, including artificial neural networks, deep learning, decision trees, and support vector machines, are widely used in biosystems engineering (LeCun et al., 2015; Maione and Barbosa, 2019; Pham et al., 2019, Zhao et al., 2019).

The model development methods described above can be used for both regression and classification problems. For regression, the final outputs are the results produced when the independent variables are input into the established models. For classification, a further operation is required to attain the final numbers for categorical representation. Normally, a rounding operation is adopted. For instance, a direct output of 1.1 from the model tends to be rounded down to 1 as the final result, which can be a label for a certain class. After such modification, the name of the regression method can be changed from PLSR to partial least squares *discriminant analysis* (PLS-DA), as an example. However, these numbers do not have actual physical meanings, and therefore they are often termed *dummy variables*.

Since a model can be established using different modeling methods, some of which are outlined above, the decision on which type of method to use is task-specific. If the objective is to achieve stable model with high precision, the one that can lead to the best model performance should be employed. However, if the main concern is simplicity and easy interpretation based on feasible application, a linear method will often be the best choice. In cases when a linear model fails to depict the correlation between X and Y, nonlinear models established by applying artificial neural networks or support vector machines could then be applied.

Model Evaluation

The full process of model development includes the *calibration*, *validation*, and *evaluation* of models. Model calibration tries to employ different modeling methods to the training data to find the best parameters for representation of samples. For example, if PLSR is applied to NIR spectral data to quantify beef adulteration with pork, the important parameters including the number of LVs and regression coefficients are determined so that when the spectra are inputted to the model, the predicted percentage of adulteration levels can be calculated. It is clear that this process simply works on the training data itself and the resultant model can best explain the data of the particular samples. However, since the modeling process is data specific, good model performance sometimes can be due to the modeling of noise and such models will fail to function with new, independent data. This problem is known as *over-fitting* and should be always avoided during modeling. Therefore, it is of crucial importance to validate the performance of the models using independent data, i.e., data that are not included in the calibration set and that are totally unknown to the established model.

Model validation is a process to verify whether similar model performance can be attained to that of calibration. There are basically two ways to conduct model validation. One is to use cross-validation, if there are not enough samples available. Cross-validation is implemented based on the training set and often a leave-one-out approach is taken (Klanke and Ritter, 2006). During leave-one-out cross-validation, one sample is left out from the calibration set and a calibration model is developed based on the remaining data. The left-out sample is then inputted to the developed model based on the other samples. This procedure terminates when all samples have been left out once. Finally, all samples will be predicted for comparison with the measured values. However, this method should be used with caution since it may lead to over-optimistic evaluation or model overfitting. Another approach, called external validation, is to introduce an independent prediction set that is not included in the calibration set and apply the model to the new, independent dataset. External validation is always preferred for model evaluation. Nevertheless, it is recommended to apply both cross-validation and external validation methods to evaluate the performance of models. This is particularly important in biosystems engineering because biological samples are very complex and their properties can change with time and environment. For meat samples, the chemical components of meat vary due to species, geographical origins, breeding patterns, and even different body portions of the same type of animal. The packaging atmosphere and temperature also have great influence on the quality variations of meat. Ideally, with a good and stable model, the results from cross-validation and external validation should be similar.

Model evaluation is an indispensable part of model development, which aims to determine the best performance of a model as well as to verify its validity for future applications by calculating and comparing some statistics (Gauch et al., 2003). For regression problems, two common parameters, coefficient of determination (R^2), and root mean squared error (RMSE), are calculated to express the performance of a model. They are defined as follows:

$$R^2 = 1 - \frac{\sum_{i=1}^{n}(Y_{i,meas} - Y_{i,pre})^2}{\sum_{i=1}^{n}(\overline{Y} - Y_{i,pre})^2} \qquad (12)$$

$$RMSE = \sqrt{\frac{1}{n}\sum_{i=1}^{n}(Y_{i,meas} - Y_{i,pre})^2} \qquad (13)$$

where $Y_{i,pre}$ and $Y_{i,meas}$, respectively, represent the predicted value and the measured value of targets for sample i; $\overline{Y}$ is the mean target value for all samples. An R^2 of 1 and RMSE of 0 for all data sets would indicate a "perfect" model. Thus, the goal is to have R^2 as close to 1 as possible and RMSE close to 0. In addition, a stable model has similar R^2 and RMSE values for calibration and validation. It should be noted that R, the square root of R^2, or correlation coefficient, is also frequently used to express the linear relationship between the predicted and measured values. Moreover, since different data sets may be used during model development, the above parameters can be modified in accordance. For example, R^2_C, R^2_{CV} and R^2_P can be used to represent the coefficients of determination for calibration, cross-validation, and prediction, respectively. Root mean squared errors for calibration, cross-validation, and prediction are denoted as RMSEC, RMSECV, and RMSEP, respectively.

For classification problems, a model's overall correct classification rate (OCCR) is an important index used to evaluate the classification performance:

$$\text{OCCR} = \frac{\text{Number of correctly classified samples}}{\text{Total number of samples}} \quad (14)$$

The number of correctly classified samples is determined by comparing the predicted classification with the known classification. To investigate the detailed classification performance, a confusion matrix can be utilized (Townsend, 1971). A confusion matrix for binary classifications is shown in Table 1. In the confusion matrix, true positive and true negative indicate samples that are predicted correctly. False positives and false negatives are encountered when what is not true is wrongly considered as true and vice versa. Based on the confusion matrix, parameters can be attained to evaluate the classification model, including the sensitivity, specificity, and prevalence, among others:

$$\text{Sensitivity} = \frac{\sum \text{True positive}}{\sum \text{Condition positive}} \quad (15)$$

$$\text{Specificity} = \frac{\sum \text{True negative}}{\sum \text{Condition negative}} \quad (16)$$

$$\text{Prevalence} = \frac{\sum \text{Condition positive}}{\sum \text{Total population}} \quad (17)$$

Table 1. Confusion matrix for binary classification.

	Condition Positive	Condition Negative
Predicted Positive	True positive (Power)	False positive (Type I error)
Predicted Negative	False negative (Type II error)	True negative

Applications

Beef Adulteration Detection

Food adulteration causes distrust in the food industry by leading to food waste due to food recall and loss of consumer trust. Therefore, it is crucial to use modern technologies to detect deliberate adulteration or accidental contamination. For example, a handheld spectrometer can be used to obtain spectra from beef samples. The raw spectra can be processed by the spectrometer to quantify the level, if any, of adulteration of each beef sample. To properly process the raw spectra, purposeful contamination experiments can be used to determine the

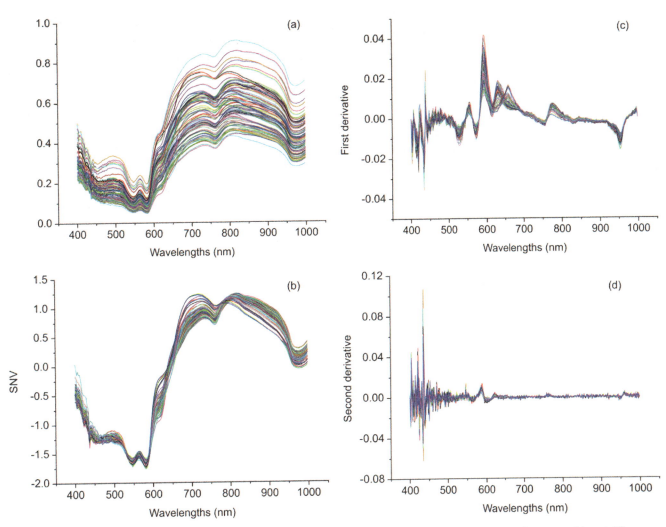

Figure 5. Preprocessing of beef spectra for adulterated beef: (a) raw spectra; (b) SNV preprocessed spectra; (c) and (d) spectra preprocessed with first and second derivatives.

appropriate pretreatment (or preprocessing) method(s) for the raw data. For example, figure 5a shows spectra corresponding to different adulteration levels. Adulteration concentration in such an experiment should range from 0% to 100% with 0% being pure fresh beef and 100% for pure spoiled beef. The experiment should include a calibration dataset to develop the predictive relationship from spectra and an independent dataset to test the validity of the prediction. The following process can be used to determine the best preprocessing method for quantification of beef adulteration.

The raw spectral data (figure 5a) have what is probably random noise with the signal, particularly at the lower wavelengths (400–500 nm). The reason for saying this is there are variations in spectral magnitude among the samples that do not change linearly with adulteration concentration. It is possible that these variations (noise in this application) are due to differences in chemical components of the samples, since spoiled meat is very different from fresh meat, so when the two are mixed in different proportions a clear signal should be visible. Noise might also be introduced due to small differences in the physical structure of samples causing variation of light scattering between the samples. Also note that there are only limited peaks and there is evident offset in the raw spectra. Therefore, different preprocessing methods including S-G smoothing, SNV, and the first and second derivatives can be applied to the raw spectra (figure 5) and their performance in terms of improving the detection of beef adulteration compared.

Table 2 shows the performance of different preprocessing methods together with PLSR in determining the adulteration concentration. All the preprocessing methods applied lead to better models with smaller RMSEs, although such improvement is not very much. The optimal model was attained by using SNV as the preprocessing method, which had coefficients of determination of 0.93, 0.92, and 0.88 as well as RMSEs of 7.30%, 8.35%, and 7.90% for calibration, cross-validation, and prediction, respectively. Though second derivative spectra have contributed to better prediction precision (7.37%), the corresponding model yielded larger RMSEs for both calibration and cross-validation. Therefore, the best preprocessing method in this case is SNV. This preprocessing method can be embedded in a handheld spectrometer, where the raw spectra of adulterated beef samples acquired can be normalized by removing the average and then dividing by the standard deviation of the spectra. The prediction model can then be applied to the SNV-preprocessed data to estimate levels of beef adulteration and to provide insights into the authenticity of the beef product.

Table 2. Comparison of different data preprocessing methods combined with PLSR for predicting beef adulteration.

Methods	RMSEC (%)	RMSECV (%)	RMSEP (%)	R^2C	R^2CV	R^2P	LV
None	8.35	9.34	7.99	0.91	0.90	0.88	4
1st Derivative	8.05	8.78	7.92	0.92	0.91	0.88	3
2nd Derivative	7.92	10.03	7.37	0.92	0.88	0.90	4
SNV	7.30	8.35	7.90	0.93	0.92	0.88	4
S-G	7.78	8.90	7.91	0.93	0.91	0.88	5

C = calibration
CV = coefficient of variation
SEP = standard error of prediction
P = prediction
LV = latent variables

Bacterial Classification

Identification and classification of bacteria are important for food safety, for the design of processes such as thermal treatment, and to help identify the causes of illness when bacterial contamination has occurred. This example outlines how a classification system can be developed (Feng et al., 2015). A spectral matrix was derived by scanning a total of 196 bacterial suspensions of various concentrations using a near infrared spectrometer over two wavelength ranges, i.e., 400-1100 nm and 1100-2498 nm. A column vector that recorded the labels for each bacterium (i.e., its name or classification) was also constructed. This dataset were used to classify different bacteria including three *Escherichia coli* strains and four *Listeria innocua* strains. Since the dataset contained a large number (>1000) of variables, it was interesting to visualize the structure of the data to investigate potential sample clustering. By using appropriate modeling methods, it was possible to establish a model for classifying bacteria at species level.

PCA can be used to understand the structure of data. Since the scores of a PCA model can be used to elucidate the distribution of samples, it is interesting to draw a score plot such as figure 6. The first two columns of the score matrix **T** are the scores for the first two PCs and is generated by using the first one as x-axis and the other as y-axis. The loading plots in figure 6 can be created by plotting the first two columns of the loading matrix $\boldsymbol{P}^T$ versus variable names (wavelengths in this case), respectively.

The first and second PCs have covered 58.34% and 35.04% of the total variance of the spectral data set, leading to 93.38% of the information explained. Based on such information, it is demonstrated clearly that the two bacteria are well separated along the first PC though very few samples mixed together. By investigating loading 1, it is found that five main wavelengths including 1392, 1450, 1888, 1950, and 2230 nm are important variables that contribute to the separation of the two bacterial species. Also, it is interesting to find that two clusters appear within either of the two bacterial species and such separation can then be explained by the four major wavelengths indicated in loading 2 (figure 6c).

The next target is to establish a classification model in the 400–1100 nm region for the classification of these bacterial species. To achieve this, PLS-DA was employed where the spectral data and the bacterial labels are used as independent and dependent variables, respectively. Figure 7 shows the performance of the established model. The optimized model takes four latent variables to produce OCCRs of 99.25% and 96.83% for calibration and prediction, respectively. To calculate OCCRs, the predicted values of individual samples are first rounded to get values of 1 or 0 and these predicted labels are then compared with the true labels, following which equation 14 is employed.

A confusion matrix showing the classification details for prediction is shown in table 3. It shows that the true positive for detecting *E. coli* and *L. innocua* are 25 and 36, respectively. Accordingly, the sensitivity for detecting *E. coli* and *L. innocua* species are 0.93 (25/27) and 1 (36/36), respectively. All the above parameters for both calibration and prediction demonstrate that the two bacterial species can be well classified.

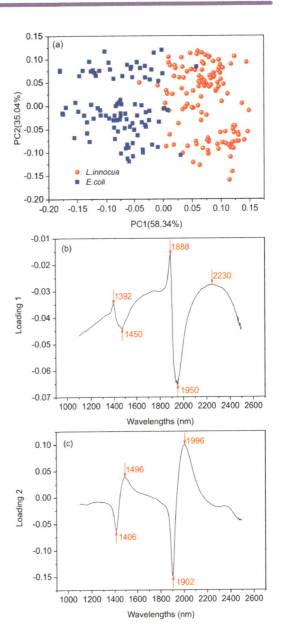

Figure 6. Score plots and loadings of the PCA model (1100–2498 nm) for *E. coli* and *L. innocua* bacterial suspensions. (a) Score plot; (b) and (c) are loadings for the first two PCs (Feng et al., 2015).

Table 3. Confusion matrix for bacterial species classification.

Actual Class	Predicted Class		
	E. coli	L. innocua	Total
E. coli	25	2	27
L. innocua	0	36	36
Total	25	38	63

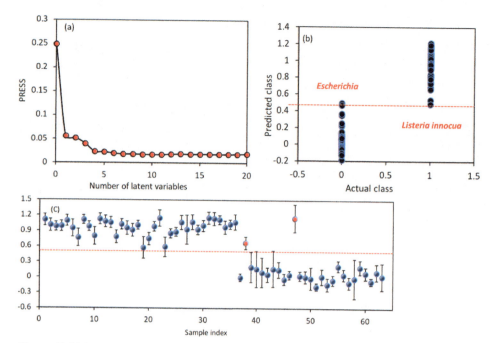

In microbial safety inspection of food products, it is important to identify the culprit pathogens that are responsible for foodborne diseases. To achieve this, bacteria on food surfaces can be sampled, cultured, isolated, and suspended, and the model can be applied to the spectra of bacterial suspensions to tell us which of those two species of bacteria are present in the food product.

Figure 7. PLS-DA classification model performance in the visible-SWNIR range (400–1100 nm). (a) Selection of the optimal number of latent variables; (b) model performance for calibration; (c) model performance for prediction. The dotted lines indicate the threshold value of 0.5 (Feng et al., 2015).

Examples

Example 1: Moving average calculation

Problem:

Fruit variety and ripeness of fruit can be determined by non-destructive methods such as NIR spectroscopy. A reflectance spectrum of a peach sample was acquired; part of the spectral data in the wavelength range of 640–690 nm is shown in table 4. Though the spectrometer is carefully configured, there still might be noise present in the spectra due to environmental conditions. Apply the moving average method to smooth the spectrum and to reduce potential noise.

Solution:

Various software, including Microsoft, MATLAB, and commercial chemometric software (the Unscrambler, PLS Toolbox etc.) are available for implementing the moving average. Taking Microsoft Excel as an example, the "average" function is required. Given a spectrum presented column-wise (for example, column B), the value for the smoothed spectrum at cell B10 can be obtained as average(B9:B11) if the window size is 3, and average(B8:B12) or average(B7:B13) if the window size is 5 or 7, respectively. For both ends of the spectrum, only the average of values present in the window of a particular size is calculated. For instance, the spectral value at 639.8 nm after moving average smoothing under the window size of 3 can be obtained as the mean values of the original spectrum at 639.8, 641.1 and 642.2 nm, that is, (0.4728 + 0.4745 + 0.4751)/3 =0.4741.

Figure 8 shows the smoothed spectrum, the result of using the moving average method. Note that the spectra are shifted 0.01, 0.02, and 0.03 unit for the

Win = 3, Win = 5, and Win = 7 spectra to separate the curves for visual presentation purposes. It is clear that for the original data, there is slight fluctuation and such variation is diminished after moving average smoothing.

Example 2: Evaluation of model performance

Problem:

As pigs cannot sweat, it is important to be able to rapidly confirm that conditions in a pig house are not causing them stress. Rectal temperature is the best indicator of heat stress in an animal, but it can be difficult to measure. A pig's surface temperature, however, can be measured easily using non-contact sensors. Table 5 shows the performance of two PLSR models used to predict the rectal temperature of pigs by using variables including surface temperature and several environmental conditions. Model 1 is a many-variable model and Model 2 is a simplified model that utilizes an optimized subset of variables. Determine which model is better. The performance of models is presented by R and RMSEs for calibration, cross-validation, and prediction.

Solution:

The first step is to check whether R is close to 1 and RMSE to 0. Correlation coefficients range from 0.66 to 0.87 (table 5), showing obvious correlation between the predicted rectal temperature and the real rectal temperature. By investigating the RMSEs, it is found that these errors are relatively small (0.25°–0.38°C) compared with the measured range (37.8°–40.2°C). Therefore, both models are useful for predicting the rectal temperature of pigs.

The second step is to check the stability of the established models by evaluating the difference among Rs or RMSEs for calibration, cross-validation, and prediction. For the specific example, although the best correlation coefficient for calibration (R_C) and root mean squared error for calibration (RMSEC) were attained for the many-variable model, its performance in cross-validation and prediction was inferior to that of the simplified model. Most importantly, the biggest difference

Table 4. Spectral data of a peach sample in the 640–690 nm range.

Wavelength (nm)	Reflectance	Wavelength (nm)	Reflectance
639.8	0.4728	665.2	0.4755
641.1	0.4745	666.5	0.4743
642.4	0.4751	667.7	0.4721
643.6	0.4758	669.0	0.4701
644.9	0.4766	670.3	0.4680
646.2	0.4777	671.5	0.4673
647.4	0.4791	672.8	0.4664
648.7	0.4807	674.1	0.4661
650.0	0.4829	675.3	0.4672
651.2	0.4850	676.6	0.4689
652.5	0.4854	677.9	0.4715
653.8	0.4854	679.2	0.4747
655.0	0.4851	680.4	0.4796
656.3	0.4838	681.7	0.4862
657.6	0.4826	683.0	0.4932
658.8	0.4814	684.3	0.5010
660.1	0.4801	685.5	0.5093
661.4	0.4789	686.8	0.5182
662.7	0.4782	688.1	0.5269
663.9	0.4765	689.3	0.5360

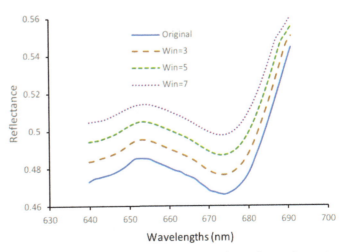

Figure 8. Example of moving average smoothing of a peach spectrum. The spectra are shifted 0.01, 0.02, and 0.03 units for Win = 3, Win = 5 and Win = 7 spectra, respectively, for better visual presentation.

Table 5. Comparison of the performance of two models, many-variable Model 1 and simplified Model 2 (Feng et al., 2019). RC, RCV, and RP are correlation coefficients for calibration, cross-validation, and prediction, respectively.

Model	RC	RCV	RP	RMSEC (°C)	RMSECV (°C)	RMSEP (°C)	LV
Model 1	0.87	0.66	0.76	0.25	0.38	0.37	4
Model 2	0.80	0.78	0.80	0.30	0.32	0.35	2

among Rs of the many-variable model was 0.21, while only a tenth of such difference (0.02) was found for the simplified model. A similar trend was also observed for the RMSEs where the maximum differences of 0.05°C and 1.3°C were yielded for the simplified and many-variable models, respectively. These results strongly demonstrate that the simplified model is much more stable than the many-variable model.

The third step can evaluate the simplicity of the model. In this example, four latent variables were employed to establish the many-variable model while only two were needed for the simplified model. Above all, the simplified model showed better prediction ability, particularly for cross-validation and prediction, with fewer latent variables. Therefore, it is considered as the better model.

Image Credits

Figure 1. Feng, Y. (CC By 4.0). (2020). S-G smoothing of a spectral signal.
Figure 2. Feng, Y. (CC By 4.0). (2020). NIR derivative spectra of bacterial suspensions.
Figure 3. Feng, Y. (CC By 4.0). (2020). SNV processing of vis-NIR spectra of beef samples adulterated with chicken meat.
Figure 4. Feng, Y. (CC By 4.0). (2020). Plot of root mean squared error (RMSE) as a function of number of latent variables (LV) for a PLSR model.
Figure 5. Feng, Y. (CC By 4.0). (2020). Preprocessing of beef spectra.
Figure 6. Feng, Y. (CC By 4.0). (2020). Score plots and loadings of the PCA model (1100-2498 nm) for E. coli and L. innocua bacterial suspension.
Figure 7. Feng, Y. (CC By 4.0). (2020). PLS-DA classification model performance in the visible-SWNIR range (400-1000 nm).
Figure 8. Feng, Y. (CC By 4.0). (2020). Example of moving average smoothing of a peach spectrum.

Acknowledgement

Many thanks to Mr. Hai Tao Zhao for his help in preparing this chapter.

References

Bai, X., Wang, Z., Zou, L., & Alsaadi, F. E. (2018). Collaborative fusion estimation over wireless sensor networks for monitoring CO_2 concentration in a greenhouse. *Information Fusion, 42*, 119-126. https://doi.org/10.1016/j.inffus.2017.11.001.

Baietto, M., & Wilson, A. D. (2015). Electronic-nose applications for fruit identification, ripeness and quality grading. *Sensors, 15*(1), 899-931. https://doi.org/10.3390/s150100899.

Dhanoa, M. S., Lister, S. J., Sanderson, R., & Barnes, R. J. (1994). The link between multiplicative scatter correction (MSC) and standard normal variate (SNV) transformations of NIR spectra. *J. Near Infrared Spectroscopy, 2*(1), 43-47. https://doi.org/10.1255/jnirs.30.

Feng, Y.-Z., & Sun, D.-W. (2013). Near-infrared hyperspectral imaging in tandem with partial least squares regression and genetic algorithm for non-destructive determination and visualization of *Pseudomonas* loads in chicken fillets. *Talanta, 109,* 74-83. https://doi.org/10.1016/j.talanta.2013.01.057.

Feng, Y.-Z., Downey, G., Sun, D.-W., Walsh, D., & Xu, J.-L. (2015). Towards improvement in classification of *Escherichia coli, Listeria innocua* and their strains in isolated systems based on chemometric analysis of visible and near-infrared spectroscopic data. *J. Food Eng., 149,* 87-96. https://doi.org/10.1016/j.jfoodeng.2014.09.016.

Feng, Y.-Z., ElMasry, G., Sun, D.-W., Scannell, A. G., Walsh, D., & Morcy, N. (2013). Near-infrared hyperspectral imaging and partial least squares regression for rapid and reagentless determination of Enterobacteriaceae on chicken fillets. *Food Chem., 138*(2), 1829-1836. https://doi.org/10.1016/j.foodchem.2012.11.040.

Feng, Y.-Z., Zhao, H.-T., Jia, G.-F., Ojukwu, C., & Tan, H.-Q. (2019). Establishment of validated models for non-invasive prediction of rectal temperature of sows using infrared thermography and chemometrics. *Int. J. Biometeorol., 63*(10), 1405-1415. https://doi.org/10.1007/s00484-019-01758-2.

Friedman, J., Hastie, T., & Tibshirani, R. (2001). *The elements of statistical learning. No. 10.* New York, NY: Springer.

Ganesh, S. (2010). Multivariate linear regression. In P. Peterson, E. Baker, & B. McGaw (Eds.), *International encyclopedia of education* (pp. 324-331). Oxford: Elsevier. https://doi.org/10.1016/B978-0-08-044894-7.01350-6.

Gauch, H. G., Hwang, J. T., & Fick, G. W. (2003). Model evaluation by comparison of model-based predictions and measured values. *Agron. J., 95*(6), 1442-1446. https://doi.org/10.2134/agronj2003.1442.

Geladi, P., & Kowalski, B. R. (1986). Partial least-squares regression: A tutorial. *Anal. Chim. Acta, 185,* 1-17. https://doi.org/10.1016/0003-2670(86)80028-9.

Gowen, A. A., O'Donnell, C. P., Cullen, P. J., Downey, G., & Frias, J. M. (2007). Hyperspectral imaging: An emerging process analytical tool for food quality and safety control. *Trends Food Sci. Technol., 18*(12), 590-598. https://doi.org/10.1016/j.tifs.2007.06.001.

Hotelling, H. (1933). Analysis of a complex of statistical variables into principal components. *J. Ed. Psychol., 24,* 417-441. https://doi.org/10.1037/h0071325.

Klanke, S., & Ritter, H. (2006). A leave-k-out cross-validation scheme for unsupervised kernel regression. In S. Kollias, A. Stafylopatis, W. Duch, & E. Oja (Eds.), *Proc. Int. Conf. Artificial Neural Networks. 4132,* pp. 427-436. Springer. doi: https://doi.org/10.1007/11840930_44.

LeCun, Y., Bengio, Y., & Hinton, G. (2015). Deep learning. *Nature, 521*(7553), 436-444. https://doi.org/10.1038/nature14539.

Maione, C., & Barbosa, R. M. (2019). Recent applications of multivariate data analysis methods in the authentication of rice and the most analyzed parameters: A review. *Critical Rev. Food Sci. Nutrition, 59*(12), 1868-1879. https://doi.org/10.1080/10408398.2018.1431763.

Mevik, B.-H., Wehrens, R., & Liland, K. H. (2011). PLS: Partial least squares and principal component regression. R package ver. 2(3). Retrieved from https://cran.r-project.org/web/packages/pls/pls.pdf.

O'Donnell, C. P., Fagan, C., & Cullen, P. J. (2014). *Process analytical technology for the food industry.* New York, NY: Springer. https://doi.org/10.1007/978-1-4939-0311-5.

Park, B., & Lu, R. (2015). *Hyperspectral imaging technology in food and agriculture.* New York, NY: Springer. https://doi.org/10.1007/978-1-4939-2836-1.

Pham, B. T., Jaafari, A., Prakash, I., & Bui, D. T. (2019). A novel hybrid intelligent model of support vector machines and the MultiBoost ensemble for landslide susceptibility modeling. *Bull. Eng. Geol. Environ., 78*(4), 2865-2886. https://doi.org/10.1007/s10064-018-1281-y.

Savitzky, A., & Golay, M. J. (1964). Smoothing and differentiation of data by simplified least squares procedures. *Anal. Chem., 36*(8), 1627-1639. https://doi.org/10.1021/ac60214a047.

Townsend, J. T. (1971). Theoretical analysis of an alphabetic confusion matrix. *Perception Psychophysics, 9*(1), 40-50. https://doi.org/10.3758/BF03213026.

Zhao, H.-T., Feng, Y.-Z., Chen, W., & Jia, G.-F. (2019). Application of invasive weed optimization and least square support vector machine for prediction of beef adulteration with spoiled beef based on visible near-infrared (Vis-NIR) hyperspectral imaging. *Meat Sci., 151,* 75-81. https://doi.org/10.1016/j.meatsci.2019.01.010.

Traction

Daniel M. Queiroz
Department of Agricultural Engineering
Universidade Federal de Viçosa
Viçosa, Minas Gerais, Brazil

John K. Schueller
Department of Mechanical and Aerospace Engineering
University of Florida
Gainesville, Florida, USA

KEY TERMS

Mechanics of traction Traction devices Tractors
Engine power Transport devices Pulled implements
Tractive force

Variables

- μ = coefficient of friction
- μ_g = gross traction ratio
- μ_n = net traction ratio
- ρ = motion resistance ratio
- ω = angular velocity of the wheel
- e = horizontal offset
- f_r = force required per row of planter
- F = gross tractive force
- F_f = friction force
- F_i = force required to pull an implement
- F_p = force required to pull a planter
- F_r = motion resistance force
- F_x = any force applied to the wheel in x direction
- F_z = any force applied to the wheel in z direction
- G = soil reaction at resistance center
- H = net tractive force
- n_r = number of rows
- P_{DB} = drawbar power

P_e = engine gross flywheel power
P_t = tractive power developed by the wheel
P_w = power transferred to the wheel axle
r = wheel rolling radius
r_t = torque radius of the wheel
R = vertical reaction force of the wheel
s = travel reduction ratio, or slip
T = torque transferred to the wheel axle
T_E = tractive efficiency of the wheel
v_a = actual velocity of the wheel
v_i = implement velocity
v_t = theoretical velocity of the wheel
W = dynamic wheel load

Introduction

Tractors were created to reduce human and animal labor inputs and increase efficiency and productivity in crop production activities (Schueller, 2000). The main use of tractors is to pull implements such as tillage tools, planters, cultivators, and harvesters in the field and, to some extent, on the road (Renius, 2020). To pull implements efficiently, a tractor needs to generate traction between the tires and the soil surface. Traction is the way a vehicle uses force to move over a surface.

Quite early in tractor development, the direct transfer of power from tractors to implements was made possible by using power take-offs (PTOs) that transfer rotary power to implements and machines and by using hydraulic systems to lift and lower implements and to move parts of attached machines. Pulling implements is still the most common use of tractor power. The field capacity of agricultural machines, i.e., the field area that can be covered per unit time, has caused bigger implements to be developed and used. The increased sizes require greater traction from the pulling tractor. More efficient systems to create the tractive force are necessary to provide the large forces necessary to pull those implements.

The efficiency of how tractors convert the power generated by an engine to the power required to pull the implements depends on many variables associated with the tractor and the soil conditions. Traction is especially important in agriculture as field soils are not as firm as the roads used by cars and trucks. This chapter presents the basic principles of traction applied to agricultural machinery.

> **Outcomes**
>
> After reading this chapter, you should be able to:
>
> - Explain how tractors develop tractive force
> - Describe the effect of some important variables on the tractive force
> - Calculate how much power a tractor can develop when pulling an implement
> - Calculate the power requirements to match tractors to implements

Concepts

Traction and Transport Devices

According to the American Society of Agricultural and Biological Engineers (ASABE Standards, 2018), there are two types of surface contact devices associated with the motion of a vehicle: traction devices and transport devices. A traction device receives power from an engine and uses the reactions of forces from the supporting surface to propel the vehicle, while a transport device does not receive power, but is needed to support the vehicle on a surface while the vehicle is moving over that surface. Wheels, tires, and tracks can be traction devices if they are connected to an engine or other power source; if not connected, they are transport devices. The main components of an agricultural tractor are presented in figure 1. In this example, the tractor is 2-wheel drive, so the large rear wheels, which receive power from the engine, are the traction devices, and the small front wheels are the transport devices. All wheels would be traction devices if the tractor were 4-wheel drive. The engine is connected to the traction device by the drive train, often consisting of a clutch, transmission, differential, axles, and other components. (The drive train is not discussed in this chapter.) The drawbar is an attachment point through which the tractor can apply pulling force to an implement.

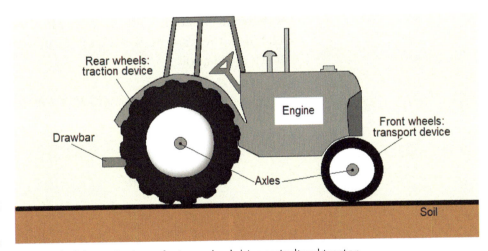

Figure 1. Schematic view of a two-wheel drive agricultural tractor.

Mechanics of Traction

The simplest way of analyzing the traction produced by a traction device, such as a wheel or track, is to consider friction forces that act at the contact between a traction device and the surface when the system is in equilibrium.

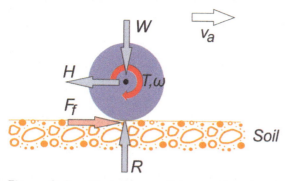

Figure 2. Simplified diagram of the variables related to a wheel developing a net tractive force.

ω = angular velocity of the wheel
F_f = friction force
H = net tractive force
R = vertical reaction force of the wheel
T = torque transferred to the wheel axle
v_a = actual velocity of the wheel
W = dynamic wheel load

For simplification the machine is assumed to be moving at a constant velocity on a non-variable surface (figure 2). A traction device (hereafter simplified to the most common implementation as a "wheel") has two main functions: to support the load acting on the wheel axle (W) and to produce a net tractive force (H). The force W is generally called the dynamic load acting on the wheel. The dynamic load depends on how the weight of the tractor at that point in time is distributed to each wheel. If the system is in equilibrium, the surface reacts to W by applying a vertical reaction force (R) to the wheel. In the contact between the surface and the wheel, a friction force (F_f) is generated. To keep equilibrium in the horizontal direction, the magnitude of the net tractive force H is equal to the magnitude of the friction force F_f. To produce a net tractive force H, the friction force needs to be overcome. This is done by applying a torque (T) to the wheel axle. This torque is proportional to the torque produced by the tractor engine according to the drive train, including the current transmission ratio.

When moving, the wheel (figure 2) rotates with a constant angular velocity (ω), and this angular speed is proportional to the engine rotation speed, depending on the gearing ratio in the drive train. The wheel has an actual velocity v_a, which is equal to the angular velocity multiplied by the wheel's rolling radius reduced by the slip (as discussed below). In an equilibrium situation, ω and v_a are constants. The power transferred to the wheel axle (P_w) can be calculated as the product of the torque (T) and the angular velocity (ω), as shown in equation 1. The tractive power developed by the wheel (P_t) is the product of the net tractive force (H) and the actual velocity (v_a), as shown in equation 2. The tractive efficiency of the wheel (T_E) can be calculated as the ratio between tractive power and the wheel axle power, as shown in equation 3.

$$P_w = T\omega \tag{1}$$

$$P_t = H v_a \tag{2}$$

$$T_E = \frac{P_t}{P_w} \tag{3}$$

where P_w = power transferred to the wheel axle (W)
T = torque transferred to the wheel axle (N m)
ω = angular velocity of the wheel (rad s^{-1})
P_t = tractive power developed by the wheel (W)
H = net tractive force (N)
v_a = actual velocity of the wheel (m s^{-1})
T_E = tractive efficiency of the wheel (dimensionless)

The friction force (F_f in figure 2) is generated by the interaction between the wheel and the surface. The friction force can be calculated by multiplying the reaction force (R) by the equivalent friction coefficient (μ). Table 1 presents some typical values. Because R is equal to the dynamic load acting on the wheel axle (W) and the net tractive force is equal to the friction force, the tractive force can be calculated as the product of the equivalent coefficient of friction and the dynamic load, as:

$$H = \mu W \quad (4)$$

where μ = coefficient of friction (dimensionless).

Table 1. Equivalent coefficient of friction for a tractor wheel working on different surfaces.

Surface type	Equivalent coefficient of friction (μ)[a]
Soft soil	0.26–0.31
Medium soil	0.40–0.46
Firm soil	0.43–0.53
Concrete	0.91–0.98

[a] These values were estimated based on data presented by Kolator and Bialobrzewski (2011).

The theoretical velocity (v_t) is determined by the wheel's rotational velocity (ω) times the rolling radius (r) as shown in equation 5, but the actual wheel velocity (v_a) is less due to the relative motion at the interface between the wheel and the surface. This relative motion is the travel reduction ratio, commonly called slip, and is defined as the ratio of the loss of wheel velocity to the theoretical velocity, that is, the velocity that wheel would have if there was no loss. Equation 6 shows how the travel reduction ratio can be estimated:

$$v_t = \omega r \quad (5)$$

$$s = \frac{v_t - v_a}{v_t} \quad (6)$$

where v_t = theoretical velocity of the wheel (m s^{-1})
r = wheel rolling radius (m s^{-1})
s = travel reduction ratio, or slip (dimensionless)

The travel reduction ratio is an important variable for wheel tractive force analysis. The travel reduction ratio of a wheel can vary from 0 to 1 depending on wheel and surface conditions. When the travel reduction ratio is equal to 0, there would be no relative motion between the periphery of the wheel and the surface. The wheel rotation causes a perfect translational motion relative to the surface. However, experience has shown that for a wheel to develop a tractive force, there must be relative motion (slip) between the wheel and the surface. Therefore, a wheel generating tractive force needs to have a travel reduction ratio greater than zero. When a wheel generates more tractive force, the travel reduction ratio increases, and the actual wheel velocity reduces. When the travel reduction ratio is equal to 1, the wheel does not move forward when it rotates. The models used to calculate the tractive force generally use the travel reduction ratio as one of the variables.

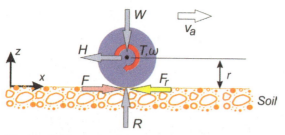

Figure 3. Diagram of the variables related to a wheel developing a net tractive force (*H*) including the gross tractive force (*F*) and the motion resistance force (F_r).

- ω = angular velocity of the wheel
- F = gross tractive force
- H = net tractive force
- R = vertical reaction force of the wheel
- r = rolling radius
- T = torque transferred to the wheel
- F_r = motion resistance force
- v_a = actual velocity of the wheel
- W = dynamic wheel load

Another important concept when analyzing the traction process of a moving wheel is the motion resistance force (F_r) (figure 3). If a wheel is moving, the wheel and the surface deform. Energy is spent to produce this deformation. The resistance produced by the wheel and surface deformations must be overcome to allow the wheel to move. Considering the existence of the motion resistance force, in the contact between the wheel and the surface, it is necessary to generate a friction force greater than the motion resistance force at the wheel-surface contact to produce a tractive force. This friction force is now termed the gross tractive force (denoted by *F*). Thus, the gross tractive force would be the net tractive force generated by the wheel if there was no motion resistance. Adding the concepts of motion resistance and gross tractive forces to figure 2 results in figure 3, which is an improved representation of forces acting on a wheel.

If the wheel represented in figure 3 has no motion in the vertical direction (z axis), the wheel is in static equilibrium in this direction. In this condition, the summation of forces in the z (vertical) direction is zero. Therefore,

$$\sum F_z = 0 \tag{7a}$$

$$R - W = 0 \tag{7b}$$

$$R = W \tag{7c}$$

where F_z = any force applied to the wheel in z direction (N)
R = vertical reaction force of the wheel (N)

If the actual speed of the wheel represented in figure 3 is constant, the horizontal forces are in static equilibrium in this direction and the sum of the horizontal forces is zero. Therefore,

$$\sum F_x = 0 \tag{8a}$$

$$F - F_r - H = 0 \tag{8b}$$

$$H = F - F_r \tag{8c}$$

where F_x = any force applied to the wheel in x direction (N)
F = gross tractive force (N)
F_r = motion resistance force (N)

Based on equation 8c, the gross tractive force (*F*) must be the net tractive force (*H*) plus the motion resistance force (F_r). If both sides of equation 8c are divided by the dynamic load (*W*) acting on the wheel, resulting in equation 9a, three dimensionless numbers, i.e., μ_n, μ_g, and ρ, are created as shown in equations 9c, 9d, and 9e. The first one is the net traction ratio (μ_n), defined as the net tractive force divided by the dynamic load. The second one is the gross traction ratio (μ_g), defined as the gross tractive force divided by the dynamic load. And the third one is the motion resistance ratio (μ), defined as the motion resistance force divided by the dynamic load.

$$\frac{H}{W} = \frac{F}{W} - \frac{F_r}{W} \tag{9a}$$

$$\mu_n = \mu_g - \rho \tag{9b}$$

$$\mu_n = \frac{H}{W} \tag{9c}$$

$$\mu_g = \frac{F}{W} \tag{9d}$$

$$\rho = \frac{F_r}{W} \tag{9e}$$

Equation 9b shows that μ_n, μ_g, and ρ are not independent. By using a technique called dimensional analysis, functions were developed to predict how μ_g and ρ change as a function of the wheel variables and soil resistance. This analysis is presented by Goering et al. (2003) and is beyond the scope of this chapter. If μ_g, ρ, and W are known, the tractive force generated by the wheel can be predicted using equation 10:

$$H = (\mu_g - \rho)W \tag{10}$$

The R, F, and F_r forces (figure 3) act on a point called the wheel resistance center. This point is not aligned with the direction of the dynamic load W but is a little bit ahead of it. This horizontal distance is called the horizontal offset (e). The static analysis of a towed wheel (figure 4) shows that the wheel resistance center is not aligned with the direction of the dynamic wheel load. In a towed wheel, there is no torque applied to its axle. The soil reaction (G) at the resistance center is the resultant of the R and F_r forces. The direction of the G force passes through the wheel center. To move the towed wheel at a constant actual velocity (v_a), a net tractive force (H) equal to the motion resistance force (F_r) needs to be applied to the wheel. For the wheel to keep an angular velocity constant, the sum of the momentums at the center of the wheels must equal zero. Goering et al. (2003) showed that the horizontal offset can be calculated with equation 11.

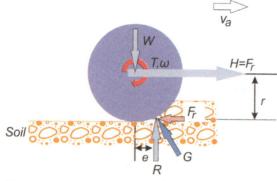

Figure 4. Diagram of forces acting on a towed wheel.

- ω = angular velocity of the wheel
- e = horizontal offset
- G = soil reaction at resistance center
- H = net tractive force
- R = vertical reaction force of the wheel
- r = rolling radius
- F_r = motion resistance force
- v_a = actual velocity of the wheel
- W = dynamic wheel load

$$Re - F_r r = 0 \tag{11a}$$

$$e = \frac{F_r}{R} r \tag{11b}$$

$$e = \frac{F_r}{W} r \tag{11c}$$

where e = horizontal offset (m).

By using equation 10, the tractive force can be predicted. The other important information in the wheel traction analysis is to predict how much torque needs to be transferred to the wheel axle to generate the tractive force (H). In equation 5, the wheel radius is used to convert the rotational angular velocity to the theoretical wheel velocity. The wheel radius can also be used to calculate the torque necessary to produce the wheel tractive force. The torque (T) necessary to keep the angular velocity of the wheel constant and produce the net tractive force is the product of the gross tractive force and the torque radius of the wheel, as given by:

$$T = F r_t \tag{12}$$

where r_t = torque radius of the wheel (m).

The wheel radius defined in equation 5 is different from the torque radius of the wheel defined in equation 12 because of the interaction of the wheel and the surface, which varies on a soft soil surface. Generally, a rolling radius based on the distance from the center of the wheel axle to a hard surface is used. Therefore, equation 13 can be used to estimate the torque acting at wheel axle:

$$T = F r \tag{13}$$

Engine Power Needed to Produce a Tractive Force

ASABE Standards (2015) presented a diagram (figure 5) of the approximate typical power relationship for agricultural tractors. Tractors can be specified by their engine gross flywheel rated power (P_e). One of the standards used to define the engine gross flywheel rated power is SAE J1995 (SAE, 1995). The rated power defined by this standard is the mechanical power produced by the engine without some of its accessories (such as the alternator, the radiator fan, and the water pump). Therefore, the engine gross flywheel rated power is greater than the net power produced by the engine. The approximate engine net flywheel power can be estimated by multiplying the gross flywheel power by 0.92. The power at the tractor PTO is about equal to the engine gross flywheel power multiplied by 0.83 or the engine net flywheel power multiplied by 0.90.

The power that the tractor can generate to pull implements, often termed drawbar power because many implements are attached to the tractor's drawbar, depends on the tractor type, i.e., 2-wheel drive (2WD), mechanical front wheel drive (MFWD), 4-wheel drive (4WD), or tracked. The surface condition where the tractor is used has an even greater effect. Using these two pieces of information, coefficients that show estimates of the relationship between the drawbar power and the PTO power is given in figure 5.

The drawbar power required to pull an implement is:

$$P_{DB} = F_i v_i \qquad (14)$$

where P_{DB} = drawbar power (W)
F_i = force required to pull an implement (N)
v_i = implement velocity (m s^{-1})

The force required to pull an implement depends on the implement. For example, the force required to pull a planter F_p is the force required per row times the number of rows:

$$F_p = f_r n_r \qquad (15)$$

where f_r = force required per row of planter (N row^{-1})
n_r = number of rows

Once the required drawbar power is determined, the values in figure 5 can be used to calculate the estimated needed gross flywheel rated power of a tractor to pull the implement.

Applications

The concepts of traction and tractor power are necessary for properly matching the tractor to an implement. Agricultural operations cannot be performed if the tractor cannot develop enough power or traction to pull the implement. As implements have increased in size over the years, it is necessary that the tractors have enough power and enough traction for the tasks they have to perform. Choosing a tractor that is too large will negatively impact agricultural profitability because larger tractors cost more than smaller tractors. An oversize tractor may also increase fuel consumption and exhaust emissions. This is significant because even the most efficient tractors get less than 4 kWh of work per liter of diesel fuel.

Tractors range greatly in size (e.g., figure 6). For example, one large contemporary manufacturer sells tractors from 17 to 477 kW. The weight of the tractor must be enough to generate sufficient traction force, as shown in equation 10. However, besides the cost

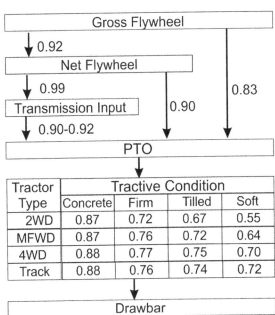

Figure 5. Diagram of the approximate power relationships in agricultural tractors (types are defined in the main text) and soil conditions (ASABE, 2015).

(a) (b)

Figure 6. Typical contemporary (a) small and (b) large tractors.

of adding weight, additional weight may increase soil compaction and depress crop yields. It is therefore necessary to understand these concepts to design tractors and implements. The capabilities of the tractor's engine, power transmission elements, and wheels need to be appropriately scaled. There needs to be a trade-off between making them large and powerful with making them compact and inexpensive. The above analyses can be used to guide tractor choice and design.

The concepts are also applied to other types of agricultural machinery, such as self-propelled harvesters and sprayers. For these machines to be able to complete their tasks, they need to be able to move across agricultural soils. The same calculations can be used to determine if there is enough power and to design the various components on those machines. The wheels, axles, and power transmission components must be able to withstand the forces, torques, and power during the machines' use.

Examples

Example 1: Tractive force

Problem:
Calculate the tractive force produced by a tractor wheel that works on a firm soil with a dynamic load of 5 kN. The wheel velocity is 2 m s^{-1}. If the tractive efficiency is 0.73, what is the power that needs to be transferred to the wheel axle?

Solution:
Assume an equivalent coefficient of friction of 0.48, the mean value for firm soil presented in table 1. Calculate the tractive force using equation 4:

$$H = \mu W = 0.48 \times 5 = 2.4 \, \text{kN}$$

Now, calculate the tractive power for the tractor wheel using equation 2:

$$P_t = H v_a = 2.4 \times 2 = 4.8 \, \text{kW}$$

Calculate the power that needs to be transferred to the wheel axle for using equation 3 with the given tractive efficiency of 0.73:

$$P_w = \frac{P_t}{T_E} = \frac{4.8}{0.73} = 6.58 \, \text{kW}$$

This value of needed power can be used to design the various power transmission components. The power consumption can also be used to calculate the power demanded of the ultimate power source, probably an engine, to calculate fuel consumption and, thereby, costs of a particular field operation.

Example 2: Torque and travel reduction ratio, or slip

Problem:
A wheel on another tractor receives 40 kW from the tractor powertrain. The wheel rotates at 25 rpm, which is an angular velocity, ω, of 2.62 rad s^{-1}. (Note: 2π rad × 25 rpm/60 min s^{-1} = 2.62 rad s^{-1}.) If the rolling radius of the wheels is 0.81 m and the speed of the tractor is 2 m s^{-1}, calculate the torque acting on the wheel and the travel reduction ratio (commonly known as slip).

Solution:
Calculate the torque acting on the wheel T for a power P_w of 40 kW using equation 1:

$$T = \frac{P_w}{\omega} = \frac{40}{2.62} = 15.28 \, \text{Nm}$$

Calculate the power to be transferred to the wheel for producing 2.4 kN of tractive force at 2 m s^{-1} of wheel speed using equation 3:

$$P_w = \frac{P_t}{T_E} = \frac{4.8}{0.73} = 6.58 \, \text{kW}$$

Calculate the theoretical velocity of the wheel v_t for a rolling radius r of 0.81 m using equation 5:

$$v_t = \omega \, r = 2.62 \times 0.81 = 2.12 \, \text{m s}^{-1}$$

Since the actual velocity of the wheel is 2 m s^{-1}, which is less than the theoretical velocity of the wheel, calculate the travel reduction ratio s using equation 6:

$$s = \frac{v_t - v_a}{v_t} = \frac{2.12 - 2.00}{2.12} = 0.057, \text{ or } 5.7\%$$

In addition to providing guidance to the design of the agricultural machine and its power consumption, calculation of the slip is useful to determine how fast the operation will be performed. Excessive slip can also have adverse effects on the soil's structure and inhibit plant growth.

Example 3: Tractive force and power

Problem:
Consider a wheel that works with a dynamic load of 10 kN, a motion resistance ratio of 0.08, and a gross traction ratio of 0.72. Find the tractive force that the wheel can develop. If this wheel rotates at 40 rpm and the rolling radius of the wheel is 0.71 m, how much power is necessary to move this wheel?

Solution:
Calculate the gross tractive force developed by the wheel F using equation 9d:

$$F = \mu_g W = 0.72 \times 10 = 7.2 \text{ kN}$$

Calculate the motion resistance F_r of this wheel using equation 9e:

$$F_r = \rho W = 0.08 \times 10 = 0.80 \text{ kN}$$

The tractive force H developed by the wheel, according to equation 8c, is the difference between the gross tractive force and the motion resistance:

$$H = F - F_r = 7.2 - 0.8 = 6.4 \text{ kN}$$

Calculate the torque necessary to move this wheel using equation 13:

$$T = F r = 7.2 \times 0.71 = 5.11 \text{ kN m}$$

Calculate the power P_w necessary to turn the wheel using equation 1:

$$P_w = T \omega = T \frac{2 \pi N}{60} = 5.11 \times \frac{2 \times \pi \times 40}{60} = 21.4 \text{ kW}$$

Example 4: Engine gross flywheel power

Problem:
Calculate the necessary power of a MFWD tractor to pull a 30-row planter. According to ASABE Standards (2015), a force of 900 N per row is required to pull a drawn row crop planter if it is only performing the seeding operation. The speed of the tractor will be 8.1 km h⁻¹ (2.25 m s⁻¹). The soil is in the tilled condition. Consider that the tractor should have a power reserve of 20% to overcome unexpected overloads.

Solution:
Calculate the drawbar force needed to pull the planter using equation 15:

$$F_p = f_r n_r = 900 \times 30 = 27,000 \text{ N}$$

Calculate the drawbar power P_{DB} needed to pull the planter using equation 14:

$$P_{DB} = F_p v_p = 27{,}000 \times 2.25 = 60{,}750 \text{ W}$$

Therefore, the tractor needs to produce a drawbar power of 60.75 kW. From figure 5, find that the coefficient that relates the drawbar power to the PTO power of the tractor for a MFWD tractor working on tilled soil condition is 0.72. Thus, the tractor PTO power P_{PTO} should be:

$$P_{PTO} = \frac{P_{DB}}{0.72} = \frac{60.75}{0.72} = 84 \text{ kW}$$

Considering that the coefficient that relates the PTO power to the engine gross flywheel power is 0.83 (figure 5), the engine gross flywheel power P_e is:

$$P_e = \frac{P_{PTO}}{0.83} = \frac{84.375}{0.83} = 102 \text{ kW}$$

Considering a reserve of power of 20% to overcome unexpected overloads, the tractor selected should have an engine gross flywheel power at least 20% greater than that needed to pull the 30-row planter, or 1.2 × 102 kW = 122 kW.

These calculations will help the farm manager select the proper tractor for the operation.

Image Credits

Figure 1. Queiroz, D. (CC By 4.0). (2020). Schematic view of a two-wheel drive agricultural tractor.
Figure 2. Queiroz, D. (CC By 4.0). (2020). Simplified diagram of the variables related to a wheel developing a net tractive force.
Figure 3. Queiroz, D. (CC By 4.0). (2020). Diagram of the variables related to a wheel developing a net tractive force (H) including the gross tractive force (F) and the motion resistance force (F_r).
Figure 4. Queiroz, D. (CC By 4.0). (2020). Diagram of forces acting on a towed wheel.
Figure 5. ASABE Standard ASAE D497.7 (CC By 4.0). (2020). Diagram of the approximate power relationships in agricultural tractors (types are defined in the main text) and soil conditions.
Figure 6. Schueller, J. (CC By 4.0). (2020). Typical contemporary (a) small and (b) large tractors.

References

ASABE Standards. (2018). ANSI/ASAE S296.5 DEC2003 (R2018): General terminology for traction of agricultural traction and transport devices and vehicles. St. Joseph, MI: ASABE.
ASABE Standards. (2015). ASAE D497.7 MAR2011 (R2015): Agricultural machinery management data. St. Joseph, MI: ASABE.
Goering, C. E., Stone, M. L., Smith, D. W., & Turnquist, P. K. (2003). Traction and transport devices. In *Off-road vehicle engineering principles* (pp. 351-382). St. Joseph, MI: ASAE.

Kolator, B., & Białobrzewski, I. (2011). A simulation model of 2WD tractor performance. *Comput. Electron. Agric.* 76(2): 231-239.

Renius, K. T. (2020). *Fundamentals of tractor design.* Cham, Switzerland: Springer Nature.

SAE. (1995). SAE J1995_199506: Engine power test code—Spark ignition and compression ignition—Gross power rating. Troy, MI: SAE.

Schueller, J. K. (2000). In the service of abundance: Agricultural mechanization provided the nourishment for the 20th century's extraordinary growth. *Mech. Eng.* 122(8):58-65.

Crop Establishment and Protection

Roberto Oberti
Department of Agricultural and Environmental Science, University of Milano, Milano, Italy

Peter Schulze Lammers
Department of Agricultural Engineering, University of Bonn, Bonn, Germany

KEY TERMS		
Tillage	Spreading	Field performance parameters
Planting	Spraying	Application rate and quality

Variables

- ε = dynamic factor
- ρ_R = rolling resistance
- A = field area receiving the material
- A_o = area of the nozzle orifice
- AR = application rate
- B = bite length
- c_{AC} = content of active compound in the raw material or solution
- c_d = discharge coefficient that accounts for the losses due to viscous friction through the orifice
- C_a = field capacity
- C_t = theoretical field capacity of the machine
- CV = coefficient of variation
- d = depth of furrow
- D = dose of application, i.e., amount of active compound per unit area
- F_v = vertical force
- F_z = draft force
- g = gravity (9.8 m s^{-2})
- h = height of cell above seed furrow

i = number of moldboards
k = static factor
k_n = nozzle-specific efflux coefficient
LAI = Leaf Area Index
M = mass of material applied
n = number of gauge wheels
n_d = the deposit density on the target surface
N = number of measured samples
p = operating pressure of the circuit
p_z = seed spacing frequency
q = liquid material or volume flow rate
q_n = flow rate discharged by the nozzle
Q = material flow rate
r = rotary speed
s = field speed
t = time of dropping
v = travel speed
V = volume of the material applied
VMD = volume median diameter of the spray
w = operating, or working, width of a machine
w_f = width of furrow
x_i = seed spacing
$\bar{x}$ = mean seed spacing
z = number of blades per tool assembly

Introduction

Field crops are most often grown to provide food for humans and for animals. Growing field crops requires a sequence of operations (figure 1) that usually starts with land preparation followed by planting. These two stages are known as crop establishment. Crop growth requires a supply of nutrients through application of fertilizers as well as protection against weeds, diseases, and pest insects using biological, chemical, and/or physical treatments. Finally, the crop is harvested and transported to processing locations. This general sequence of operations can be more complex or specifically modified for a particular crop or cropping system. For example, crop establishment is only required once, while crop protection and fertilization may be repeated multiple times annually.

Engineering is integral to maximize the productivity and efficiency of

Figure 1. Typical operations involved in growing field crops.

these operations. This chapter introduces some of the engineering concepts and equipment used for crop establishment and crop protection in arable agriculture.

> **Outcomes**
>
> After reading this chapter, you should be able to:
>
> Describe the fundamental principles of agricultural mechanization
>
> Apply physics concepts to some aspects of crop establishment and protection equipment
>
> Calculate field performance for land preparation, planting, fertilizing, and plant protection based on operating parameters

Concepts

Field Performance Parameters

Regardless of the specific operation, the work of a field machine is evaluated through some fundamental parameters: the operating width and the field capacity of the machine.

Operating Width

The operating, or working, width w of a machine is the width of the portion of field worked by each pass of the machine. In field work, especially with large equipment, the effective operating width can be less than the theoretical width due to unwanted partial overlaps between passes.

Field Capacity of a Machine

An important parameter to be considered when selecting a machine for an operation is the field capacity, which represents the machine's work rate in terms of area of land or crop processed per hour. The theoretical field capacity of the machine (also called the area capacity) can be computed as:

$$C_t = w\,s \qquad (1)$$

where C_t = theoretical field capacity (m² h⁻¹)
w = operating width (m)
s = field speed (m h⁻¹)

C_t is typically expressed in ha h⁻¹. Figure 2 illustrates the field area worked during a time interval, t.

In actual working conditions, this theoretical capacity is reduced by idle times (e.g., turnings, refills, transfers, or breaks) and possible

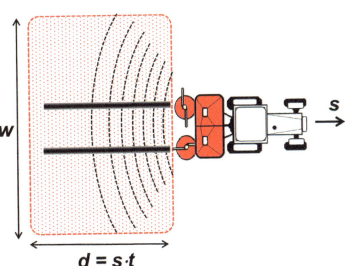

Figure 2. The area-rate worked by a machine, i.e., its field capacity (C_a) originates from its working width (w) and field speed (s).

reductions in working width or in nominal field speed due to operational considerations, resulting in an actual field capacity:

$$C_a = e_f C_t \qquad (2)$$

where e_f is the field efficiency (decimal). Its value largely depends on the operation, which can be estimated for given operations and working conditions.

Tillage

Definition of Tillage

Soil preparation by mechanical interventions is called *tillage*. The major function of tillage is loosening the soil to create pores so they can contain air and water to enable the growth of roots. Other main tasks of tillage are crushing soil aggregates to required sizes, reduction or elimination of weeds, and admixing of plant residues. Tillage needs to be adapted to the soil type and condition (such as soil water or plant residue content) and conducted at a proper time.

Soil Mechanics

Soil is classified by grain sizes into sand, silt, and clay categories. Loam is a mixture of these soil types. Soil is subjected to shear stress and reacts by strain when it is tilled. The tillage tool moving through the soil causes a force which causes a stress between adjacent soil grains. This leads to a deformation or strain of the soil. Sandy soil is characterized by low shear strength and high friction, while clay is characterized by high cohesion and, after cracking, by low friction. Tillage tools usually act as wedges. They engage with the soil and cause relative movement in a shear plane where some of the soil moves with the tool while the adjacent soil stays in place. Energy is expended in shearing and lifting of the soil and overcoming the friction on the tool.

Primary Tillage

Primary tillage tools or implements are designed for loosening the soil and mixing or incorporating crop residues left on the field surface after harvest. Subsequent soil treatment to prepare a seedbed is secondary tillage. A typical implement for primary tillage is the plow (spelled "plough" in some countries), which is used for deep soil cultivation.

The three most common kinds of plow are moldboard, chisel, and disc (figure 3). The moldboard plow body and its action are shown in figure 3a. A plow share cuts the soil horizontally and the attached moldboard upturns the soil strip and turns it almost upside down in the furrow made by the previous plow body. The heel makes sure the plow follows the proper path. These parts are connected by a supporting part (breast) which is connected by a leg to the frame of the plow. Chisel plows do not invert all the soil, but they mix the top soil layer, including residues, into deeper portions of the soil. Chisel plows use heavy tines with shares on the bottom of the tines (figure 3b). A disc plow uses concave round discs (figure 3c) to cut the soil on the furrow bottom and turns the soil with the rotating motion of the disc.

The draft forces to pull a plow are provided by a tractor to which it is attached. Equation 3 is one way used to calculate the draft force needed to pull a moldboard plow. This calculation follows Gorjachkin (1968):

$$F_z = n F_v \rho_R + i k w_f d + i \varepsilon w_f d v^2 \tag{3}$$

where F_z = draft force (N)
n = number of gauge wheels
F_v = vertical force (N)
ρ_R = rolling resistance
i = number of moldboards or shares
k = static factor (N cm^{-2}) ranging from 2 to 14 depending on soil type
w_f = width of furrow (cm)
d = depth of furrow (cm)
ε = dynamic factor (N s^2 m^{-2} cm^{-2}), ranging from 0.15 to 0.36 depending on soil type and moldboard design
v = traveling speed (m s^{-1})

About half the energy consumed in plowing, the effective energy, does the work of cutting (13–20%), elevating and accelerating the soil (13–14%), and deformation (14–15%). The remaining energy is spent on noneffective losses (e.g., friction), which do not contribute to tillage effectiveness.

Secondary Tillage (Seedbed Preparation)

Secondary tillage prepares the seedbed after primary tillage. Implements for secondary tillage are numerous and of many different designs. A harrow is the archetype of secondary tillage; it consists of tines fixed in a frame. Cultivators are heavier, with longer tines formed as chisels in a rigid frame or with a flexible suspension. Soil is opened by shares and the effect is characterized by the tine spacing, depth of furrow, and speed.

Most tillage implements are pulled by tractors and limited by traction, as discussed in another chapter. (Power tillers exist as rotary harrows with vertical axles or rotary cultivators with horizontal axles; these are discussed below.) The mechanical connection of the tractor power take-off (PTO) to the tillage implement provides the power to drive the implement's axles, which are equipped with

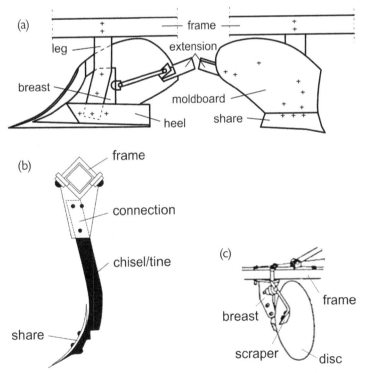

Figure 3. (a) moldboard plow body; (b) chisel (tine) of a cultivator; (c) disc plow.

blades, knives, or shares. Each blade cuts a piece of soil (Schilling, 1962) and the length of that bite is determined as a function of the tractor travel speed and axle rotation speed according to:

$$B = \frac{v}{r\,z}\frac{10{,}000}{60} \qquad (4)$$

where B = bite length (cm)
v = travel speed (km h^{-1})
r = rotary speed (min^{-1})
z = number of blades per tool assembly

The cut soil is often flung against a hood or cover which helps crush soil agglomerations, with the axle rotation speed affecting the impact force.

Planting

Crops are sown (planted) by placing seeds in the soil (or in some cases, discussed below, by transplanting). Basic requirements are:

- equal distribution of seeds in the field,
- placing the seeds at the proper depth of soil, and
- covering the seeds.

The seed bed should be prepared by tilling the soil to aggregates the size of the seed grain, and gently compacting the soil at the placement depth of the seed, e.g., 2–4 cm for wheat and barley. Germination is triggered by soil temperature (e.g., 2–4°C for wheat) and soil water content. The seeding rate of cereal grains is in the range of 200 to 400 grains per m² resulting in 500 to 900 heads per m² as a single plant may produce multiple heads. As the metering of seeds is mass-based, commercial seed is indexed by the mass of one thousand grains (table 1).

The appropriate distribution of seeds is a fundamental condition for a successful crop yield. Seeds can be broadcasted, which means they are scattered randomly (figure 4). This is commonly done for crops such as grasses and alfalfa. But the seeds for most crops are deposited in rows. Common distances between

Table 1. Characteristics of seeds.

	Thousand Grain Mass (g)	Bulk Density (kg L^{-1})	Seed Rate (kg ha^{-1})	Area per Grain (cm²)
Wheat (*Triticum aestivum*)	25–50	0.76	100–250	22–37
Barley (*Hordeum vulgare*)	24–48	0.64	100–180	27–48
Maize (*Zea mays*)	100–450	0.7	50–80	-
Peas (*Pisum sativum*)	78–560	0.79	120–280	-
Rapeseed (*Brassica napus*)	3.5–7	0.65	6–12	-

rows for cereals are 12 to 15 cm. For crops commonly called row crops (e.g., soybean or maize), the row spacing is 45 to 90 cm. If the rows are at a suitable distance, the wheels of farm machines can avoid driving on the plants as they grow from the seeds.

Within rows, seeds have a random spacing if seed drills are used, or have a fixed distance between seeds if precision seeders (discussed below) are used. Seed drills are commonly used for small grains and consist of components that:

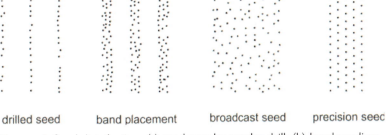

Figure 4. Seed distribution: (a) seed row by regular drill; (b) band seeding; (c) broadcasted seeds; (d) precision seed.

- hold the seeds to be planted,
- meter (singulate) the seed,
- open a row furrow in the soil,
- transport the seeds to the soil,
- place the seeds in the soil, and then
- cover the open furrow with soil.

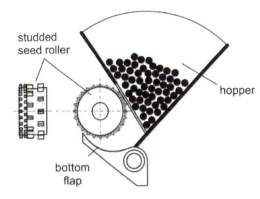

Figure 5. Studded seed wheel for metering seeds, with bottom flap for adjustment for seed size.

To extend the area per seed when sowing in rows, a wider opening of the furrow allows band seeding (figure 4).

Regular seed drills often meter the seeds with a studded roller as seen in figure 5. The ideal is to have a uniform distribution, but there will be variation, including infrequent longer distances, as shown by figure 6 (Heege, 1993).

A second very common metering device is a cell wheel, which is used for central metering in pneumatic seed drills. A rotating cell wheel is filled by the seeds in the hopper and empties into an air stream via a venturi jet (figure 7). The grains are entrained by air and collide with a plate. A relatively uniform distribution of grains occurs along the circumference of the plate where pipes for transporting the grains to the coulters are arranged.

The frequency of the seed distances (figure 6) as metered by the feed cells or studded rollers corresponds to an exponential function (Heege, 1993):

$$p_z = \frac{1}{\overline{x}} e^{-\frac{x_i}{\overline{x}}} \qquad (5)$$

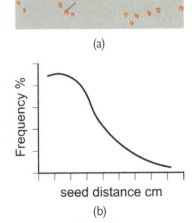

Figure 6. (a) Seed deposition in a row, drilled; (b) frequency of seed distances, drilled.

where p_z = seed spacing frequency
x_i = seed spacing (cm)
$\overline{x}$ = mean seed spacing (cm)

The accuracy of the longitudinal seed distribution is indicated by the coefficient of variation (Müller et al., 1994):

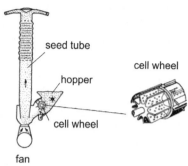

Figure 7. Fluted pipe with plate for distributing the seeds along the circumference into the seed tubes, and cell wheel in detail.

$$CV = \frac{\sqrt{\frac{\sum (x_i - \bar{x})^2}{N-1}}}{\bar{x}} \cdot 100\% \qquad (6)$$

where CV = coefficient of variation (%)
N = number of measured samples
x_i = spacing (cm)
$\bar{x}$ = mean spacing (cm)

A CV smaller than 10% is considered good, while above 15% is unsatisfactory.

Precision Seed Drills

Crops such as maize, soybeans, sugar beets, and cotton have higher yields if seeded as single plants. Precision seed drills singulate the seeds and place them at constant separation distances in the row. Cell wheels single the seeds out of the bulk, move them by constant rotational speed to the coulter, and drop them in the seed furrow with a target distance between each seed. Filling of the cells is a crucial function of precision seed drills. Every cell needs to be filled with one, and only one, seed. The cell wheels rotate through the bulk of seeds in the hopper and are filled by gravity, or the filling is accomplished by an air stream sucking the grains to the holes of the cell wheels (figure 8). Then the seeds are released from the cell wheels and drop onto the bottom of the seed furrow. The trajectory of a single seed grain is affected by gravity (including the time for the seed to drop). To avoid rolling of the seeds in the furrow, the backward speed of the seed relative to the drill should match the forward speed of the seed drill.

The time required for the seed to drop is:

$$t = \sqrt{\frac{2h}{g}} \qquad (7)$$

where t = time of dropping (s)
h = height of cell above seed furrow (m)
g = acceleration due to gravity (9.8 m s^{-2})

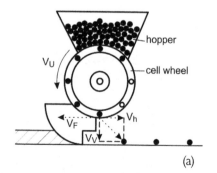
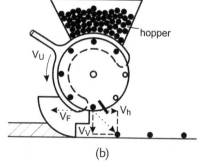

placement optimal:
$V_F = V_U \approx 0$
V_h = min.
$V_V \approx 0$

V_F = machine travel speed
V_h = horizontal speed
V_U = angular speed of cell wheel
V_V = vertical speed

(a)　　　(b)

Figure 8. Precision seeder singling seed grains for seed placement with definite spacing: (a) mechanical singling by cell wheel, (b) pneumatic singling device with cell wheel.

Transplanting

In the case of short vegetation periods or intensive agriculture (e.g., crops that require production in a short time), some crops are not direct-seeded into fields but instead are transplanted.

Seeds may be germinated under controlled conditions, such as in greenhouses (glasshouses). The small plant seedlings may be grown on trays or in pots then transplanted into fields where they will produce a harvestable crop. In the case of rice, which is often not direct-seeded, the plants are grown in trays and then transplanted. When transplanting rice, an articulated mechanical device punches the root portion together with the upper parts of the plants out of the tray and presses it into the soil, while keeping the plants fully saturated with water.

Other plants are propagated by vegetative methods (cloning or cuttings). Special techniques are required for potatoes. "Seed" tubers are put in the hopper of the planter and planted in ridges. Row distances are commonly 60–90 cm, which produce 40,000–50,000 potatoes per ha.

Compared to seeds, small plants (whether seedlings, tubers, cuttings, etc.) are easily damaged. The fundamental requirements of transplanting, both manual (which is labor-intensive) and mechanized, are:

- no damage to the seedlings,
- upright positioning of the seedlings in the soil at a target depth,
- correct spacing between plants in a row, and
- close contact of the soil with the roots.

Fertilization

Crop yield is strictly related to the availability of nutrients that are absorbed by the plant during growth. As crops are harvested, nutrients are removed from the soils. The major nutrients (nitrogen (N), phosphorus (P), and potassium (K)) usually have to be replaced by the application of fertilizers in order to maintain the soil's productivity. Minor nutrients are also sometimes needed.

Organic fertilizers are produced within the farming process, and their use can be seen as a conservation recycling of nutrients. The concentration of nutrient elements in organic fertilizers is low (e.g., 1 kg of livestock slurry can contain 5 g of N or less), requiring very large structures for long-term storage, suitable protection from nutrient volatilization or dilution, large machines for field distribution, and the application of large amounts for sufficient fertilization of crops. Based on their solid contents, organic fertilizers are classified as slurry (solid content less than 14%) that can be pumped and managed as fluids, or as manure (solid content at least 14%) that are managed as solids with scrapers and forks.

Mineral fertilizers are produced by industrial processes and are characterized by a high concentration of nutrient chemicals, prompt availability for plant uptake, ease of storage and handling, and stability over time. The most used form of mineral fertilizers worldwide is solid granules (e.g., urea prills, calcium ammonium nitrate, potassium chloride, and N-P-K compound fertilizers). Other techniques rely on the distribution of liquid solutions or suspensions of mineral fertilizers or on soil injection of anhydrous ammonia.

Application Rate

During distribution operations (including fertilization as well as distribution of other inputs such as pesticides), the application rate is the amount of material distributed per unit of surface area, i.e., for solids by mass:

$$AR = M/A \qquad (8a)$$

and for liquids by volume:

$$AR = V/A \qquad (8b)$$

where AR = application rate (kg ha^{-1} or L ha^{-1})
M = mass of material distributed (kg)
V = volume of material distributed (L)
A = field area receiving the material (ha)

Dose of Application

The dose of application, D, refers to the amount of active compound (e.g., chemical nutrient, pesticide ingredient) distributed per unit of surface area:

$$D = c_{AC} \, AR \qquad (9)$$

where D = dose of application (kg$_{AC}$ ha^{-1})
c_{AC} = content of active compound in the raw material or solution distributed at the application rate (g kg^{-1} or g L^{-1})

Longitudinal and Lateral Uniformity of Distribution

The uniformity of the application rate during distribution is of fundamental importance for the agronomic success of the operation. The machine must be able to guarantee a suitable uniformity along both the traveling (longitudinal uniformity) and the transversal (lateral uniformity) directions.

The *longitudinal uniformity* of the distribution to the field is obtained by appropriate metering of the mass (or volume) flow out of the machine, by means of control devices such as adjustable discharge gates or valves. The rate of material flow to be set depends on the desired application rate, the traveling speed, and the working width of the distributing machine. This can be seen by dividing by time both the numerator and denominator of equation 8a (and similarly for 8b), which leads to:

$$AR = (M/t)/(A/t)$$

The numerator of the right side of the equation is the mass outflow Q (typically expressed in kg min^{-1}), and the denominator is the theoretical field capacity of the machine C_t, or $0.1\,w\,s$ (see equation 1).

Then, it follows that, for AR in units of kg ha^{-1}:

$$AR = \frac{(Q\,\text{kg min}^{-1})(60\,\text{min h}^{-1})}{0.1ws} = \frac{600Q}{ws}$$

Rearranging:

$$Q = AR\,w\,s / 600 \qquad (10\text{a})$$

where Q is the value of material flow (kg min^{-1}) to be set in order to obtain a desired application rate AR (kg ha^{-1}), when the distributing machine works at a speed s (km h^{-1}) and with a working width w (m).

Similarly, for liquid material distributed, the volume rate q (L min^{-1}) is calculated as follows:

$$q = AR\,w\,s / 600 \qquad (10\text{b})$$

The *lateral uniformity* of distribution along the working width is obtained by ensuring two conditions: a controlled distribution pattern and an appropriate overlapping of swath distance between the adjacent passes of machine. A properly operating distribution system can maintain the regular shape of the distribution pattern, which can be triangular, trapezoidal, or rectangular, depending on the distribution system. The overall lateral distribution is obtained by the proper overlapping of the individual pattern produced by each pass of the equipment (the working width). The uniformity of distribution can be tested by travelling past trays placed on the ground and measuring the amount of fertilizer deposited in the individual trays. Coefficient of variation analyses similar to that discussed above for seeding (equation 6) can be performed to evaluate the uniformity.

Fertilizer Spreader Types and Functional Components

A fertilizer spreader is a machine that carries, meters, and applies fertilizer to the field. There are many types of fertilizer spreaders with different characteristics, depending on the fertilizer material and local farming needs.

Slurry tankers are often used for spreading organic fertilizers that can be pumped, while manure spreaders are used for drier materials with higher solids content, often including straw or plant residues in addition to animal waste. Granular mineral fertilizers are distributed by centrifugal spreaders or by pneumatic or auger spreaders. Liquid fertilizers are usually distributed by boom sprayers or by micro-irrigation systems, and anhydrous ammonia by pressure injectors.

All fertilizer spreaders include three main functional components: the hopper or tank, the metering system, and the distributor. A hopper (for solid materials) or a tank (for liquids and slurries) is the container where the fertilizer is loaded. In tractor-mounted spreaders, the hopper capacity is generally below 1000–1500 kg, while for trailed equipment the capacity can reach 5000 kg. The load capacity of slurry tankers and manure spreaders is much higher (from 3 m^3 to more than 25 m^3), since the application rates for organic fertilizers are very high to compensate for their low concentration of nutrients. Hoppers and tanks are treated to be corrosion resistant, while slurry tankers are typically made of stainless steel for similar reasons.

The fertilizers are fed from the hopper or tank either by gravity (centrifugal spreader), a mechanical conveyor (pneumatic spreader or manure spreader), or by pressure (slurry tanker), through the metering system toward the distribution system. The mass outflow Q (kg min^{-1}) in fertilizer spreaders is often metered by an adjustable gate, which can change the outlet's opening area to set the fertilizer application rate (equation 10a). Since the flow characteristics of granular material through a given opening depends on particle size, shape, density, friction, etc., a calibration procedure is necessary to establish the mathematical relationship between gate opening and mass flow Q for a specific fertilizer. This is generally carried out by disabling the distributor system, setting the metering gate in a defined position, collecting the fertilizer discharged during a given time (e.g., 30 s) with a bucket, and finally computing the mass flow obtained. This procedure may be repeated for multiple metering gate positions, although the manufacturer usually provides instructions to extrapolate from a single measurement point (calibration factor) to a full relationship between gate opening and flow.

In a manure spreader, mass outflow is metered by varying the speed of a floor conveyor or of a hydraulic push-gate in the case of very large machines. Flow control in pressurized slurry tankers is made through a metering valve or by varying the pump speed for machines with direct slurry pumping.

The metered flow is then spread by the distributor across the distribution width. In centrifugal spreaders, the distribution is produced by two (occasionally one in small machines) rotating discs powered by the tractor PTO or by hydraulic or electric motors. On each disc, two or more radial vanes impress a centrifugal acceleration to fertilizer granules that are propelled away with velocities ranging between 15 m s^{-1} and more than 50 m s^{-1}, within a certain direction angle resulting from the combination of tangential and radial components of the velocity. The granules then follow an almost parabolic (drag friction decelerates the particle) trajectory in the air, obtaining a very large distribution width.

In addition to the rotational speed of the discs, a crucial parameter in defining the spreading pattern in a centrifugal spreader is the feeding position, i.e., the dropping position of the granules on the disc that, in turn, defines the time during which each particle is accelerated by the vanes and hence its launching velocity. By changing the feeding position, together with the metering gate opening, the distribution pattern and width can be kept uniform for different fertilizer granules or it can be used to obtain specific distribution patterns, such as for spreading near field borders.

In pneumatic spreaders, the fertilizer granules are fed into a stream of carrier air generated by a fan. The air stream transports the fertilizer through pipes mounted on a horizontal boom and the fertilizer is finally distributed by hitting deflector plates. The spacing between plates is about 1–2 m producing a small overlapping of spreading, which results in uniform transversal distribution across the whole working width.

Manure spreaders usually have two or more rotors mounted on the back of the spreader. The rotors are equipped with sharp paddle assemblies that shred and spread manure particles over a distributing width of 5–8 m. Slurry is spread in similar widths by a pressurized flow into a deflector plate or by means of soil applicators that deposit the slurry directly on or into the ground.

Crop Protection

The development and productivity of crops require protection against the competition by undesired plants (weeds), against infestations by diseases (fungi, viruses, and bacteria), and damage from pest insects. This can be obtained through the integration of one or more different approaches, including rotation of crops and selection of resistant varieties, crop management techniques, distribution of beneficial organisms, and application of physical (e.g., mechanical or thermal) or chemical treatments.

The current primary method of crop protection is the use of chemical protection products, commonly pesticides, which play a vital role in securing worldwide food and feed production. Pesticide formulations are sometimes distributed as fumigation, powder, or solid granules, such as during seeding. But the technique most used is liquid application, after dilution in water, by means of a pressurized liquid sprayer.

Droplet Size

To optimize the biological efficacy of pesticides, the liquid is atomized into a spray of droplets. The number of droplets and their size affect the spray's ability to cover a larger surface, to hit small targets, and to penetrate within foliage. Each spray provides a range, or distribution, of droplet sizes. The droplet size is usually represented as a *volume median diameter* (VMD or $D_{v0.5}$) in μm and is classified as in table 2. Crop protection applications mostly use droplets ranging from fine to very coarse diameters.

Droplet Drift vs. Adhesion and Coverage

The effect of drag and buoyancy forces increases as droplet size decreases. This makes finer sprays more prone to drift, i.e., to be transported out of the target zone by air convection. Moreover, in dry air, evaporation of water reduces the droplet size during transport, especially of small droplets, further amplifying drift risks. Besides the reduced crop protection efficacy, spray drift is a major concern for pesticide deposition on unintended targets, contamination of surface water and surrounding air, and risks due to over-exposure for operators and other people.

On the other hand, coarse droplets cover less target area with the same liquid volume (figure 9), and their adhesion on target surfaces after impact can be problematic. If the kinetic energy at impact overcomes capillary forces, the

Table 2. Droplet size classifications in accordance with ANSI/ASABE S572.1 (ASABE Standards, 2017), and typical use in crop protection applications. CP and SP refer to the action mode of the product: CP = contact product; SP = systemic product.

Droplet Size Category	Symbol	VMD (µm)	Typical Use
Very fine	VF	<140	Greenhouse fogging
Fine	F	140–210	CP on tree crops
Medium	M	210–320	CP on arable crops
Coarse	C	320–380	SP on crops; CP on soil
Very coarse	VC	380–460	CP on soil; anti-drift applications
Extremely coarse	EC	460–620	Anti-drift applications; liquid fertilizers
Ultra coarse	UC	>620	Liquid fertilizers

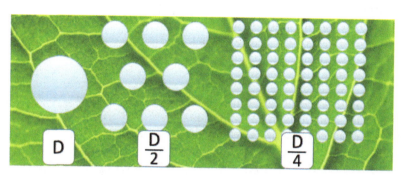

Figure 9. By reducing droplet size D, a larger target surface can be covered with the same liquid volume.

droplet shatters or bounces, resulting in runoff instead of adhering to the surface as liquid.

As a consequence, the optimal droplet size distribution is a matter of careful optimization: while a fine spray can take advantage of air turbulence and be beneficial for improving coverage in a dense canopy, medium-coarse spray is preferred to decrease drift risks with product losses in air, water, and soil. Coarse to very coarse sprays need to be used when wind velocity is above the optimal range (1–3 m s^{-1}) and treatments cannot be postponed.

Action Mode and Application Parameters

Spray characteristics have to be adapted to the features of the target and crop and to the pesticide action mode. There are mainly two broad groups of pesticide modes of action: *contact pesticides*, with a protection efficacy restricted to the areas directly reached by the chemical in a sufficient amount; and *systemic pesticides*, with a protection efficacy depending on the overall absorption by the plant of a sufficient amount of chemical and its internal translocation to the site of action.

Contact products generally require high deposit densities (75–150 droplets cm^{-2}) for a dense coverage of the target surface, as obtained with closely spaced droplets of finer sprays. On the other hand, for systemic products, coverage of the surface is less important provided that a sufficient dose of pesticide is delivered to, and absorbed by, the plant. Hence, lower deposit densities (20–40 droplets cm^{-2}) are used, associated with coarser sprays.

Application Rate

By combining the droplet size and deposit density chosen for a pesticide treatment, the application rate AR, i.e., the liquid volume per unit of sprayed area, can be computed as:

$$AR = \frac{\text{liquid volume}}{\text{sprayed area}} = (\text{mean drop volume})\left(\frac{\text{number of drops}}{\text{sprayed area}}\right)$$

The determination of the sprayed area should take into account that, for soil treatments, the target surface is the field area, whereas for plant treatments, it is the total vegetation surface of the plant. The relationship between the two is usually expressed as leaf area index, LAI, which is the ratio between the leaf surfaces of the target and the surface of the field in which it is growing. At early stages of growth, an LAI of about 1 is usually assumed (as for soil), while with further development LAI increases to 5 or more, depending on the crop. The previous expression can then be rewritten as:

$$AR = \frac{4}{3}\pi\left(\frac{VMD}{2}\right)^3 \times n_d \times LAI = \frac{\pi}{6}VMD^3\left(\mu m^3\right)\left(10^{-15} L\ \mu m^{-3}\right) \times n_d\ \left(cm^{-2}\right)\left(10^8\ cm^2\ ha^{-1}\right) \times LAI$$

that is,

$$AR = 10^{-7} \times \frac{\pi}{6}VMD^3 \times n_d \times LAI \tag{11}$$

where AR = application rate (L ha^{-1})
 VMD = volume median diameter of the spray (μm)
 n_d = the deposit density on the target surface (number of droplets cm^{-2})
 LAI = leaf area index of sprayed plants (decimal; = 1 for soil and early growth stages)

Functional Components of the Sprayer

The sprayer is the machine that carries, meters, atomizes, and applies the spray material to the target. The main functional components of a sprayer are shown in figure 10.

The *tank* contains the water-pesticide mixture to be applied, with capacities that vary from 10 L for human-carried knapsack models to more than 5 m³ for large self-propelled sprayers. Tanks are made of corrosion-resistant and tough material, commonly polyethylene plastic, suitably shaped for easy

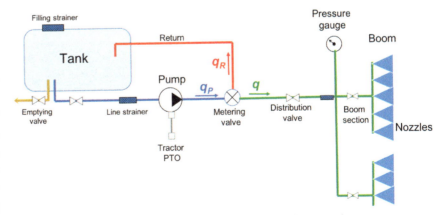

Figure 10. Schematic diagram of a sprayer with main functional components: q_P is the flow rate produced by the pump, q_R is the portion returned by the valve to the tank, and q is the flow rate metered for distribution.

filling and cleaning. To keep uniform mixing of the liquid in the tank, suitable agitation is provided by the return of a part of the pumped flow or, more rarely, by a mechanical mixer.

The *pump* produces the liquid flow in the circuit, working against the resistance generated by the components of the system (valves, filters, nozzles, etc.) and by viscous friction. The higher the resistance that the pump must overcome, the greater the pressure of liquid in the circuit.

Diaphragm pumps (figure 11) are the most common type used in sprayers, because they are lightweight, low cost, and can handle abrasive and corrosive chemicals. The pumping chamber is sealed by a flexible membrane (diaphragm) connected to a moving piston. When the piston moves to increase the chamber volume (figure 11, left), liquid enters by suction through the inlet valve. As the piston returns, the diaphragm reduces the chamber volume (figure 11, right), propelling the liquid through the outlet valve.

As for any positive displacement pump, diaphragm pumps deliver a constant flow for each revolution of the pump shaft, regardless of changes of pressure (within the working range):

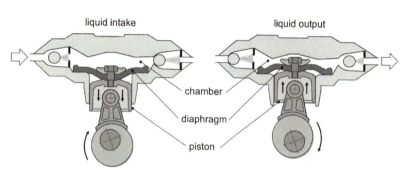

Figure 11. Diaphragm pump: intake (left) and output stroke (right) of the liquid.

$$q_p = 10^{-3} V_p \, n_p \qquad (12)$$

where q_p = flow rate delivered by the pump (L min^{-1})
V_p = pump displacement (cm^3)
n_p = rotational speed of the pump shaft (min^{-1})

In tractor-coupled sprayers, the pump is actuated by the tractor PTO shaft to provide the spraying liquid hydraulic power necessary to operate the circuit. The hydraulic power is:

$$P_{hyd} = p \, q_p / 60000 \qquad (13)$$

where p = pressure of the circuit (kPa)
q_p = flow rate produced by the pump (L min^{-1})

Some sprayers use centrifugal pumps. In these cases, the flow from the pump will not be positive displacement and will depend upon the pressure the pump has to pump against.

Control valves in the circuit enable the desired functioning of the sprayer, by controlling flow direction and volume in the different sections, and by maintaining a desired liquid pressure that, in turn, defines the spray characteristics and the distributed volume.

Since pressure is a fundamental parameter for spray distribution, a *pressure gauge* with appropriate accuracy and measurement range (e.g., two times the expected maximum pressure) is always installed in the sprayer circuit.

Nozzles are the core component of the sprayer that atomizes the pesticide-water mixture into droplets, producing a spray with a specific pattern to cover the target. The most common atomizing technology in sprayers is the hydraulic nozzle (figure 12), which breaks up the stream of liquid as it emerges by pressure from a tiny orifice into spray droplets.

For a given liquid (i.e., for a given density and surface tension), the operating pressure and the orifice area directly determine the size of the droplets of the produced spray. In particular, by increasing the pressure with a specific nozzle, the size of the droplets decreases. Conversely, for a given pressure the size of the droplets increases with the area of the nozzle orifice.

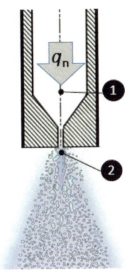

Figure 12. Hydraulic nozzle operation: the liquid flowing in the nozzle body (1) at flow rate q_n is atomized into droplets by forcing it through a tiny orifice (2).

Flow Rate Metering by Pressure Control

The discharge flow rate through a nozzle with a given orifice size can be metered by setting the liquid pressure in the circuit before the nozzle. The Bernoulli equation, which describes the conservation of energy in a flowing liquid, can be applied to the liquid flow at two points of the nozzle body: one in the nozzle chamber before entering the nozzle orifice (point 1 in figure 12) and the other at the outlet of the orifice (point 2 in figure 12). Neglecting the energy losses due to viscous friction, the Bernoulli equation gives:

$$p_1 + \frac{1}{2}\rho v_1^2 + \rho g z_1 = p_2 + \frac{1}{2}\rho v_2^2 + \rho g z_2$$

where p_1 = absolute pressure of the liquid in the circuit
p_2 = atmospheric pressure
ρ = density of the liquid
v_1 and v_2 = mean velocities of the liquid before entering the orifice and just after it
g = acceleration due to gravity
z_1 and z_2 = vertical positions of the two considered points

From flow continuity, it is also obvious that:

$$q_n = A_1 \cdot v_1 = A_2 \cdot v_2$$

where q_n = flow rate through the nozzle
A_1 and A_2 = area of sections of the nozzle chamber and orifice, respectively

Due to the tiny diameter of the orifice, the fluid velocity v_2 in the orifice is much larger than that in the chamber v_1, which can be neglected in the equation. Moreover, due to the small distance between the two points, we can consider $z_1 \cong z_2$. The Bernoulli equation for the nozzle then simplifies as:

$$p \cong \frac{1}{2} \rho \left(\frac{q_n}{A}\right)^2$$

that can be rearranged, leading to the nozzle equation:

$$q_n = 1.9 \, c_d A_o \sqrt{\frac{2p}{\rho}} \qquad (14)$$

where q_n = flow rate discharged by the nozzle (L min^{-1})
    1.9 = a constant resulting from units adjustment
    c_d = discharge coefficient that accounts for the losses due to viscous friction through the orifice = <1 (decimal) (typically proportional to v_2^2)
    A_o = area of the nozzle orifice (mm^2)
    p = operating pressure of the circuit (kPa), i.e., $p = p_2 - p_1$ the differential pressure to the atmosphere

In practical applications, equation 14 is used in the form:

$$q_n = k_n \sqrt{p} \qquad (15)$$

where k_n is a nozzle-specific efflux coefficient that incorporates its construction characteristics and viscous losses. The value of k_n (commonly in the range of 0.03 to 0.2 L min^{-1} kPa$^{-1/2}$) can be derived from flow-pressure tables provided by the nozzle manufacturer.

Equation 15 shows that the discharged volume rate of pesticide-water mixture can be varied by adjusting the circuit pressure p. Increasing the pressure will increase the flow rate and decrease the spray droplet size simultaneously. However, there is usually a limited working range of pressure (depending on nozzle type, this can be from 150 kPa up to 800 kPa, rarely above) because outside that range, the spray droplets will be either too large or too small. In this working range of pressure, the flow rate increases proportionally to the square root of pressure; if larger changes in discharge rate are needed, a nozzle with a different orifice area (i.e., different k_n) has to be selected.

Sprayer Application Rate Metering

For a required application rate AR (corresponding to a defined dose of pesticide), the sprayer volume rate, q, to be discharged in the field has to be set to a value computed by applying equation 10b, which includes the operating speed and width of the machine. By dividing the total outflow rate q

by the number of nozzles equipping the sprayer, the nozzle flow rate q_n is obtained.

Once an appropriate nozzle is chosen (i.e., a nozzle able to deliver q_n within the usual working range of pressures), the circuit pressure has to be fine tuned, by means of the control valve, until the liquid pressure value (read on the pressure gauge) is the one obtained by solving equation 15 for p and using the k_n value from the nozzle manufacturer, i.e.:

$$p = \left(\frac{q_n}{k_n}\right)^2 \tag{16}$$

This relationship is also used in sprayer electronic controllers to achieve a constant application rate as the sprayer speed varies, or to adjust the application rate for different areas in the fields.

Applications

The concepts and calculations discussed above are widely used to design crop production equipment, and also for the adjustment and management of equipment to suit local conditions on individual farms.

Tillage Equipment

Plows are used for deep tillage operations and are unique in soil movement as they invert soil to be almost upside down, as shown in figure 13. Disc plows cultivate the soil in shallow layers aiming at weed elimination, loosening the soil and uprooting crop plants remaining after harvest.

In contrast, PTO-driven implements (such as shown in figure 14) cultivate the soil more intensively, breaking it into smaller pieces. The intensity is controlled by the axle rotation speed and the tractor speed resulting in a bite length as pointed out by equation 4. Powered implements use the engine power of the tractor more efficiently because slip of wheels due to non-optimal track conditions in the fields are avoided. They are smaller than primary tillage machines in length and weight and are therefore appropriate for combining with other tillage tools (e.g., rollers) or seed drills.

Some other implements used on farms for primary and secondary tillage are illustrated in figure 15.

Figure 13. Tractor-mounted moldboard plow working in the field.

Figure 14. Rotary power tiller as an example of a PTO-driven tillage implement.

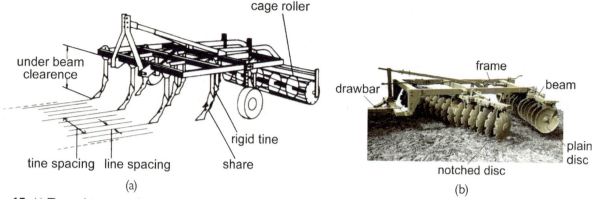

Figure 15. (a) Tine cultivator with tines line spacing and tine spacing; (b) disc cultivator in A-type formation to compensate lateral forces.

Planting Equipment

The most common technique in sowing seeds is drilling with metering by wheels for each row as shown in figure 16. Each row needs a metering wheel and a share. The hopper supplies all rows and extends over the entire working width. The metering wheels grasp seeds from the hopper bottom and transport them over a bottom flap to drop them via the seed tubes into the share and from there into the soil. As a result, the seeds are not distributed in the row with constant distances but are placed randomly. The function of the marker (figure 16a) is to guide the tractor in the subsequent path.

Centralized hoppers can increase capacity of seeding machines. The centralized hopper has only one metering wheel, under the conical hopper bottom. An air stream conveys the grain via a distributor to the shares using flexible pipes.

Seeds can be sown in well prepared soil (after secondary tillage), which is the regular case (figure 17), or under other soil conditions when minimum tillage or no till is applied. Minimum tillage cultivates the soil without deep intervention (e.g., plowing) and no tillage means seeding without any tillage manipulation of the soil. A typical part of the seeding machine is the hopper for fertilizer (between tractor and seeder; figure 17b). This combination offers fertilization and seeding in the same pass.

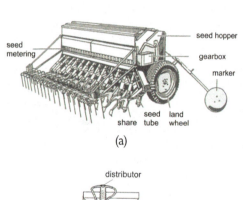

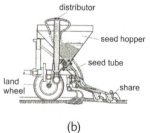

Figure 16. (a) Regular seed drill; (b) pneumatic seed drill.

Machines that plant potatoes, transplant seedlings, and so on, have mechanisms that are quite varied, depending upon what is to be planted into the soil. One example for potatoes is displayed in figure 18a. A chain with catch cups passes through a pile of potatoes in the hopper and picks up potatoes. If there is more than one potato on a cup, the excess potatoes fall off in the horizontal section. The potatoes are then transported down and placed at a constant separation distance in the open furrow.

A rice transplanter is displayed in figure 18b. The seedlings are kept in trays gliding down to the transplanting mechanism. A crank arm for each row will pick out a single seedling and place it in the soil.

(a)   (b)

Figure 17. (a) Tractor-mounted seed drill working in a well prepared soil with small wheels for recompaction of the cover soil after embedding the seeds (wheat); (b) precision seeder sowing maize (corn) in well prepared soil with larger recompaction wheels.

Fertilizer Distribution Equipment

The most common fertilization equipment is the centrifugal spreader (figure 19). This machine is often powered by a tractor through the PTO shaft and often mounted on the tractor's three point-hitch. Large units (hopper capacity above 1500 kg) can have their own wheels and be pulled by a tractor, or be mounted on trucks. The fertilizer granules flow by gravity, with aid of an agitator, from two outlets on the hopper bottom. The outlets area is adjustable through a sliding gate, which meters the mass outflow Q (kg min^{-1}) to control the rate of fertilizer application to the field.

Under the outlets, the metered fertilizer drops on rotating discs (30 to 50 cm in diameter) that impart a centrifugal force on the fertilizer granules, thus distributing them at distances that can reach 50 m. Nevertheless, the working width of centrifugal spreaders is typically 18–24 m and rarely higher than 30 m. Centrifugal spreaders do not uniformly deposit the fertilizer across the working width, but rather with a triangular pattern that requires a partial overlap between two adjacent passes to obtain a uniform transversal distribution within the field. Field speeds typically range from 8 to 12 km h^{-1}, but smooth ground conditions can enable applications faster than 15 km h^{-1}.

Liquid Fertilizer Distributors

Use of liquid mineral fertilizers is rather limited in Europe, except in vegetable crops where the nutrient solution is distributed by a sprayer (see next section) or, much more frequently, in association with irrigation through micro-irrigation systems (fertigation). On the other hand, in North America, fertilization with liquid anhydrous ammonia is very common due to its high nitrogen content (82%) and low cost. Non-refrigerated anhydrous ammonia is applied from high pressure vessels, and it has to be handled with care to prevent hazardous situations. Equipment for application (figure 20) includes

Figure 18. (a) Potato transplanter with device to complete the filling of cups; (b) rice transplanter with device moving the seedlings from the tray in the soil.

Figure 19. Centrifugal fertilizer spreader mounted on the three-point hitch of a tractor.

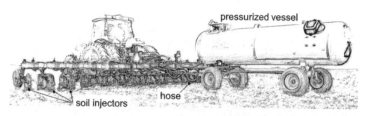

Figure 20. Trailed equipment for anhydrous ammonia soil injection.

Figure 21. Double-axle slurry tanker with soil applicators.

Figure 22. Boom sprayer mounted on the three point-hitch of a tractor

injectors mounted on soil-cutting knives spaced 20–50 cm apart that reach a depth of 15–25 cm in the soil. When delivered in the soil, the ammonia turns from liquid into gas that reacts with water and rapidly converts to ammonium made available to plant roots. This molecule strongly adheres to mineral and organic matter particles in the soil, helping to prevent gaseous or leaching losses.

Slurry Tankers

A slurry tanker (figure 21) is commonly used for distributing organic fertilizer in areas with livestock. The tanker is a trailed, massive piece of equipment mounted on a single or double axle frame (or three axles for tank capacity above 20 m^3) equipped with wide wheels (up to 800 mm) to reduce soil compaction. In vacuum tankers, the stainless steel tank is pressurized at 150–250 kPa for spreading by compressed air, which is pumped in. During tank filling, the pump produces a negative pressure difference (vacuum) with the atmospheric pressure, enabling the slurry to be sucked into the tank by a flexible pipe. Slurry flow can also be obtained with direct slurry pumping by a multiple-lobe pump.

Traditional distribution from a slurry tanker involves a deflector or splash plate mounted on the back of the tank. The slurry impacts the plate and thus is spread over an umbrella pattern covering a width of 4–8 m. Splash plates have been banned by legislation in some countries due to odor emission and nutrient losses (e.g., by ammonia volatilization), so they have been replaced by soil applicators. Soil applicators have multiple hoses mounted on a horizontal boom ending with trailed openings, spaced about 20–30 cm apart, that deposit the slurry flowing through the hose directly on the soil. The soil applicator can also be an injector, made of a tine or a vertical disc tool, that makes a groove in the ground where the slurry is injected at depths ranging from 5 cm (for meadows) to 15–20 cm (tilled soil).

To obtain a uniform distribution of slurry flow among the multiple hose lines, soil applicators require the adoption of a homogenizer. This is a hydraulically-driven shredding unit that processes the slurry with rotating blades to cut fibers and clogs to ensure the regular and even feeding of all the hoses connected to the injectors.

Sprayers

In addition to the common functional parts of sprayers (tank, pump, valves, boom, nozzles), sprayers are manufactured in a wide variety of types for specific crops, various application techniques, environmental regulations, purchase costs, etc. CIGR (1999) provides information about various types of sprayers.

Boom sprayers (figure 22) are the main type used for protection treatments on field crops (e.g., cereals, vegetables, and leguminous crops). They are named for the wide horizontal boom where nozzles are mounted. Booms often range from 8 to 36 m (and sometimes more) in width, with a height from the soil adjustable from 30 cm to more than 150 cm to ensure a good spray pattern at the level of the target. The boom is generally self-leveling to reduce travel undulation and provide more uniform spray application.

Nozzles are mounted on the booms with a typical spacing of 50 cm, although the spacing may range from 20–150 cm depending upon the specific application and the type of nozzle. The most used nozzle on boom sprayers is the fan type that can produce a wide spectrum of droplet size, from medium-fine to coarse spray, at low pressure (150–500 kPa), meeting most field crop spraying requirements.

A boom sprayer is commonly mounted on a tractor by the three point-hitch, or in the case of sprayers with large capacity tanks (above 1 m³), may be a trailed unit pulled by a tractor or self-propelled. Operating speed can vary, largely with field conditions and type of treatment, but during accurate protection treatments a speed range of 7–10 km h⁻¹ is typical.

Examples

Example 1: Work rate and timeliness of row-crop planting

Problem:
A farmer has a six-row planter for planting maize with a row spacing of 75 cm. The farmer wants to know the field capacity of the planter and whether it can successfully plant 130 ha within five working days. If not, what size planter could do this task?

Assumptions:

- Forward speed, s, = 9 km h⁻¹. This is a typical value that depends on the seedbed (firmness, levelness, residue, etc.) and the characteristics of the equipment
- Field efficiency, e_f, = 0.65. This typical value allows for non-planting times, such as filling the planter with seeds and turning at the end of rows.
- Five working days. This is given, but is very dependent upon the weather.
- Eight hours per day of effective field time. This is the time that the planter is available for field work and does not include time for machine preparation, transfer to fields, operator breaks, and other non-planting activities.

Solution:
The first step is to calculate the field capacity, C_a, using equation 2:

$$C_a = 0.1\, e_f\, w\, s \qquad (2)$$

We are given e_f and s. The planter's operating width, w, can be calculated as:

$$\text{number of rows} \times \text{width per row} = 6 \text{ rows} \times 75 \text{ cm row}^{-1} \times (\text{m} / 100 \text{ cm}) = 4.5 \text{ m}$$

Substituting the values into equation 2:

$$C_a = 0.1 \times 0.65 \times 4.5 \text{ m} \times 9 \text{ km h}^{-1} = 2.63 \text{ ha h}^{-1}$$

Therefore, the planter is capable of planting 2.63 ha every hour. If the planter is used to plant on five days for eight hours on each day, the area planted in that time is:

$$A = (2.63 \text{ ha h}^{-1}) \times (5 \text{ days}) \times (8 \text{ h day}^{-1}) = 105.2 \text{ ha}$$

That is less than the required 130 ha. Perhaps the farm staff will have to work more hours, but one option for the farmer would be to get a larger planter, which may require a larger tractor. The following calculations help the farm manager select equipment and manage its use.

The field capacity of the new planter needs to be:

$$C_a \geq (130 \text{ ha}) / (40 \text{ h}) = 3.25 \text{ ha h}^{-1}$$

Then, by rearranging equation 2 the minimum operating width can be computed:

$$w \geq (3.25 \text{ ha h}^{-1})(10) / (0.65 \times 9 \text{ km h}^{-1}) = 5.56 \text{ m}$$

This width corresponds to a number of rows:

$$Nr \geq (5.56 \text{ m}) / (0.75 \text{ m row}^{-1}) = 7.41 \text{ rows}$$

Therefore, the farmer should get an 8-row planter (i.e., the next market size ≥ 7.41) to accomplish the planting of 130 ha within 40 hours of work.

Example 2: Draft force while plowing

Problem:
When designing the frame and hitch of a plow, an engineer needs to know the draft force to ensure that the frame and hitch have enough strength. The draft force also affects tractor selection, since the draft force and speed determine the required pulling power. Determine the draft force needed to pull the plow at a speed of 7 km h^{-1} given the following information about the plow:

- 4-share plow
- 1 gauge wheel
- 5000 N weight on gauge wheel
- 0.15 gauge wheel rolling resistance factor
- 40 cm furrow width

- 30 cm furrow depth
- 5 N cm⁻² static factor
- 0.21 N s² m⁻² cm⁻² dynamic factor

Solution:

Calculate the draft force using equation 3:

$$F_z = n F_v \rho_R + i k w_f d + i \varepsilon w_f d v^2 \tag{3}$$

where F_z = draft force (N)
n = number of gauge wheels = 1
F_v = vertical force = 5000 N
ρ_R = rolling resistance = 0.15
i = number of moldboards or shares = 4
k = static factor = 5 N cm⁻²
w_f = width of furrow = 40 cm
d = depth of furrow = 30 cm
ε = dynamic factor = 0.21 N s² m⁻² cm⁻²
v = traveling speed = 7 km h⁻¹

Draft force F_z = 1 × 5000 × 0.15 + 4 × 5 × 40 × 30 + 4 × 0.21 × 40 × 30 × 7 = 31,806 N

Example 3: Length of a rotovator (rotary tiller) bite

Problem:

Determine the bite taken by each blade on the rotary tiller with these characteristics:

- rotary tiller turning at 240 revolutions per minute
- travelling at 5 km h⁻¹
- 4 blades on each tool assembly

Solution:

Use equation 4:

$$B = \frac{v \; 10{,}000}{n \; z \; 60} \tag{4}$$

where B = bite length
v = travel speed = 5 km h⁻¹
n = rotary speed = 240 min⁻¹
z = number of blades per tool assembly = 4

Bite length, B = 5 × 1000 / (240 × 4 × 60) = 8.68 cm

Each blade takes an 8.68 cm bite. The size of this bite will affect the properties of the tilled soil.

Example 4: Nitrogen fertilization with a centrifugal spreader

Problem:

A test was conducted to determine if nitrogen fertilizer was being applied uniformly at the target application rate. The situation is described by the following:

- centrifugal spreader with working width of 18 m
- travel speed of 9 km h^{-1}
- desired nitrogen dose of 70 kg$_N$ ha^{-1}
- calcium ammonium nitrate is 27% nitrogen
- spreader hopper holds 1000 kg of calcium ammonium nitrate
- spreader tested with 50 cm by 50 cm trays collecting applied fertilizer
- figure below shows the amount of fertilizer that was collected in each tray while testing the spreader

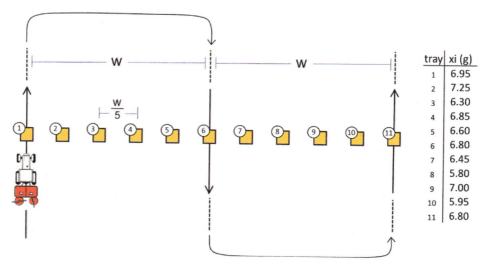

tray	xi (g)
1	6.95
2	7.25
3	6.30
4	6.85
5	6.60
6	6.80
7	6.45
8	5.80
9	7.00
10	5.95
11	6.80

Analyze the collected data to determine the following:

(a) spreader flow rate (kg/min) of calcium ammonium nitrate to achieve desired nitrogen dose
(b) time between fillings of the hopper
(c) average application rate and coefficient of variation from the test

Solution:

(a) By applying equation 9, the amount of calcium ammonium nitrate needed to achieve 70 kg$_N$ ha^{-1} is:

$$D = c_{AC} \, AR \qquad (9)$$

where D = dose of application = 70 kg$_{AC}$ ha^{-1}
c_{AC} = content of active compound in the raw material = 0.27 kg$_N$ kg^{-1}
AR = application rate

Rearranging and using the given information,

$$AR = (70 \text{ kg}_N \text{ ha}^{-1}) / (0.27 \text{ kg}_N \text{ kg}^{-1}) = 259.3 \text{ kg ha}^{-1}$$

This corresponds to a flow rate of the fertilizer (equation 10a):

$$Q = AR\, w\, s / 600 \qquad (10a)$$

$$Q = (259.3 \text{ kg ha}^{-1}) \times 18 \text{ m} \times 9 \text{ km h}^{-1} / 600 \text{ (min h}^{-1}\text{) (m km ha}^{-1}\text{)} = 70 \text{ kg min}^{-1}$$

Therefore, the flow out of the spreader must be adjusted to 70 kg min^{-1}.

(b) The time between fillings is the time it takes to spread all the fertilizer from the spreader:

$$t = 1000 \text{ kg} / (70 \text{ kg min}^{-1}) = 14.3 \text{ minutes}$$

The hopper must be refilled every 14.3 minutes.

(c) The average amount applied is found by summing the amounts of fertilizer in the trays and dividing by the number of trays:

$$\bar{x} = (6.95 + 7.25 + 6.30 + \ldots + 6.80)/11 = 6.62 \text{ g}$$

The mean application rate can be found by dividing that amount by the surface area of a tray:

$$\text{Mean } AR = \bar{x} / (\text{area of tray})$$

$$= (6.62 \text{ g}) \times (\text{kg}/1000 \text{ g}) / [(0.5 \text{ m}) \times (0.5 \text{ m}) \times (\text{ha}/10000 \text{ m}^2)] = 264.8 \text{ kg ha}^{-1}$$

That is close to the desired rate, but represents an error of:

$$[(264.8 - 259.3)/259.3] \times 100\% = 2.1\% \text{ error}$$

The uniformity of distribution is quantified by the coefficient of variation (CV) of the collected material as shown by equation 6:

$$CV = \frac{\sqrt{\dfrac{\sum (x_i - \bar{x})^2}{N-1}}}{\bar{x}} 100\% \qquad (6)$$

where CV = coefficient of variation (%)
  N = number of measured samples
  x_i = amount of fertilizer collected in each tray (g)
  $\bar{x}$ = mean amount (g)

That is, by using the appropriate values given in the fertilizer test figure:

$$CV = \sqrt{\frac{(6.95-6.62)^2 + (7.25-6.62)^2 + (6.30-6.62)^2 + ... + (6.80-6.62)^2}{11-1}}$$

$$\times \frac{100\%}{6.62} = 6.2\%$$

Since a coefficient of variation under 10% is considered good, the field test shows that the spreader is performing satisfactorily.

Example 5: Sprayer pressure setting

Problem:

A fungicide treatment has to be sprayed to a crop at an application rate, $AR = 250$ L ha^{-1}. For this kind of treatment the farmer mounts nozzles that deliver a flow $q_n = 1.95$ L min^{-1} at a circuit pressure of 400 kPa. (This information is provided by the nozzle manufacturer.) Determine the proper pressure to be set in the sprayer circuit in order to distribute the fungicide at the desired application rate.

Assumptions:

- boom width, w, = 24 m with a typical nozzle spacing, d, = 50 cm
- forward speed, s, = 8 km h^{-1}, usual for fungicide treatments (depends on wind conditions)

Solution:

The first step is to calculate the volume rate q (L min^{-1}) required to distribute the application rate by applying equation 10b:

$$q = AR\ w\ s\ /\ 600 \tag{10b}$$

Substituting the given values into the equation:

$$q = 250\ \text{L ha}^{-1} \times 24\ \text{m} \times 8\ \text{km h}^{-1}\ /\ 600 = 80\ \text{L min}^{-1}$$

The number of nozzles equipping the boom is:

$$(24\ \text{m}) / (0.50\ \text{m nozzle}^{-1}) = 48\ \text{nozzles}$$

The required flow rate per nozzle q_n is:

$$q_n = (80\ \text{L min}^{-1}) / (48\ \text{nozzles}) = 1.67\ \text{L min}^{-1}$$

In order to choose the pressure setting to obtain the desired flow of 1.67 L min^{-1} we need to calculate, for these nozzles, the discharge coefficient k_n of equation 15:

$$q_n = k_n \sqrt{p} \tag{15}$$

By substituting the values provided by the nozzle manufacturer ($q_n = 1.95$ L min^{-1}, $p = 400$ kPa) we find:

$$k_n = \frac{1.95 \text{ L min}^{-1}}{\sqrt{400 \text{ kPa}}} = 0.0975$$

Then, by equation 16 we compute the set value of the sprayer circuit pressure:

$$p = \left(\frac{q_n}{k_n}\right)^2 \tag{16}$$

$$p = \left(\frac{1.67}{0.0975}\right)^2 = 293$$

Thus, the metering valve has to be adjusted until the circuit pressure reads 293 kPa (2.93 bar) on the pressure gauge.

Image Credits

Figure 1. Oberti, R. (CC By 4.0). (2020). Typical operations involved in growing field crops.
Figure 2. Oberti, R. (CC By 4.0). (2020). The field capacity of a machine.
Figure 3. Schulze Lammers, P. (CC By 4.0). (2020). (a) Moldboard plow body. (b) Chisel tine of a cultivator. (c) Disc plough.
Figure 4. Schulze Lammers, P. (CC By 4.0). (2020). Seed distribution (a) drilled seed, (b) band seed, (c) broadcasted seed, (d) precision seed.
Figure 5. Schulze Lammers, P. (CC By 4.0). (2020). Studded seed wheel for metering seeds, with bottom flap for adjustment to seed size.
Figure 6. Schulze Lammers, P. (CC By 4.0). (2020). (a) Seed deposition, drilled. (b) Frequency of seed distances, drilled.
Figure 7. Schulze Lammers, P. (CC By 4.0). (2020). Fluted pipe with plate for distribution of seeds along the circumference into the seed tubes, and cell wheel in detail.
Figure 8. Schulze Lammers, P. (CC By 4.0). (2020). Precision seeder singling seed grains for seed placement with definite spacing, (a) mechanical singling by cell wheel. (b) pneumatic singling device with cell wheel.
Figure 9. Oberti, R. (CC By 4.0). (2020). Reducing droplet size.
Figure 10. Oberti, R. (CC By 4.0). (2020). Schematic diagram of a sprayer.
Figure 11. Mancastroppa, S. (CC By 4.0). (2020). Diaphragm pump.
Figure 12. Oberti, R. (CC By 4.0). (2020). Hydraulic nozzle operation.
Figure 13. Schulze Lammers, P. (CC By 4.0). (2020). Tractor mounted moldboard plow working in the field.
Figure 14. Schulze Lammers, P. (CC By 4.0). (2020). Rotary tiller as an example of a PTO-driven tillage implement.
Figure 15. Schulze Lammers, P. (CC By 4.0). (2020). (a) Tine cultivator with tine line spacing and tine spacing. (b) Disc cultivator in A-type formation to compensate lateral forces.
Figure 16. Schulze Lammers, P. (CC By 4.0). (2020). (a) Regular seed drill. (b) Pneumatic seed drill.
Figure 17. Schulze Lammers, P. (CC By 4.0). (2020). (a) Tractor mounted seed drill working in a well prepared with small wheels for recompaction of cover soil after embedding the seeds (wheat). (b) Precision seeder sowing maize in well prepared soil with larger recompaction wheels.

Figure 18. Schulze Lammers, P. (CC By 4.0). (2020). (a) Potato transplanter with device to complete the filling of cup grippers. (b) Rice transplanter with device moving seedlings from the tray in the soil.
Figure 19. Mancastroppa, S. (CC By 4.0). (2020). Centrifugal fertilizer spreader.
Figure 20. Mancastroppa, S. (CC By 4.0). (2020). Equipment for anhydrous ammonia.
Figure 21. Mancastroppa, S. (CC By 4.0). (2020). Slurry tanker.
Figure 22. Mancastroppa, S. (CC By 4.0). (2020). Boom sprayer.
Example 4. Oberti, R. (CC By 4.0). (2020).

References

ASABE Standards. (2017). ANSI/ASABE S572.1: Spray nozzle classification by droplet spectra. St. Joseph, MI: ASABE.

CIGR. (1999). Plant production engineering. In B. A. Stout & B. Cheze (eds). *CIGR handbook of agricultural engineering* (vol. 3). St. Joseph, Michigan: ASAE.

Gorjachkin, W. P. (1968). *Sobranie socinenij* (vol. 2). Moscow: Kolow Press.

Heege, H. J. (1993). Seeding methods performance for cereals, rape, and beans. *Trans. ASAE*, 36(3): 653-661. https://doi.org/10.13031/2013.28382.

Müeller, J., Rodriguez, G., & Koeller, K. (1994). Optoelectronic measurement system for evaluation of seed spacing. AgEng '94 Milano Report N. 94-D-053.

Schilling, E. (1962). *Landmaschinen* (2nd ed., p. 288).

Grain Harvest and Handling

Tim Stombaugh
Biosystems and Agricultural Engineering
University of Kentucky
Lexington, Kentucky, USA

KEY TERMS		
Performance	Efficiency	Dissociation
Productivity	Functional processes	Separation
Quality	Engagement	Transport

Variables

Note about units: In this list of variables, dimensions of variables are given. In the text, variable definitions include dimensions as well as example SI units for illustration.

ε = strain
η_v = volumetric efficiency of the conveyor
θ = angular position of rotating support, or connecting, arms
ρ_{air} = density of the air in mass per volume
ρ_{grain} = density of the grain
σ = stress in units of force per unit area
σ_b = bending stress in the connective tissue in force per unit area
σ_t = tensile stress in the connective tissue in force per unit area
σ_y = yield stress in force per unit area
ω = rotational speed
a = acceleration of the particle in the direction of the resultant force in length per time squared
A = cross-sectional area of connective tissue or structural member
A_{part} = characteristic area of the particle
C_a = effective field capacity in area per unit time
C_d = drag coefficient of the particle

C_m = material capacity in weight or volume per unit time
d = diameter of shaft
D = outside diameter of the flighting
E = modulus of elasticity in force per unit area
E_f = field efficiency
E_h = harvest efficiency
F = force in the member
F_d = drag force
F_g = gravitational force
F_r = resultant force on the particle
F_t = tensile force acting on connective tissue
g = gravitational constant in length per unit time squared
h = height that the material is lifted
I = moment of inertia of the connective tissue cross section in mass x length squared
L = length of a structural member
L_h = total harvest loss
L_s = separation loss
L_{sh} = shatter losses
L_{th} = threshing loss
m = mass of the particle
M = bending moment in force x distance
n = number of paddles that are discharged per unit time
Q_a = actual volumetric flowrate of grain in volume per unit time
Q_t = theoretical flowrate in volume per unit time
P = pitch length of the flighting
P_g = power required to overcome gravity in (force x distance) per unit time
r = length of rotating support, or connecting, arms
R_w = total weight that the cart wheels and axles must support
s = field speed in distance per unit time
T_a = actual completion time
T_t = theoretical completion time
v = velocity in length per time
v_f = forward speed
v_{part} = velocity of the particle relative to the air
v_h = horizontal component of tangential velocity
v_t = tangential velocity
v_v = vertical component of tangential velocity

V = volume of material carried by a single paddle

w = machine width

x_g = grain center of gravity

y = perpendicular distance to neutral axis

y_a = actual harvested, or recovered, yield of grain in weight or volume per unit area

y_p = potential yield of grain in the field in weight or volume per unit area

y_{ph} = pre-harvest yield loss in weight or volume per unit area

y_t = total yield that the plants actually produced in weight or volume per unit area

Introduction

One unique skill that biosystems engineers must develop is the ability to understand how mechanical systems interact with biological systems. This interaction is very prevalent in the design of machinery and systems for harvesting grains such as corn (maize, *Zea mays*), soybean (*Glycine max*), wheat (*Triticum*), or canola (*Brassica napus*). The machines must traverse through a field on a biologically active soil to engage the plants growing in that field. The variability in plant and soil properties (e.g., maturity, moisture content, and structural integrity) within a field can be extensive. This variability presents a challenge to design engineers to conceive machines that can accommodate this variability and provide the machine operator with the flexibility needed to properly engage the plants. The goal of this chapter is to lay the engineering foundation needed to design machinery systems for harvesting grain crops.

Outcomes

After reading this chapter, you should be able to:

- Identify the basic functional processes needed to harvest grain
- Describe the basic engineering principles governing grain harvest machinery design
- Quantify machine performance and design basic machine components using basic engineering principles

Concepts

One key to becoming a great engineer is the ability to identify and understand the core problem to be solved. Too often, engineers focus on improving current solutions to problems rather than looking for better solutions. In the case of grain harvest, the engineer might be tempted to focus on ways to improve the grain combine (figure 1), which is the machine most commonly used to harvest grain. The most unique and creative engineering solutions will often come

Figure 1. Typical grain combine with grain table header harvesting a soybean crop.

only when the engineer focuses on identifying the fundamental problem to be solved.

With grain harvest, the core challenge is to recover a certain fraction or fractions of the plants in a grain crop that is grown in large fields. The fraction that is to be retrieved may vary by plant and by situation. In corn (maize, *Zea mays*) harvest, for example, the most commonly harvested plant fraction is the kernel, which is used in a variety of products including food, sugar, and biofuel production. For fresh market sweet corn harvest, the whole ear is recovered with the husks intact. In some animal production systems, the whole ear without husks is recovered for animal feed; in other animal production systems, the entire plant is harvested and ensiled for feed. In these examples, the maturity and moisture content of the plant material may be drastically different if the corn is being recovered for sugar production, animal feed, or human consumption. A further challenge may exist where multiple plant fractions are harvested for different purposes. In industrial hemp (cannabis) production, for example, the seeds of the plant might be recovered for oil or food production, and the plant stems might be recovered separately for fiber production. The plurality of production streams will not be independent of each other and must be considered in the design of the mechanization solution. This chapter focuses on systems where only the grain is recovered.

Performance

The performance of a grain harvesting machine or system can be measured using three general metrics: productivity, quality, and efficiency.

Productivity

The productivity of a harvesting machine or operation is a measure of how much useful work is accomplished. As described in ASABE Standard EP496.3 (2015a), it can be quantified using two primary metrics. First, it can be measured on an area basis indicating how much of the area of a field was covered per unit time. This metric is expressed as the effective field capacity (C_a) and can be calculated as:

$$C_a = s w E_f \qquad (1)$$

where C_a = effective field capacity in area per unit time (m² h⁻¹)
s = field speed in distance per unit time (m h⁻¹)
w = machine width (m)
E_f = field efficiency (decimal)

Second, productivity can be measured on a material basis indicating how much of the grain is recovered per unit time. Material capacity (C_m) is related to the area capacity by:

$$C_m = C_a y \qquad (2)$$

where C_m = material capacity in weight or volume per unit time (m³ h⁻¹)
y_a = recovered (harvested) crop yield in weight or volume per unit area (m³ m⁻²)

Material capacity can be reported on either a volume or mass basis with appropriate density conversion. In international trade, grain quantity is typically reported in metric tons. In U.S. grain production, grain quantity is commonly measured using units of bushels. While a bushel is technically a volume measurement equaling 35.239 L, in grain production it is a unit that reflects a standardized weight of grain at a particular moisture content specified for that grain. The standardized weights of a grain bushel for some common crops are listed in table 1.

Table 1. Bushel weight of common grain crops at standardized moisture content.

Commodity	Moisture Content (%)	Weight (lb/bushel)	Weight (kg/bushel)
Corn *(Zea mays)*	15.5	56	25.40
Soybean *(Glycine max)*	13	60	27.22
Sunflower *(Helianthus annuus)*	10	100	45.36
Wheat *(Triticum aestivum)*	13.5	60	27.22

Quality

The second measure of performance of a mechanized grain harvest system is product quality. Ideally, the product (grain) that is recovered is free from any foreign matter and damage, but this is rarely the case. Small pieces of plant material and other foreign matter are often captured with the grain. The machinery can also cause physical damage to the grains as they pass through the different mechanisms. Machine design, as well as crop and operating conditions, can have an effect on foreign matter and damage, which are often referred to collectively as dockage. The term dockage is used because producers generally incur a financial penalty (docked some amount) from the market value of the grain by the buyer if the grain is damaged, contains excessive foreign matter, or is not at the proper moisture content.

Efficiency

The third measure of performance, efficiency, can be quantified in two ways. The first is a time-based field efficiency (E_f) that relates the actual time required to complete a field operation to the theoretical completion time had there been no delays, such as turning around at the ends of the field, machine repair, and operator breaks. Field efficiency is calculated as:

$$E_f = \frac{T_t}{T_a} \tag{3}$$

where T_t = theoretical completion time
T_a = actual completion time

Efficiency can also be measured based on the completeness of the harvest operation. This harvest efficiency (E_h) is a measure of the amount of desirable product that is actually recovered relative to the amount of product that was originally available to the harvesting machine. It is calculated as:

$$E_h = \frac{y_a}{y_p} \tag{4}$$

where y_a = actual yield of grain recovered measured in weight per unit area (kg m^{-2})
y_p = potential yield of grain in the field measured in weight per unit area (kg m^{-2})

The antithesis of harvest efficiency is harvest loss (L_h), which is the amount of grain lost by the harvesting machine per unit area expressed as a percentage of the potential yield. It can be calculated as:

$$L_h = 1 - E_h \tag{5}$$

When focusing on the harvesting operation, the potential yield of the crop is considered to be the harvestable grain that is still attached to the plants. Potential yield does not consider the grains that have fallen from the plants before the machine engages them. If harvest is delayed after the optimum time, potential yield will often decrease due to natural forces causing seeds to fall from the plants. This pre-harvest yield loss (y_{ph}) is the amount of grain that is lost before harvest expressed on a per area basis. It is the difference between the total yield that the plants actually produced (y_t) and the potential harvestable yield:

$$y_{ph} = y_t - y_p \tag{6}$$

There are often strong interrelationships between productivity, grain quality, and efficiency of grain harvest. For example, an increase in productivity could be realized by an increase in speed or field efficiency; however, grain quality and harvest efficiency may be compromised. Finding the fiscally optimum operation point is a challenge to be addressed by both the design engineer and the machine operator. The engineer must understand the needs of the operator and incorporate the appropriate flexibility of control into the design of the machine.

Functional Processes

When considering the design of any mechanization system, the engineer should first carefully consider the potential processes that will have to be undertaken to complete the task. Srivastava et al. (2006) expand on a number functional process that could occur in grain harvest. These processes can be simplified to four main processes:

- Engage the crop to establish mechanical control of the grain.
- Dissociate or break the connection between the individual grains and the plant.
- Separate the grain from all of the other plant material.
- Transport the grain to the proper receiving facility.

Depending on the specific system, these functions could take place in varying order, and some processes may be repeated multiple times. Historically, before mobile grain combines were developed, a harvesting process involved gathering the whole plant from a field and transporting it to a central location where the grains were separated (threshed) from the plant material either by hand or with a stationary machine. With modern harvesting machines, the grain dissociation and separation is accomplished as the machine moves through the field, and the only material that is transported away from the field is the grain. Likewise, some functions may be accomplished multiple times. For instance, there are often several separation stages in a single machine, and the product might be transported multiple times between different mechanisms, temporary storage units, and transportation vehicles before reaching the final destination.

Engagement

The process of engaging the crop can occur in many different ways depending on the particular crop and machine configuration. Often there is some type of mechanism that will grasp or pull the standing crop toward the machine as it moves forward. The grasping mechanism could be mechanical, such as a rotating arm or chain, or it could involve other forces such as pneumatics or gravity. Engagement may include a cutting action that severs the part of the plant containing the grain from the rest of the plant. The grasping and cutting actions usually result in the material being caught on some surface where it can then be moved into the machine.

Dissociation

The dissociation function of a grain harvest process involves physically breaking the connection between the desired particle and the plant. The term threshing is often used to describe the dissociation function, but in some contexts, threshing could also include some separation and cleaning functions. The dissociation performance of a harvesting machine is quantified by the threshing loss, L_{th}, which is the amount of grain that remains connected to the plants expressed as a percentage of the total amount of grain that was presented to the threshing mechanism.

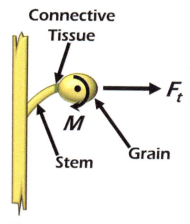

Figure 2. Dissociation forces acting on a grain. F_t is the tensile force and M is the bending force, or moment.

Engineers designing mechanisms to dissociate grain need to understand the basic principles of tensile and bending failure. Most grains are attached to the plant by some kind of stem and/or connective tissue. The connective tissue can often be understood as a cylindrical bundle of connective tissue. Most dissociation mechanisms apply a tensile force, a bending force, or a combination of the two on the seed relative to the plant (figure 2). The goal of the force application is to cause a failure of the connective tissue between the grain and the stem or plant. Failure will occur when the stress in the connective tissue exceeds its ultimate yield stress, which is the stress at which the material will break. Obviously, it is desirable for the dissociation failure to occur as close to the individual grain as possible so that there is no stem or other plant material captured with the grain.

Tensile failure is the mode where the grain is pulled straight away from the stem until the connection fails. The linear tensile force induces stress in the connective tissue that can be calculated by the following equation:

$$\sigma_t = \frac{F_t}{A} \quad (7)$$

where σ_t = tensile stress in the connective tissue measured in force per unit area (N m^{-2})
F_t = tensile force acting on connective tissue (N)
A = cross-sectional area of the connective tissue (m^2)

Bending failure occurs when the grain is rotated relative to the stem inducing bending stress in the connective tissue. Bending force or moment (M) can also be thought of as a rotational torque applied to the grain. Bending stress is described by the following equation:

$$\sigma_b = \frac{M\,y}{I} \quad (8)$$

where σ_b = bending stress in the connective tissue measured in force per unit area (N m^{-2})
M = bending force acting on the connective tissue measured in force × distance (N m)
y = perpendicular distance to neutral axis (m)
I = area moment of inertia of the connective tissue cross section in units of length to the fourth power (m^4)

The moment of inertia is a quantity that is based on the shape of the cross section of the member and is used to characterize the member's ability to resist bending.

Bending stress is a bit more complicated than tensile stress because the stress in the member is not constant across its cross section. Fibers farther from the neutral axis of bending will experience higher stress, which can actually enhance dissociation.

One of the challenges with describing failure of plant materials mathematically is the wide variability that can occur. The strength of the connective tissue is affected by three main factors based on biological properties of the plant. The first factor is plant size. Some plants within a single crop may grow larger than others and may have more connective tissue between the stem and grain. This would correspond to larger area and moment of inertia in equations 7 and 8, which would require more force to reach the ultimate stress.

The second factor affecting connective tissue strength is plant maturity. Most plants will lose their grains naturally when they reach maturity as a mechanism for propagation. Mathematically, this natural dissociation is described by a reduction in the yield strength of the connective tissue. Quite often, this natural maturity state corresponds with the optimum time for grain harvest; however, it is not always possible to harvest at that exact time. Therefore, the failure strength could vary significantly based on actual harvest time.

The third factor affecting the strength of the connective tissue is moisture content. The strength properties of plant material vary greatly with moisture content. Engineers need to understand a number of biological properties of the plant as they are affected by moisture content. For instance, the turgor pressure in a plant is the pressure exerted on the walls of the cells within the plant by the moisture in the cells. As turgor pressure decreases, which is caused by a reduction in moisture content, plants become less rigid. In some plants, this might weaken the plant structure, making dissociation easier, but in others it could make it more difficult to dissociate a grain because the plant material would be more elastic. Depending on weather conditions and solar intensity, turgor pressure can vary greatly throughout a single day, affecting the dissociation of the crop.

Further complicating the mathematical representation of plant strength is the fact that there can be significant variability of plant size, maturity, and moisture content between different regions of a field, between different plants, and even between different grains on a single plant. Engineers need to develop mechanisms that accommodate this variability and give machine operators the flexibility to adapt the machine to the various conditions.

Separation

Once the grains are dissociated from the plant, they must be separated from the rest of the plant material and any other undesirable material. This undesirable material is called material other than grain (MOG). The separation performance of a harvesting machine is quantified by the separation loss (L_s), which is the amount of free grain that cannot be separated from the MOG expressed as a percentage of the total amount of grain that was dissociated from the plants by the machine.

There are two main principles that are typically used to separate grain from MOG. The first is mechanical separation through sieving. A sieve is simply a barrier with holes of a correct size that allows the desired particles to pass through while preventing larger particles from passing, or vice versa, allow smaller undesirable particles to pass through while retaining the desired particles. Some grain sieving mechanisms rely on gravitational forces on the particles to cause them to pass downward through the sieve openings; others utilize centripetal forces of rotating mechanisms to force particles outward through the sieves. Most gravitational sieving mechanisms induce a shaking or bouncing motion on the material to enhance the separation process by facilitating particle motion downward through the mat of material as well as causing motion of the material across the sieve.

Consider the sieve plate connected to the parallel rotating bars as shown in figure 3. This is a classic four-bar linkage mechanism. The sieve plate moves in a circular pattern while maintaining its horizontal orientation. If the design of the length of the rotating arms along with the rotational speed is correct, the material is bounced laterally across the plate. As it bounces, the grains move downward through the mat of material, then through appropriately-sized holes in the plate. The MOG travels across the plate and is deposited off the end of the sieve.

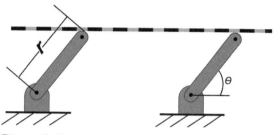

Figure 3. Simple sieve mechanism.

The velocity of the sieve plate (v) can be calculated from the following equation:

$$v = r\omega \qquad (9)$$

where v = velocity of the sieve plate (m s^{-1})
r = length of the rotating support, or connecting, arms (m)
ω = rotational speed of the arms (radians s^{-1})

The velocity of the sieve plate is actually the tangential velocity of the rotating support arms. The direction of this velocity changes sinusoidally as the bars rotate. The vertical (v_v) and horizontal (v_h) components of the velocity can be described with the following equations:

$$v_v = v \cos\theta \qquad (10)$$

$$v_h = v \sin\theta \qquad (11)$$

where θ = the angular position of the arms.

The bouncing motion of a particle is analyzed by considering the momentum of the particle in relation to the upward moving, but decelerating, plate to determine if and when the particle will leave the plate.

The second principle that is used to separate grains from MOG is aerodynamic separation. Quite often, the grain particles are denser and have significantly different aerodynamic properties than the MOG, especially the lighter leaf and hull particles. These differences are exploited to separate the MOG from the grain.

A particle that is moving through any fluid, including air, is subjected to gravity and drag forces (figure 4). Gravity acts downward on the particle and produces the force represented by:

$$F_g = mg \qquad (12)$$

where F_g = gravitational force (N, or m kg s^{-2})
m = mass of the particle (kg)
g = gravitational constant in units of length per unit time squared (9.81 m s^{-2})

Figure 4. Forces affecting particle motion when it is free falling (left) and when subjected to a directed air flow (right).

The drag force acts in the opposite direction of the particle's motion relative to the air. The drag force is calculated by:

$$F_d = 0.5 \rho_{air} v_{part}^2 C_d A_{part} \qquad (13)$$

where F_d = drag force (N, or m kg s^{-2})
ρ_{air} = density of the air in units of mass per unit of volume (kg m^{-3})
v_{part} = velocity of the particle relative to the air in units of length per time (m s^{-1})
C_d = unitless drag coefficient of the particle
A_{part} = characteristic area of the particle (m^2).

The motion of the particle is determined by the vector sum of the two forces and the fundamental motion equation:

$$a = \frac{F_r}{m} \qquad (14)$$

where a = acceleration of the particle in the direction of the resultant force (m s^{-2})
F_r = resultant force on the particle (N, or m kg s^{-2})

The particle trajectory can be described mathematically by integrating the acceleration equation once to get the velocity equation, then a second time to find position as a function of time.

The drag coefficient is a function of many particle factors including its shape and surface texture. Many MOG particles, such as seed hulls and stem particles, have a more irregular shape and surface texture than the grains and, thus, will have a higher drag coefficient. Aerodynamic separation occurs by capitalizing on these drag differences as well as differences in mass between the grain and MOG particles. Particles can be separated if an air stream is directed through the mat of grain and MOG in such a manner that the MOG is directed in a different trajectory than the grains.

Consider an air stream created by a fan blowing straight upward at a falling particle. If the air speed is increased to the point that the drag force equals the gravity force, the particle will be suspended in the air stream. The velocity of the air at this point is, by definition, the terminal velocity of the particle. If the air speed is increased, the particle will move upward; if the air speed is decreased, the particle will move downward.

Consider a mixture of grain and MOG particles being dropped through a directed air stream as illustrated in figure 4. If the air velocity is set slightly below the terminal velocity of the grains, their trajectory will be altered to the right somewhat, but they will continue to move downward. MOG particles that have a much higher drag force and correspondingly lower terminal velocity will be carried more upward and to the right by the air stream moving them out of the grain flow.

Transport

Once the grain is dissociated and separated from the MOG, it must be transported to a receiving station. This is usually accomplished in several steps or stages using a variety of mechanisms. Various types of conveyors are used to move the grain from one part of a machine to another or from one machine to another. At different stages of the process, the grain might be stored or carried in various bulk containers.

The principles involved with designing or analyzing bulk storage or transportation containers are primarily strength of materials. The designer first needs to determine what forces will be produced on the structure by the grain. Free body diagrams are analyzed to determine the magnitude and direction of all forces. One challenge in designing grain harvesting machinery is that the machines are often mobile. As the machines move across the rough terrain typically encountered in agricultural fields, dynamic forces are induced as the grain load bounces. Designers typically utilize a variety of techniques to predict the maximum dynamic loads that could be induced on a structure.

Once the forces are known, the designer then determines what stresses are induced in each structural member by the grain load. Stress (σ) describes the amount of force (F) being carried per unit area (A) of a given structural member:

$$\sigma = \frac{F}{A} \tag{15}$$

where σ = stress in units of force per unit area (N m^{-2})
F = force in the member (N)
A = characteristic area of the member (m^2)

The stress in any part of a structural member cannot exceed the yield stress of the material or permanent damage (deformation) will be incurred. But even if permanent deformation is not induced in a structural member, engineers still need to be concerned about how much a structural member flexes or deflects. The deflection in a member is calculated from strain, which is:

$$\varepsilon = \frac{dL}{L} = \frac{\sigma}{E} \qquad (16)$$

where ε = strain (dimensionless)
L = length of the member (m)
E = modulus of elasticity reported in force per unit area (N m^{-2})

Stress and strain are related by the modulus of elasticity, also known as Young's modulus. The lower the modulus of elasticity, the more deflection a given force will cause in a member. Some deflection can be good in a structure, especially when dynamic forces are involved, because it helps to absorb energy without causing high peak loads. In the case of a machine moving across a rough field, for example, some deflection in the structure can absorb some of the energy caused by uneven terrain and prevent structural failure.

For shorter distance transportation, several different conveying devices can be employed. When designing conveying devices, the designer is primarily concerned about the capacity of the device and the power required to convey the material. Some of the simplest conveying devices utilize paddles or buckets connected to chains (figure 5) to drag or convey the grain. The capacity of paddle conveyors, which is the flow rate of material through the conveyor, is calculated simply by the amount of material carried by each paddle and the number of paddles that pass a point in a given amount of time:

$$Q_a = V\,n \qquad (17)$$

where Q_a = actual flowrate in volume per unit time (m^3 s^{-1})
V = volume of material carried by a single paddle (m^3)
n = number of paddles that are discharged per unit time (s^{-1})

The volume of material that can be carried by the paddles is affected by a number of parameters. Grain properties such as particle shape, size, surface friction and moisture content affect the shape of the pile of grain on each paddle. The slope of the conveyor limits the size of the piles before the grain runs over the top of the paddle and back down the conveyor.

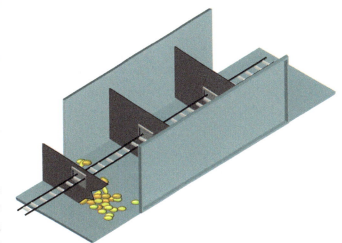

Figure 5. Simple paddle conveyor.

Another conveying device commonly used in grain harvest and handling is a screw conveyor, commonly known as an auger (figure 6). Screw conveyors utilize a continuous helicoid plate, called flighting, attached to a rotating shaft. The capacity of a horizontal screw conveyor that is completely full of grain is the volume displaced by a single rotation of the shaft times the number of rotations in a given unit of time, which can be calculated by:

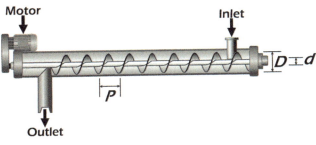

Figure 6. Simple screw conveyor.

$$Q_t = \frac{\pi}{4}\left(D^2 - d^2\right)P\omega \tag{18}$$

where Q_t = theoretical flowrate in volume per unit time (m³ s⁻¹)
D = outside diameter of the flighting (m)
d = diameter of shaft (m)
P = pitch length of the flighting (m)
ω = rotational speed of the shaft (radians s⁻¹)

When the conveyor is inclined, the flighting will no longer be full as the grain will tend to slide down around the flighting. The actual volumetric flowrate (Q_a) can be calculated by:

$$Q_a = Q_t \eta_v \tag{19}$$

where η_v is the volumetric efficiency of the conveyor. Predicting the volumetric efficiency can be very challenging because it is affected by numerous factors, including conveyor slope, rotational speed, grain moisture content, particle size, particle to conveyor friction, and particle-to-particle friction. Because of this complexity, mathematical prediction is usually accomplished with empirical relationships.

The power required to convey the material is affected by gravitational and friction forces. If the grain is lifted any vertical distance, the conveyor must overcome the gravitational force opposing that lift. Power is defined as a force applied over a given distance in a given amount of time (force × distance/time). The force and time components of the gravitational power calculation come from the flow rate of the grain through the conveyor expressed in units of weight per unit of time. The density of the grain can be used to convert volumetric flow rate into a weight flow rate. The distance component of power is simply the vertical distance that the grain is lifted. The gravimetric component of power is:

$$P_g = Q_a \rho_{grain} h \tag{20}$$

where P_g = power required to overcome gravity (W or J/s)
Q_a = actual volumetric flow rate of grain (m³ s⁻¹)
ρ_{grain} = density of the grain (kg m⁻³)
h = height that the material is lifted (m).

The friction component of power can be more complicated to compute. In the case of a paddle conveyor, the grain must be slid along the bottom of the conveyor surface. This friction force can sometimes be predicted quite well from the coefficient of friction between the grains and the surface of the conveyor. If that coefficient of friction gets too large, the forces due to friction on the grains at the interface between the grain pile and the conveyor surface will cause the grains within the pile to begin to move relative to each other. At this point, it becomes more difficult to mathematically describe the energy necessary to overcome these internal friction forces as well as the surface friction.

Frictional forces in screw conveyors are similar. The grain in a full horizontal conveyor slides along the outside wall of the conveyor tube as well as the flighting but does not move as much within the grain mass. As the conveyor is inclined and it is no longer completely full, the amount of motion within the grain mass increases and becomes more critical to the evaluation.

Applications

The most common machine used for grain harvest is the modern grain combine (figure 1). Combines typically have an interchangeable attachment on the front called a header that engages the crop and passes certain fractions of that plant into the combine. The material then passes through a threshing mechanism that dissociates the grains from the plant stems and also performs some separation of grain from MOG. The grain and MOG are then passed through various separating and cleaning mechanisms. The MOG is generally passed longitudinally through the machine and expelled from the back. The clean grain generally moves downward through the machine to a catch reservoir on the bottom. From there, it is moved upward with paddle and/or screw conveyors to a holding tank on the top of the machine. A large screw conveyor is then used to periodically empty the contents of the holding tank into a truck or other vehicle, which transports the grain to a receiving station.

Engagement

Header attachments are used to engage the crop. The two most common types of header attachments on grain combines are the grain table and the corn or row head. Grain tables (figure 1) are typically used in small grain crops such as wheat and soybean. They generally include a large gathering reel to engage the crop and pull it into the header. A cutting mechanism, typically a sickle bar, cuts the plant as it is pulled into the header to gain mechanical control of the grain. Since grain tables can be 12 m wide or wider, cross conveyors move the crop material from the ends of the header to the center where it is fed into the combine.

The height of the cut depends on the crop and the cultural practice of the operation. The threshing and separation processes in the combine are most efficient when the MOG entering the combine is minimized. In soybean, for example, the seed pods can grow very low on the plant stem; therefore, the crop must be cut near the ground to prevent losses. In crops like wheat where the grains grow in a head on the top of the plant stem, the cut height could be just low enough to capture the entire head but minimize the amount of MOG passed into the machine. In some production systems, though, the MOG might be used for animal bedding or bioenergy. In these cases, the header is operated lower so

Figure 7. Combine with attached baler to collect biomass.

Figure 8. Typical row-crop header on a combine.

that more MOG is gathered, passed through the combine and deposited in a narrow line, called a windrow, behind the combine. The windrow can easily be gathered by another biomass harvesting machine in a separate operation. Occasionally, the MOG is gathered by another machine, such as a baler, attached directly to the combine (figure 7).

In corn (maize), the grains are produced toward the middle of the plant. Cutting the plant to capture the ears for threshing would mean the introduction of large amounts of MOG into the harvest stream, hampering threshing and separation performance. Since corn is typically grown in rows spaced 0.5–0.75 m, corn heads are constructed with fingers that pass between the rows so that each row of corn can be engaged individually (figure 8). Long parallel rollers on each side of the row grab the plant stems below the ear and pull them downward as the machine moves forward. Stripper plates above each roller are spaced such that the plants pass down between them, but the corn ears do not. As the plants are pulled downward, the ears are stripped off the plants. Ideally, all of the plant material is pulled down through the header and does not pass into the combine. Depending on stalk condition, some stalk breakage and leaf removal will occur, and that MOG will have to be separated in the combine. Chains with fingers above each stripper plate move the ears and MOG back into the header. Cross conveyors then move material from the edge of the header into the center where it is fed into the combine.

One of the performance measures of any header is its effectiveness in gathering all of the grain from the field into the combine. This engagement process is complicated by the fact that the grains tend to naturally fall off of the plants more easily when the crop is in its optimum harvest condition. Losses by the header are called shatter losses (L_{sh}). They are quantified as a percentage of the potential yield (y_p), i.e., the available yield of the plants.

Shatter losses are affected by crop conditions, including maturity and moisture content. They are also affected by the design and operation of the header. For example, the speed of the gathering reel on grain tables must be matched to the forward speed of the machine. If the reel speed is too high, the plants are beaten aggressively, causing grains to fall off the plants before they can be caught on the header platform. If the reel speed is too slow, the plants could be pushed forward, again knocking grains onto the ground before they enter the header. Depending on crop conditions, the tangential speed of the reel is typically operated at least 25–50% faster than the forward speed of the combine to pull the plants into the header. Some machines utilize sensors and electronic controls to automatically adjust the reel speed to match machine forward speed.

Dissociation

The dissociation function in combines is generally accomplished by rotating cylinders called threshing cylinders. There are two basic configurations of threshing cylinders, distinguished by the direction the material moves through the cylinder. Some cylinders are mounted with their axis of rotation horizontal and perpendicular to the longitudinal axis of the machine. The material enters from one side of the cylinder and exits the other (figure 9). Bars oriented along the outside of the cylinder rub the plant material against the stationary housing around the outside of the cylinder, which is called the concave. The rubbing action affects the dissociation of the grain from the plants. Holes in the concave facilitate a sieving action to separate some of the grain from the longer plant material.

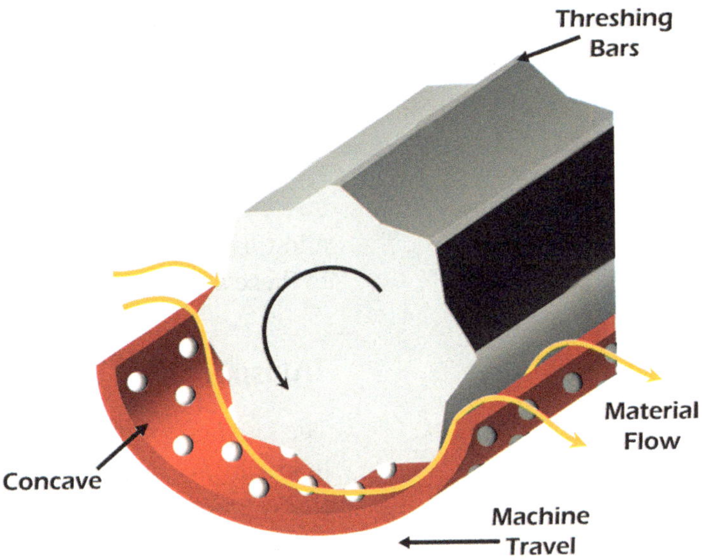

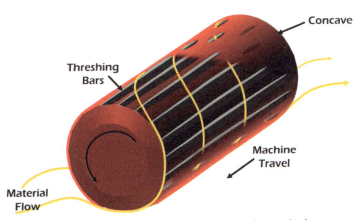

Figure 9. Transverse-mounted (conventional) threshing cylinder.

The other common threshing configuration has the rotating threshing cylinder mounted parallel to the longitudinal axis of the machine. Material enters the end of the cylinder and moves in a helical pattern around and past the cylinder (figure 10). Similar concave structures around the cylinder provide resistance to the flow to induce the dissociation and separation functions.

Threshing effectiveness is measured by the percentage of grains that are dissociated from the plants, the percentage of grains that are damaged during the threshing process, and the amount of MOG break-up. Excess amounts of small MOG particles can hamper separation efficiency since they can be indistinguishable from grains in the separation process. Threshing efficiency and grain damage are affected by plant properties, the design of the cylinder and concaves, and operational adjustments. Machine operators often have real-time control of the cylinder speed as well as the clearance between the cylinder and concave.

Figure 10. Longitudinal-mounted (rotary) threshing cylinder.

Separation

There are two different types of separation systems used in grain combines that are generally associated with the two types of threshing devices. Laterally oriented threshing cylinders generally feed the material stream onto a vibrating separator platform commonly called a straw walker. The oscillating plate is essentially a sieve allowing the smaller particles, including the grains, to fall through the sieve as the MOG is moved back through the machine.

On combines with axially oriented threshing cylinders, the latter portion of the cylinder and concave accomplish initial separation. These rotary separators utilize centripetal forces to separate grains outward through concave openings.

Regardless of the initial separator configuration, most combines pass the grain stream captured from the threshing unit and initial separation unit through an additional multi-stage cleaning sieve. Pneumatic separation is also applied in these sieves to enhance separation of grain from MOG.

Transport

The cleaned grain stream is conveyed from the bottom of the combine to a holding tank on the top of the machine using a combination of paddle and screw conveyors. The holding tanks vary in size with the size of the machine. Depending on the crop and operating conditions, the combine tank could be filled in as little as 3–4 minutes. In some operations, the combine is driven to the edge of the field when the tank is full so that it can be emptied into a truck for transport to a receiving station. This is often considered an inefficient use of a very expensive harvesting machine. Productivity of the harvesting operation is maximized if the combine can be operated as close to continuously as possible.

Combines can be unloaded while they are harvesting if a receiving vehicle can be driven alongside the combine. Over-the-road trucks are not typically used for this operation because of their relatively small tires. Traction in potentially soft soil conditions limits their mobility. Also, there is a concern regarding compaction of the soil in the field. Heavy loads on small tires will compact the soil under the tires causing damage that will affect the performance of future crops in the field.

In-field transport of grain is often accomplished with a grain cart (figure 11). Grain carts are large transport tanks typically pulled by large agricultural tractors. Both the cart and tractor will be equipped with large tires or tracks to reduce the pressure on the soil.

With the use of grain carts, a logistical challenge arises around the best way to get the grain away from the combine to keep it harvesting. Many operations use multiple combines in a field simultaneously. Managers must decide

Figure 11. Typical grain cart receiving clean grain from a combine.

how many grain carts are needed, how big those carts need to be, and how many trucks are needed to get the grain away from the field. Operationally, vehicle scheduling is a challenge to anticipate which combines in a multiple combine fleet must be emptied so that they do not fill up and become unproductive.

Examples

Example 1: Combine harvest efficiency

Problem:
One way to evaluate the harvest efficiency of a combine is to measure losses that occur as a combine moves through the field. This can be done by physically gathering and counting or weighing the grains found at different locations in the combine's operating space.

Consider the combine in figure 12 that was stopped while harvesting a very uniform crop of wheat. Field measurements were taken at three different locations as shown. At each point, a 1 m square area was selected as a representative test area. At point A in front of the combine, all of the standing plants in the test area were carefully cut and hand harvested to determine how much grain was available in the field. After that, the grains that were laying on the ground in that test area were gathered and weighed. At point B, all of the grains found within the test area were gathered and weighed. At point C, which is beyond the discharge trajectory of material being expelled from the back of the combine when it was stopped, all of the grains were collected and weighed separately by those that were still attached to the plants and those that were free. The following are the data collected at each location.

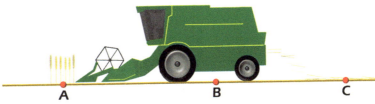

Figure 12. Locations of performance test measurements for a grain combine harvesting wheat.

    Point A:
        335 g unharvested grain
        15 g free grains (grains laying on the ground)
    Point B:
        40 g free grain
    Point C:
        63 g free grain
        14 g grain attached to plant

    Determine the gathering, threshing, and separating efficiencies of this harvest operation.

Solution:

The theoretical or potential yield, y_p, of the crop is the harvestable grain that is still attached to the plants when the combine engages it. In this example, the potential yield is based on the unharvested seed at point A.

$$y_p = \frac{0.335 \text{ kg}}{\text{m}^2} \cdot \frac{10000 \text{ m}^2}{\text{ha}} = \frac{3350 \text{ kg}}{\text{ha}} = \frac{3.35 \text{ T}}{\text{ha}}$$

A simple unit conversion can be performed to convert the metric yield into common U.S. yield units of bu/acre as:

$$y_p = \frac{3350 \text{ kg}}{\text{ha}} \cdot \frac{1 \text{ bu}}{27.22 \text{ kg}} \cdot \frac{1 \text{ ha}}{2.47 \text{ acre}} = \frac{49.8 \text{ bu}}{\text{acre}}$$

Note that the potential yield calculation does not consider the grains that had fallen from the plants before the machine engaged them. In this example, the pre-harvest yield loss, y_{ph}, was:

$$y_{ph} = \frac{0.015 \text{ kg}}{\text{m}^2} \cdot \frac{10000 \text{ m}^2}{\text{ha}} = \frac{150 \text{ kg}}{\text{ha}}$$

As a percentage of the total available grain, the pre-harvest yield loss was:

$$L_{ph} = \frac{150}{3350 + 150} \cdot 100 = 4.3\%$$

The grain that was collected at point B under the combine includes the shatter losses as well as the pre-harvest losses. The pre-harvest losses are subtracted from the total grain at point B to determine grain lost as the header engaged the crop. The shatter loss, L_{sh}, is calculated as a percentage of the theoretical yield as follows:

$$L_{sh} = \frac{(40 \text{ g} - 15 \text{ g})}{335 \text{ g}} \cdot 100 = 7.5\%$$

Threshing loss is a quantification of the grains that did not get dissociated from the plant. These grains are found at point C still attached to plant material. The threshing loss percentage is based on the total grain that actually enters the combine. In this example, the shatter loss is removed from the total available grain in calculating the threshing loss, L_{th}, as follows:

$$L_{th} = \frac{14 \text{ g}}{(335 \text{ g} - 25 \text{ g})} \cdot 100 = 4.5\%$$

The separation loss is threshed grain that is not removed from the MOG stream and is lost out the back of the combine. The free grain collected at point C includes the separation loss as well as the shatter and yield losses. Therefore, the loss due only to separation, L_s, is:

$$L_s = \frac{(63\text{ g} - 15\text{ g} - 25\text{ g})}{(335\text{ g} - 25\text{ g})} \cdot 100 = 7.4\%$$

The total harvest loss, L_h, is based on all the grain lost by the combine, which would be:

$$L_h = \frac{(63\text{ g} + 14\text{ g} - 15\text{ g})}{335\text{ g}} \cdot 100 = 19\%$$

The actual harvested yield, y_a, then, is:

$$y_a = \frac{3.35\text{ T}}{\text{ha}} - \frac{0.19(3.35\text{ T})}{\text{ha}} = \frac{2.71\text{ T}}{\text{ha}}$$

The harvest efficiency is (equation 6):

$$E_h = 1 - L_h = 1 - 0.19 = 0.81$$

This can be verified by equation 4:

$$E_h = \frac{y_a}{y_t} = \frac{2.71\text{ T}}{\text{ha}} \cdot \frac{\text{ha}}{3.35\text{ T}} = 0.81$$

The prudent manager would scrutinize these harvest efficiency numbers to determine if improvements are merited. For wheat harvest, these losses would probably be considered quite large. The manager may consider adjustment and/or operational changes to the combine that might reduce harvest losses.

Example 2: Reel speed

Problem:
One of the causes of shatter loss with grain tables is improper speed of the reel. The designers of a grain table need to provide ample adjustability in the rotational speed of the reel so that the operator can compensate for crop conditions and forward speed. Specifically, the designer needs to determine the range of speeds that the design must be able to achieve. Consider the grain table in figure 13 that has a 1.3 m diameter reel. Determine the range of reel speeds that the design must be able to achieve.

Figure 13. Gathering reel on grain table.

Solution:

As mentioned earlier, the tangential speed of the engaging devices on the end of the reel should typically be 25–50% greater than the combine forward speed, v_f. ASAE Standard D497.7 is a great resource for operating parameters of common agricultural machinery. Table 3 of that standard (reprinted in part as table 2 of this chapter) indicates that the typical forward speed of a self-propelled combine ranges from 3.0 to 6.5 km/hr. The minimum rotational speed of the

Table 2. Field efficiency and field speed for common harvesting machinery (excerpt from table 3 in ASABE Standard D497.7, 2015b).

Harvesting Machine	Field Efficiency		Field Speed			
	Range %	Typical %	Range mph	Typical mph	Range km/h	Typical km/h
Corn picker sheller	60–75	65	2.0–4.0	2.5	3.0–6.5	4.0
Combine	60–75	65	2.0–5.0	3.0	3.0–6.5	5.0
Combine (SP)	65–80	70	2.0–5.0	3.0	3.0–6.5	5.0
Mower	75–85	80	3.0–6.0	5.0	5.0–10.0	8.0
Mower (rotary)	75–90	80	5.0–12.0	7.0	8.0–19.0	11.0
Mower-conditioner	75–85	80	3.0–6.0	5.0	5.0–10.0	8.0
Mower-conditioner (rotary)	75–90	80	5.0–12.0	7.0	8.0–19.0	11.0
Windrower (SP)	70–85	80	3.0–8.0	5.0	5.0–13.0	8.0
Side delivery rake	70–90	80	4.0–8.0	6.0	6.5–13.0	10.0
Rectangular baler	60–85	75	2.5–6.0	4.0	4.0–10.0	6.5
Large rectangular baler	70–90	80	4.0–8.0	5.0	6.5–13.0	8.0
Large round baler	55–75	65	3.0–8.0	5.0	5.0–13.0	8.0
Forage harvester	60–85	70	1.5–5.0	3.0	2.5–8.0	5.0
Forage harvester (SP)	60–85	70	1.5–6.0	3.5	2.5–10.0	5.5
Sugar beet harvester	50–70	60	4.0–6.0	5.0	6.5–10.0	8.0
Potato harvester	55–70	60	1.5–4.0	2.5	2.5–6.5	4.0
Cotton picker (SP)	60–75	70	2.0–4.0	3.0	3.0–6.0	4.5

reel would occur with the reel tangential speed 25% greater than the slowest forward speed of 3.0 km/hr. Conversely, the maximum speed would occur at 150% of 6.5 km/hr. The tangential speed, v_t, is calculated using equation 9:

$$v_t = r\omega$$

At the minimum rotational speed, the tangential speed should be:

$$v_t = v_f (1.25)$$

Combining the equations, the minimum rotational speed is:

$$\omega = \frac{v_t}{r} = \frac{v_f(1.25)}{r} = \frac{3 \text{ km}}{\text{hr}} \cdot \frac{1.25}{1} \cdot \frac{2}{1.3 \text{ m}} \cdot \frac{1000 \text{ m}}{\text{km}} \cdot \frac{1 \text{ hr}}{60 \text{ min}} \cdot \frac{1 \text{ rev}}{2\pi} = 15.4 \text{ rpm}$$

It follows, then, that the maximum rotational speed is:

$$\omega = \frac{v_t}{r} = \frac{6 \text{ km}}{\text{hr}} \cdot \frac{1.5}{1} \cdot \frac{2}{1.3 \text{ m}} \cdot \frac{1000 \text{ m}}{\text{km}} \cdot \frac{1 \text{ hr}}{60 \text{ min}} \cdot \frac{1 \text{ rev}}{2\pi} = 36.7 \text{ rpm}$$

The revolution units are added to the calculations by noting that radians are considered unitless, and there are 2π radians in one complete revolution. The conclusion is that the drive system for the reel on that grain table must be able to achieve speeds varying from 15.4 to 36.7 rpm, so the drive mechanism for the reel should be designed accordingly.

Example 3: Axle loads

Problem:

The design of the structure of a vehicle relies heavily on understanding the effects of all the forces on the machine. Consider a two-wheeled grain cart pulled by a tractor as shown in figure 14. The task is to calculate the required size (diameter) of the cylindrical axles to support the cart wheel assembly. Assume that the grain load is evenly distributed in the tank of the cart and that the tank is laterally symmetrical, which means that the loads are evenly distributed between the left and right wheels of the cart. Besides the dimensions shown in figure 15, the following data are given by a manufacturer for a very similar cart:

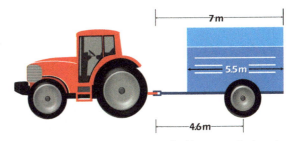

Figure 14. Basic grain cart pulled by an agricultural tractor.

Cart capacity: 850 bushels of corn (maize)
Empty cart weight: 54 kN
Tongue weight of empty cart: 11 kN

Solution:

Because of the left/right symmetry of the cart, the free body analysis can be conducted in two dimensions looking at the side of the machine (figure 15). The two cart wheels will have identical loads. Since the tractor supports 11 kN of the empty cart weight from the tongue at the hitch point (F_{ct}), the rest of the empty cart weight, which is 54 kN − 11 kN = 43 kN, must be supported by the cart wheels (F_{cw}). Given the symmetry and uniform loading assumptions, the center of gravity of the grain load will be at the geometric center of the bin on the cart. The distance from the hitch point to the grain center of gravity, x_g, is:

$$x_g = 7 - \frac{5.5}{2} = 4.25 \text{ m}$$

The weight of the grain is:

$$F_g = \frac{850 \text{ bu}}{1} \cdot \frac{25.4 \text{ kg}}{\text{bu}} \cdot \frac{9.81 \text{ N}}{\text{kg}} = 212 \text{ kN}$$

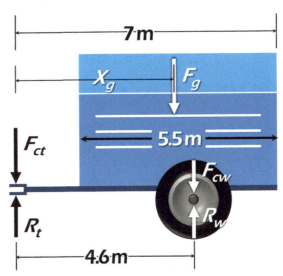

Figure 15. Free body forces acting on a grain cart.

The cart and grain must be supported by the cart wheels and by the tractor at the hitch point. These forces are represented as reaction forces R_w and R_t (figure 15). R_w is the total weight that the cart wheels and axles must support, which can be calculated by summing the moments about the hitch point between the cart and tractor. If counter-clockwise rotation is positive, that moment equation is:

$$R_w(4.6) - F_{cw}(4.6) - F_g(4.25) + F_{ct}(0) - R_t(0) = 0$$

Note that because the tongue load and reaction force both pass through the hitch point; their moment arm distances are zero and they fall out of the equation. The moment equation is now solved for R_w:

$$R_w = \frac{F_{cw}(4.6 \text{ m}) + F_g(4.25 \text{ m})}{4.6 \text{ m}} = \frac{43 \text{ kN}(4.6\text{m}) + 212 \text{ kN}(4.25 \text{ m})}{4.6 \text{ m}} = 240 \text{ kN}$$

Since there are two wheels, each wheel must support 120 kN.

With the load known, it is possible to calculate the diameter of the cylindrical axle required to support the wheel. The axle (figure 16) is a cantilever configuration since it is rigidly fixed to the frame on one end. The simplified configuration of the axle (figure 16) shows that the reaction force from the wheel is applied 30 cm out from the base of the axle. The downward force of the cart and grain at the base of the axle and the upward reaction force from the tire will cause bending stress in the axle. The bending moment is:

$$M = Fd = 120{,}000 \text{ N} \cdot 0.3 \text{ m} = 3600 \text{ Nm}$$

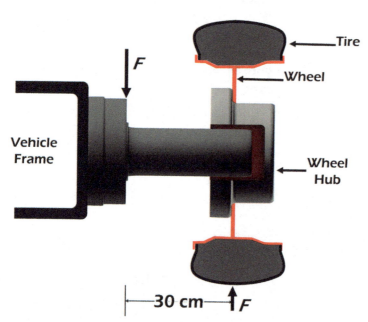

Figure 16. Configuration of axle on grain cart.

The maximum stress in the axle cannot exceed the yield stress of the material, which would cause permanent deformation in the axle, compromising its functionality and strength. The designer needs to know what material will be used to manufacture the axle and then determine the yield stress for that particular material. A number of material engineering handbooks and other resources can be consulted to find the yield stress of different materials. For this example, assume that mild steel would be used. The yield stress for mild steel (σ_y) can be found from a number of resources to be 250 MPa. Note that a Pa is defined as a N/m^2. The bending stress in the axle is (equation 8):

$$\sigma_b = \frac{My}{I}$$

Note that y is the distance from the neutral axis, which is the center of the circular shaft. The maximum stress will occur at the top and bottom of the axle.

The equations for moment of inertia for different cross-sectional shapes can be found in a number of engineering handbooks or strength of materials text books. For a circular cross section, the moment of inertia is:

$$I = \frac{1}{4}\pi r^4 \qquad (21)$$

Substituting equation 21 into equation 8, the bending stress equation becomes:

$$\sigma_b = \frac{4My}{\pi r^4}$$

Since the critical failure point will be at the outermost fibers of the circular cross section, the stress will be calculated at $y = r$. Also, the stress at those outermost fibers should not exceed the yield stress; therefore,

$$\sigma_y = \frac{4M}{\pi r^3}$$

Now solve for r:

$$r^3 = \frac{4M}{\pi \sigma_y} = \frac{4}{\pi} \cdot \frac{3600 \text{ Nm}}{1} \cdot \frac{\text{m}^2}{350 \times 10^6 \text{ N}}$$

The calculated minimum radius is 2.6 cm.

Any calculated number or computer output should always be scrutinized to make sure that it represents a reasonable conclusion. In this case, an experienced engineer should be concerned that a 2.6-cm radius axle seems unusually small for a large grain cart. There are several factors that were not considered in the analysis. First, the load on the axle was the static weight of the cart and grain. There was no consideration for peak dynamic loads that would be induced as the vehicle moved across the terrain of a farm field. The dynamic analysis would also need to consider fatigue stress in the material due to repeated loading. There was no safety factor considered to compensate for inconsistencies in material properties of the axle or overloading of the cart by the operator. Depending on the method of attachment of the axle to the frame, there could be significant stress concentrations at sharp corners or weldments. These stress concentrations are usually identified with a finite element analysis of the structure. But even if catastrophic failure did not occur in the mechanism, the engineer should consider the effects of the elasticity of the axle. In this case, excessive elastic deflection in the axle could cause the tire to become misaligned, which could cause adverse tracking of the cart or unacceptable wear of the tire. All these factors would need to be addressed to achieve a final design that prevents failure and assures proper operation.

Image Credits

Figure 1. Stombaugh, T. (CC By 4.0). (2020). Typical grain combine.
Figure 2. Stamper, D. & Stombaugh, T. (CC By 4.0). (2020). Forces acting on grain.
Figure 3. Stamper, D. & Stombaugh, T. (CC By 4.0). (2020). Simple sieve mechanism.

Figure 4. Stamper, D. & Stombaugh, T. (CC By 4.0). (2020). Drag forces.
Figure 5. Stamper, D. & Stombaugh, T. (CC By 4.0). (2020). Paddle conveyor.
Figure 6. Stamper, D. & Stombaugh, T. (CC By 4.0). (2020). Screw conveyor.
Figure 7. Stombaugh, T. (CC By 4.0). (2020). Combine and baler.
Figure 8. Stombaugh, T. (CC By 4.0). (2020). Row crop head.
Figure 9. Stamper, D. & Stombaugh, T. (CC By 4.0). (2020). Conventional threshing cylinder.
Figure 10. Stamper, D. & Stombaugh, T. (CC By 4.0). (2020). Rotary threshing cylinder.
Figure 11. Stombaugh, T. (CC By 4.0). (2020). Combine and grain cart.
Figure 12. Stamper, D. & Stombaugh, T. (CC By 4.0). (2020). Combine test locations.
Figure 13. Stamper, D. & Stombaugh, T. (CC By 4.0). (2020). Gathering reel.
Figure 14. Stamper, D. & Stombaugh, T. (CC By 4.0). (2020). Tractor and grain cart.
Figure 15. Stamper, D. & Stombaugh, T. (CC By 4.0). (2020). Forces on grain cart.
Figure 16. Stamper, D. & Stombaugh, T. (CC By 4.0). (2020). Axle configuration.

References

ASABE Standards. (2015a). ASAE EP496.3 FEB2996 (R2015): Agricultural machinery management. St. Joseph, MI: ASABE.

ASABE Standards. (2015b). ASAE D497.7 MAR2011 (R2015): Agricultural machinery management data. St. Joseph, MI: ASABE.

Srivastava, A. K., Goering, C. E., Rohrbach, R. P., & Buckmaster, D. R. (2006). *Engineering principles of agricultural machines* (2nd ed.). St. Joseph, MI: ASABE.

Mechatronics and Intelligent Systems in Agricultural Machinery

Francisco Rovira-Más
Agricultural Robotics Laboratory, Universitat Politècnica de València, Valencia, Spain

Verónica Saiz-Rubio
Agricultural Robotics Laboratory, Universitat Politècnica de València, Valencia, Spain

Qin Zhang
Center for Precision & Automated Agricultural Systems, Washington State University, Prosser, Washington

KEY TERMS		
Control systems	Analog and digital data	Auto-guided tractors
Actuators	Positioning	Variable-rate application
Sensors	Vision and imaging	Intelligent machinery

Variables

- θ_t = angle at time t
- $\dot{\theta}_t$ = angular rate at time t measured by a gyroscope
- λ = latitude
- φ = longitude
- a = semi-major axis of WGS 84 reference ellipsoid
- a_t = linear acceleration recorded by an accelerometer or inertial measurement unit
- A = horizontal dimension of the imaging sensor
- d_1 = distance between the imaging sensor and the optical center of the lens
- d_2 = distance between the optical center of the lens and the target object

e = eccentricity of WGS 84 reference ellipsoid
f = lens focal length
FOV = horizontal field of view covered in the images
h = altitude
L = number of digital levels in the quantization process
n = number of bits
(N, E, D) = LTP coordinates north, east, down
N_0 = length of the normal
Δt = time interval between two consecutive measurements
T = ambient temperature
V = speed of sound through air
V_t = velocity of a vehicle at time t
(X, Y, Z) = ECEF coordinates
(X_0, Y_0, Z_0) = user-defined origin of coordinates in ECEF format

Introduction

Visitors to local farm fairs have a good chance of seeing old tractors. Curious visitors will notice that the oldest ones, say, those made in the first three decades of the 20th century, are purely mechanical. As visitors observe newer tractors, they may find that electronic and fluid powered components appeared in those machines. Now, agricultural machinery, such as tractors and combines, are so sophisticated that they are fully equipped with electronic controls and even fancy flat screens. These controls and screens are the driver interface to electromechanical components integrated into modern tractors.

The term *mechatronics* is used to refer to systems that combine computer controls, electrical components, and mechanical parts. A mechatronics solution is not just the addition of sensors and electronics to an already existing machine; rather, it is the balanced integration of all of them in such a way that each individual component enhances the performance of the others. This outcome is achieved only by considering all subsystems simultaneously at the earliest stages of design (Bolton, 1999). Thus, mechatronics unifies the technologies that underlie sensors, automatic control systems, computing processors, and the transmission of power through mechanisms including fluid power actuators.

During the 20th century, agricultural mechanization greatly reduced the drudgery of farm work while increasing *productivity* (more land farmed by fewer people), *efficiency* (less time and resources invested to farm the same amount of land), and *work quality* (reduced losses at harvesting, more precise chemical applications, achieving uniform tillage). The Green Revolution, led by Norman Borlaug, increased productivity by introducing region-adapted crop varieties and the use of effective fertilizers, which often resulted in yields doubling, especially in developing countries. With such improvements initiated by the Green Revolution, current productivity, efficiency, and quality food crops may be sufficient to support a growing world population projected to surpass 9.5 billion by 2050, but the actual challenge is to do it in a sustainable way by means of a regenerative agriculture (Myklevy et al., 2016). This challenge is further complicated by the continuing decline of the farm workforce globally.

Current agricultural machinery, such as large tractors, sprayers, and combine harvesters, can be too big in practice because they must travel rural roads, use powerful diesel engines that are subjected to restrictive emissions regulations, are difficult to automate for liability reasons, and degrade farm soil by high wheel compaction. These challenges, and many others, may be overcome through the adoption of mechatronic technologies and intelligent systems on modern agricultural machinery. Mechanized farming has been adopting increased levels of automation and intelligence to improve management and increase productivity in field operations. For example, farmers today can use auto-steered agricultural vehicles for many different field operations including tilling, planting, chemical applications, and harvesting. Intelligent machinery for automated thinning or precise weeding in vegetable and other crops has recently been introduced to farmers.

This chapter introduces the basic concepts of mechatronics and intelligent systems used in modern agricultural machinery, including farming robots. In particular, it briefly introduces a number of core technologies, key components, and typical challenges found in agricultural scenarios. The material presented in this chapter provides a basic introduction to mechatronics and intelligent technologies available today for field production applications, and a sense of the vast potential that these approaches have for improving worldwide mechanization of agriculture in the next decades.

Outcomes

After reading this chapter, you should be able to:

- Explain the purpose of intelligent machinery for agricultural operations

- Describe common sensing devices for intelligent agricultural machines, such as inertial measuring units, rangefinders, digital cameras, and global navigation satellite system positioning receivers

- Apply important concepts of mechanized and robotic farming operations to relevant use cases

Concepts

The term *mechatronics* applies to engineering systems that combine computers, electronic components, and mechanical parts. The concept of mechatronics is the seamless integration of these three subsystems; its embodiment in a unique system leads to a *mechatronic system*. When the mechatronic system is endowed with techniques of artificial intelligence, the mechatronic system is further classified as an *intelligent system*, which is the basis of robots and intelligent farm machinery.

Automatic Control Systems

Machinery based on mechatronics needs to have control systems to implement the automated functions that accomplish the designated tasks. Mechatronic systems consist of electromechanical hardware and control software encoding the algorithm or model that automate an operation. An automatic control system obtains relevant information from the surrounding environment to manage (or regulate) the behavior of a device performing desired operations. A good example is a home air conditioner (AC) controller that uses a thermostat to determine the deviation of room temperature from a preset value and turn the AC on and off to maintain the home at the preset temperature. An example in agricultural machinery is auto-steering. Assume a small utility tractor has been modified to steer automatically between grapevine rows in a vineyard. It may use a camera looking ahead to detect the position of vine rows, such that deviations of the tractor from the centerline between vine rows are related to the proper steering angle for guiding the tractor in the vineyard without hitting a grapevine. From those two examples, it can be seen that a *control system*, in general, consists of *sensors* to obtain information, a *controller* to make decisions, and an *actuator* to perform the actions that automate an operation.

Actuation that relies on the continuous tracking of the variable under control (such as temperature or wheel angle) is called *closed-loop control* and provides a stable performance for automation. Closed-loop control allows the real-time estimation of the error (which is defined as the difference between the desired output of the controlled variable and the actual value measured by a feedback sensor), and calculates a correction command with a control function—the controller—for reducing the error. This command is sent to the actuator (discussed in the next section) for automatically implementing the correction. This controller function can be a simple proportion of the error (*proportional controller, P*), a measure of the sensitivity of change (*derivative controller, D*), a function dependent on accumulated (past) errors (*integral controller, I*), or a combination of two or three of the functions mentioned above (*PD, PI, PID*). There are alternative techniques for implementing automated controls, such as *intelligent systems* that use artificial intelligence (AI) methods like neural networks, fuzzy logic, genetic algorithms, and machine learning to help make more human-like control decisions.

Actuators

An electromechanical component is an integrated part that receives an electrical signal to create a physical movement to drive a mechanical device performing a certain action. Examples of electromechanical components include electrical motors that convert input electrical current into the rotation of a shaft, and pulse-width modulation (PWM) valves, such as variable rate nozzles and proportional solenoid drivers, which receive an electrical signal to push the spool of a hydraulic control valve to adjust the valve opening that controls the amount of fluid passing through. Because hydraulic implement systems are widely used on agricultural machinery, it is common to see many more electrohydraulic components (such as proportional solenoid drivers and servo drivers) than electrical motors on farm machines. However, as robotic solutions become increasingly more available in agriculture, applications of electrical motors on modern agricultural machinery will probably increase, especially on intelligent and robotic versions. The use of mechatronic components lays the foundation for adopting automation technologies to agricultural machinery, including the conversion of traditional machines into robotic ones capable of performing field work autonomously.

Intelligent Agricultural Machinery and Agricultural Robots

For intelligent agricultural machinery to be capable of performing automated field operations, it is required that machines have the abilities of: (1) becoming aware of actual operation conditions; (2) determining adaptive corrections suitable for continuously changing conditions; and (3) implementing such corrections during field operations, with the support of a proper mechanical system. The core for achieving such a capability often rests on the models that govern intelligent machinery, ranging from simple logic rules controlling basic tasks all the way to sophisticated *AI algorithms* for carrying out complex operations. These high-level algorithms may be developed using popular techniques such as *artificial neural networks*, *fuzzy logic*, *probabilistic reasoning*, and *genetic algorithms* (Russell and Norvig, 2003). As many of those intelligent machines could perform some field tasks autonomously, like a human worker could do, such machinery can also be referred to as robotic machinery. For example, when an autonomous lawn mower (figure 1a) roams within a courtyard, it is typically endowed with basic navigation and path-planning skills that make the mower well fit into the category of robotic machinery, and therefore, it is reasonable to consider it a field robot. Though these robotic machines are not presently replacing human workers in field operations, the introduction of robotics in agriculture and their widespread use is only a matter of time. Figure 1b shows an autonomous rice transplanter (better called a rice transplanting robot) developed by the National Agriculture and Food Research Organization (NARO) of Japan.

Many financial publications forecast that there will be a rapid growth of the market for service robots in the next two decades, and those within agricultural applications will play a significant role. Figure 2 shows the expected growth of the U.S. market for agricultural robots by product type. Although robots for milking and dairy management have dominated the agricultural robot market in the last decade, crop production robots are expected to increase their presence commercially and lead the market in the coming years, particularly for specialty crop production (e.g., tree fruit, grapes, melons, nuts, and vegetables). This transformation of the 21st century farmer from laborer to digital-age manager may be instrumental in attracting younger generations to careers in agricultural production.

(a)

(b)

Figure 1. (a) Autonomous mower (courtesy of John Deere); (b) GPS-based autonomous rice transplanter (courtesy of NARO, Japan).

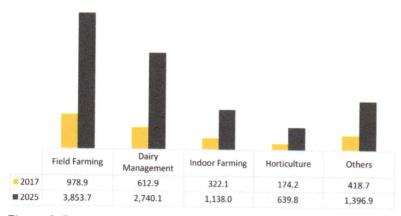

Figure 2. Expected growth of agricultural robot market in the U.S. for 2016-2025. Amounts are millions of U.S. dollars (Verified Market Research, 2018).

Sensors in Mechatronic Systems

Sensors are a class of devices that measure significant parameters by using a variety of physical phenomena. They are important components in a mechatronic system because they provide the information needed for supporting automated operations. While the data to be measured can be in many forms, sensors output the measured data either in analog or digital formats (described in the next section). In modern agricultural machinery, sensor outputs are eventually transformed to digital format and thus can be displayed on an LCD screen or fed to a computer. This high connectivity between sensors and computers has accelerated the expansion of machinery automation. An intelligent machine can assist human workers in conducting more efficient operations: in some cases, it will simply entail retrieving clearer or better information; in other cases, it will include the automation of physical functions. In almost all situations, the contribution of reliable sensors is needed for machines to interact with the surrounding environment. Figure 3 shows the architecture of an intelligent tractor, which includes the typical sensors onboard intelligent agricultural machinery.

Even though sensors collect the data required to execute a particular action, that may not be enough because the environment of agricultural production is often complicated by many factors. For example, illumination changes throughout the day, adverse weather conditions may impair the performance of sensors, and open fields are rife with uncertainty where other machines, animals, tools, and even workers may appear unexpectedly in the near vicinity. Sensed data may be insufficient to support a safe, reliable, and efficient automated operation, and therefore data processing techniques are necessary to get more comprehensive information until it becomes sufficient to support automated operations. As a rule of thumb, *there is no sensor that provides all needed information, and there is no sensor that never fails*. Depending on specific needs, engineers often use either *redundancy* or *sensor fusion* to solve such a problem. The former acquires the same information through independent sources in case one of them fails or decreases in reliability, and the latter combines information of several sources that are complementary. Once the sensed information has

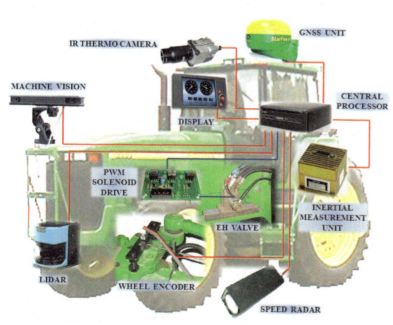

Figure 3. Sensor architecture for intelligent agricultural vehicles.

been processed using either method, the actuation command can be calculated and then executed to complete the task.

Analog and Digital Data

As mentioned above, mechatronic systems often use sensors to obtain information to support automated operations. Sensors provide measurements of physical magnitudes (e.g., temperature, velocity, pressure, distance, and light intensity) represented by a quantity of electrical variables (such as voltages and currents). These quantities are often referred to as *analog data* and normally expressed in base 10, the decimal numbering system. In contrast, electronic devices such as controllers represent numbers in base 2 (the binary numbering system or simply "binary") by adopting the on-off feature of electronics, with a numerical value of 1 assigned to the "on" state and 0 assigned to the "off" state.

A binary system uses a series of digits, limited to zeros or ones, to represent any decimal number. Each of these digits represents a bit of the binary number; a binary digit is called a bit. The leftmost 1 in the binary number 1001 is called the most significant bit (MSB), and the rightmost 1 is the least significant bit (LSB). It is common practice in computer science to break long binary numbers into segments of 8 bits, known as *bytes*. There is a one-to-one correspondence between binary numbers and decimal numbers. For instance, a 4-bit binary number can be used to represent all the positive decimal integers from 0 (represented by 0000) to 15 (represented by 1111). Signal *digitization* consists of finding that particular correspondence.

The process of transforming binary numbers to decimal numbers and vice versa is straightforward for representing positive decimal integers. However, negative and floating-point numbers require special techniques. While the transformation of data between two formats is normally done automatically, it is important to know the underlying concept for a better understanding of how information can be corrected, processed, distributed, and utilized in intelligent machinery systems. The resolution of digital data depends on the number of bits, such that more bits means more precision in the digitized measurement. Equation 1 yields the relationship between the number of bits (n) and the resulting number of digital levels available to code the signal (L). For example, using 4 bits leads to $2^4 = 16$ levels, which implies that an analog signal between 0 V and 2 V will have a resolution of $2/15 = 0.133$ V; as a result, quantities below 133 mV will not be detected using 4-bit numbers. If more accuracy is necessary, digitization will have to use numbers with more bits. Note that equation 1 is an exponential relationship rather than linear, and quantization grows fast with the number of bits. Following with the previous example, 4 bits produce 16 levels, but 8 bits give 256 levels instead of 32, which actually corresponds to 5 bits.

$$L = 2^n \quad (1)$$

where L = number of digital levels in the quantization process
n = number of bits

Position Sensing

One basic requirement for agricultural robots and intelligent machinery to work properly, reliably, and effectively is to know their location in relation to the surrounding environment. Thus, positioning capabilities are essential.

Global Navigation Satellite Systems (GNSS)

Global Navigation Satellite System (GNSS) is a general term describing any satellite constellation that provides positioning, navigation, and timing (PNT) services on a global or regional basis. While the USA Global Positioning System (GPS) is the most prevalent GNSS, other nations are fielding, or have fielded, their own systems to provide complementary, independent PNT capability. Other systems include Galileo (Europe), GLONASS (Russia), BeiDou (China), IRNSS/NavIC (India), and QZSS (Japan).

When the U.S. Department of Defense released the GPS technology for civilian use in 2000, it triggered the growth of satellite-based navigation for off-road vehicles, including robotic agricultural machinery. At present, most leading manufacturers of agricultural machinery include navigation assistance systems among their advanced products. As of 2019, only GPS (USA) was fully operational, but the latest generation of receivers can already expand the GPS constellation with other GNSS satellites.

GPS receivers output data through a serial port by sending a number of bytes encoded in a standard format that has gained general acceptance: NMEA 0183. The NMEA 0183 interface standard was created by the U.S. National Marine Electronics Association (NMEA), and consists of GPS messages in text (ASCII) format that include information about time, position in *geodetic coordinates* (i.e., latitude (λ), longitude (φ), and altitude (h)), velocity, and signal precision. The World Geodetic System 1984 (WGS 84), developed by the U.S. Department of Defense, defines an ellipsoid of revolution that models the shape of the earth, and upon which the geodetic coordinates are defined. Additionally, the WGS 84 defines a Cartesian coordinate system fixed to the earth and with its origin at the center of mass of the earth. This system is the *earth-centered earth-fixed* (ECEF) coordinate system, and it provides an alternative way to locate a point on the earth surface with the conventional three Cartesian coordinates X, Y, and Z, where the Z-axis coincides with the earth's rotational axis and therefore crosses the earth's poles.

The majority of the applications developed for agricultural machinery, however, do not require covering large surfaces in a short period of time. Therefore, the curvature of the earth has a negligible effect, and most farm fields can be considered flat for practical purposes. A *local tangent plane* coordinate system (LTP), also known as NED coordinates, is often used to facilitate such small-scale operations with intuitive global coordinates north (N), east (E), and down (D). These coordinates are defined along three orthogonal axes in a Cartesian configuration generated by fitting a tangent plane to the surface of the earth at an arbitrary point selected by the user and set as the *LTP origin*. Given that standard receivers provide geodetic coordinates (λ, φ, h) but practical field operations require a local frame such as LTP, a fundamental operation for mapping applications in agriculture is the real-time transformation between the two coordinate systems (Rovira-Más et al., 2010). Equations 2 to 8 provide the step by step procedure for achieving this transformation.

$$a = 6378137 \tag{2}$$

$$e = 0.0818 \tag{3}$$

$$N_0(\lambda) = \frac{a}{\sqrt{1 - e^2 \cdot \sin^2 \lambda}} \tag{4}$$

$$X = (N_0 + h) \cdot \cos \lambda \cdot \cos \varphi \tag{5}$$

$$Y = (N_0 + h) \cdot \cos \lambda \cdot \sin \varphi \tag{6}$$

$$Z = \left[h + N_0 \cdot (1 - e^2) \right] \cdot \sin \lambda \tag{7}$$

$$\begin{bmatrix} N \\ E \\ D \end{bmatrix} = \begin{bmatrix} -\sin \lambda \cdot \cos \varphi & -\sin \lambda \cdot \sin \varphi & \cos \lambda \\ -\sin \varphi & \cos \varphi & 0 \\ -\cos \lambda \cdot \cos \varphi & -\cos \lambda \cdot \sin \varphi & -\sin \lambda \end{bmatrix} \cdot \begin{bmatrix} X - X_0 \\ Y - Y_0 \\ Z - Z_0 \end{bmatrix} \tag{8}$$

where a = semi-major axis of WGS 84 reference ellipsoid (m)
e = eccentricity of WGS 84 reference ellipsoid
N_0 = length of the normal (m)

Geodetic coordinates:
λ = latitude (°)
φ = longitude (°)
h = altitude (m)
(X, Y, Z) = ECEF coordinates (m)
(X_0, Y_0, Z_0) = user-defined origin of coordinates in ECEF format (m)
(N, E, D) = LTP coordinates north, east, down (m)

Despite the high accessibility of GPS information, satellite-based positioning is affected by a variety of errors, some of which cannot be totally eliminated. Fortunately, a number of important errors may be compensated by using a technique known as *differential correction*, lowering errors from more than 10 m to about 3 m. Furthermore, the special case of *real-time-kinematic* (RTK) differential corrections may further lower error to just centimeter level.

Sonar Sensors

In addition to locating machines in the field, another essential positioning need for agricultural robots is finding the position of surrounding objects during farming operations, such as target plants or potential obstacles. *Ultrasonic rangefinders* are sensing devices used successfully for this purpose. Because they measure the distance of target objects in terms of the speed of sound, these sensors are also known as *sonar* sensors.

The underlying principle of sonars is that the speed of sound is known (343 m s^{-1} at 20°C), and measuring the time that the wave needs to hit an obstacle and return to the sensor—the echo—allows the estimation of an object's distance. The speed of sound through air, V, depends on the ambient temperature, T, as:

$$V \text{(m s}^{-1}) = 331.3 + 0.606 \times T \text{(°C)} \tag{9}$$

The continuously changing ambient temperature in agricultural fields is one of many challenges to sonar sensors. Another challenge is the diversity of target objects. In practice, sonar sensors must send out sound waves that hit an object and then return to the sensor receiver. This receiver must then capture the signal to measure the elapsed time for the waves to complete the round trip. Understanding the limitations posed by the reflective properties of target objects is essential to obtain reliable results. The distance to materials that absorb sound waves, such as stuffed toys, will be measured poorly, whereas solid and dense targets will allow the system to perform well. When the target object is uneven, such as crop canopies, the measurements may become noisy. Also, sound waves do not behave as linear beams, but propagate in irregular cones that expand in coverage with distance. When objects are outside the cone, they may be undetected. Errors will often vary with ranges such that farther ranges lead to larger errors.

An important design feature to consider is the distance between adjacent ultrasonic sensors, as echo interference is another source of unstable behavior. Overall, sonar rangefinders are helpful to estimate short distances cost-efficiently when accuracy and reliability are not critical, as when detecting distances to the canopy of trees for automated pesticide spraying.

Light Detection and Ranging (Lidar) Sensors

Another common position-detecting sensor is *lidar*, which stands for *light detection and ranging*. Lidars are optical devices that detect the distance to target objects with precision. Although different light sources can be used to estimate ranges, most lidar devices use laser pulses because their beam density and coherency result in high accuracy.

Lidars possess specific features that make them favorable for field robotic applications, as sunlight does not affect lidars unless it hits their emitter directly, and they work excellently under poor illumination.

Machine Vision and Imaging Sensors

One important element of human intelligence is *vision*, which gives farmers the capability of visual perception. A basic requirement for intelligent agricultural machinery (or agricultural robots) is to have surrounding awareness capability. *Machine vision* is the computer version of the farmer's sight; the cameras function as eyes and the computers as the brain. The output data of vision systems are digital images. A *digital image* consists of little squares called *pixels* (picture elements) that carry information on their level of light intensity. Most of the digital cameras used on agricultural robots are CCD (charge coupled devices), which are composed of a small rectangular sensor made of a grid of tiny light-sensitive cells, each of them producing the information of its corresponding pixel in the image. If the image is in black and white (technically called monochrome), the intensity level is represented in a gray scale between a minimum value (0) and a maximum value (i_{max}). The number of levels in the gray scale depends on the number of bits in which the image is coded. Most of the images used in agriculture are 8 bits, which means that the image can distinguish 256 gray levels (2^8), where the minimum value is 0 representing complete black, and the maximum value is 255 representing pure white. In practical terms, human eyes cannot distinguish so many levels, and 8 bits are many times more than enough. When digital images reproduce a scene in color, pixels carry information of intensity levels for the three channels of red (R), green (G), and blue (B), leading to RGB images. The processing of RGB images is more complicated than monochrome images and falls outside the scope of this chapter.

> In general, precision is a measure of the difference between a specific reading and the average of all readings of the same value being measured, and accuracy is defined as the closeness of a measured value to its actual value.

Monocular cameras (which have one lens) constitute simple vision systems, yet the information they retrieve is powerful. When selecting a camera, engineers must choose important technical parameters such as the focal length of the lens, the size of the sensor, and optical filters when there are spectral ranges (colors) that need to be blocked from the image. The focal length (*f*) is related to the scope of scene that fits into the image, and is defined in equation 10. The geometrical relationship described by figure 4 and equation 11 determines the resulting field of view (FOV) of any given scene. The design of a machine vision system, therefore, must include the right camera and lens parameters to assure that the necessary FOV is covered and the target objects are in focus in the images.

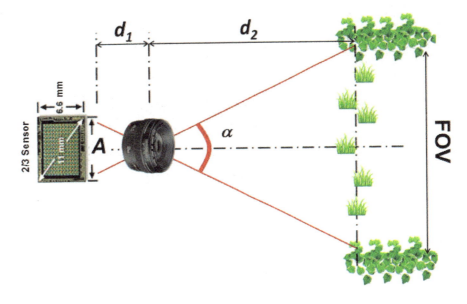

Figure 4. Geometrical relationship between an imaging sensor, lens, and FOV; α is the angular equivalent of the horizontal field of view.

$$\frac{1}{f} = \frac{1}{d_1} + \frac{1}{d_2} \qquad (10)$$

$$\frac{d_1}{d_2} = \frac{A}{FOV} \qquad (11)$$

where f = lens focal length (mm)
d_1 = distance between the imaging sensor and the optical center of the lens (mm)
d_2 = distance between the optical center of the lens and the target object (mm)
A = horizontal dimension of the imaging sensor (mm)
FOV = horizontal field of view covered in the images (mm)

After taking the images, the first step of the process (image acquisition) is complete. The second step, analysis of the images, begins with image processing, which involves the delicate task of extracting the useful information from each image for its later use. Figure 5 reproduces the results of a color-based segmentation algorithm to find the position of mandarin oranges in a citrus tree.

Even though digital images reproduce scenes with great detail, the representation is flat, that is, in two dimensions (2D). However, real scenes are in three dimensions (3D), with the third dimension being the *depth*, or distance between the camera and the objects of interest in the scene. In the image shown in figure 5, for instance, a specific orange can be located with precision in the horizontal and vertical axes, but how far it is from the sensor cannot be known. This information would be essential, for example, to program a robotic arm to retrieve the oranges. Stereo cameras (which are cameras with at least two lenses that meet the principles of stereoscopy) allow the acquisition of two (or more) images in a certain relative position to which the principles of *stereoscopic vision* apply. These principles mimic how human vision works, as the images captured by human eyes in the retinas are slightly offset, and this offset (known as disparity) is what allows the brain to estimate depth.

Figure 5. Color-based segmentation of mandarin oranges.

Estimation of Vehicle Dynamic States

The parameters that help understand a vehicle's dynamic behavior are known as the *vehicle states*, and typically include velocity, acceleration, sideslip, and angular rates yaw, pitch, and roll. The sensors needed for such measurements are commonly assembled in a compact motion sensor called an *inertial measurement unit* (IMU), created from a combination of accelerometers and gyroscopes. The accelerometers of the IMU detect acceleration as the change in velocity of the vehicle over time. Once the acceleration is known, its mathematical integration gives an estimate of the velocity, and integrating again gives an estimate of the position. Equation 12 allows the calculation of instantaneous velocities from the acceleration measurements of an IMU or any individual accelerometer. Notice that for finite increments of time Δt, the integral function is replaced by a summation. Similarly, *gyroscopes* can detect the angular rates of the turning vehicle; integrating these values leads to *roll*, *pitch*, and *yaw* angles, as specified by equation 13. A typical IMU is composed of three accelerometers and three gyroscopes assembled along three perpendicular axes that reproduce a Cartesian coordinate system. With this physical configuration, it is possible to calculate the three components of acceleration and speed in Cartesian coordinates as well as Euler angles roll, pitch, and yaw. Current IMUs on the market are small and inexpensive, favoring the accurate estimation of vehicle states with small devices such as microelectromechanical systems (MEMS).

> *Euler angles* are three angles that provide the orientation of a rigid body with respect to a Cartesian frame of reference given in three dimensions.

$$V_t = V_{t-1} + a_t \cdot \Delta t \tag{12}$$

$$\theta_t = \theta_{t-1} + \dot{\theta}_t \cdot \Delta t \tag{13}$$

where V_t = velocity of a vehicle at time t (m s^{-1})
a_t = linear acceleration recorded by an accelerometer (or IMU) at time t (m s^{-2})
Δt = time interval between two consecutive measurements (s)
θ_t = angle at time t (rad)
$\dot{\theta}_t$ = angular rate at time t measured by a gyroscope (rad s^{-1})

Applications

Mechatronic Systems in Auto-Guided Tractors

As mentioned early in this chapter, mechatronic systems are now playing an essential role in modern agricultural machinery, especially on intelligent and robotic vehicles. For example, the first auto-guided tractors hit the market at the turn of the 21st century; from a navigation standpoint, farm equipment manufacturers have been about two decades ahead of the automotive industry. Such auto-guided tractors would not be possible if they had not been upgraded to state-of-the-art mechatronic systems, which include sensing, controlling, and electromechanical (or electrohydraulic) actuating elements. One of the most representative components never seen before on conventional mechanical tractors as an integrated element is the high-precision GPS receiver, which furnishes tractors with the capability to locate themselves in order to guide them following designated paths.

Early navigation solutions that were commercially available did not actually control the tractor steering system; rather, they provided tractor drivers with lateral corrections in real time, such that by following these corrections the vehicle easily tracked a predefined trajectory. This approach is easy to learn and execute, as drivers only need to follow a light-bar indicator, where the number of lights turned on is proportional to the sidewise correction to keep the vehicle on track. In addition to its simplicity of use, this system works for any agricultural machine, including older ones. Figure 6a shows a light-bar system mounted on an orchard tractor, where the red light signals the user to make an immediate correction to remain within the trajectory shown on the LCD screen.

Another essential improvement of modern tractors based on mechatronics technology is the electrohydraulic system that allows tractors to be maneuvered by wire. This means that an operation of the tractor, such as steering or lowering the implement installed on the three-point hitch, can be accomplished by an electronically controlled electrohydraulic actuating system in response to control signals generated by a computer-based controller. An electrohydraulic steering system allows a tractor to be guided automatically, by executing navigation commands calculated by an onboard computer based on received GPS positioning signals. One popular auto-steering application is known as *parallel tracking*, which allows a tractor being driven automatically to follow desired pathways in parallel to a reference line between two points, say A-B line, in a field recorded by the onboard GPS system. These reference lines can even include curved sectors. Figure 6b displays the control screen of a commercial auto-guidance system implemented in a wheel-type tractor. Notice the magnitude of the tractor deviation (the off-track error) from the predefined trajectory is shown at the top bar, in a similar fashion as the corrections conveyed through light-bars. The implementation of automatic guidance has reduced pass-to-pass overlaps,

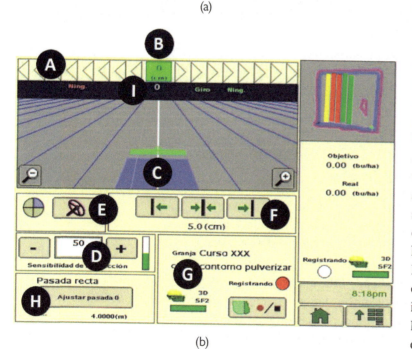

Figure 6. Auto-guidance systems: (a) Light-bar kit; (b) Parallel tracking control screen, where A is the path accuracy indicator, B is the off-track error, C represents the guidance icon, D provides the steering sensitivity, E mandates steer on/off, F locates the shift track buttons, G is the GPS status indicator, H is the A-B (0) track button, and I shows the track number.

especially with large equipment, resulting in significant savings in seeds, fertilizer, and phytosanitary chemicals as well as reduced operator fatigue. Farmers are seeing returns on investment in just a few years.

Automatic Control of Variable-Rate Applications

The idea of variable rate application (VRA) is to apply the right amount of input, i.e., seeds, fertilizers, and pesticides, at the right time and at sub-plot precision, moving away from average rates per plot that result in economic losses and environmental threats. Mechatronics enables the practical implementation of VRA for precision agriculture (PA). Generally speaking, state-of-the-art VRA equipment requires three key mechatronic components: (1) sensors, (2) controllers, and (3) actuators.

Sub-plot precision is feasible with GPS receivers that provide the instantaneous position of farm equipment at a specific location within a field. In addition, vehicles require the support of an automated application controller to deliver the exact amount of product. The specific quantity of product to be applied at each location is commonly provided by either a *prescription map* preloaded to the vehicle's computer, or alternatively, estimated in real time using onboard crop health sensors.

There are specific sensors that must be part of VRA machines. For example, for intelligent sprayers to be capable of automatically adapting the rate of pesticide to the properties of trees, global and local positioning in the field or related to crops is required. Fertilizers, on the other hand, may benefit from maps of soil parameters (moisture, organic matter, nutrients), as well as vegetation (vigor, stress, weeds, temperature). In many modern sprayers, pressure and flow of applied resources (either liquid or gaseous) must be tracked to support automatic control and eventually achieve a precise application rate. Controllers are the devices that calculate the optimal application rate on the fly and provide intelligence to the mechatronics system. They often consist of microcontrollers reading sensor measurements or loaded maps to calculate the instantaneous rate of product application based on internal algorithms. This rate is continuously sent to actuators for the physical application of product. Controllers may include small monitoring displays or switches for manual actuation from the operator cabin, if needed. Actuators are electromechanical or electrohydraulic devices that receive electrical signals from the controllers to regulate the amount of product to apply. This regulation is usually achieved by varying the rotational speed of a pump, modifying the flow coming from a tank, or changing the settings of a valve to adjust the pressure or flow of the product. Changing the pressure of sprayed liquids, however, results in a change of the droplet size, which is not desirable for pest control. In these cases, the use of smart nozzles that are controlled through PWM signals is recommended.

As VRA technology is progressing quickly, intelligent applicators are becoming available commercially, mainly for commodity crops. An intelligent system can automatically adjust the amount of inputs dispersed in response to needs, even permitting the simultaneous use of several kinds of treatments, resulting in new ways of managing agricultural production. For example, an intelligent VRA seeder has the ability to change the number of seeds planted in the soil according to soil potential, either provided by prescription maps or detected using onboard sensors. Control of the seeding rate is achieved by actuating the opening of the distributing device to allow the desired number of seeds to go through.

In many cases, a feedback control system is required to achieve accurate control of the application rate. For example, in applying liquid chemicals, the application rate may be affected by changes in the moving speed of the vehicle, as well as the environmental conditions. Some *smart sprayers* are programmed to accurately control the amount of liquid chemical by adjusting the nozzles in response to changes of sprayer forward speed. This is normally accomplished using electronically controlled nozzle valves that are commanded from the onboard processor. Such a mechatronic system could additionally monitor the system pressure and flow in the distribution circuit with a GPS receiver, and even compensate changes of the amount of liquid exiting the nozzles resulting from pressure or flow pattern changes in the circuit.

Redesigning a Tractor Steering System with Electrohydraulic Components

Implementing auto-guidance capabilities in a tractor requires that the steering system can be controlled electrically for automated turning of the front wheels. Therefore, it is necessary to replace a traditional hydraulic steering system with an electrohydraulic system. This could be accomplished simply by replacing a conventional manually actuated steering control valve (figure 7a) by an electrohydraulic control system. Such a system (figure 7b) consists of a rotary potentiometer to track the motion of the steering wheel, an electronic controller to convert the steering signal to a control signal, and a solenoid-driven electrohydraulic control valve to implement the delivered control signal.

The upgraded electrohydraulic steering system can receive control signals from a computer controller enabled to create appropriate steering commands in terms of outputs from an auto-guided system, making navigation possible without the input of human drivers to achieve autonomous operations with the tractor. As the major components of an electrohydraulic system are connected by wires, such an operation is also called "actuation by wire."

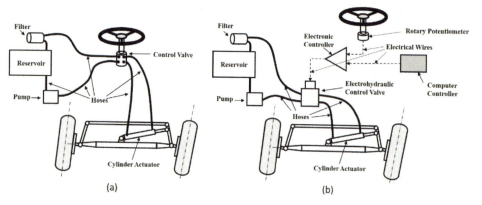

Figure 7. Tractor steering systems: (a) traditional hydraulic steering system; and (b) electrohydraulic steering system.

Use of Ultrasonic Sensors for Measuring Ranges

Agricultural machinery often needs to be "aware" of the position of objects in the vicinity of farming operations, as well as the position of the machinery. Ultrasonic sensors are often used to perform such measurements.

In order to use an ultrasonic (or sonar) sensor, a microprocessor is often needed to convert the analog signals (which are in the range of 0–5 V) from the ultrasonic sensor to digital signals, so that the recorded data can be further used by other components of automated or robotic machinery. For an example consider the HC-SR04, which consists of a sound emitter and an echo receiver such that it measures the time elapsed between a sound wave being sent by the emitter and its return back from the targeted object. The speed of sound is approximately 330 m·s^{-1}, which means that it needs 3 s for sound to travel 1,000 m. The HC-SR04 sensor can measure ranges up to 4.0 m, hence the time measurements are in the order of milliseconds and microseconds for very short ranges. The sound must travel through the air, and the speed of sound depends on environmental conditions, mainly the ambient temperature. If this sensor is used on a hot summer day with an average temperature of 35°C, for example, using equation 9, the corrected sound speed will be slightly higher, at 352 m·s^{-1}.

Figure 8 shows how the sensor was connected to and powered by a commercial product (Arduino Uno microprocessor, for illustration purposes) in a laboratory setup (also for illustration). After completing all the wiring of the system as shown in figure 8, it is necessary to select an unused USB port and any of the default baud rates in the interfacing computer. If the baud rate and serial port are properly set in a computer with a display console, and the measured ranges have been set via software at an updating frequency of 1 Hz, the system could then perform one measurement per second. After the system has been set up, it is important to check its accuracy and robustness by moving the target object in the space ahead of the sensor.

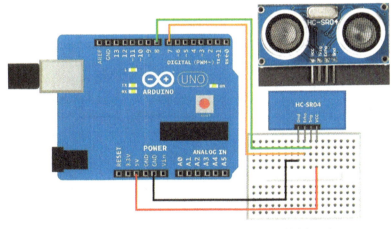

Figure 8. Assembly of an ultrasonic rangefinder HC-SR04 with an Arduino microprocessor.

Examples

Example 1: Digitization of analog signals

Problem:
Mechatronic systems require sensors to monitor the performance of automated operations. Analog sensors are commonly used for such tasks. A mechatronics-based steering mechanism uses a linear potentiometer to estimate the steering angle of an auto-guided tractor, outputting an analog signal in volts as the front wheels rotate. To make the acquired data usable by

a computerized system to automate steering, it is necessary to convert the analog data to digital format.

Given the analog signal coming from a steering potentiometer, digitize the signal using 4 bits of resolution, by these steps.

1. Calculate the number of levels coded by the 4-bit signal taking into account that the minimum voltage output by the potentiometer is 1.2 V and the maximum voltage is limited to 4.7 V, i.e., any reading coming from the potentiometer will belong to the interval 1.2 V–4.7 V. How many steps comprise this digital signal?
2. Establish a correspondence between the analog readings within the interval and each digital level from 0000 to 1111, drafting a table to reflect the correlation between signals.
3. Plot both signals overlaid to graphically depict the effect of digitizing a signal and the loss of accuracy behind the process. According to the plot, what would be the digital value corresponding to a potentiometer reading of 4.1 V?

Solution:

The linear potentiometer has a rod whose position varies from retraction (1.2 V) to full extension (4.7 V). Any rod position between both extremes will correspond to a voltage in the range 1.2 V–4.7 V. The number of levels L encoded in the signal for $n = 4$ bits is calculated using equation 1:

$$L = 2^n = 2^4 = \mathbf{16\ levels}$$

Thus, the number of steps between the lowest digital number 0000 and the highest 1111 is **15 intervals**. Table 1 specifies each digital value coded by the 4-bit signal, taking into account that the size of each interval ΔV is set by:

$$\Delta V = (4.7 - 1.2)/15 = 3.5/15 = 0.233\ V$$

A potentiometer reading of **4.1 V** belongs to the interval between [4.000, 4.233], that is, greater or equal to 4 V and less than 4.233 V, which according to table 1 corresponds to **1101**. Differences below 233 mV will not be registered with a 4-bit signal. However, by increasing the number of bits, the error will be diminished and the "stairway" profile of figure 9 will get closer and closer to the straight line joining 1.2 V and 4.7 V.

Table 1. Digitization of an analog signal with 4 bits between 1.2 V and 4.7 V.

Bit 1	Bit 2	Bit 3	Bit 4	4-Bit Digital Signal	Analog Equivalence (V)
1	1	1	1	1111	4.70000
1	1	1	0	1110	4.46666
1	1	0	1	1101	4.23333
1	1	0	0	1100	4.00000
1	0	1	1	1011	3.76666
1	0	1	0	1010	3.53333
1	0	0	1	1001	3.30000
1	0	0	0	1000	3.06666
0	1	1	1	0111	2.83333
0	1	1	0	0110	2.60000
0	1	0	1	0101	2.36666
0	1	0	0	0100	2.13333
0	0	1	1	0011	1.90000
0	0	1	0	0010	1.66666
0	0	0	1	0001	1.43333
0	0	0	0	0000	1.20000

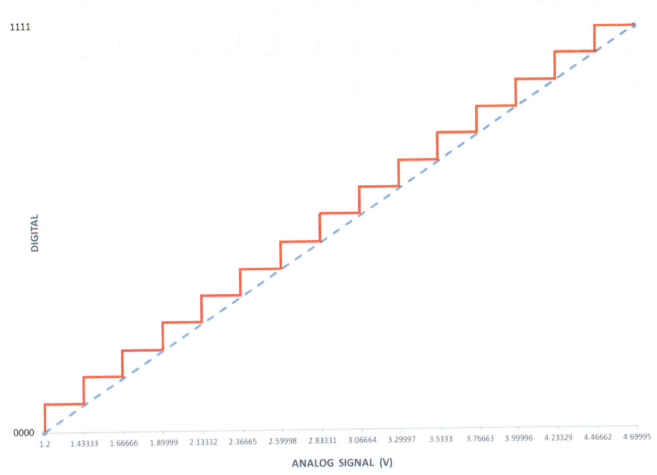

Figure 9. Digitization of an analog signal with 4 bits between 1.2 V and 4.7 V.

Example 2: Transformation of GPS coordinates

Problem:
A soil-surveying robot uses a GPS receiver to locate sampling points forming a grid in a field. Those points constitute the reference base for several precision farming applications related to the spatial distribution of soil properties such as compactness, pH, and moisture content. The location data (table 2) provided by the GPS receiver is in a standard NMEA code format. Transform the data (i.e., the geodetic coordinates provided by a handheld GPS receiver) to the local tangent plane (LTP) frame to be more directly useful to farmers.

Solution:
The first step in the transformation process requires the selection of a reference ellipsoid. Choose the WGS 84 reference ellipsoid because it is widely used for agricultural applications. Use equations 2 to 7 and apply the transform function (equation 8) to the 23 points given in geodetic coordinates (table 2) to convert them into LTP coordinates. For that reference ellipsoid,

a = semi-major axis of WGS 84 reference ellipsoid = 6378137 m

e = eccentricity of WGS 84 reference ellipsoid = 0.0818

$$N_0(\lambda) = \frac{a}{\sqrt{1-e^2 \cdot \sin^2 \lambda}} \tag{4}$$

$$(N_0 + h) \cdot \cos\lambda \cdot \cos\varphi \tag{5}$$

$$Y = (N_0 + h) \cdot \cos\lambda \cdot \sin\varphi \tag{6}$$

$$Z = \left[h + N_0 \cdot (1-e^2) \right] \cdot \sin\lambda \tag{7}$$

$$\begin{bmatrix} N \\ E \\ D \end{bmatrix} = \begin{bmatrix} -\sin\lambda \cdot \cos\varphi & -\sin\lambda \cdot \sin\varphi & \cos\lambda \\ -\sin\varphi & \cos\varphi & 0 \\ -\cos\lambda \cdot \cos\varphi & -\cos\lambda \cdot \sin\varphi & -\sin\lambda \end{bmatrix} \cdot \begin{bmatrix} X - X_0 \\ Y - Y_0 \\ Z - Z_0 \end{bmatrix} \tag{8}$$

The length of the normal N_0 is the distance from the surface of the ellipsoid of reference to its intersection with the rotation axis and $[\lambda, \varphi, h]$ is a point in geodetic coordinates recorded by the GPS receiver; $[X, Y, Z]$ is the point transformed to ECEF coordinates (m), with $[X_0, Y_0, Z_0]$ being the user-defined origin of coordinates in ECEF; and $[N, E, D]$ is the point being converted in LTP coordinates (m).

MATLAB® can provide a convenient programming environment to transform the geodetic coordinates to a flat frame, and save them in a text file. Table 3 summarizes the results as they would appear in a MATLAB® (.m) file.

These 23 survey points can be plotted in a Cartesian frame East-North (namely in LTP coordinates) to see their spatial distribution in the field, with East and North axes oriented as shown in figure 10.

A crucial advantage of using flat coordinates like LTP is that Euclidean geometry

Table 2. GPS geodetic coordinates of field points.

Point	Latitude (°)	Latitude (min)	Longitude (°)	Longitude (min)	Altitude (m)
Origin	39	28.9761	0	−20.2647	4.2
1	39	28.9744	0	−20.2539	5.1
2	39	28.9788	0	−20.2508	5.3
3	39	28.9827	0	−20.2475	5.9
4	39	28.9873	0	−20.2431	5.6
5	39	28.9929	0	−20.2384	4.8
6	39	28.9973	0	−20.2450	5.0
7	39	28.9924	0	−20.2500	5.2
8	39	28.9878	0	−20.2557	5.2
9	39	28.9832	0	−20.2593	5.4
10	39	28.9792	0	−20.2626	5.2
11	39	28.9814	0	−20.2672	4.8
12	39	28.9856	0	−20.2638	5.5
13	39	28.9897	0	−20.2596	5.5
14	39	28.9941	0	−20.2542	5.0
15	39	28.9993	0	−20.2491	5.0
16	39	29.0024	0	−20.2534	5.1
17	39	28.9976	0	−20.2590	4.9
18	39	28.9929	0	−20.2643	4.9
19	39	28.9883	0	−20.2695	4.9
20	39	28.9846	0	−20.2738	4.8
21	39	28.9819	0	−20.2770	4.7
22	39	28.9700	0	−20.2519	4.5

can be extensively used to calculate distances, areas, and volumes. For example, to calculate the total area covered by the surveyed grid, split the resulting trapezoid into two irregular triangles (figure 11), one defined by the points A-B-C, and the other by the three points A-B-D. Apply Euclidean geometry to calculate the area of an irregular triangle from the measurement of its three sides using the equation:

$$\text{Area} = \sqrt{K \cdot (K-a) \cdot (K-b) \cdot (K-c)} \quad (14)$$

where, a, b, and c are the lengths of the three sides of the triangle, and $K = \dfrac{a+b+c}{2}$.

The distance between two points A and B can also be determined by the following equation:

$$L_{A-B} = \sqrt{(E_A - E_B)^2 + (N_A - N_B)^2} \quad (15)$$

where L_{A-B} = Euclidean distance (straight line) between points A and B (m)
 $[E_A, N_A]$ = the LTP coordinates east and north of point A (m)
 $[E_B, N_B]$ = the LTP coordinates east and north of point B (m), calculated in table 3.

Using the area equation, the areas of the two triangles presented in figure 11 are determined as 627 m² for the yellow triangle (ADB) and 1,054 m² for the green triangle (ABC), with a total **area of 1,681 m²**. The corresponding Euclidean distances are 50.9 m, 42.1 m, 60.0 m, 27.8 m, and 46.6 m, respectively, for L_{A-C}, L_{C-B}, L_{A-B}, L_{A-D}, and L_{D-B}, as:

$$L_{A-B} = \sqrt{(E_A - E_B)^2 + (N_A - N_B)^2} =$$

$$\sqrt{(16.2 - 18.3)^2 + (48.7 - (-11.3))^2} = 60.0$$

We have not said anything about the Z direction of the field, but the Altitude column in table 2 and the Down column in table 3 both suggest that the field is quite flat, as the elevation of the points over the ground does not vary by much along the 22 points.

Figure 12 shows the sampled points of figure 10 overlaid with a satellite image that allows users to know additional details of the field such as crop type, lanes, surrounding buildings (affecting GPS performance), and other relevant information.

Table 3. LTP coordinates for the field surveyed with a GPS receiver.

Point	East (m)	North (m)	Down (m)
Origin	0	0	0
1	15.5	−3.1	−0.9
2	19.9	5.0	−1.1
3	24.7	12.2	−1.7
4	31.0	20.7	−1.4
5	37.7	31.1	−0.6
6	28.2	39.2	−0.8
7	21.1	30.2	−1.0
8	12.9	21.6	−1.0
9	7.7	13.1	−1.2
10	3.0	5.7	−1.0
11	−3.6	9.8	−0.6
12	1.3	17.6	−1.3
13	7.3	25.2	−1.3
14	15.1	33.3	−0.8
15	22.4	42.9	−0.8
16	16.2	48.7	−0.9
17	8.2	39.8	−0.7
18	0.6	31.1	−0.7
19	−6.9	22.6	−0.7
20	−13.0	15.7	−0.6
21	−17.6	10.7	−0.5
22	18.3	−11.3	−0.3

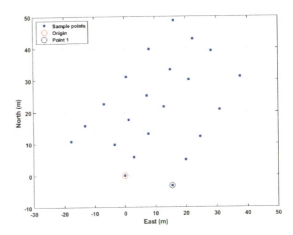

Figure 10. Planar representation of the 23 points sampled in the field with a local origin.

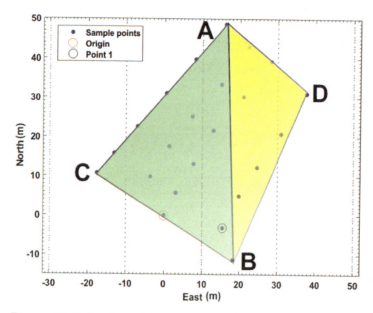

Figure 11. Estimation of the area covered by the sampled points in the surveyed field.

Figure 12. Sampled points over a satellite image of the surveyed plot (origin at point 23).

Example 3: Configuration of a machine vision system for detecting cherry tomatoes on an intelligent harvester

Problem:

Assume you are involved in designing an in-field quality system for the on-the-fly inspection of produce on board an intelligent cherry tomato harvester. Your specific assignment is the design of a machine vision system to detect blemishes in cherry tomatoes being transported by a conveyor belt on the harvester, as illustrated in figure 13. You are required to use an existing camera that carries a CCD sensor of dimensions 6.4 mm × 4.8 mm. The space allowed to mount the camera (camera height h) is about 40 cm above the belt. However, you can buy any lens to assure a horizontal FOV of 54 cm to cover the entire width of the conveyor belt. Determine the required focal length of the lens.

Solution:

The first step in the design of this sensing system is to calculate the focal length (f) of the lens needed to cover the requested FOV. Normally, the calculation of the focal length requires knowing two main parameters of lens geometry: the distance between the CCD sensor and the optical center of the lens, d_1, and the distance between the optical center of the lens and the conveyor belt, d_2. We know d_2 = 400 mm, FOV = 540 mm, and A, the horizontal dimension of the imaging sensor, is 6.4 mm, so d_1 can be easily determined according to equations 10 and 11:

$$\frac{d_1}{d_2} = \frac{A}{FOV} \tag{11}$$

Thus,

$$d_1 = \frac{A \cdot d_2}{FOV} = \frac{6.4 \cdot 400}{540} = 4.74 \text{ mm}$$

The focal length, f, can then be determined using equation 10:

$$\frac{1}{f} = \frac{1}{d_1} + \frac{1}{d_2} \tag{10}$$

Thus,

$$f = \frac{d_1 \cdot d_2}{d_1 + d_2} = \frac{4.74 \cdot 400}{4.74 + 400} = 4.68 \text{ mm}$$

No lens manufacturer will likely offer a lens with a focal length of 4.68 mm; therefore, you must choose the closest one from what is commercially available. The lenses commercially available for this camera have the following focal lengths: 2.8 mm, 4 mm, 6 mm, 8 mm, 12 mm, and 16 mm. A proper approach is to choose the best lens for this application, and readjust the distance between the camera and the belt in order to keep the requested FOV covered. Out of the list offered above, the best option is choosing a lens with $f = $ **4 mm**. That choice will change the original parameters slightly, and you will have to readjust some of the initial conditions in order to maintain the same FOV, which is the main condition to meet. The easiest modification will be lowering the position of the camera to a distance of **34 cm** to the conveyor (d_2 = 340 mm from the focal length equation). If the camera is fixed and d_2 has to remain at the initial 40 cm, the resulting field of view would be larger than the necessary 54 cm, and applying image processing techniques would be necessary to remove useless sections of the images.

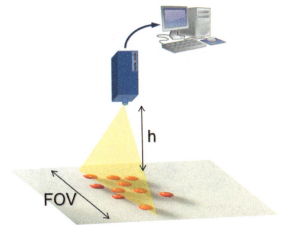

Figure 13. Geometrical requirements of a vision-based inspection system for a conveyor belt on a harvester.

Example 4: Estimation of a robot velocity using an accelerometer

Problem:

The accelerometer of figure 14a was installed in the agricultural robot of figure 14c. When moving along vineyard rows, the output measurements from the accelerometer were recorded in table 4, including the time of each measurement and its corresponding linear acceleration in the forward direction given in g, the gravitational acceleration.

1. Calculate the instantaneous acceleration of each point in m·s^{-2}, taking into account that one g is equivalent to 9.8 m·s^{-2}.
2. Calculate the time elapsed between consecutive measurements Δt in s.
3. Estimate the average sample rate (Hz) at which the accelerometer was working.
4. Calculate the corresponding velocity for every measurement with equation 12, taking into account that the vehicle started from a resting position ($V_0 = 0$ m s^{-1}) and always moved forward.
5. Plot the robot's acceleration (m s^{-2}) and velocity (km h^{-1}) for the duration of the testing run.

Table 4. Acceleration of a vehicle recorded with the accelerometer of figure 14a.

Data point	Time (s)	Acceleration (g)
1	7.088	0.005
2	8.025	0.018
3	9.025	0.009
4	10.025	0.009
5	11.025	0.008
6	12.025	0.009
7	13.025	0.009
8	14.025	0.009
9	15.025	0.008
10	16.025	0.008
11	17.025	0.009
12	18.025	0.009
13	19.025	0.008
14	20.088	0.009
15	21.088	−0.009
16	21.963	−0.019
17	23.025	−0.001

Solution:

Table 4 can be completed by multiplying the acceleration expressed in g by 9.8, and by applying the expression $\Delta t = t_k - t_{k-1}$ to every point of the table, except for the first point t_1 that has no preceding measurement.

Data point	Time (s)	Acceleration (g)	Acceleration (m s^{-2})	Δt (s)
1	7.088	0.005	0.050	0
2	8.025	0.018	0.179	0.938
3	9.025	0.009	0.091	1.000
4	10.025	0.009	0.085	1.000
5	11.025	0.008	0.083	1.000
6	12.025	0.009	0.088	1.000
7	13.025	0.009	0.085	1.000
8	14.025	0.009	0.084	1.000
9	15.025	0.008	0.080	1.000
10	16.025	0.008	0.081	1.000
11	17.025	0.009	0.086	1.000
12	18.025	0.009	0.084	1.000
13	19.025	0.008	0.083	1.000
14	20.088	0.009	0.089	1.063
15	21.088	−0.009	−0.092	1.000
16	21.963	−0.019	−0.187	0.875
17	23.025	−0.001	−0.009	1.063
			Average	0.996

According to the previous results, the average time elapsed between two consecutive measurements Δt is 0.996 s, which corresponds to approximately one measurement per second, or 1 Hz. The velocity of a vehicle can be estimated from its acceleration with equation 12. Table 5 specifies the calculation at each specific measurement.

Figure 15 plots the measured acceleration and the calculated velocity for a given time interval of 16 seconds. Notice that there are data points with a negative acceleration (deceleration) but the velocity is never negative because the vehicle always moved forward or stayed at rest. Accelerometers suffer from noisy estimates, and as a result, the final velocity calculated in table 5 may not be very accurate. Consequently, it is a good practice to estimate vehicle speeds redundantly with at least two independent sensors working under different principles. In this example, for instance, forward velocity was also estimated with an onboard GPS receiver.

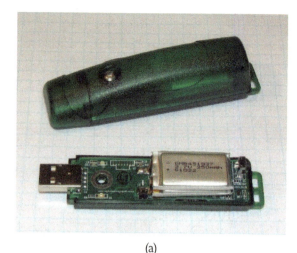

(a)

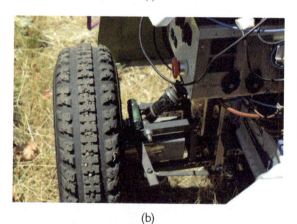

(b)

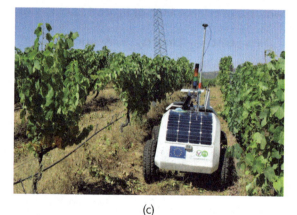

(c)

Figure 14. (a) Accelerometer Gulf Coast X2-2; (b) sensor mounting; (c) in an agricultural robot.

Table 5. Velocity of a robot estimated with an accelerometer.

Data point	Acceleration (m s^{-2})	Δt (s)	Velocity (m s^{-1})	V (km h^{-1})
1	0.050	0	$V_1 = V_0 + a_1 \cdot \Delta t = 0 + 0.05 \cdot 0 = 0$	0.0
2	0.179	0.938	$V_2 = V_1 + a_2 \cdot \Delta t = 0 + 0.179 \cdot 0.938 = 0.17$	0.6
3	0.091	1.000	$V_3 = V_2 + a_3 \cdot \Delta t = 0.17 + 0.091 \cdot 1 = 0.26$	0.9
4	0.085	1.000	0.34	1.2
5	0.083	1.000	0.43	1.5
6	0.088	1.000	0.51	1.9
7	0.085	1.000	0.60	2.2
8	0.084	1.000	0.68	2.5
9	0.080	1.000	0.76	2.7
10	0.081	1.000	0.84	3.0
11	0.086	1.000	0.93	3.3
12	0.084	1.000	1.01	3.7
13	0.083	1.000	1.10	3.9
14	0.089	1.063	1.19	4.3
15	-0.092	1.000	1.10	4.0
16	-0.187	0.875	0.94	3.4
17	-0.009	1.063	0.93	3.3

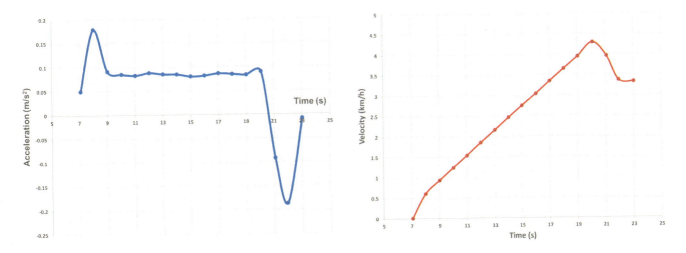

Figure 15. Acceleration and velocity of a farm robot estimated with an accelerometer.

Image Credits

Figure 1a. John Deere. (2020). Autonomous mower. Retrieved from https://www.deere.es/es/campaigns/ag-turf/tango/. [Fair Use].

Figure 1b. Rovira-Más, F. (CC BY 4.0). (2020). (b) GPS-based autonomous rice transplanter.

Figure 2. Verified Market Research (2018). (CC BY 4.0). (2020). Expected growth of agricultural robot market.

Figure 3. Rovira-Más, F. (CC BY 4.0). (2020). Sensor architecture for intelligent agricultural vehicles.

Figure 4. Rovira-Más, F. (CC BY 4.0). (2020). Geometrical relationship between an imaging sensor, lens, and FOV.

Figure 5. Rovira-Más, F. (CC BY 4.0). (2020). Color-based segmentation of mandarin oranges.

Figure 6. Rovira-Más, F. (CC BY 4.0). (2020). Auto-guidance systems: (a) Light-bar kit; (b) Parallel tracking control screen.

Figure 7. Zhang, Q. (CC BY 4.0). (2020). Tractor steering systems: (a) traditional hydraulic steering system; and (b) electrohydraulic steering system.

Figure 8. Adapted from T. Karvinen, K. Karvinen, V. Valtokari (*Make: Sensors*, Maker Media, 2014). (2020). Assembly of an ultrasonic rangefinder HC-SR04 with an Arduino processor. [Fair Use].

Figure 9. Rovira-Más, F. (CC BY 4.0). (2020). Digitalization of an analog signal with 4 bits between 1.2 V and 4.7 V.

Figure 10. Rovira-Más, F. (CC BY 4.0). (2020). Planar representation of the 23 points sampled in the field with a local origin.

Figure 11. Rovira-Más, F. (CC BY 4.0). (2020). Estimation of the area covered by the sampled points in the surveyed field.

Figure 12. Saiz-Rubio, V. (CC BY 4.0). (2020). Sampled points over a satellite image of the surveyed plot (origin in number 23).

Figure 13. Rovira-Más, F. (CC BY 4.0). (2020). Geometrical requirements of a vision-based inspection system for a conveyor belt on a harvester.

Figure 14a. Gulf Coast Data Concepts. (2020). Accelerometer Gulf Coast X2-2. Retrieved from http://www.gcdataconcepts.com/x2-1.html. [Fair Use].

Figure 14b & 14c. Saiz-Rubio, V. (CC BY 4.0). (2020). (b) sensor mounting. (c) mounting an agricultural robot.

Figure 15. Rovira-Más, F. (CC BY 4.0). (2020). Acceleration and velocity of a farm robot estimated with an accelerometer.

References

Bolton, W. (1999). *Mechatronics* (2nd ed). New York: Addison Wesley Longman Publishing.

Myklevy, M., Doherty, P., & Makower, J. (2016). *The new grand strategy.* New York: St. Martin's Press.

Rovira-Más, F., Zhang, Q., & Hansen, A. C. (2010). *Mechatronics and intelligent systems for off-road vehicles.* London: Springer-Verlag.

Russell, S., & Norvig, P. (2003). *Artificial Intelligence: a modern approach* (2nd ed). Upper Saddle River, NJ: Prentice Hall.

Verified Market Research. (2018). Global agriculture robots market size by type (driverless tractors, automated harvesting machine, others), by application (field farming, dairy management, indoor farming, others), by geography scope and forecast. Report ID: 3426. Verified Market Research Inc.: Boonton, NJ, USA, pp. 78.

Water Budgets for Sustainable Water Management

Stacy L. Hutchinson
Kansas State University
Biological and Agricultural Engineering
Manhattan, Kansas, USA

KEY TERMS		
Hydrologic cycle	Infiltration and runoff	Water balance
Soil water relationships	Evapotranspiration	Agricultural water management
Precipitation	Water storage	Urban stormwater management

Variables

- θ_g = gravimetric water content
- θ_v = volumetric water content
- ρ_b = bulk density of the soil
- AW = available water
- E = Evaporation
- ET = evapotranspiration
- DS = deep seepage
- FC = field capacity
- In = Infiltration
- Ir = irrigation
- PWP = permanent wilt point
- P = precipitation
- R = net runoff ($R_{in} - R_{out}$)
- R_{in} = runoff into the area of interest
- R_{out} = runoff out of the area of interest
- T = Transpiration
- ΔS = change in soil water storage

Introduction

Water is central to many discussions regionally, nationally, and globally—be it the lack of water, the overabundance of water, or poor water quality—pushing us to seek answers on how to ensure we can maintain a safe, reliable, adequate water supply for human and environmental well-being. The United Nations (2013) defines water security as

> ...the capacity of a population to safeguard sustainable access to adequate quantities of acceptable quality water for sustaining livelihoods, human well-being, and socio-economic development, for ensuring protection against water-borne pollution and water-related disasters, and for preserving ecosystems in a climate of peace and political stability...

This highlights the need to understand the complex relationships associated with water and the need to research, develop, and implement engineered systems that assist with enhancing water security across the nation and world. This chapter focuses on understanding the system water balance, which is the fundamental basis for all water management decisions.

Outcomes

After reading this chapter, you should be able to:

- Describe the basic components of a water balance, including precipitation, infiltration, evaporation, evapotranspiration, and runoff

- Calculate a water budget

- Use a water budget for the design and implementation of a simple water management system for irrigation or sustainable stormwater management

Concepts

Hydrologic Cycle

Hydrology is the study of how water moves around the Earth in continuous motion, cycling through liquid, gaseous, and solid phases. This cycle is called the hydrologic cycle or water cycle (figure 1). At the global scale, the hydrologic cycle can be thought of as a closed system that obeys the conservation law; a closed system has no external interactions. The vast majority of water in the system continues to cycle through the three states of matter: liquid, gaseous, and solid.

Key processes in the hydrologic cycle are:

- Precipitation (P), which is the primary input into a water budget. Precipitation describes all forms of water (liquid and solid) that falls or condenses from the atmosphere and reaches the earth's surface (Huffman

et al., 2013), including rainfall, drizzle, snow, hail, and dew.
- Infiltration (In), which is the movement of water into the soil. Infiltrated water from precipitation and irrigation are the primary sources of water for plant growth.
- Evaporation (E), which is the conversion of liquid or solid water into water vapor (gaseous water).
- Transpiration (T), which is the process through which plants use water. Water is absorbed from the soil, moved through the plant, and evaporated from the leaves.
- Evapotranspiration (ET), which is the combination of evaporation and transpiration to describe the water use, or output, from vegetated surfaces.
- Runoff (R), which is the precipitation water that does not infiltrate into the soil. Runoff is generally an output, or loss of water, from the system or area of interest, but can also be an input if this water runs into the system or area of interest.
- Deep seepage (DS), which is water that infiltrates below the root zone, which is the depth of the area of interest for the water budget.

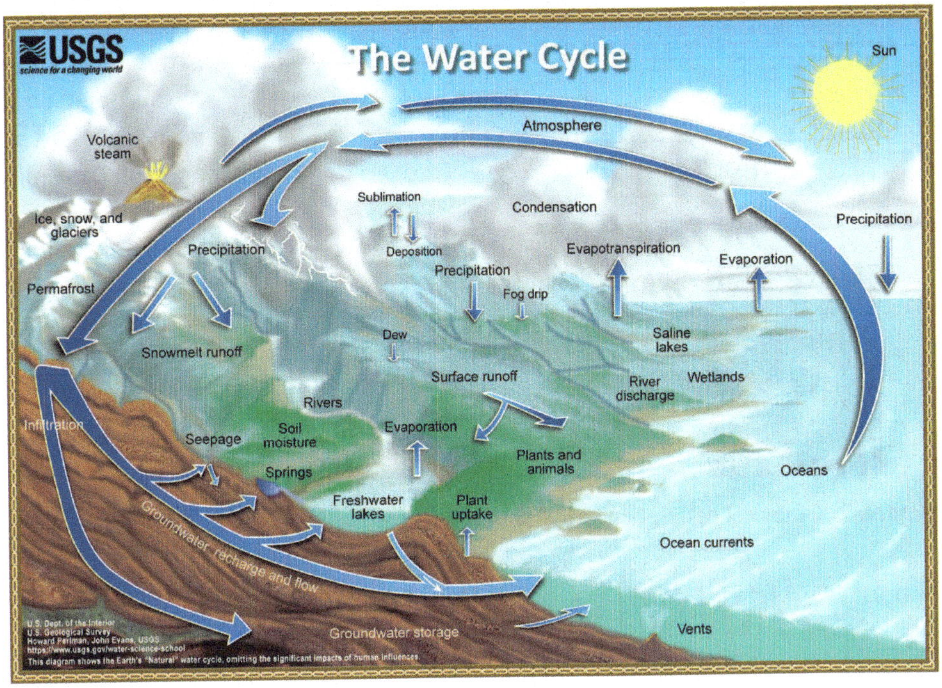

Figure 1. The water cycle, hydrologic cycle (USGS, 2020a).

Soil Water Relationships

The design of water management systems by biosystems engineers involves water moving through or being held in the soil. Thus, soil-water dynamics are a critical factor in the design process.

A volume of soil is comprised of solids and voids, or pore space. Porosity is the volume of voids as compared to the total volume of the soil. The proportions of solids and voids depend on the soil particles (sand, slit, clay, organic matter) and structure (known as peds), with coarse-textured soils (i.e., dominated by sand) having approximately 30% voids and finer-textured soils (i.e., containing more silt or clay) having as much as 50% void space. Water that infiltrates into the soil profile is stored in the soil voids. When all the void space is filled with water, the soil is at saturation water content.

Gravitational forces remove water up to 33 kPa (1/3 bar) of tension; this is drainage or gravitational water. The soil water content after gravitational drainage for

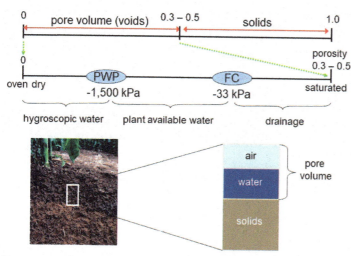

Figure 2. Graphical description of soil water.

approximately 24 hrs is called field capacity (*FC*) and is the maximum water available for plant growth. At this point, water is held in the soil in the smaller pore spaces by capillary action and surface tension, and can be removed from the soil profile by plant roots extracting the water from those pores. The water tension of the soil at field capacity (the suction pressure required to extract water from the pore spaces) is around 10 kPa (0.1 bars) for sands to 30 kPa (0.3 bars) for heavier soils containing more silt or clay. Plants are able to extract water from the soil profile with up to 1,500 kPa (15 bars) of tension; the water content at 1,500 kPa (15 bars) of tension is called the permanent wilt point (*PWP*). The total plant available water (*AW*) for a given soil profile is the difference between *FC* and *PWP*:

$$AW = FC - PWP \qquad (1)$$

Plant water use is the primary means of removing water from the soil profile. Once all of the available water has been taken up by plants, some water remains in the soil, in very small pore spaces where very high suction pressure would be needed to remove that water. This film of water is more tightly bound to soil particles than can be extracted by plants and is called *hygroscopic water*. These relationships are shown in figure 2.

The amount of water in the soil is called the soil water content or soil moisture content, and is often indicated using the symbol θ. When soil water content is expressed on a mass basis, i.e., mass of water in the soil compared to total mass of soil, it is called gravimetric water content (θ_g). Gravimetric water content can be calculated by expressing mass of water as a proportion of total wet mass of soil, known as wet weight water content, or as a proportion of total dry mass of soil, known as dry weight water content (equation 2). When expressed on a volume basis, i.e., volume of water as a proportion of the total volume, with units $cm_w^3\ cm^{-3}$, the value is called volumetric water content (θ_v) (equation 3). Gravimetric (dry weight) and volumetric water content are related through the bulk density of the soil (ρ_b), which is the mass of soil particles in a given volume of soil, expressed in g cm^{-3}.

$$\text{gravimetric water content }(\theta_g) = \frac{\text{mass water}}{\text{mass dry soil}} = \frac{\text{mass total soil} - \text{mass dry soil}}{\text{mass dry soil}} \qquad (2)$$

$$\text{volumetric water content }(\theta_v) = \frac{\text{volume water}}{\text{total volume soil}} = \theta_g \frac{\text{soil bulk density }(\rho_b)}{\text{density of water }(\rho_w)} \qquad (3)$$

As collection of a known volume of soil is more difficult than collecting a simple grab sample of soil, it is much easier to determine gravimetric water content by mass. However, volumetric soil moisture is much easier to use in calculations because it can be expressed as an equivalent depth of rainfall (mm) over a given area and, thus, be directly related to rainfall, which is most commonly reported in units of depth (such as mm) and never in mass. The volume of water input to the water budget can be calculated by multiplying the depth of rainfall by the area receiving rain.

Water Budget Calculation

A water budget, or water balance, is a measure of all water flowing into and out of an area of interest, along with the change of water storage in the area. This could be an irrigated field, a lake or pond, or green infrastructure such as a stormwater management system like a bioretention cell or a rain garden (a vegetated area to absorb and store stormwater runoff). At smaller scales, the hydrologic cycle is characterized using a water budget, which is the primary tool used for designing and managing water resources systems, including stormwater runoff management and irrigation systems. Water budgets are calculated for a defined system or area (e.g., field, pond) over a specified time period (e.g., rainfall event, growing season, month, year).

The water budget is calculated by quantifying the inputs, outputs, and change in water storage (ΔS) of the system or project (equation 4; figure 3). While precipitation is the primary input to the water budget, others include runoff into the system (R_{in}) and water added through irrigation (Ir). Outputs from the system include runoff (R_{out}), deep seepage (DS), and evapotranspiration (ET). The change in system storage (ΔS) may be positive, such as an increase in pond water level after a rainfall event, or negative, such as the decline in soil moisture from plant water use.

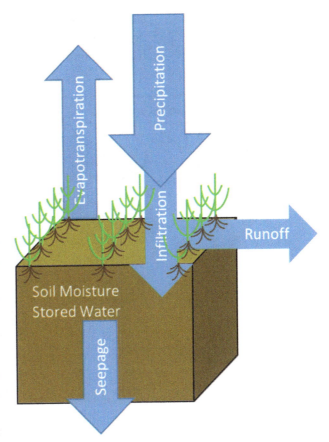

Figure 3. Water balance.

$$P + Ir \pm R - ET - DS = \Delta S \qquad (4)$$

where P = precipitation
Ir = irrigation
R = net runoff ($R_{in} - R_{out}$)
ET = evapotranspiration
DS = deep seepage
ΔS = change in soil water storage

The first step in developing a water budget for water resources management is to collect input, output, and storage data. There are many sources of data for this, including the National Oceanic and Atmospheric Administration

(NOAA) National Centers for Environmental Information (NCEI) (NOAA, 2019), which includes data from around the world. Within the U.S., the U.S. Geological Survey (USGS) real-time water data (USGS, 2020b) and the U.S. Department of Agriculture (USDA) Natural Resources Conservation Service (NRCS) Web Soil Survey (USDA-NRCS, 2019a) are also available. Each of these sources has extensive sets of data available to assist with management of water resources. Local-level data from state climatologists, research farms, and project gages should also be considered when available. High quality precipitation data are particularly valuable, and necessary, when designing and implementing water management systems including irrigation systems and green infrastructure.

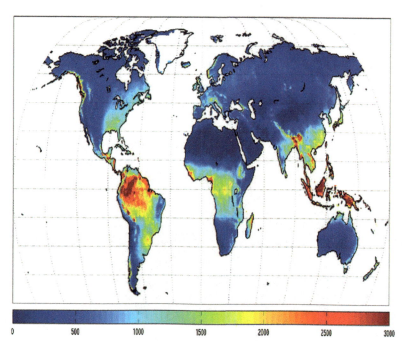

Figure 4. Mean annual precipitation (mm year^{-1}) for the 1980-2010 period from AgMERRA (from Ruane et al., 2015).

Precipitation

Precipitation is the primary input into a water budget. The most important form of precipitation for agricultural and biological engineering applications in most parts of the world is rainfall. Precipitation varies significantly across regions (figure 4) and throughout the year, but is relatively consistent over long time periods at a given location. This means long-term precipitation records can be used to calculate a water budget and for planning water resource management systems.

In addition to data available from NOAA NCEI and local sources mentioned above, long-term design storm information is available for the U.S. from the NOAA Atlas 14 point precipitation frequency estimates (HDSC, 2019). Statistical analysis of historic precipitation data is used to determine the magnitude and annual probability for various rainfall events. These values are used to size stormwater systems to deal with the majority of events, but not the most extreme.

Infiltration

Precipitation starts moving into the soil profile, i.e., infiltrating, as soon as it reaches the ground surface. Initial infiltration rates depend on the initial soil water content, but if the soil is drier than field capacity, can be very high as water fills depressions on the soil surface and begins moving into the soil. As the surface depressions fill and the surface soil becomes saturated, infiltration slows to steady-state and excess precipitation begins to run off the surface. Depending on the type of precipitation, initial soil water content, duration, and intensity (depth per time), precipitation may infiltrate into the soil profile and become soil water, or it will not infiltrate and will become runoff. The infiltration rate is related to soil bulk density, porosity, pore size distribution, and pore connectivity.

Runoff

Water that does not infiltrate into the soil is an output from the water budget (a loss from the water management system) unless the system is designed to collect it. The amount of runoff depends on land cover type, soil type, initial soil water content, and rainfall intensity; these four factors all interact and are not independent of each other. Land cover plays a significant role in determining the amount of runoff. Natural vegetated land usually has the highest infiltration rates. Infiltration rates decrease as natural land is converted to production land, due to increased compaction and changes in soil structure, and to developed land, due to impervious surfaces such as buildings and roads. Soil type also influences whether water can infiltrate. Soils with greater connected pore space will tend to generate less runoff. Initial soil water content is important because, if the soil is at or near saturation water content, the infiltration rate is more likely to be less than the rainfall rate. If rainfall rate is very high, runoff can occur even if the soil is quite dry. One of the most common methods for estimating likely surface runoff is the curve number method (USDA-NRCS, 2004). More information on the curve number method can be found in Chapter 10, Part 630, of the USDA-NRCS National Engineering Handbook (USDA-NRCS, 2004).

Evapotranspiration

Evapotranspiration (ET) is the primary means for removing water from the top of the soil system when plants are present. If there are no plants this loss will be confined to evaporation only. The amount of ET depends on the type and growth stage of the vegetation, the current weather, and soil water content at the location. The most accurate ET calculations involve an energy balance associated with incoming solar radiation and the mass transfer of water from moist vegetation surfaces to a drier atmosphere (Allen et al., 2005). ET is driven by solar radiation, but as the soil becomes drier, ET decreases. The relationship depends on the pore size distribution of the soil, so is regulated by the interaction of soil texture (sand, silt, and clay content) and structure (presence and type of soil peds).

Storage

Soils can store a significant amount of water and play an important role in controlling the rainfall-runoff processes, stream flow, and ET (Dobriyal et al., 2012; Meng et al., 2017). Available soil water influences productivity of natural and agricultural systems (Dobriyal et al., 2012). Understanding the available water stored in surface soils allows effective design and implementation of irrigation and stormwater management systems. The maximum storage is determined by the total porosity of the soil, which will be almost the same as the saturated water content. The ability of the soil to store more water at any given time is determined by the difference in current water content and saturated water content. These factors are influenced by soil texture, structure, porosity, and bulk density. More information about soils can be found from sources such as the FAO Soils Portal of the Food and Agriculture Organization of the United Nations (FAO, 2020), the European Soil Data Centre (https://esdac.jrc.ec.europa.eu/), and on the Web Soil Survey maintained by the USDA-NRCS (2019b).

Applications

Water is essential for life and critical for both food and energy production. With a finite supply of available fresh water and increasing global population demanding access to clean water for drinking, energy, and food production, proper water management is of the utmost importance. It is imperative to research, develop, and implement engineered systems that enhance local, regional, national, and global water security. Water balances are used in the design of systems to manage excess stormwater and to manage scarce available water in low-rainfall areas. Excess water management systems include components to retain and regulate runoff safely, while limited water management systems include components to provide irrigation water during dry seasons.

Urban Stormwater Management

As urbanization and land development occurs, the addition of impervious structures (roads, sidewalks, buildings) dramatically changes the hydrology of an area. Runoff volume linearly increases with increases in the impervious surface area. Hydrologic models and long-term stream flow monitoring show that, compared to pre-development, developed suburban areas have 2 to 4 times more annual runoff, and high-density areas have 15 times more runoff (Sahu et al., 2012; Suresh Babu and Mishra, 2012; Christianson et al., 2015). As runoff travels over pavement, rooftops, and fertilized lawns that make up much of the urban landscape, it picks up contaminants such as pathogens, heavy metals, sediments, and excess nutrients (Davis et al., 2001), creating both water quantity and water quality concerns.

Increases in the total volume of runoff from urban areas are caused not only by impervious structures. Even in low-density suburban areas where individual lots have large lawns and large public parks are created, current methods of construction and development greatly reduce the infiltration capacity of soils. Development typically involves stripping native vegetation from large areas of land, accelerating soil erosion rates 40,000 times (Gaffield et al., 2003). During this process, construction equipment compacts soil, reducing its ability to absorb runoff. Developed lawns can generate 90% as much runoff as pavement (Gaffield et al., 2003). Urban runoff may also include dry-weather flows from the irrigation of lawns and public parks. The final result of urban development is a significant increase in runoff that transports pollutants from dense urban centers to receiving water bodies, and small changes in land use can relate to large increases in flood potential and pollutant loading.

Over the past three decades, urban and urbanizing areas have started to increase the use of green infrastructure, or natural-based systems, for stormwater management. Green infrastructure works to reduce stormwater runoff and increase stormwater treatment on-site for floodwater and nonpoint, or diffuse, source pollution control using infiltration and biologically based treatment in the root zone. Green infrastructure is very different from traditional grey stormwater management systems, such as storm sewers. The more traditional grey systems use a centralized approach to water management, designed to quickly move runoff off the land and into nearby surface water with little to no storage or treatment. Green infrastructure is more resilient and can offer additional benefits, such as habitat, in developed and developing areas. For more information about green infrastructure see the U.S. Environmental Protection Agency (USEPA) green infrastructure webpage (USEPA, 2020).

Bioretention cells are one of the most effective green infrastructure systems. Bioretention cells are designed to infiltrate, store, and treat runoff water from impervious surfaces such as parking lots and roadways. Bioretention cells can be lined to prevent contaminants from moving into groundwater or unlined when there is no concern of groundwater contamination. Ideally, bioretention cells are designed to infiltrate and store the "first flush" rainfall event. (The first flush rainfall is the first 13-25 mm of rain that removes the majority of accumulated pollution from surfaces.) However, in many cases, there is not enough space to install such a large bioretention cell and the system is designed to fit the space in order to treat as much stormwater as possible.

Initial design and assessment of bioretention cell function is completed using a system water balance. Runoff from the impervious parking is directed toward stormwater management practices, like bioretention cells, where water infiltrates and is stored until removed from the soil (or growing media) through evapotranspiration.

Agricultural Water Management

Changing climate conditions, population growth, and urbanization present challenges for food supply. Agricultural intensification impacts local resources, particularly usable freshwater. This vulnerability is amplified by a changing climate, in which drought and variability in precipitation are becoming increasingly common. As a result, 52% of the world's population is projected to live in regions under water stress by 2050 (Schlosser et al., 2014). Concurrently, varying water availability, along with limiting nutrients, will constrain future food and energy production.

Agricultural water management is key to optimizing crop production. The amount of water needed for crop growth depends on the crop type, location, and time of year, but averages roughly 6.5 mm day^{-1} during the growing season (Brouwer and Heibloem, 1986). Depending on location, installation of runoff management systems such as terraces (ASAE S268.6 MAR2017) and grassed waterways (ASABE EP 464.1 FEB2016) (ASABE Standards, 2016, 2017) may be needed to manage excess rainfall and reduce flooding and erosion within fields. Design standards with specific design criteria are available from the ASABE Standards (https://asabe.org/Publications-Standards/Standards-Development/National-Standards/Published-Standards).

In addition to managing runoff from rainfall events, supplementing natural rainfall with irrigation water for crop production may also be required (sometimes in the same places). Maintaining available water for crop growth and development has a significant impact on crop yields. Irrigation system design and management rely on the use of detailed soil water balances to minimize water losses and optimize crop production. A daily water budget to account for all inputs and outputs from the system can be calculated.

Examples

Example 1: Calculating soil water content

Problem:
The following information is provided to determine the amount of available water storage in a soil profile. A soil sample collected in the field has a wet weight of 238 g. After drying at 105°C for 24 hrs, the soil sample dry weight is 209 g. Careful measurement of the soil sample determines the volume of the soil core is 135 cm³. Determine the available water storage, in both mass and volume basis, in the soil profile. (a) What was the original water content of the soil on a gravimetric basis (mass of water per total mass) and (b) on a volumetric basis (volume of water per volume of bulk soil)?

Solution:
(a) Calculate the gravimetric water content using equation 2:

$$\text{gravimetric water content } (\theta_g) = \frac{\text{mass water}}{\text{mass dry soil}} = \frac{\text{total mass soil} - \text{dry soil mass}}{\text{dry soil mass}}$$

$$\text{mass water (g)} = 238 \text{ g total} - 209 \text{ g dry} = 29 \text{ g water}$$

$$\text{gravimetric water content } (\theta_g) = \frac{29 \text{ g water}}{209 \text{ g dry}}$$

$$= 0.139 \frac{\text{g water}}{\text{g dry}} = 13.9\% \text{ water by mass}$$

(b) Calculate the volumetric water content using equation 3:

$$\text{volumetric water content } (\theta_v) = \frac{\text{volume water}}{\text{total volume}}$$

$$\text{volume of water} = \frac{\text{mass water}}{\text{density of water}} = \frac{29 \text{ g water}}{1 \text{ g per cm}^3} = 29 \text{ cm}^3 \text{ water}$$

$$\text{volumetric water content } (\theta_v) = \frac{29 \text{ cm}^3 \text{ water}}{135 \text{ cm}^3 \text{ dry}}$$

$$= 0.215 \frac{\text{cm}^3 \text{ water}}{\text{cm}^3 \text{ dry}} = 21.5\% \text{ water by volume}$$

Example 2: Determining plant available water

Problem:
Plant available water is one factor that helps to determine the need for irrigation as well as the available water storage in bioretention cells. A field has an established grass cover. The grass has an effective root zone depth of 0.90 m. The soil is a fine sandy loam with $FC = 23\%$ (vol) and $PWP = 10\%$ (vol), as shown in the water balance diagram.

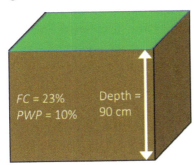

(a) If the soil is at field capacity, how much available water (cm) is in the effective root zone? (b) If the field water content averages 18% (vol) in the root zone, what is the available water storage depth for rainfall?

Solution:
(a) Calculate the available water using equation 1:

$$AW = FC - PWP = 23\% \text{ (vol)} - 10\% \text{ (vol)} = 13\% \text{ (vol)}$$

Thus, 13% of the soil volume is available water for plant use. When the volumetric water content is considered on a per unit area of soil, e.g., (cm³ water cm⁻² soil)/(cm³ soil cm⁻² soil), the units become depth water/depth soil, e.g. cm water cm⁻¹ soil profile. Thus, consider a unit area of soil and calculate the depth of available water:

$$\text{available water} = \frac{13 \text{ cm water}}{100 \text{ cm soil profile}} \times 90 \text{ cm}$$

$$= 11.7 \text{ cm water available in root zone}$$

(b) If the soil water content is 18% (vol), calculate the available storage as the difference between FC and volumetric water content:

$$\text{available storage} = FC - \theta_v = 23\% - 18\% = 5\%$$

And the depth of available storage in the root zone is:

$$\text{depth of available storage} = 0.05 \times 90 \text{ cm} = 4.5 \text{ cm}$$

Thus, the soil profile would be able to store 4.5 cm in the 90 cm root zone.

Example 3: Using a water balance to design a simple pond to store runoff

Problem:
A developer is planning the layout of a small housing development on 16.2 ha (40 ac) of land near Manhattan, KS, USA. According to local ordinance, the developer must retain any increased runoff due to development from the 2-yr, 24-hr rainfall event (86 mm) (HDSC, 2019). Prior to development, the area was able to infiltrate and store approximately 50 mm of this rainfall event. With the increase of impervious land cover (e.g. houses and roads), it is expected that infiltration and storage will be reduced to 30 mm. Determine the pond volume required to store the difference in expected runoff.

Solution:
Apply equation 4 to the 16.2-ha site to determine the expected increase in runoff from the site to due development:

Water balance equation:

$$\text{inputs} - \text{outputs} = \text{change in storage}$$

$$P + Ir \pm R - ET - DS = \Delta S$$

Assumptions:

$$\text{Pond is dry prior to rain}$$

$$Ir = 0$$

$$ET = 0 \text{ for short duration events}$$

$$DS = 0$$

$$\text{Therefore, } P \pm R = \Delta S$$

Pre-development:

$$P = 86 \text{ mm}$$

$$R = ?$$

$$\Delta S = 50 \text{ mm}$$

$$R = 86 \text{ mm} - 50 \text{ mm} = 36 \text{ mm of runoff}$$

Post-development:

$$P = 86 \text{ mm}$$

$$R = ?$$

$$\Delta S = 30 \text{ mm}$$

$$R = 86 \text{ mm} - 30 \text{ mm} = 56 \text{ mm of runoff}$$

Change in runoff:

$$\Delta S = 56 \text{ mm} - 36 \text{ mm} = 20 \text{ mm runoff}$$

$$= 20 \text{ mm} \times 16.2 \text{ ha}$$

$$= 0.02 \text{ m} \times 162{,}000 \text{ m}^2 = 3{,}240 \text{ m}^3$$

The pond must be designed to detain, or slow down, 20 mm of runoff from the developed land. This equates to 3,240 m³ of runoff water from the entire development of 16.2 ha.

Example 4: Estimate the amount of storage available in a bioretention cell

Problem:
Consider a bioretention cell located in the center of a parking lot. The parking lot, an area of 26 m by 12 m, is sloped to direct runoff into the bioretention cell. The cell contains an engineered growing media that is 60% sand and 40% organic compost, with a porosity of 45% by volume, planted with native grasses and forbs. The cell is 2.0 m wide, 1.2 m deep, and 12 m long.

(a) What is the maximum water storage volume of the bioretention cell?
(b) What is the largest storm (maximum precipitation depth) the cell can infiltrate if all storage is available?

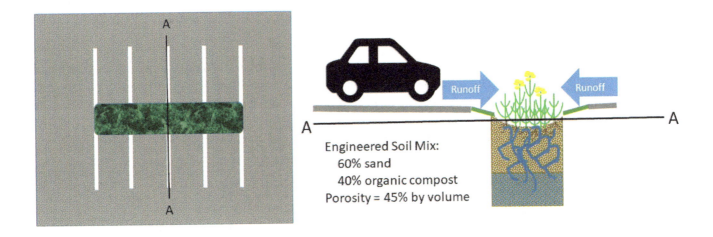

Solution:

(a) Calculate the total volume of the bioretention cell:

$$\text{volume of cell} = \text{length} \times \text{width} \times \text{depth}$$

$$= 12 \text{ m} \times 2 \text{ m} \times 1.2 \text{ m} = 28.8 \text{ m}^3$$

The maximum storage is equal to the total void space (porosity):

$$\text{void space} = \text{volume of cell} \times \text{porosity}$$

$$= 28.8 \text{ m}^3 \times 0.45 = 12.96 \text{ m}^3$$

(b) Assuming all rainfall will run off the parking lot, the maximum storm depth that can be stored is:

$$\text{rainfall} = \frac{\textit{volume of cell void space}}{\textit{area of parking lot}} = \frac{12.96 \text{ m}^3}{12 \text{ m} \times 26 \text{ m}} = 0.042 \text{ m} = 42 \text{ mm}$$

Thus, the maximum depth of precipitation the cell can store is 42 mm if all voids are available. If the cell did not have an underdrain, water would be removed from the cell through ET. The ET rate would depend on the time of year, the type of plant/vegetation, and the weather (e.g., temperature, solar radiation, humidity, and wind). During the summer, plants would evapotranspire about 5-10 mm day^{-1}, preparing the cell to store the next rainfall event.

Example 5: Development of an irrigation schedule for corn in a water limited area

Problem:

Given the following information, determine the daily changes in the soil water content. How much irrigation water should be added on the 10th day to raise the water content of the root zone back to the initial water content? The root zone is 1 m and the initial soil moisture content is 20% by volume. Assume all seepage passes through the root zone and is not stored.

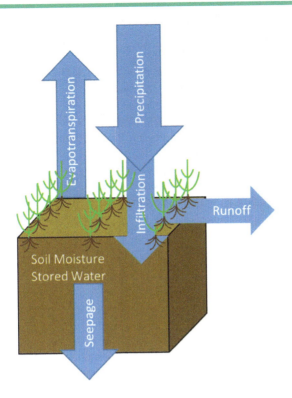

Solution:

Convert the initial soil amount in the profile at the beginning of day 1, $\theta_v = 20\%$ of root zone, to units of depth of water in the root zone:

$$\theta_{vi} = 0.2 \times 1000 \text{ mm} = 200 \text{ mm of water}$$

Apply equation 4 to the root zone for each day:

$$\text{Daily water balance} = P + Ir \pm R - ET - DS = \Delta S$$

$$\text{Day 1 inputs} = P + Ir \pm R = 0 + 0 + 0 = 0 \text{ mm}$$

$$\text{Day 1 outputs} = R + ET + \text{seepage} = 0 + 7 + 0 = 7 \text{ mm}$$

$$\text{Day 1 } \Delta S = \text{inputs} - \text{outputs} = 0 \text{ mm} - 7 \text{ mm} = -7 \text{ mm}$$

$$\text{Day 1 } \theta_{v1} = \theta_{vi} - \Delta S = 200 \text{ mm} - 7 \text{ mm} = 193 \text{ mm}$$

There is 193 mm of water in the soil profile at the end of day 1.

It is easiest to complete the rest of these calculations by setting up a spreadsheet. The following table shows the daily water budget components, the summed inputs and outputs as calculated above, and the resulting soil water depth (mm) in the right-most column. Initial water = 200 mm. At water content is 183 mm. Thus, to return the amount of water in the profile to in the initial soil water content of 200 mm, 17 mm would need to be added on day 10.

Image Credits

Figure 1. USGS. (CC By 1.0). (2020). The water cycle. Retrieved from https://www.usgs.gov/media/images/water-cycle-natural-water-cycle

Figure 2. Hutchinson, S. L. (CC By 4.0). (2020). Soil water.

Figure 3. Hutchinson, S. L. (CC By 4.0). (2020). Water Balance.

Figure 4. Ruane et. Al. (2018). Annual Precipitation. Retrieved from https://data.giss.nasa.gov/impacts/agmipcf/

Figure 5. Hutchinson, S. L. (CC By 4.0). (2020). Example 2.

Figure 6. Hutchinson, S. L. (CC By 4.0). (2020). Example 4.

Figure 7. Hutchinson, S. L. (CC By 4.0). (2020). Example 5.

References

Allen, R. G., Walter, I. A., Elliott, R. L., Howell, T. A., Itenfisu, D., Jensen, M. E., & Snyder, R. L. (2005). The ASCE standardized reference evapotranspiration equation. Reston, VA: American Society of Civil Engineers.

ASABE Standards. (2016) EP 464.1: Grassed waterway for runoff control. St. Joseph, MI: ASABE.

ASAE Standards. (2017). S268.6: Terrace systems. St. Joseph, MI: ASABE.

Brouwer, C., & Heibloem, M. (1986). Irrigation water management: Irrigation water needs. Food and Agriculture Organization of the United Nations. Retrieved from http://www.fao.org/3/s2022e/s2022e00.htm#Contents

Christianson, R., Hutchinson, S., & Brown, G. (2015). Curve number estimation accuracy on disturbed and undisturbed soils. *J. Hydrol. Eng.*, *21*(2). http://doi.org/10.1061/(ASCE)HE.1943-5584.0001274.

Davis, A. P., Shokouhian, M., Sharma, H., & Minami, C. (2001). Laboratory study of biological retention for urban stormwater management. *Water Environ. Res.*, *73*(1), 5-14. http://doi.org/10.2175/106143001x138624.

Dobriyal, P., Qureshi, A., Badola, R., & Hussain, S. A. (2012). A review of the methods available for estimating soil moisture and its implications for water resource management. *J. Hydrol.*, *458-459*, 110-117. https://doi.org/10.1016/j.jhydrol.2012.06.021.

FAO. (2020). FAO soils portal. Food and Agriculture Organization of the United Nations. Retrieved from http://www.fao.org/soils-portal/en/.

Gaffield, S. J., Goo, R. L., Richards, L. A., & Jackson, R. J. (2003). Public health effects of inadequately managed stormwater runoff. *Am. J. Public Health*, *93*(9), 1527-1533. https://doi.org/10.2105/ajph.93.9.1527.

HDSC. (2019). NOAA atlas 14 point precipitation frequency estimates. Hydrometeorological Design Studies Center. Retrieved from https://hdsc.nws.noaa.gov/hdsc/pfds/pfds_map_cont.html.

Huffman, R. L., Fangmeier, D. D., Elliot, W. J., Workman, S. R., & Schwab, G. O. (2013). *Soil and water conservation engineering* (7th ed.). St. Joseph, MI: ASABE.

Meng, S., Xie, X., & Liang, S. (2017). Assimilation of soil moisture and streamflow observations to improve flood forecasting with considering runoff routing lags. *J. Hydrol.*, *550*, 568-579. http://doi.org/10.1016/j.jhydrol.2017.05.024.

NOAA. (2019). Climate data online. National Oceanic and Atmospheric Administration, National Centers for Environmental Information. Retrieved from https://www.ncdc.noaa.gov/cdo-web/.

Ruane, A. C., Goldberg, R., & Chryssanthacopoulos, J. (2015) AgMIP climate forcing datasets for agricultural modeling: Merged products for gap-filling and historical climate series estimation, *Agr. Forest Meteorol.*, *200*, 233-248, http://doi.org/10.1016/j.agrformet.2014.09.016.

Sahu, R. K., Mishra, S., & Eldho, T. (2012). Improved storm duration and antecedent moisture condition coupled SCS-CN concept-based model. *J. Hydrol. Eng. 17*(11). https://doi.org/10.1061/(ASCE)HE.1943-5584.0000443.

Schlosser, C. A., Strzepek, K., Gao, X., Fant, C., Blanc, E., Paltsev, S., . . . Gueneau, A. (2014). The future of global water stress: An integrated assessment. *Earth's Future, 2*(8), 341-361. https://doi.org/10.1002/2014EF000238.

Suresh Babu, P., & Mishra, S. (2012). Improved SCS-CN–inspired model. *J. Hydrol. Eng. 17*(11). https://doi.org/10.1061/(ASCE)HE.1943-5584.0000435

United Nations. (2013). Water security and the global water agenda. A UN-Water Analytical Brief. Retrieved from https://www.unwater.org/publications/water-security-global-water-agenda/.

USDA-NRCS. (2004). Chapter 10: Estimation of direct runoff from storm rainfall. In *National engineering handbook, part 630 hydrology*. U.S. Department of Agriculture Natural Resources Conservation Service. Retrieved from https://www.nrcs.usda.gov/wps/portal/nrcs/detailfull/national/water/manage/hydrology/?cid=STELPRDB1043063.

USDA-NRCS. (2019a). Web soil survey. USDA Natural Resources Conservation Service. Retrieved from https://websoilsurvey.sc.egov.usda.gov/App/HomePage.htm.

USDA-NRCS. (2019b). Soils. U.S. Department of Agriculture Natural Resources Conservation Service. Retrieved from https://www.nrcs.usda.gov/wps/portal/nrcs/main/soils/survey/.

USEPA (2020). Green infrastructure. U.S. Environmental Protection Agency. Retrieved from https://www.epa.gov/green-infrastructure.

USGS (2020a) The water science school. U.S. Geological Survey. Retrieved from https://water.usgs.gov/edu/watercycle.html.

USGS (2020b). National Water Information System: Web interface. U.S. Geological Survey. Retrieved from https://waterdata.usgs.gov/nwis.

Water Quality as a Driver of Ecological System Health

Leigh-Anne H. Krometis
Biological Systems Engineering
Virginia Tech
Blacksburg, Virginia, USA

KEY TERMS		
Water pollution	Source control	Water quality standards
Ecological and ecosystem services	Delivery control	Nutrient management
	Assimilative capacity	Urban stormwater planning
Pollutant budget		

Variables

- μ = viscosity of water
- ρ_p = density of particle
- ρ_w = density of the fluid (water)
- b = benefit
- BCR = benefit to cost ratio
- c = cost
- d = particle diameter
- D = depth
- g = acceleration due to gravity
- L = length
- M_{in} = mass of pollutant entering the system of interest (e.g., field, structure)
- M_{out} = mass of pollutant leaving the system of interest (e.g., field, structure)
- MOS = margin of safety
- NPS = nonpoint source contributions of the targeted pollutant

PS = point source contributions of the targeted pollutant
Q = inflow rate
Q_R = overflow rate
ΔS = mass of pollutant retained or treated by the system of interest (e.g., field, structure)
$TMDL$ = maximum permissible total quantity of targeted pollutant that can be added to the receiving water each day
V_y = average vertical ("fill") velocity
V_s = settling velocity
W = width

Introduction

Water is critical to all known forms of life, human and non-human. Poor management of water resources can result in risks to human health through the spread of toxic chemical and pathogenic microorganisms, reduction in species diversity through changes in water chemistry and/or habitat loss, economic hardship due to a failure to meet industrial, agricultural, and energy needs and political conflict or instability as neighboring states or nations struggle to equitably distribute water to their people.

Globally, 70% of freshwater withdrawals are used by the agricultural sector (World Bank, 2017). It is important to recognize, however, that these consumptive values can vary considerably by nation or global region depending on the local population size, ecology, climate, and primary industries present. In the United States, the U.S. Geological Survey (USGS) estimated that in 2011, 41% of consumptive water use (water that is not returned quickly to the same source from which it was taken) was devoted to hydroelectric power generation, 40% was used to support various forms of agriculture (aquaculture, livestock, crop irrigation), 13% supported domestic household use, and the remaining 6% was used for industrial purposes or in extractive industries (e.g., mining) (USGS, 2018). In contrast, the United Nations Food and Agriculture Organization (UNFAO) estimated in 2015 that over 64% of water in China and nearly 80% of water in Egypt supported agriculture (UNFAO, 2019). Although per capita water use has declined in recent years, the human population and its attendant need for clean water, affordable energy, and nutritious food continues to increase. Concurrently, non-human species diversity continues to decline as forest, soil, and water resources are increasingly exploited (MEA, 2005; Raudsepp-Hearne et al., 2010). More explicit consideration of the intricate feedback of food-energy-water systems within human populations and their impacts on other ecological services are needed to ensure sustainability.

This chapter introduces basic concepts related to water management and provides examples of best management practices that can be used to preserve and improve water quality. Here, we define ecosystem health as the capacity of a natural system to support human and non-human needs. In this chapter particular focus is on chemical, microbial, and physical constituents in water as drivers of ecosystem health.

Outcomes

After reading this chapter, you should be able to:

- Define pollution in terms of assimilative capacity and use of water body
- Explain the concept of ecological, or ecosystem, services and their relationship to water quality
- Describe strategies for water pollution control, including pollutant budgets and stormwater best management practices
- Calculate a range of water quality impairment and cost parameters

Concepts

Definition and Description of Water Pollution

The U.S. Environmental Protection Agency defines water pollution as "human-made or human-induced alteration of chemical, physical, biological, and radiological integrity of water" (USEPA 2018a). These alterations include the addition of specific pollutants (e.g., chemicals, microorganisms, sediment) to an aquatic system or the alteration of natural conditions, such as pH or temperature. In this context, the "integrity" of the water refers to the ability of the water to continue to perform appropriate human or ecological functions. These functions are spelled out explicitly by the European Environment Agency's definition of pollution as "the introduction of substances or energy into the environment, resulting in deleterious effects of such a nature as to endanger human health, harm living resources and ecosystems, and impair or interfere with amenities and other legitimate uses of the environment" (EEA, 2019). While the two definitions are similar, it is critical to note that the EEA does not specify that pollution needs to be human-made.

A place where pollutants directly enter a receiving water such as a stream, river, or lake through an identifiable pipe or culvert (e.g., industrial outfall or wastewater treatment plant effluent) is referred to as a point source (PS) of pollution. Point sources are generally reasonably constant in flow and concentration (i.e., the pattern, type, and amount of pollution being discharged is consistent), because they tend to be governed by predictable or controlled processes. Places from which pollutants are transported to receiving waters via stormwater runoff (e.g., eroded sediment from construction sites and leachate from septic drainfields), or are more diffuse and less predictable in nature, are referred to as nonpoint sources (NPS) of pollution. NPS pollution is sometimes referred to as diffuse pollution, as the sources are distributed throughout the catchment, rather than originating from a distinct location. NPS discharges are generally highly variable and much more severe following significant weather events such as rainfall or seasonal events such as snowmelt. Consequently, NPS pollution is often more serious during high flows when greater quantities of pollutants are being transported to receiving waters, while PS pollution is more serious during low flows when there is less dilution of constant discharges (Novotny, 2003).

Although any changes to water through PS or NPS contributions may meet the technical definition of pollution, pollution is only considered a concern if it exceeds the receiving water's waste assimilative capacity so that the water no longer supports its human or ecological purpose. Waste assimilative capacity is defined as the natural ability of a water body to absorb pollutants without harmful effects. Receiving waters can naturally process some level of pollution through dilution, photodegradation, and bioremediation. For example, native aquatic plants use nutrients including nitrogen and phosphorus to grow; however, very high nutrient contributions from anthropogenic sources can stimulate algal overgrowth, leading to harmful blooms, eutrophication and aquatic ecosystem collapse (Withers et al., 2014).

Ecological Services and Water Quality Decisions

Historically, human and ecological uses of water resources were sometimes regarded as separate or even competing aims. There is an increasing effort to acknowledge the inherent linkages and interdependence of human and ecological well-being through the concept of *ecological*, or ecosystem, *services*. Rather than promoting conservation of habitat and non-human species diversity solely for the sake of nature, the concept of ecological services recognizes that preservation and restoration of natural ecosystems also protects functions that ensure the sustainability of human communities. Ecological or ecosystem services are classified into four categories: *regulating services* (climate, waste, disease, buffering); *provisioning services* (food, fresh water, raw materials, genetic resources); *cultural services* (inspiration, spiritual, recreational, educational, scientific); and *supporting services* (nutrient cycling, habitats, primary production). Making ecological services (such as supporting fish populations, carbon sequestration, nutrient cycling, and flood mitigation) explicit allows for their quantification and inclusion in cost-benefit analyses associated with future land use planning and the allocation of funds for water quality improvements (Keeler et al., 2012; APHA, 2013; Hartig et al., 2014). Continuing research aims to uncover and quantify additional linkages between human health and well-being and ecosystem integrity, including promotion of mental health and community cohesiveness (Sandifer et al., 2015). This is in keeping with the mission of the American Society of Agricultural and Biological Engineers, whose members "ensure that we have the necessities of life: safe and plentiful food to eat, pure water to drink, clean fuel and energy sources, and a safe, healthy environment in which to live" (ASABE, 2020).

Designated Uses and Water Quality Standards

Water quality standards vary around the world. In some jurisdictions (e.g., countries, regions), minimum water quality standards are set for all water bodies regardless of use; in other jurisdictions, appropriate levels of different water quality constituents are generally determined based on chosen, intended, or planned uses of a water body. For example, in the U.S., states, tribes, and territories assign "designated" uses to surface waters to protect human health following water contact (e.g., drinking water reservoirs, recreation, fishing), to preserve ecological integrity (e.g., trout stream, biological integrity), and for economic or industrial use (e.g., navigation, sufficient flow for hydroelectric power). Acceptable levels of critical pollutants are then established to ensure the water body can continue to meet these designated uses. For instance, a water body used only for irrigation may have concentrations of nitrate (NO_3^-), a soluble form of an important plant nutrient, that exceed safe levels for drinking water, without interfering with its use as irrigation water. Basing water quality standards on the designated use of the water body allows for these differences in quality requirements by use category to be taken into account.

An Example of Water Pollution Regulation: U.S.A.

The U.S. Clean Water Act, introduced in 1972, remains the primary regulatory mechanism to ensure surface waters in the U.S. continue to meet the designated uses while protecting human and ecological health. At its most basic, the Clean Water Act regulates point sources through the National Pollutant Discharge Elimination System (NPDES), which requires permits for discrete discharges to ensure implementation of best practicable technology and appropriate monitoring.

NPS pollution is primarily regulated through the Total Maximum Daily Load (TMDL) program, which requires states to monitor surface water and compile lists of waters that do not meet standards applicable for their designated uses, which are then classified as impaired and require TMDL development (Keller and Cavallaro, 2008; USEPA, 2019). The acronym TMDL has two distinct definitions: (1) the mathematical quantity of a targeted pollutant that a receiving water can absorb without harmful effects (equation 1); and (2) the restoration process developed to bring that water body back into compliance with water quality standards (Freedman et al., 2004). Through this restoration process, acceptable levels of pollutant discharges are identified that will not exceed the waste assimilative capacity of the water body so that it can maintain pollutant levels appropriate to its designated use. Mathematically, TMDL is defined as:

$$TMDL = PS + NPS + MOS \qquad (1)$$

where *TMDL* = maximum permissible total quantity of targeted pollutant that can be added to the receiving water each day (mass day^{-1})

PS = all point source contributions of the targeted pollutant (mass day^{-1}), regulated through the NPDES process

NPS = all nonpoint source contributions of the targeted pollutant (mass day^{-1})

MOS = a margin of safety (mass day^{-1})

TMDL is calculated on a load (mass) basis, e.g., mg day^{-1}, and for each individual pollutant that is compromising the use of the water body in question. Margins of safety are included to account for future land development, changes in climate, and uncertainties in measurements and modeling used in TMDL development. Once a total maximum daily load is determined for a water body to meet the relevant water quality standards (including how much pollutant can be allowed from PS and NPS), then treatment systems and land use changes can be designed to meet that maximum daily load.

Determining the allowable total maximum daily load combines mass balance and concentration information, which are described in more detail later in this chapter. While this calculation is simple, the most important part of solving the problem is keeping track of units and identifying the necessary data and information needed to complete the task. Occasionally, there may be an abundance of data but not all of it is valuable to the engineer, thus it is essential for an engineer to master the skills to identify exactly what data are needed to complete the calculation.

Engineering Strategies for Water Quality Protection

Strategies to preserve surface water integrity from degradation or to address water quality impairments are often referred to collectively as best management practices (BMPs). The USEPA defines a BMP as "a practice or combination of practices determined by an authority to be the most effective means for preventing or reducing pollution to a level compatible with water quality goals" (USEPA, 2018b). This term is more broadly encompassing than the National Academies' Stormwater Control Measure (SCM), which primarily refers to structural practices implemented in urban areas to intercept stormwater (NRC, 2009). In addition to structural practices, the term BMP can be used to describe non-structural efforts to protect water quality, including public participation, community education, and pollutant budgeting, and is used to describe these efforts in a variety of land-water environments, including urban, agricultural, and industrial (e.g., mining, forestry) landscapes.

Water quality protection strategies can be broadly categorized as implementing either source control or delivery control. The function of many strategies for water quality protection can be described very simply using a mass balance:

$$M_{in} = \Delta S + M_{out} \qquad (2)$$

where M_{in} = mass of pollutant entering the system of interest (e.g., field, structure) (kg)
ΔS = mass of pollutant retained or treated by the system of interest (kg)
M_{out} = mass of pollutant leaving the system of interest (kg)

This simple relationship is the foundation for the design, performance evaluation, and costing of BMPs. Application of equation 2 to different types of strategies is described in the following sections.

Source Control

Source control refers to efforts to reduce the presence or availability of the pollutant in the land-water system (e.g., eliminating pesticide use) or to prevent transport of the pollutant from its original source (e.g., discouraging erosion by managing tillage in a field). Widespread use of chemical fertilizers has facilitated a more than doubling of cereal grain production globally over the past 50 years, allowing for the feeding of an ever-increasing population (Tilman et al., 2002). However, overuse of fertilizers can lead to nutrient losses via runoff to surface water and/or leaching to groundwater following precipitation events if soil amendments are not applied via an appropriate method at the time of year best suited to promote plant growth. Excessive nutrient loadings can result in eutrophication and aquatic biology impairments, as well as difficulty in meeting municipal drinking water needs. Use of fertilizers in excess of crop needs also represents an unnecessary and nontrivial expenditure for the producer. When applied to a source control practice such as nutrient management, the variables in equation 2 are defined as:

M_{in} = mass of nutrient applied to the crop
ΔS = mass of nutrient taken up by the crop + mass of nutrient adsorbed by the soil
M_{out} = mass of nutrient leaving the field in runoff, lateral flow through the soil, and in deep seepage.

Delivery Control

Delivery control refers to efforts to reduce pollutant movement to source waters after pollutants are moved from their point of origin. Often, delivery control efforts involve interception, treatment, and/or storage of pollutants in water (e.g., riparian buffer, detention basin) prior to their discharge into a receiving water. When applied to a source control practice such as a detention basin, the variables in equation 2 are defined as:

M_{in} = mass of pollutant in inflow
ΔS = mass of pollutant treated or retained
M_{out} = mass of pollutant in outflow

An example of delivery control is a detention basin in which runoff water is collected and the particles are allowed to settle out before the water flows out of the basin. A detention basin can be placed at the outlet of a watershed in which soil erosion is occurring (the NPS) to reduce the mass load of sediment flowing out of the watershed as part of a plan to meet the TMDL. Theoretical particle removal by size class in the detention basin can be calculated by assuming a theoretical stormwater basin of depth D, width W, and length L (figure 1):

Assuming a constant inflow rate of Q, the average vertical ("fill") velocity is approximated as flow divided by cross-sectional area of the basin, or:

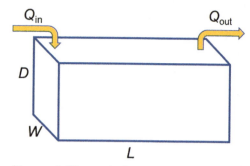

Figure 1. Theoretical stormwater basin dimensions.

$$V_y = \frac{Q}{W \times L} \quad \text{for depth of water in the basin} < D \qquad (3)$$

where V_y = average vertical ("fill") velocity (m h^{-1})
Q = inflow rate (m^3 h^{-1})
W = width (m)
L = length (m)

When the basin is full of water and inflow continues, water will flow out of the basin, and:

$$Q_R = \frac{Q}{W \times L} \quad \text{when depth of water in the basin} = D \qquad (4)$$

where Q_R = overflow rate (m h^{-1})

Under laminar flow conditions (smooth flow at low velocity), the theoretical rate at which a particulate will settle from the runoff water is governed by Stokes' law:

$$V_s = \frac{gd^2(\rho_p - \rho_w)}{18\mu} \qquad (5)$$

where V_s = settling velocity (m s^{-1})
ρ_p = density of particle (kg m^{-3})
ρ_w = density of the fluid (water), (kg m^{-3})
g = acceleration due to gravity (m s^{-1} s^{-1})
d = particle diameter (m)
μ = viscosity of water, 10^{-3} N s m^{-2} at 20°C

Equations 3 and 5 can be used together to determine dimensions of the basin that will allow even small and light particles, such as fine silts, to settle to the bottom of the basin for a given inflow rate. Notice that both equations 3 and 5 represent a velocity in the vertical direction; equation 3 describes the change in water depth over time as the inflow fills the basin, and equation 5 describes the vertical velocity of the particles in the water. When these are equal to each other, the dimensions of the basin are sufficient to allow the sediment particles to settle to the bottom of the basin before the now-clean water begins to leave the basin from the top (figure 1). Detention basins can be sized to completely remove particles of a minimum size and density by setting overflow rate (equation 4) equal to the theoretical settling velocity for that design size particle, e.g.:

$$Q_R = \frac{Q}{W \times L} = \frac{gd^2(\rho_p - \rho_w)}{18\mu} \qquad (6)$$

Given a design inflow rate Q, and size and density of the small particles carried in with the flow, combinations of W and L (the dimensions of the basin) can be found that will accomplish the objective of settling the small particles to the bottom. This reduces the load carried in the outflow, reducing the consequences of sediment pollution downstream and helping to meet targets for maximum allowable load.

Cost-Benefit Analysis

Cost-benefit analysis (CBA) at its simplest uses an estimate of the monetary value of the benefits of a project (*b*, any currency, e.g., $ or €) divided by the costs (*c*, must be same currency as b), as:

$$BCR = \frac{b}{c} \qquad (7)$$

where *BCR* is the benefit-to-cost ratio (unitless).

In practice, it requires more detailed considerations, particularly concerning when the costs are incurred and when the benefits will accrue, because the value of a unit of money changes through time. To ensure *BCR* is meaningful, all costs should be adjusted to a reference time period, using inflation data for these adjustments. At its simplest, to calculate *BCR* for an engineered structure for water quality protection, it is necessary to know who has a vested interest (the stakeholders) and what benefits they want to prioritize. Once this is done, the production costs can be estimated; the benefits, expressed as monetary values, can be estimated; all costs converted to the same currency; and all adjusted to represent the same time period. In general, it is relatively straightforward to estimate the cost of a project because a design can be converted into a bill of materials and a construction schedule, and the operating cost can be estimated from current practice. The benefits can be much more difficult to cost but could be estimated from the medical costs for human illness or the willingness of people to pay for a cleaner environment. Putting a price on non-human ecosystem services that can be damaged by poor water quality requires ingenuity. For example, the cost of eutrophication (a result of excess nutrient loadings) on local ecology, recreation, and aesthetics could be quantified by loss of fish yields related to tourism or fishing permits, local home prices, or tourism revenue, but quantifying the cost of a lost species that most people are not even aware of is much more difficult.

Applications

Designated Uses and Water Quality Standards

As noted earlier, the concept of basing water quality standards on the designated use of water bodies allows for differences in water quality requirements by use category to be taken into account. Establishing water quality standards based on designated uses provides challenges to policymakers. They must consider common designated uses for surface freshwaters and decide which uses should have more stringent standards. For example, in considering standards for drinking water reservoirs vs. fishing, drinking water might generally be expected to require more stringent standards, as this implies direct human contact. However, it is worth noting that a drinking water reservoir is directed into a treatment plant, which may remove pollutants of concern (though at some expense); and for some contaminant and species interactions, human drinking water standards are insufficient to provide protection, e.g., the selenium drinking water standard set by the USEPA is 50 ppb, but there is research suggesting selenium levels over 5 ppb may be toxic to some freshwater fish due to bioaccumulation. Considerations to deliberate upon when thinking about irrigation vs. navigation include the fact that irrigation involves potential application to plants that then could be consumed by humans and so this water would likely need to be of higher quality. However, it is also useful to think about water quality issues that might impede navigation, for example, extreme eutrophication. Filamentous algae can tangle motors and docks (and irrigation intake pumps). One very difficult use comparison is habitat maintenance vs. recreation. Full-body-contact recreation by humans could involve ingestion and/or submersion in water. Habitat maintenance could involve water chemistry and habitat not compatible with human submersion.

Source Control: Nutrient Management

Nutrient management, which is the science and practice of managing the application of fertilizers, manure, amendments, and organic materials to agricultural landscapes as a source of plant nutrients, is a source-control BMP designed to simultaneously support water quality protection and agroeconomic goals. This pollution control strategy sets a nutrient budget whereby primary growth-limiting nutrients (generally nitrogen, phosphorus, and potassium, or N-P-K) are applied only in amounts to meet crop growth needs. Fertilizer application is intentionally timed to coincide with times of maximum crop need (e.g., prior to or just after germination) and to avoid high-risk transport periods (e.g., avoiding prior to large rainfall events or when the ground is frozen). In minimizing the amount of fertilizer applied, the risk of loss to the environment and the cost of production are also minimized.

In its simplest form, nutrient management planning can be thought of in terms of a mass balance (equation 2). Using a mass balance approach also requires deciding on the appropriate scale of analysis; it may be appropriate to consider inputs and outputs on a per-unit-land-area basis, and/or it may be appropriate to consider a whole farm. The latter can be especially useful for managing nutrients in a combined animal and plant production system, where the animals generate waste that contains a concentration of nutrients, and where the animal waste (manures) can be applied to an area of land to meet the nutrient demand of the plants. Nutrient concentration information can be converted to nutrient mass information for use in a mass-balance approach by multiplying the concentration by the relevant total area or volume:

$$\text{Mass (in an area or volume)} = \text{concentration (per unit area or volume)} \times \text{total area or volume} \quad (8)$$

Units in equation 8 will vary depending on the specific application, so it is important to keep track of the units and convert units as needed. Common units for concentration on a per unit volume basis are mg L^{-1} and g cm^{-3}. On a per unit area basis, common units are kg ha^{-1}.

It is relatively straightforward to estimate nutrient application rates. For example, if N demand of a crop is known, and the available N in a wastewater or manure is known, it is possible to calculate whether field application to manage the wastewater or manure is likely to exceed crop demand and thus cause pollution. Nitrogen needed for a field (kg) can be calculated as

$$\text{Nitrogen needed by the crop} = \text{area} \times \text{crop nitrogen demand per unit area} \quad (9)$$

where area (ha) can be determined from maps or farm records and crop demand (kg ha^{-1}) can be taken from advisory/extension service or agronomy guidelines. If the N content of a wastewater is known, equation 8 can be used to calculate the available supply of nitrogen. The difference between amount spread and amount needed indicates whether polluting losses are likely.

While nutrient management is relatively simple conceptually and practically in terms of chemical fertilizers, the practice becomes much more complicated when animal wastes such as manure are used as a source of crop nutrients and soil organic matter. The use of manure as a fertilizer and soil conditioner has proven a successful agricultural strategy since the Neolithic Revolution and continues to be recommended as a sustainable means of recycling nutrients in agricultural systems today. However, because manures are quite heterogeneous in composition, matching manure nutrient content with crop needs can be quite complex.

Other complicating factors in nutrient management plans, particularly those reliant on manure, include the impacts of historical land uses on soil nutrient levels and additional potentially harmful components of animal wastes. Years of fertilization with manure have resulted in P saturation of many agricultural soils (Sims et al., 1998). Given that P is generally the growth-limiting nutrient for freshwater systems (i.e., additional P is likely to result in eutrophication), many agricultural nutrient management guidelines are P-based and so do not permit addition of fertilizer beyond crop P needs. This can render disposal of manures difficult if surrounding croplands have P-saturated soils. Manures and agricultural wastes can also contain additional contaminants of human health concern, including pathogenic microorganisms and antibiotics. Consequently, crops for human consumption cannot be fertilized with animal manures unless there is considerable oversight and pre-treatment (e.g., composting) (USFDA, 2018).

While the previous examples have focused primarily on agricultural landscapes, nutrient management is also widely applied in urban landscapes as well to minimize nutrient loss following fertilization of ornamental plants, lawns, golf courses, etc. (e.g., Chesapeake Stormwater Network, 2019).

Delivery Control: Detention Basins and Wetlands

One example of a common pollutant in water systems is excess sediment that arrives in the water body with surface runoff, and which carries eroded particles from the soil over which the water has moved. Human activities, including agriculture, urban development, and resource extraction, have been estimated to move up to 4.0 to 4.5×10^{13} kg yr^{-1} of soil globally (Hooke, 1994, 2000). Given the sheer magnitude of earth-moving activities involved, it is perhaps inevitable that these activities accelerate erosion, i.e., the wearing away and loss of local soils. Erosion represents a significant concern as it results in the degradation of soil quality and the contamination of local receiving waters. Eroded sediments alone can threaten aquatic ecology through sedimentation of habitat, physical injury to aquatic animals, and disruption of macroinvertebrate biological processes (Govenor et al., 2017). In addition, these sediments can carry with them additional adsorbed pollutants, including bacteria (Characklis et al., 2005), metals (Herngren et al., 2005), nutrients (Vaze and Chiew, 2004), and some emerging organic contaminants (Zgheib et al., 2011). Eroded sediments can also compromise storage capacity of lakes and reservoirs. Detention, or "settling," basins (also called ponds) are a popular BMP in the USA and beyond that are implemented in a variety of landscapes to prevent eroded soils from contaminating local waterways.

In recent decades, low impact development (LID) practices have started to emerge as BMPs. LID "refers to systems and practices that use or mimic natural processes that result in the infiltration, evapotranspiration or use of stormwater in order to protect water quality and associated aquatic habitat" (USEPA, 2018c). LID is a design approach to managing stormwater runoff in urban and suburban environments, both in new developments and retrofitting older developments. Although the term LID was first coined in the U.S., this paradigm is now widely practiced elsewhere (Saraswat et al., 2016; Hager et al., 2019). Specific BMPs used to support LID include wetlands, which rely on both physical (e.g., settling) and biological (e.g., denitrification) processes to remove water quality pollutants, and bioretention cells, which use infiltration through a bioactive media to remove contaminants and decrease peak flows (figure 2). Selection of an appropriate BMP requires knowledge of the specific target pollutants requiring treatment, available land and land cost, and stakeholder preferences and capacity for continuing maintenance. LID approaches also consider the broader ecological impacts beyond the reduction of a target pollutant by the BMPs employed, including habitat restoration and carbon/nutrient cycling.

The advent of these strategies to manage stormwater has partially led to the creation of a new subdiscipline of agricultural and biological engineering during the past few decades, known as ecological engineering. Ecological engineering is defined as "the design of sustainable ecosystems that integrate human society with its natural environment for the benefit of both" (Mitsch, 2012). As with any emerging discipline, there is substantial current research codifying ecological engineering design guidelines and quantifying expected outcomes of relevant BMP implementation (Hager et al., 2019).

Figure 2. Bioretention cells for urban stormwater control in Brazil (left) and the USA (right). These cells are designed to temporarily store water, allowing sediments to settle, and using plants for nutrient uptake. Note the use of local native vegetation.

Urban Stormwater Planning

An important aspect of urban planning is effective stormwater control. The selection of appropriate BMPs for each urban setting depends on the specifics of the situation. For example, consider an urban community that is particularly concerned about maintaining a small downstream reservoir for aquatic recreation. Samples from this reservoir must occasionally be tested for levels of fecal coliform bacteria. The presence of fecal coliform indicates that the water has been contaminated with human or other animal fecal material and that it is possible other pathogenic organisms are present. To ensure fecal coliform values are lower than the recommended levels, specific BMPs can be implemented. Implementing source control practices, such as dog waste collection stations, could be part of the solution. In addition, one or more delivery control practices, such as bioretention cells, detention basins, or a wetland basin, would be required to remove fecal coliforms from stormwater flows. The design of these urban features to reduce coliform transport to local streams and the reservoir requires knowledge of local climate, specifically rainfall patterns and some idea of the loading that might be expected, specifically, the number and magnitude of sources of coliforms. To evaluate which BMP is most appropriate and to obtain design guidelines, a tool such as the International Stormwater BMP Database (Clary et al., 2017) can be used. The database includes data and statistical analyses from over 700 BMP studies, performance analysis results, and other information (International Stormwater BMP Database, 2020). Interpretation of the results of the statistical analysis have to consider issues such as whether the magnitude of the average decrease or the reliability of the BMP is most important, whether the BMP might actually export bacteria, how location specific the data might be, and how useful a particular BMP might be for related pollutants, in this case for something like *E. coli*. Ultimately size and cost calculations need to be used to select a specific design.

Examples

Example 1: Quantifying ecosystem services

Problem:

Presently in the U.S. Midwest there is concern that the use of fertilizer on agricultural lands to maximize crop production may result in downstream concentrations of nitrate that render the water more difficult and costly to treat for human consumption. The current maximum permissible concentration of nitrate in drinking water is 10 mg L^{-1}. Assume that the average nitrate concentration in a drinking water treatment plant intake is 12.3 mg L^{-1}. The plant must treat and distribute 1.5×10^8 L of water per day to meet consumer demand. Treating water to remove nitrate costs \$2 kg^{-1}. What is the minimum cost of nitrate treatment per year?

Solution:

The cost of nitrate treatment is expressed in units of \$ kg^{-1} of nitrate. Thus, to determine the total cost, determine the mass of nitrate treated using equation 2:

$$\text{mass nitrate in inflow} = \text{mass nitrate treated} + \text{mass nitrate in outflow}$$

In this case, the concentration of nitrate in the inflow is 12.3 mg L^{-1}. The concentration of nitrate in the outflow should not exceed 10 mg L^{-1}. The difference can be used to estimate the minimum amount of nitrate that must be treated:

$$\text{mass nitrate treated} = (\text{concentration in inflow} - \text{concentration in outflow}) \times \text{volume}$$

$$= (12.3 \text{ mg } L^{-1} - 10.0 \text{ mg } L^{-1}) \times 1.5 \times 10^8 \text{ L day}^{-1}$$

$$= 3.45 \times 10^8 \text{ mg day}^{-1} \times (1 \text{ kg}/10^6 \text{ mg}) = 345 \text{ kg day}^{-1}$$

The annual cost of treatment can then be calculated as:

$$\$2 \text{ kg}^{-1} \times 345 \text{ kg day}^{-1} \times 365 \text{ days year}^{-1} = \$251,850$$

This calculation provides no contingency for inefficiency in the plant. If a safety margin of 1 mg L^{-1} were included, the outflow concentration would be 9 mg L^{-1}, and the calculation would be:

$$\text{mass nitrate treated} = (12.3 \text{ mg L}^{-1} - 9.0 \text{ mg L}^{-1}) \times 1.5 \times 10^8 \text{ L day}^{-1}$$

$$= 4.95 \times 10^8 \text{ mg day}^{-1} \times (1 \text{ kg}/10^6 \text{ mg}) = 495 \text{ kg day}^{-1}$$

and

$$\$2 \text{ kg}^{-1} \times 495 \text{ kg day}^{-1} \times 365 \text{ days year}^{-1} = \$361{,}350$$

A cost benefit analysis would have to be used to decide whether it was worth paying $109,500 per year for what might be seen as greater certainty that outflow water quality would be better than the permissible limit.

Example 2: Calculating a TMDL

Problem:

You are a water quality manager tasked with ensuring that a stream within a small, rapidly urbanizing watershed remains in compliance with applicable state standards. At present, water quality monitoring indicates that nitrate-nitrogen (NO_3-N) levels (mg 100 mL^{-1}) in grab samples are just below the state standard. Knowing that future development will likely increase nutrient discharges, you decide to calculate a current TMDL value for future reference based on a current inventory of loadings to the stream. An inventory of local NPDES permits provides the loadings in table 1; water quality models estimate that nonpoint sources contribute roughly 2.3×10^9 g month^{-1} of NO_3-N. Prior experience indicates that the margin of safety should be equivalent to 35% of total current nonpoint and point source loadings in order to account for errors, growth, and missing data. What TMDL value (in Mg day^{-1}) do you report for this stream under the current conditions?

Table 1. Average daily discharge and NPDES permitted loading from local point sources.

Source	Average Daily Discharge, L day^{-1}	Permitted Loading (per day)
Wastewater treatment plant	6.4×10^6	5.6×10^6 E. coli; 0.7 Mg NO_3-N
Mid-sized concentrated animal feeding operation (CAFO)	1.0×10^4	4.4×10^5 E. coli; 0.2 Mg sediment
City storm sewer 1	5.3×10^5	10.4 Mg sediment
City storm sewer 2	0.13×10^5	3.2×10^7 Mg NO_3-N

Solution:

Calculate the TMDL using equation 1; specifically, sum the point (*PS*) and nonpoint (*NPS*) source loads of NO_3-N and add a margin of safety (*MOS*):

$$TMDL = PS + NPS + MOS$$

Point sources of NO_3-N, based on the inventory of local NPDES permits, are a wastewater treatment plant and city storm sewer #2. The total *PS* loadings per day are:

$$PS = 0.7 \text{ Mg} + 3.2 \times 10^7 \text{ Mg} = 3.2 \times 10^7 \text{ Mg } NO_3\text{-N}$$

The loading from the wastewater treatment plant is negligible compared to that of the city storm sewer.

Nonpoint sources of NO_3-N are 2.3×10^9 g month^{-1}. Assuming 30 days per month yields the *NPS* loading per day:

$$NPS = 2.3 \times 10^9 \text{ g month}^{-1} / (30 \text{ days month}^{-1}) = 77 \text{ Mg } NO_3\text{-N}$$

Since the specified margin of safety is 35% of the total *PS* and *NPS* loadings, the TMDL is:

$$\text{TMDL} = PS + NPS + 0.35 \, (PS + NPS) = 1.35 \times (PS + NPS) = 1.35 \times [(3.2 \times 10^7) + 77]$$

$$= 4.32 \times 10^7 \text{ Mg } NO_3\text{-N day}^{-1}$$

Example 3: Nutrient management to meet crop needs

Problem:
You are advising a producer who is managing 30.3 ha in continuous cultivation for corn (maize; *Zea mays*) silage. You have determined from agronomic advice that for the soil type and cultivar the crop needs 326 kg ha^{-1} of nitrogen after initial planting. An adjacent dairy has a slurry (mixture of manure and milking parlor wastewater) that could be used as a source of nitrogen. Laboratory analyses indicate that the slurry contains 15.6 kg available nitrogen per 1000 L of slurry.

(a) How much slurry would be required to completely fertilize the field to meet crop needs?
(b) Assuming the available slurry spreader can spread no less than 47,000 L ha^{-1}, what is the minimum quantity of slurry that can be applied?
(c) Is the application of slurry to the field likely to cause pollution?

Solution:
(a) To calculate the total amount of slurry needed to provide the needed amount of nitrogen to the cropped area, first, calculate the total amount of nitrogen needed in the field:

$$\text{N needed in the field} = 30.3 \text{ ha} \times 326 \text{ kg N ha}^{-1} = 9{,}877.8 \text{ kg N}$$

Then, calculate the amount of slurry needed to provide the needed N, based on the N content of the slurry:

slurry needed in the field = 9,877.8 kg N × (1,000 L/15.6 kg N) = 633,192 L

(b) The machine can apply a minimum of 47,000 L ha⁻¹. Using the available slurry, the amount of nitrogen that would be applied at this rate is:

15.6 kg N/1,000 L × 47,000 L ha⁻¹ × 30.3 ha = 22,216 kg N in the field.

(c) As the minimum application rate would result in 22,216 kg N applied to the field, and the crop only needs 9,877.8 kg N, there will be an excess of 12,338.2 kg N applied to the field, so it is likely to cause pollution. The producer could consider several options: dilute the available slurry; find another source of slurry with lower concentration of available nitrogen; or find a slurry spreader with a lower minimum spreading rate.

Example 4: Calculating theoretical detention basin removals by particle size class

Problem:
Assuming theoretical conditions as described above, what is the surface area of a detention basin required to remove 100% of particulates greater than 0.1 mm in size and with a density of 2.6 g cm⁻³? Given the size of the watershed and typical design storm, the basin will need to be designed to treat 10×10^6 m³ of water over 24 hours.

Solution:
Detention basins can be sized to completely remove particles of a minimum size and density by setting the overflow rate equal to the theoretical settling velocity for that design size particle.

Calculate overflow rate, Q_R, as expressed by equation 4:

$$Q_R = \frac{Q}{W \times L} \qquad (4)$$

$$Q_R = \frac{Q}{W \times L} = \frac{\frac{(10 \times 10^6 \text{ m}^3)}{24 \text{ hr}(3600 \text{ s hr}^{-1})}}{W \times L} = \frac{11.57 \text{ m}^3\text{s}^{-1}}{W\ L}$$

Calculate the settling velocity (equation 5):

$$V_s = \frac{gd^2(\rho_p - \rho_w)}{18\mu} \qquad (5)$$

where V_s = settling velocity (m s⁻¹)
ρ_p = density of particle = 2.6 g cm⁻³ = 2,600 kg m⁻³
ρ_w = density of the fluid (water) = 1,000 kg m⁻³
g = acceleration due to gravity = 9.81 m s⁻²
d = particle diameter = 0.1 mm = 0.0001 m
μ = viscosity of water, 10⁻³ N s m⁻² at 20°C

$$V_s = \frac{(9.81 \text{ m s}^{-2})(0.0001\text{m})^2(2{,}600 \text{ kg m}^{-3} - 1{,}000 \text{ kg m}^{-3})}{18(10^{-3}\text{N s m}^{-2})} = 0.00872 \text{ m s}^{-1}$$

Set overflow rate equal to settling velocity and solve for the required surface area, or $W \times L$, of the detention basin:

$$\frac{11.57 \text{ m}^3\text{s}^{-1}}{W \times L} = 0.00872 \text{ m s}^{-1}$$

$$W \times L = \frac{11.57 \text{ m}^3\text{s}^{-1}}{0.00872 \text{ m s}^{-1}} = 1{,}327 \text{ m}^2$$

The required surface area of the detention basin is 1,327 m².

Image Credits

Figure 1. Krometis, Leigh-Anne. H. (CC By 4.0). (2020). Theoretical stormwater basin dimensions.

Figure 2. Krometis, Leigh-Anne. H. (CC By 4.0). (2020). Bioretention cells for urban stormwater control in Brazil (left) and the USA (right). These cells are designed to temporarily store water, allowing sediments to settle, and using plants for nutrient uptake. Note the use of local native vegetation.

References

APHA. (2013). Improving health and wellness through access to nature. APHA Policy Statement 20137. American Public Health Association. Retrieved from https://www.apha.org/policies-and-advocacy/public-health-policy-statements/policy-database/2014/07/08/09/18/improving-health-and-wellness-through-access-to-nature

ASABE (2020). About the profession. https://asabe.org/About-Us/About-the-Profession

Characklis, G. W., Dilts, M. J., Simmons, III, O. D., Likirduplos, C. A., Krometis, L. A., & Sobsey, M. D. (2005). Microbial partitioning to settleable solids in stormwater. *Water Res.* 39(9), 1773-1782.

Chesapeake Stormwater Network. (2019). Chesapeake Stormwater Network's urban nutrient management guidelines. Retrieved from https://chesapeakestormwater.net/bmp-resources/urban-nutrient-management/

Clary, J., Strecker, E., Leisenring, M., & Jones, J. (2017). International stormwater BMP database: New tools for a long-term resource. *Proc. Water Environment Federation WEFTEC 2017*, Session 210-219, pp. 737-746.

EEA. (2019). European Environment Agency. Retrieved from https://www.eea.europa.eu/archived/archived-content-water-topic/wise-help-centre/glossary-definitions/pollution

Freedman, P. L., Nemura, A. D., & Dilks, D. W. (2004). Viewing total maximum daily loads as a process, not a singular value: Adaptive watershed management. *J. Environ. Eng.*, 130, 695-702. https://doi.org/10.1061/(ASCE)0733-9372(2004)130:6(695)

Govenor, H., Krometis, L., & Hession, W. C. (2017). Invertebrate-based water quality impairments and associated stressors identified through the US Clean Water Act. *Environ. Manag.* 60(4), 598-614.

Hager, J., Hu, G., Hewage, K., & Sadiq, R. (2019). Performance of low-impact development best management practices: A critical review. *Environ. Rev.*, 27(1), 17-42. https://doi.org/10.1007/s00267-017-0907-3.

Hartig, T., Mitchell, R., de Vries, S., & Frumkin, H. (2014). Nature and public health. *Ann. Rev. Public Health, 35:* 207-228. https://doi.org/10.13140/RG.2.2.15647.61600.

Herngren, L., Goonetilleke, A., & Ayoko, G. A. (2005). Understanding heavy metal and suspended solids relationships in urban stormwater using simulated rainfall. *J. Environ. Manag., 76*(2), 149-158. https://doi.org/10.1016/j.jenvman.2005.01.013.

Hooke, R. L. (1994). On the efficacy of humans as geomorphic agents. *GSA Today 4.* Retrieved from https://www.geosociety.org/gsatoday/archive/4/9/pdf/i1052-5173-4-9-sci.pdf.

Hooke, R. L. (2000). On the history of human as geomorphic agent. *Geol.*, 28, 843-846.

International Stormwater BMP Database (2020). http://www.bmpdatabase.org/.

Keeler, B. L., Polasky, S., Brauman, K. A., Johnson, K. A., Finlay, J. C., O'Neill, A., . . . Dalzell, B. (2012). Linking water quality and well-being for improved assessment and valuation of ecosystem services. *Proc. Natl. Acad. Sci. USA 109:* 18619-18624. http://doi.org/10.1073/pnas.1215991109.

Keller, A. A., & Cavallaro, L. (2008). Assessing the US Clean Water Act 303(d) listing process for determining impairment of a waterbody. *J. Environ. Manag.*, 86, 699-711. http://doi.org/10.1016/j.jenvman.2006.12.013.

MEA. (2005). Ecosystems and human well-being: Biodiversity synthesis. Millennium Ecosystem Assessment. Washington, DC: World Resources Institute. Retrieved from https://www.millenniumassessment.org/documents/document.354.aspx.pdf.

Mitsch, W. 2012. What is ecological engineering? *Ecol. Eng.*, 45, 5-12. https://doi.org/10.1016/j.ecoleng.2012.04.013.

Novotny, V. (2003). *Water quality: Diffuse pollution and watershed management.* New York, NY: J. Wiley & Sons.

NRC. (2009). *Urban stormwater management in the United States.* National Research Council. Washington, DC: The National Academies Press. https://doi.org/10.17226/12465.

Raudsepp-Hearne, C., Peterson, G. D., Tengö, M., Bennett, E. M., Holland, T., Benessaiah, K., . . . Pfeifer, L. (2010). Untangling the environmentalist's paradox: Why is human well-being increasing as ecosystem services degrade? *BioSci.* 60, 576-589. https://doi.org/10.1525/bio.2010.60.8.4.

Sandifer, P. A., Sutton-Grier, A. E., & Ward, B. P. (2015). Exploring connections among nature, biodiversity, ecosystem services, and human health and well-being: Opportunities to enhance health and biodiversity conservation. *Ecosyst. Services* 12, 1-15. https://doi.org/10.1016/j.ecoser.2014.12.007.

Saraswat, C., Kumar, P., & Mishra, B. (2016). Assessment of stormwater runoff management practices and governance under climate change and urbanization: An analysis of Bangkok, Hanoi and Tokyo. Environ. Sci. *Policy 64*, 101-117. https://doi.org/10.1016/j.envsci.2016.06.018.

Sims, J. T., Simard, R. R., & Joern, B. C. (1998). Phosphorus loss in agricultural drainage: Historical perspective and current research. *J. Environ. Qual., 27*(2), 277-293. https://doi.org/10.2134/jeq1998.00472425002700020006x.

Tilman, D., Cassman, K. G., Matson, P. A., Naylor, R., & Polasky, S. (2002). Agricultural sustainability and intensive production practices. *Nature 418*, 671-677. https://doi.org/10.1038/nature01014.

UNFAO. (2019). United Nations Food and Agriculture Organization Aquastat database. Retrieved from http://www.fao.org/nr/water/aquastat/water_use/index.stm.

USEPA. (2018a.) Section 404 of the Clean Water Act. U.S. Environmental Protection Agency. Retrieved from https://www.epa.gov/cwa-404/clean-water-act-section-502-general-definitions.

USEPA. (2018b). Terms and acronyms. U.S. Environmental Protection Agency. Retrieved from https://iaspub.epa.gov/sor_internet/registry/termreg/searchandretrieve/termsandacronyms/search.do.

USEPA. (2018c). Urban runoff: Low impact development. U.S. Environmental Protection Agency. Retrieved from https://www.epa.gov/nps/urban-runoff-low-impact-development.

USEPA. (2019). US EPA's national summary webpage on water quality impairments and TMDL development. U.S. Environmental Protection Agency. Retrieved from https://ofmpub.epa.gov/waters10/attains_index.home.

USFDA. (2018). Food Safety Modernization Act. U.S. Food and Drug Administration. Retrieved from https://www.fda.gov/food/guidanceregulation/fsma/.

USGS. (2018). Water use in the United States. U.S. Geological Survey. Retrieved from https://water.usgs.gov/watuse/.

Vaze, J., & Chiew, F. H. S. (2004). Nutrient loads associated with different sediment sizes in urban stormwater and surface pollutants. *J. Environ. Eng.*, *130*(4), 391-396. https://doi.org/10.1061/(ASCE)0733-9372(2004)130:4(391).

Withers, P., Neal, C., Jarvie, H., & Doody, D. (2014). Agriculture and eutrophication: Where do we go from here? *Sustainability 6*(9), 5853-5875. https://doi.org/10.3390/su6095853.

World Bank. (2017). Globally, 70% of freshwater is used for agriculture. Retrieved from https://blogs.worldbank.org/opendata/chart-globally-70-freshwater-used-agriculture.

Zgheib, S., Moilleron, R., Saad, M., & Chebbo, G. (2011). Partition of pollution between dissolved and particulate phases: What about emerging substances in urban stormwater cathments? *Water Res.*, *45*(2), 913-925. http://doi.org/10.1016/j.watres.2010.09.032.

Quantifying and Managing Soil Erosion on Cropland

Jaana Uusi-Kämppä
Water Quality Impacts, Natural Resources
Natural Resources Institute Finland (Luke)
Tietotie 4, FI-31600 Jokioinen, Finland

KEY TERMS		
Detachment	Water erosion	Universal Soil Loss Equation
Transport	Wind erosion	Measurement
Deposition	Tillage erosion	Monitoring

Variables

θ = angle of slope
λ = slope length
a = organic matter content
A = computed average annual soil loss from sheet and rill erosion
b = soil structure code
c = soil profile permeability class
C = crop management factor
K = soil erodibility factor
LS = topographic factor
m = exponent that is a function of slope steepness
M = particle size parameter
P = conservation practice factor
R = rainfall erosivity factor

Introduction

Soil is a major natural resource in food production and therefore it is important to take care of soil in a sustainable manner. In cropland areas, topsoil is degraded by depleting available nutrients and by the removal of soil material from the soil surface via erosion caused by water or wind. Erosion usually occurs more rapidly when the soil is disturbed by human activity or during extreme weather conditions such as high precipitation or drought. Soil loss from a field decreases soil fertility and hence crop yield because of depletion of nutrients, reduction in soil organic carbon, and weakening of soil physical properties (Zhang and Wang, 2006). Recent global estimates suggest that soil erosion removes between 36 and 75 billion tonnes of fertile soil every year (Borelli et al., 2017; Fulajtar et al., 2017) causing adverse impacts to agricultural land and the environment.

In addition to the loss of fertile soil from cropland, erosion processes cause burying of crops and many environmental problems, such as siltation and pollution of receiving watercourses and degradation of air quality. Agrichemicals such as phosphorus and some pesticides adsorbed to eroded soil particles may be transported from croplands. In receiving water bodies, the chemicals may desorb and cause algal blooms or damage the local ecosystems. Due to the many harmful effects caused by soil erosion, it is important to understand erosion processes and how to monitor and prevent them, as well as how to reduce harmful environmental impacts both in the source and impacted (or target) areas. These topics are explored in more detail in this chapter.

Outcomes

After reading this chapter, you should be able to

- Define soil erosion and explain erosion mechanics and transport mechanisms
- Describe measurement and monitoring methods for quantifying erosion
- Explain and apply the Universal Soil Loss Equation (USLE) to estimate soil loss by water
- Calculate average annual soil loss and the effect of different tillage practices on erosion rates

Concepts

What is Soil Erosion?

Soil erosion is a natural geomorphological process by which surface soil is loosened and carried away by an erosive agent such as water or wind. Other agents, such as freezing and thawing, gravity, tillage, and biological activity cause soil movement. Human activity has accelerated erosion for many years, with changes in land use making soil prone to accelerated erosion so that loss is more rapid than replenishment. Tillage, and especially plowing, generally keeps the soil surface bare during winter. Bare soil is prone to erosion, whereas permanent grass or winter plant cover (i.e., cover crop or stubble) on the soil surface protects soils from erosion. Soil erosion is a local, national, and global problem. In the future, erosion processes may be intensified due to the increase in extreme weather events predicted with climate change. New erosion areas also appear due to deforestation, clearing land for cultivation, and global warming.

Soil Erosion Processes

The process of soil erosion consists of three different parts: detachment, transport, and deposition. First, soil particles are *detached* by the energy of falling raindrops, running water, or wind. Soil particles with the least cohesion are easiest to be loosened. The detached soil particles are then *transported* by surface runoff (also known as overland flow) or wind. Finally, the soil particles start to settle out, or deposit, when the velocity of overland flow or wind and sediment transport capacity decrease. *Deposited* particles are called sediment. Heavier particles, such as gravel and sand, deposit first, whereas fine silt and clay particles can generally be carried for a longer distance and time before deposition. Although particles of fine sand are more easily detached than those of a clay soil, clay particles are more easily transported than the sand particles in water (Hudson, 1971).

In addition to the energy of water or wind used in both detachment and transport of soil particles, gravity may impact erosion either directly, i.e., soil moving downhill without water (e.g., slump mass-movement), or indirectly (e.g., pulling rain to the Earth or drawing floodwaters downward). Bioturbation, which is reworking of soils and sediments by animals or plants, may also play an important role in sediment transport. For example, uprooted trees, invertebrates living underground and moving through the soil (e.g., earthworms), and many mammals burrowing into soil (e.g., moles) can cause soil transport downslope (Gabet et al., 2003).

In some other erosion processes, cycles of freezing and thawing or wetting and drying of clay soils weaken or break down soil aggregates and make the soil more susceptible to erosion. In boreal areas (i.e., northern areas with long winters and short, cool to mild summers), soil erosion may be high during snowmelt periods as a result of soils saturated by water, limited vegetation cover, and high overland flow (Puustinen et al., 2007). Soil erodibility is high in recently thawed soils, since high water content decreases the cohesive strength of soil aggregates (Van Klaveren and McCool, 1998).

Tillage Erosion

Soil erosion caused by tillage has also become more important with the development of mechanized agriculture, while soil erosion caused by water and wind has moved the Earth for millions of years. Tillage erosion has intensified with increased tillage speed, depth, and size of tillage tools, and with the tillage of steeper and more undulating lands (Lindstrom et al., 2001). The amount of soil moved by tillage can exceed that moved by interrill and rill erosion (Lindstrom et al., 2001). In agricultural areas, tillage is the main contributor to accelerated erosion rates. In certain areas, e.g., the U.S. and Belgium, tillage erosion has created soil banks of several meters high near field borders (Lindstrom et al., 2001). The net soil movement by tillage is generally presented as units of volume, mass, or depth per unit of tillage width (e.g., liter m^{-1}, kg m^{-1}, or cm m^{-1}, respectively).

Types of Soil Erosion Caused by Water on Cropland

Soil erosion caused by water can be classified into several forms including splash, sheet, interrill, rill, gully and bank (Toy et al., 2002). *Splash* erosion is caused by raindrop impact (Fernández-Raga et al., 2017). Small soil particles are broken off of the aggregate material by the energy of falling drops and are splashed into the air (figure 1). Particles may deposit on the soil surface nearby or on flowing water.

Figure 1. Splash erosion. (Photo courtesy of USDA Natural Resources Conservation Service.)

Sheet erosion occurs when a thin layer of soil is evenly removed from a large area by raindrop splash and runoff water moving as a thin layer of overland flow. It occurs generally on uniform slopes. Sheet erosion is assumed to be the first phase of the erosion process, and the soil losses are assumed to be rather small (Toy et al., 2002).

Rills are small channels, less than 5 cm deep. They exist when overland flow (or surface runoff) begins to concentrate in several small rivulets of water on the soil surface. Detachment of soil particles is caused by surface runoff (Toy et al., 2002). In general, if a small channel can be obliterated with normal farming operations, it is a rill rather than a channel. After obliteration, rills tend to form in a new location.

The areas between rills are called interrill areas, and the erosion there is defined as *interrill erosion* (Toy et al., 2002). Interrill erosion is a type of sheet erosion because it is uniform over the interrill area. Detachment occurs by raindrop impact, and both surface runoff and detached soil particles tend to flow into adjacent rills.

Gullies are large, wide channels that are carved by running water (figure 2). Ephemeral gullies may occur on croplands and they are able to be filled with soil during tillage operations (Toy et al., 2002). The macrotopography of the surface allows the formation of ephemeral gullies after refilling by tillage. Gullies may sometimes be large enough to prevent soil cultivation. These gullies are called permanent, or classic, gullies. This kind of gully erosion causes severe damage to a field and produces high sediment loads to water.

Figure 2. Gully erosion. (Photo courtesy of USDA Natural Resources Conservation Service.)

Bank erosion is direct removal of soil particles from a streambank by flowing water. Bank erosion is the progressive undercutting, scouring and slumping of the sides of natural stream channels and constructed drainage channels (OMAFRA, 2012b).

Types of Wind Erosion

Suspension, saltation, and surface creep are three types of soil movement during wind erosion (figure 3). The dominant manner of erosion depends principally on soil type and particle size. Pure sand moves by surface creep and saltation. Soils with high clay content move under saltation. The sediment moved by creep and saltation may deposit very near the source area, along a fence, in a nearby ditch, or a field (Toy et al., 2002). In suspension, fine particles (diameter less than 0.1 mm) are moved into the atmosphere by strong winds or through impact with other particles. They can be carried extremely long distances before returning to earth via rainfall or when winds subside. In saltation, bouncing soil particles (diameter 0.1–0.5 mm) move near the soil surface. A major fraction of soil moved by wind is through the saltation process. In surface creep, large soil particles (diameter 0.5–1 mm), which are too heavy to be lifted into the air, roll and slide along the soil surface. Particles can be trapped by a furrow or a vegetated area.

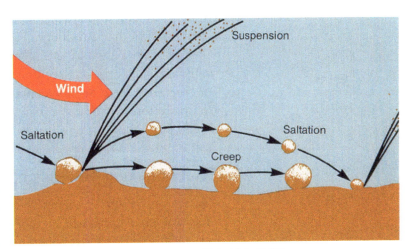

Figure 3. Wind erosion process (USDA ARS, 2020).

Factors Influencing Water and Wind Erosion

Soil erosion is affected by several factors such as climate, rainfall, runoff, slope, topography, wind speed and direction, soil characteristics, soil cover like vegetation or mulch, and farming techniques. For example, in arid climates with steep slopes without good plant cover, during heavy rains the soil erosion is much higher than in level fields with robust plant cover in a mild climate. As another example, soils with high organic matter are naturally more cohesive and, thus, less susceptible to detachment than soils with low organic matter.

Water erosion occurs in areas where rainfall intensity, duration, and frequency are high enough to cause runoff. Wind erosion is most common in arid and semi-arid areas where dry and windy conditions occur. When rainfall water exceeds infiltration (i.e., permeation of water into soil by filtration) into the soil surface, runoff starts to occur. Infiltration capacity depends on soil type. For example, water infiltrates more rapidly into sandy soils than into clay soils; however, water infiltration can be improved in clay-textured soil by aggregate formation. The aggregates, consisting of fine sand, silt, and clay, are typically formed together with a mixed adhesive including organic matter, clays, iron (Fe) and aluminum (Al) oxides, and lime. At first, rainwater runoff has an impact on light materials (i.e. silt, organic matter, and fine sand particles) in soil, whereas during heavy rainfalls, larger particles are also carried by runoff water. Topography (i.e., slope length and gradient) is also an important factor for water erosion, with longer or steeper slopes being associated with greater erosion rates.

Soil surfaces covered by dense vegetation or mulches are less prone to water erosion due to their protection against the erosive power of raindrops and runoff water. Plants also use water, and their roots bind soil particles. Wind erosion can be counteracted by vegetation, which provides shelter from wind, intercepts wind-borne sediment, and keeps the soil surface moist.

Mechanical disturbance (e.g., soil tillage) buries vegetation or residues that would ordinarily serve as protection from erosion. Anthropogenic, i.e., human-induced, influences, such as changes in land management (animal production vs. crop production) and crop pattern (crop rotation vs. monoculture), use of heavier agricultural machinery, and soil compaction, increase the water and wind erosion potential of soils. Reduced tillage and no-till practices on croplands have been successful in reducing erosion. Globally, intensive deforestation causes soil erosion in new agricultural areas, increasing the net erosion rate.

Estimation and Modeling of Soil Erosion

The average annual erosion rate can be estimated using mathematical models. One of the most widely used models for estimating soil loss by water erosion is the Universal Soil Loss Equation, USLE (Wischmeier and Smith, 1978), and its update the Revised Universal Soil Loss Equation (RUSLE) or Modified Universal Soil Equation (MUSLE). According to the USLE, the major factors affecting erosion are local climate, soil, topography (length and steepness of cropland), cover management, and conservation practices.

The standard erosion plot is 22.13 m long and 4.05 m wide, with a uniform 9% slope in continuous fallow, tilled up and down the slope (Wischmeier and Smith, 1978), and is the experimental basis for the development of the empirical USLE model. The soil loss is evaluated as follows by the USLE:

$$A = R\,K\,LS\,C\,P \qquad (1)$$

where A = computed average annual soil loss (Mg ha^{-1} yr^{-1}) from sheet and rill erosion

R = rainfall erosivity factor (MJ mm ha^{-1} h^{-1} yr^{-1})

K = soil erodibility factor (Mg ha h ha^{-1} MJ^{-1} mm^{-1})

LS = topographic factor (combines the slope length and the steepness factors L and S) (dimensionless)

C = crop management factor (dimensionless, ranging between 0 and 1)

P = conservation practice factor (dimensionless, ranging between 0 and 1; the high value, 1, is assigned to areas with no conservation practices)

Each value on the right can be estimated from figures or tables. To minimize soil loss (A), any one value on the right needs to decrease. The units of R and K in equation 1 are a result of adapting the USLE to use in SI units. The USLE was derived using customary U.S. units (e.g., tons, inches, acres). With international application of USLE, adoption of SI units was important. Several authors (e.g., Foster et al., 1981) have described approaches for use of the USLE in SI units.

Rainfall Erosivity Factor (R)

The rainfall and runoff factor (R), is related to the energy intensity of annual rainfall, plus a factor for runoff from snowmelt or applied water (irrigation) (Wischmeier and Smith, 1978). Rainfall erosivity defines the potential ability of the rain to produce erosion. Erosivity depends solely on rainfall properties (e.g., drop velocity, drop diameter, rainfall rate and duration) and frequency of a rainstorm. The greatest erosion occurs when rainfall with high intensity beats a bare soil surface without any plant cover. Plants or stubble are good cover against rainfall erosivity.

The National Soil Erosion Research Laboratory has presented a figure of the aerial erosion index for different areas of the U.S. varying from <200 to 10,000 (Foster et al., 1981). Several regional and global rainfall erosivity maps (e.g., ESDAC, 2017) are available. Erosivity also varies according to the season (Toy et al., 2002), being highest during winter and early spring in boreal areas.

Soil Erodibility Factor (K)

The soil erodibility factor is the soil loss rate per erosion index unit for a specified soil as measured on a standard erosion plot. It is based on the soil texture, soil structure, percent organic matter, and profile-permeability class (Wischmeier and Smith, 1978; Foster et al., 1981) and reflects the susceptibility of a soil type to erosion. Soils high in clay content have low K factor values because the clay soils are highly resistant to detachment of soil particles. In general, there is little control over the K factor since it is largely influenced by soil genesis. However, some management choices can result in small changes to the K factor. For example, by increasing the percent of organic carbon in soil, the K factor can be decreased, since organic matter increases soil cohesion.

The K factor in SI units (Mg ha h ha^{-1} MJ^{-1} mm^{-1}) can be estimated using a regression equation that considers soil texture, organic matter content, structure, and permeability (Mohtar, n.d.):

$$K = 2.8 \times 10^{-7} \times M^{1.14} (12 - a) + 4.3 \times 10^{-3} (b - 2) + 3.3 \times 10^{-3} (c - 3) \quad (2)$$

where M = particle size parameter = (% silt + % very fine sand) × (100 − % clay)
 a = organic matter content (%)
 b = soil structure code (very fine granular = 1; fine granular = 2; medium or coarse granular = 3; blocky, platy, or massive = 4)
 c = soil profile permeability class (rapid = 1; moderate to rapid = 2; moderate = 3; slow to moderate = 4; slow = 5; very slow = 6)

The *K* factor can also be read from nomographs, e.g., Foster et al. (1981) provided a nomograph in SI units.

In reality, soil erodibility is more complicated than equation 2 suggests. How erodible a soil is depends not only on the physical characteristics of the soil but also its treatment, which effects how cohesive the soil aggregates are. Some variations of the USLE, such as the Second Revised USLE (RUSLE2) use a more complicated and dynamic *K* factor to account for management effects.

Topographic Factor (LS)

The topographic factor (called also slope length factor) describes the combined effect of slope length and slope gradient. This factor represents a ratio of soil loss under given conditions to that on the standard plot with 9% slope. Thus, *LS* = 1 for slope steepness of 9% and slope length of 22.13 m (Wischmeier and Smith, 1978); *LS* > 1 for steeper, longer slopes than that, and <1 for gentler, shorter slopes. For example, *LS* factor values for a 61 m long slope with steepness of 5%, 10%, 14%, and 20% are 0.758, 1.94, 3.25, and 5.77, respectively. The *LS* factors for 122 m and 244 m long slopes with constant steepness of 10% are 2.74 and 3.87, respectively. The steeper and longer the slope, the higher the erosion risk. The *LS* factor can be determined from a chart or tables in standard references (Wischmeier and Smith, 1978), or from equations where both slope length and steepness have been taken into consideration, e.g., Wischmeier and Smith (1978):

$$LS = \left(\frac{\lambda}{22.13}\right)^m \left(65.41 \sin^2\theta + 4.56\sin\theta + 0.065\right) \qquad (3)$$

where λ = slope length (m)

θ = angle of slope

m = 0.5 if the slope is 5% or more, 0.4 on slopes of 3.5 to 4.5%, 0.3 on slopes of 1 to 3%, and 0.2 on uniform gradients of less than 1%

Equations such as equation 3 were derived for specific conditions, so care must be taken in using the appropriate equation for the given situation. These equations can be found in various USLE references. There is limited ability to change the *LS* factor, except for, notably, breaking a long slope into shorter slope lengths through the installation of terraces.

Cover Management Factor (C)

The cover management factor is a ratio that compares the soil loss from an area with specified cover and management to that from an identical area in tilled continuous fallow. The value of *C* on a certain field is determined by several variables, such as crop canopy, residue mulch, incorporated residues, tillage, and land use residuals (Wischmeier and Smith, 1978).

The factor may roughly be determined by selecting the cover type and tillage method that corresponds to the field and then multiplying these factors together (OMAFRA, 2012a). The height and density of a canopy reduces the rainfall energy. Residue mulch near the soil surface is more effective to reduce soil loss than equivalent percentages of canopy cover (Wischmeier and Smith, 1978).

For example, incorporating plant residue at the soil surface by shallow tillage offers a greater residual effect than moldboard plowing. The C factor for crop type varies from 0.02 (hay and pasture) to 0.40 (grain corn). The C factor for tillage method varies from 0.25 (no-till or zone tillage) to 1.0 (fall plow). However, local investigation of the C factor is highly recommended because of varying cultivation practices, and because of the interaction of the timing of crop cover development and the timing of rainfall energy, which varies from place to place. Selection of crops and tillage systems can have a huge impact on the C factor.

Conservation Practice Factor (P)

The conservation (also support practice or erosion control) factor reflects the effects of various practices that will reduce the amount and rate of water runoff and, thus, reduce the erosion rate (Wischmeier and Smith, 1978). The most commonly used supporting cropland practices are cross-slope cultivation, contour farming, and strip cropping. The highest P factor value of 1 is given in the case when no influences from conservation practices are considered. The value of 1 is also given to "up and down slopes," while "strip cropping, contour" gets the lowest value of 0.25 in the factsheet of Ministry of Agriculture, Food and Rural Affairs Ontario (OMAFRA, 2012a).

Measurement and Monitoring

Scientific research and erosion measurements are needed to understand erosion processes. Erosion is measured for three principal reasons (1) erosion inventories, (2) scientific erosion research, and (3) development and evaluation of erosion control practices (Toy et al., 2002). Measurements are also needed for the development of erosion prediction technology and implementation of conservation resources and development of conservation regulations, policies, and programs (Stroosnijder, 2005). Erosion measurements are used for development, calibration, and validation of methods of erosion prediction.

Temporal and Spatial Measurements

Erosion measurements are made at various temporal and spatial scales (Toy et al., 2002). For example, sampling duration can vary from a single rainstorm or windstorm to several years.

Spatially, water erosion measurements can range from interrill and rill sediment sources on hillslope or experimental plots to sediment discharge from watersheds. The presence of rills gives evidence of the possible erosion problems on the field. Sediment discharge from watersheds is used in reservoir design. Wind erosion measurements range also from small plots to agricultural fields and to entire regions.

Erosion Inventories

In planning erosion inventory measurements, the following issues should be included (Toy et al., 2002): selection of measurement site(s), measurement frequency and duration at the sites, and suitable measurement techniques. The selection of sites is made according to a sampling strategy. The measurement duration should be long enough to capture the temporal variability of erosion processes. The measurement technique is selected according to erosion type and study question.

How to Measure?

Erosion research is possible in the field (outdoor) or in the laboratory (Toy et al., 2002). Stroosnijder (2005) presents the following five fundamental ways to measure erosion: (1) sediment collection from erosion plots and watersheds, (2) change in surface elevation, (3) change in channel cross section dimensions, (4) change in weight, and (5) the radionuclide method. Both direct measurements and erosion prediction technology are used in erosion inventories. Commonly used erosion measurement techniques are cheap and fast but not very accurate. More accurate methods are costly and beyond the budget of many projects.

Experimental Fields and Catchments

In outdoor research settings, experimental plots, cropland fields or catchments are in use and runoff may be caused by natural or artificial rainfall. Temporal surface runoff (overland flow movement of water exclusively over the soil surface, down slope, during heavy rain) and subsurface discharge (drainage flow) from these sites can be measured and water sampled for sediment analyses. Sampling can be done automatically according to water volume or time. For indoor studies, soil blocks under a rainfall simulator (e.g., stationary drip-type rainfall simulator) can be used (Uusitalo and Aura, 2005). In both cases, representative water samples are collected for the sediment concentration analyses in the laboratory.

To predict the sediment load for a certain study area and time period, the concentration of analyzed water samples is multiplied by the water volume of the sampling period. Water flow ($L\ s^{-1}$) can be measured in stream with a flow meter or V-notch weir, and on croplands with tipping buckets. Erosion amount ($kg\ ha^{-1}$) is estimated by multiplying water flow ($L\ s^{-1}$) by the time (s) and sediment concentration ($g\ L^{-1}$) and finally dividing by the size of the study area (ha).

Also, continuously operating sensors for turbidity measurements from water can be used for measuring erosion from a study area. Turbidity is the degree to which water loses transparency due to suspended particles like sediment; the murkier the water, the more turbid it is. Turbidity sensors need good calibration and control water samples to evaluate sediment content. They must also be equipped with an automatic cleaning mechanism.

Change in Surface Elevation (Hillslope Scale)

The change in elevation is based on the principle that erosion and deposition by water or wind change the elevation of the land surface (Toy et al., 2002). The difference between the two measurements indicates the effect of erosion and deposition during that time interval. A lower elevation indicates erosion and higher elevation at the end of time interval indicates deposition.

One approach to measure change in elevation is to implant stakes or pins that remain in place in the soil for the duration of the study. The distance from the top of the stake or pin to the ground is measured at set time intervals. A decrease in distance corresponds to sedimentation whereas an increase means erosion (Stroosnijder, 2005). By multiplying the change in elevation by the soil bulk density, it is possible to convert the measurement to a mass of soil (Toy et al., 2002). In figure 4, a soil roughness meter is used to measure changes in the surface of soil. The soil roughness meter has a set of pins that sit on the surface, so that soil surface position measurements can be made relative to the top of the structure of the roughness meter. By making repeated measurements at the same location, small changes in the surface elevation can be measured. It may also be used to determine soil erosion in rills.

Figure 4. A soil roughness meter is used to measure changes in the surface of soil. Photo: Risto Seppälä, Natural Resources Institute Finland (Luke).

Change in Channel Cross Section

Channel erosion can be estimated by measuring cross sections at spaced intervals, repeating this after some time and comparing and determining the change in volume of soil. The measurement can be done either manually or using airborne laser scanners (Stroosnijder, 2005). This technique is also suitable for estimating rill or gully erosion on croplands.

Change in Weight (Collected by Splash Cups and Funnels)

This method is based on the principle that the erosion process removes material from the source area (Toy et al., 2002). Test soil, packed in a cup or funnel placed in the soil, is weighed before and after an erosion event, and loss of weight is the erosion measurement. This technique is used in studies of soil detachment and transport by raindrop force (Stroosnijder, 2005). While affordable, and accurate at a small scale, the results using this method are representative of only a very small area, and may not scale well to the field level.

Radionuclide Method

Environmental radionuclides can be used as tracers to estimate soil erosion rates (Stroosnijder, 2005). A human-induced radionuclide of cesium (Cs137) was released into the atmosphere during nuclear weapon tests in the 1950s and 1960s. It spread to the stratosphere and gradually deposited on the land surface. In studies using this method, an undisturbed reference site, on which no or minimal erosion or sedimentation occurs, is needed (Fulajtar et al., 2017). The Cs137 concentration in the study soil is compared to the concentration in the reference site. If the study site contains less Cs137 than the reference site, erosion occurs there. If the study site has more Cs137 than the reference, sedimentation (deposition of soil particles) has occurred. In radionuclide studies, the time scale is usually much longer than in agronomic or environmental studies (Stroosnijder, 2005).

Wind Erosion Measurements

Wind erosion measurements require different measurement plans and equipment than water erosion measurements (Toy et al., 2002). While water erosion follows topography and water flow paths, windblown sediment cannot be collected at a single point (Stroosnijder, 2005). Soil gains and losses due to wind erosion require a number of measurement points, followed by geostatistical analyses. Since wind blows from various directions during the year and during a storm, sediment samplers must rotate with changing wind directions. Measurements must be made at various heights to determine the vertical distribution of the sediment load (Toy et al., 2002).

Impacts of Soil Erosion In-Field and Downstream

Soil erosion has impacts both on croplands where the erosion process starts (detachment) and in the place where it ends up (deposition, sedimentation) (figure 5).

Impacts in Fields

In fields, fertile top soil material can be lost due to erosion processes. The finest particles from topsoil are generally transported from field areas under convex slopes making the areas less productive. The loss of finest particles reduces further the physical structure and fertility of soils (Hudson, 1971). Removal of fine particles or entire layers of soil or organic matter can weaken the structure and even change the texture, which can in turn affect the water-holding capacity of the soil, making it more susceptible to extreme conditions such as drought (OMAFRA, 2012b). Erosion of fertile topsoil results in lower yields and higher production costs.

Figure 5. Sediment chokes this stream due to many years of erosion on nearby unprotected farmland. (Photo courtesy of USDA Natural Resources Conservation Service.)

Sediment may either increase fertility of soil or impair its productivity on productive land. For example, in Egypt, the fields along the Nile River are very productive due to nutritious sediments from the river water. In some cases, the sediment deposited on croplands may inhibit or delay the emergence of seeds, or bury small seedlings (OMAFRA, 2012b). Dredging of open ditches, sedimentation ponds, and waterways, in which sediment is mechanically removed, is becoming more common. However, it is questionable whether dredged sediment can be recycled back to agricultural fields (Laakso, 2017). The sediment may contain substances that are harmful to crops (herbicides) or decrease soil fertility (e.g., aluminum and iron hydroxides).

Impacts Downstream and in Air

In streams and watercourses, sediment can prevent water flow, fill in water reservoirs, damage fish habitats, and degrade downstream water quality. With an enrichment of nutrients, pesticides, salts, trace elements, pathogens, and toxic substances in soil particles in the field, soil erosion causes contamination of downstream water sources, wetlands, and lakes (OMAFRA, 2012b; Zhang and Wang, 2006). Because of the potential harmful impacts of deposited soil particles in water, the control of soil erosion in the field is important. Siltation of watercourses and water storages decreases the storage capacity of water reservoirs.

In addition, fine particles (<0.1 mm) transported by wind may also cause visibility problems on roads. They may also penetrate into respiratory ducts causing health problems.

Applications

For best results, erosion control should begin at the source area, by preventing detachment of soil particles. One of the most effective ways to prevent erosion is through crop and soil management. Detached particles can be trapped by different tools both on cropped field, field edges, and outside fields.

Decreasing the Effects of Erosivity (R) and Erodibility (K)

Erosivity is rather difficult to decrease since there are no tools to affect rainfall. Soil erodibility can be decreased by increasing soil organic matter in soil, e.g., by adding manure or other carbon sources to soil. Practices that reduce or mitigate loss of soil carbon in cropped land can also decrease erodibility. These methods include managing residue to return carbon to the soil and minimizing tillage to reduce the conversion of soil carbon to carbon dioxide gas. Decreasing soil erosion caused by water on highly erodible soils requires additional methods such as permanent grass cover or zero tillage.

The addition of manure, compost, or organic sludge into soil increases aggregate stability, porosity, and water-holding capacity (Zhang, 2006). Both inorganic (stone, gravel, and sand) and organic mulches (crop residue, straw, hay, leaves, compost, wood chips, and saw dust) are used to absorb the destructive forces of raindrops and wind. All these materials also obstruct overland flow and increase infiltration (Zhang, 2006). Mulch reduces erosion until the seedlings mature to provide their own protective cover. In addition, soils treated with amendments like gypsum or structure lime are more durable against erosion than untreated soils (Zhang, 2006). The effect of these soil amendments lasts for a certain period depending on soil and environmental conditions. To maintain the effect, the amendment must be reapplied at intervals.

Soil moisture can prevent erosion. A moist soil is more stable than a dry one, since the soil water keeps the soil particles together. Soil moisture is higher in untilled soils due to a higher percent of organic carbon and minimal evaporation from the soil covered by plant residues. For example, wind erosion can be controlled by wetting the soil.

Reducing the Effect of Topography

Long slopes can be shortened by establishing terraces, but it is difficult to make steep slopes gentler. Reducing the field width (e.g., by windbreaks) protects cropped land against the effects of wind (figure 6).

Increasing the Effect of Cover and Management

Plants are excellent in the protection of soil. They keep the soil in place with their roots, intercept rainfall, provide cover from wind and runoff, increase water infiltration into soil, increase soil aggregation, and provide surface roughness that reduces the speed of water or air movement across the surface. Dense perennial grasses are the most effective erosion controlling plants.

Soil management techniques that disrupt the soil surface as little as possible are excellent at maintaining soil cover and structure. For example, eliminating tillage (called no-till, e.g., direct drilling) keeps the soil surface covered all year round (figure 7). This method, where seed is placed without any prior soil tillage in the stubble, has become common in many dry growing regions to decrease erosion potential. In winter, the stubble remaining after harvest effectively reduces soil erosion compared to bare fields (e.g., plowed in fall). Reduced, or conservation, tillage is also a better choice than fall plowing that leaves the soil surface uncovered. Tillage decreases the organic matter in soils and, thus, has a negative effect on the aggregate stability of clay soils (Soinne et al., 2016). Tillage also disturbs soil structure and, thus, reduces infiltration capacity.

Controlled grazing causes less erosion than tilled croplands; however, the number of grazing animals must be kept low enough to prevent erosion caused by over-grazing. Crop rotation and use of cover crops also maintain soil fertility and, thus, help control erosion. Cover management affects soil erosion in increasing order: meadows < grass and legume catch crops turned under in spring < residue mulch on soil surface < small grain or vetch on fall-plowed seedbed and turned at a spring planting time < row crop canopy < shallow tillage < moldboard plow < burning / removing residues < short period rough fallow in rotation < continuous fallow.

Figure 6. Field windbreaks in North Dakota (U.S.) protect the soil against wind erosion. (Photo courtesy of USDA Natural Resources Conservation Service.)

Figure 7. No-till drilling of soybeans into wheat stubble (Louisiana, USA). (Photo courtesy of USDA Natural Resources Conservation Service.)

Increasing the Effect of Support Practices

On steep slopes, erosion can be controlled by support practices like contour tillage (figure 8), strip cropping on contour, and terrace systems (Wischmeier and Smith, 1978). Strip cropping protects against surface runoff on sloping fields and decreases the transport capacity of soil.

Tillage and planting on the contour is generally effective in reducing erosion. Contouring appears to be most effective on slopes in the 3–8% range (Wischmeier and Smith, 1978). On steeper slopes, more intervention is usually needed. Contour strip cropping (figure 9) is a practice in which contoured strips of dense vegetation, e.g., grasses, legumes, or corn with alfalfa hay, are alternated with equal-width strips of row crops (e.g., soybeans, cotton, sugar beets), or small grain (Wischmeier and Smith, 1978). In erodible areas, grass strips usually 2 to 4 m wide are placed at distances of 10 to 20 m (figure 10). They can be placed on critical areas of the field and the main purpose of these strips is to protect the land from soil erosion. Terracing can be combined with contour farming and other conservation practices making them more effective in erosion control (Wischmeier and Smith, 1978).

In terrace farming, plants may be grown on flat areas of terraces built on steep slopes of hills and mountains. Terracing can reduce surface runoff and erosion by slowing rainwater to non-erosive velocity. Every step (terrace) has an outlet which channel water to the next step.

If soil detachment and transport have taken place, the next consideration is to control deposition before the runoff enters a receiving watercourse. Narrow, 1 to 5 m wide, buffer strips under perennial grasses and wider buffer zones under perennial grasses and trees (figure 11) have been established along rivers to prevent sediment transport to watercourses (Haddaway et al., 2018, Uusi-Kämppä et al., 2000). Grassed waterways (figure 10) are established on concentrated water flows in fields to decrease water flow and, thus, decrease the erosion process in a channel.

Figure 8. Contoured field in southwest Iowa, USA. (Photo courtesy of USDA Natural Resources Conservation Service.)

Figure 9. Alternating strips of alfalfa with corn on the contour protects this crop field in northeast Iowa, USA, from soil erosion. (Photo courtesy of USDA Natural Resources Conservation Service.)

Figure 10. Grass helps protect this western Iowa, USA, cropland with practices including contour buffer strips, field borders, grassed waterways, and grass on terraces. (Photo courtesy of USDA Natural Resources Conservation Service.)

Figure 11. Multiple rows of trees and shrubs, as well as a native grass strip, combine in a riparian buffer to protect Bear Creek, in Iowa, USA. (Photo courtesy of USDA Natural Resources Conservation Service.)

Sediment basins, ponds and wetlands are also used to trap sediment (Uusi-Kämppä et al., 2000). Large particles and aggregates settle over short transport distances, while small clay and silt particles can be carried over long distances in water before their sedimentation.

Country-Specific Perspectives on Soil Erosion

Due to climatic factors (R), soil characteristics (K), landscape features (LS) and cropping practices (C), soil erosion varies geographically. Soil erosion by water is highest in agricultural areas with high rainfall intensity (R factor). In the U.S., the erosion index is great (1200–10,000 MJ mm ha^{-1} h^{-1} yr^{-1}) in eastern, southern, and central parts where tropical storms and hurricanes occur. In Europe, the R factor is highest in the coastal area of the Mediterranean, from 900 to >1300 MJ mm ha^{-1} h^{-1} yr^{-1} (Panagos et al., 2015). In addition to climate, changes in cropping systems (C factor) influence the amount of erosion.

In northern Europe, the most erodible agricultural areas exist in southeast Norway (soil types are silty clay loams or silty clay), southern and central Sweden, and in southwestern Finland (with clay) due to the K factor. In these boreal areas, erosion risk is highest during late fall, winter, and spring due to surface runoff in frozen soil. Soil was previously covered by snow in winter; however, these areas have more frequently been subject to melting and runoff in winter during the last centuries due to climate change (R factor).

In the 1900s, global cropland area increased causing a similar reduction in grassland area (C factor). In Norway, the change in land use doubled soil erosion by water. In addition, extensive land levelling and putting brooks into pipes increased agricultural area in the same region in the 1970s and led to a two-to-three fold increase in erosion (Lundekvam et al., 2003), because levelling, i.e., creating smooth slopes instead of undulating ones, tended to increase the LS factor. Intensive erosion research started in the 1980s and since then Norwegian farmers have received national payments to implement erosion reducing methods, e.g., zero-tillage and growing cover crops in fall (C factor), or establishment of grassed waterways, buffer strips, and sedimentation ponds (P factor). Also, re-opening of piped brooks (decrease in *L* factor), and conversion of fall-tilled fields with high erosion risk into permanent grassland (*C* factor) have been subsidized.

In Finland, typical soil erosion processes in field are sheet erosion, rill erosion, and tillage erosion. Although the mean arable soil loss rate is low (460 kg ha^{-1} yr^{-1}) according to estimations of the RUSLE2015 model (Lilja et al., 2017), there are areas where the erosion risk is higher than this. These high risk areas, with steep slopes and high percent crop production, exist in southwestern parts of the country. In Finland, erosion is mitigated to decrease losses of phosphorus, which can be desorbed from detached soil particles into receiving water bodies where it may cause eutrophication and harmful algal blooms. To decrease soil erosion, some agri-environmental measures are subsidized by the European Union and Finland. For example, fall plowing has been replaced by conservation tillage practices, e.g. no-tillage and direct drilling (*C* factor) or fields may be left under green cover crops for the winter (*C* factor). Grass buffer zones, erosion ponds, or wetlands may be installed and maintained between fields and water bodies to trap soil particles rich in phosphorus (*P* factor).

Examples

Example 1: Calculate average annual soil loss

Problem:
Use the USLE model to calculate the annual soil loss from a Finnish experimental site (slope steepness 6%, length 61 m, 60°48′N and 23°28′E). Annual rainfall is 660 mm, and erosivity is 311 MJ mm ha^{-1} h^{-1} yr^{-1} (Lilja et al., 2017). The site is plowed (up and down slope) in the fall and sown with spring wheat. Particle distribution: clay (<0.002 mm) 30%, silt (0.002–0.02 mm) 40%, very fine sand (0.02–0.1 mm) 25%, and sand (>0.1 mm) 5%. Organic matter in the soil is 2.8%. Soil structure is fine granular, and permeability is slow to moderate.

Solution:
Determine the value of each factor in equation 1:

$$A = R\,K\,LS\,C\,P \qquad (1)$$

R = rainfall erosivity factor; given in problem statement = 311 MJ mm ha^{-1} h^{-1} yr^{-1}

K = soil erodibility factor; calculate using equation 2:

$$K = 2.8 \times 10^{-7} \times M^{1.14}(12 - a) + 4.3 \times 10^{-3}(b - 2) + 3.3 \times 10^{-3}(c - 3) \qquad (2)$$

where M = particle size parameter
= (% silt + % very fine sand) × (100 − %clay) = 65% × (100 − 30%) = 4550
a = organic matter content (%) = 2.8
b = soil structure code = 2 (fine granular)
c = soil profile permeability class = 4 (slow to moderate)

Thus, substituting values in equation 2 yields:

$K = 0.041$ Mg ha h ha^{-1} MJ^{-1} mm^{-1}

$LS =$ topographic factor; find from a published table, e.g., table 3 (Wischmeier and Smith, 1978) or the following excerpt from Factsheet table 3A (OMAFRA, 2012a):

Slope Length (m)	Slope (%)	LS Factor
61	10	1.95
	8	1.41
	6	0.95
	5	0.76
	4	0.53

For a slope length of 61 m and a slope steepness of 6%, $LS = 0.95$, or calculate LS using equation 3:

$$LS = \left(\frac{\lambda}{22.13}\right)^m \left(65.41 \sin^2\theta + 4.56 \sin\theta + 0.065\right) \quad (3)$$

$$LS = \left(\frac{61}{22.13}\right)^{0.5} \left(65.41 \sin^2(6\%) + 4.56 \sin(6\%) + 0.065\right) = 0.95$$

$C =$ crop management factor $= 0.35$ for cereals
$P =$ conservation practice factor $= 1.0$ for fall plowing up and down slope (OMAFRA, 2012a).

Substitute the values for each factor in equation 1:

$$A = R\,K\,LS\,C\,P \quad (1)$$

$$= 311 \times 0.041 \times 0.95 \times 0.35 \times 1 \text{ Mg ha}^{-1} \text{ yr}^{-1} = 4.24 \text{ Mg ha}^{-1} \text{ yr}^{-1}$$

Example 2: Effect of different tillage practices on erosion rates

Problem:
Use the USLE model to evaluate the change in erosion rate in the field runoff of the previous example when fall plowing (up and down slope) is changed (a) to spring plowing (cross slope) or (b) to no-till (up and down slope).

Solution:

(a) Using equation 1 with:

$R = 311$ MJ mm ha^{-1} h^{-1} yr^{-1}
$K = 0.041$ Mg ha h ha^{-1} MJ^{-1} mm^{-1}
$LS = 0.95$
$C = 0.35$ (cereals) × 0.9 (spring plow) = 0.315
$P = 0.75$ (cross slope)
$A = R\,K\,LS\,C\,P = 2.86$ Mg ha^{-1} yr^{-1}

The erosion rate is 32% less due to cross slope plowing in spring compared to up and down plowing in fall.

(b) Using equation 1 with:

$R = 311$ MJ mm ha^{-1} h^{-1} yr^{-1}
$K = 0.041$ Mg ha h ha^{-1} MJ^{-1} mm^{-1}
$LS = 0.95$
$C = 0.35$ (cereals) × 0.25 (no-till) = 0.0875
$P = 1$ (up and down slope)
$A = R\,K\,LS\,C\,P = 1.06$ Mg ha^{-1} yr^{-1}

The erosion rate is 75% less due to direct drilling compared to up and down plowing in fall.

Image Credits

Figure 1. USDA Natural Resources Conservation Service. (CC By 1.0). (2020). Splash erosion. Retrieved from https://photogallery.sc.egov.usda.gov/photogallery/#/

Figure 2. USDA Natural Resources Conservation Service. (CC By 1.0). (2020). Gully erosion Retrieved from https://photogallery.sc.egov.usda.gov/photogallery/#/

Figure 3. USDA ARS. (CC By 1.0). (2020). Wind erosion process. Retrieved from https://infosys.ars.usda.gov/WindErosion/weps/wepshome.html

Figure 4. Risto Seppälä / Luke. (CC By 4.0). (2020). Soil roughness meter.

Figure 5. USDA Natural Resources Conservation Service. (CC By 1.0). (2020). Sediment chokes. Retrieved from https://photogallery.sc.egov.usda.gov/photogallery/#/

Figure 6. USDA Natural Resources Conservation Service. (CC By 1.0). (2020). Field windbreaks. Retrieved from https://photogallery.sc.egov.usda.gov/photogallery/#/

Figure 7. USDA Natural Resources Conservation Service. (CC By 1.0). (2020). No-till drilling. Retrieved from https://photogallery.sc.egov.usda.gov/photogallery/#/

Figure 8. USDA Natural Resources Conservation Service. (CC By 1.0). (2020). Contoured field. Retrieved from https://photogallery.sc.egov.usda.gov/photogallery/#/

Figure 9. USDA Natural Resources Conservation Service. (CC By 1.0). (2020). Alternating strips. Retrieved from https://photogallery.sc.egov.usda.gov/photogallery/#/

Figure 10. USDA Natural Resources Conservation Service. (CC By 1.0). (2020). Grass helps. Retrieved from https://photogallery.sc.egov.usda.gov/photogallery/#/

Figure 11. USDA Natural Resources Conservation Service. (CC By 1.0). (2020). Multiple rows of trees. Retrieved from https://photogallery.sc.egov.usda.gov/photogallery/#/

References

Borrelli, P., Robinson, D. A., Fleischer, L. R., Lugato, E., Ballabio, C., Alewell, C., . . . Panagos, P. (2017). An assessment of the global impact of 21st century land use change on soil erosion. *Nature Commun.*, 8(1), 1-13. https://doi.org/10.1038/s41467-017-02142-7.

ESDAC. (2017). European soil data centre, Global rainfall erosivity. Retrieved from https://esdac.jrc.ec.europa.eu/content/global-rainfall-erosivity.

Fernandez-Raga, M., Palencia, C., Keesstra, S., Jordan, A., Fraile, R., Angulo-Martinez, M., & Cerda, A. (2017). Splash erosion: A review with unanswered questions. *Earth-Sci. Rev.*, 171, 463-477. https://doi.org/10.1016/j.earscirev.2017.06.009.

Foster, G. R., McCool, D. K., Renard, K. G., & Moldenhauer, W. C. (1981). Conversion of the universal soil loss equation to SI metric units. *JSWC*, 36(6), 355-359.

Fulajtar, E., Mabit, L., Renschler, C. S., & Lee Zhi Yi, A. (2017). Use of 137Cs for soil erosion assessment. Rome, Italy: FAO/IAEA.

Gabet, E. J., Reichman, O. J., & Seabloom, E. W. (2003). The effects of bioturbation on soil processes and sediment transport. *Ann. Rev. Earth Planetary Sci.*, 31(1), 249-273. https://doi.org/10.1146/annurev.earth.31.100901.141314.

Haddaway, N. R., Brown, C., Eales, J., Eggers, S., Josefsson, J., Kronvang, B., . . . Uusi-Kämppä, J. (2018). The multifunctional roles of vegetated strips around and within agricultural fields. *Environ. Evidence*, 7(14), 1-43. https://doi.org/10.1186/s13750-018-0126-2.

Hudson, N. (1971). *Soil conservation.* London, U.K.: B. T. Batsford.

Laakso, J. (2017). Phosphorus in the sediments of agricultural constructed wetlands. Doctoral thesis in Environmental Science. Helsinki, Finland: University of Helsinki, Department of Food and Environmental Sciences. Retrieved from https://helda.helsinki.fi/bitstream/handle/10138/224575/phosphor.pdf?sequence=1&isAllowed=y.

Lilja, H., Hyväluoma, J., Puustinen, M., Uusi-Kämppä, J., & Turtola, E. (2017). Evaluation of RUSLE2015 erosion model for boreal conditions. *Geoderma Regional*, 10, 77-84. https://doi.org/10.1016/j.geodrs.2017.05.003.

Lindstrom, M. J., Lobb, D. A., & Schumacher, T. E. (2001). Tillage erosion: An overview. *Ann. Arid Zone*, 40(3), 337-349.

Lundekvam, H. E., Romstad, E., & Øygarden, L. (2003). Agricultural policies in Norway and effects on soil erosion. *Environ. Sci. Policy*, 6(1), 57-67. https://doi.org/10.1016/S1462-9011(02)00118-1.

Mohtar, R. H. (no date). Estimating soil loss by water erosion. https://engineering.purdue.edu/~abe325/week.8/erosion.pdf.

OMAFRA. (2012a). Universal soil loss equation (USLE). Factsheet Order No. 12-051. Ontario Ministry of Agriculture, Food and Rural Affairs. Retrieved from http://www.omafra.gov.on.ca/english/engineer/facts/12-051.htm#1.

OMAFRA. (2012b). Soil erosion—Causes and effects. Factsheet Order No. 12-053. Ontario Ministry of Agriculture, Food and Rural Affairs. Retrieved from http://www.omafra.gov.on.ca/english/engineer/facts/12-053.htm.

Panagos, P., Ballabio, C., Borrelli, P., Meusburger, K., Klik, A., Rousseva, S., . . . Alewell, C. (2015). Rainfall erosivity in Europe. *Sci. Total Environ*, 511, 801-814. https://doi.org/10.1016/j.scitotenv.2015.01.008.

Puustinen, M., Tattari, S., Koskiaho, J., & Linjama, J. (2007). Influence of seasonal and annual hydrological variations on erosion and phosphorus transport from arable areas in Finland. *Soil Tillage Res.*, 93(1), 44-55. https://doi.org/10.1016/j.still.2006.03.011.

Soinne, H., Hyväluoma, J., Ketoja, E., & Turtola, E. (2016). Relative importance of organic carbon, land use and moisture conditions for the aggregate stability of post-glacial clay soils. *Soil Tillage Res.*, 158, 1-9. https://doi.org/10.1016/j.still.2015.10.014.

Stroosnijder, L. (2005). Measurement of erosion: Is it possible? *CATENA*, 64(2), 162-173. https://doi.org/10.1016/j.catena.2005.08.004.

Toy, T. J., Foster, G. R., & Renard, K. G. (2002). *Soil erosion: Processes, prediction, measurement, and control.* Hoboken, NJ: John Wiley & Sons.

USDA ARS. (2020). United States Department of Agriculture, Agricultural Research Service. Washington, DC: USDA ARS. Retrieved from https://infosys.ars.usda.gov/WindErosion/weps/wepshome.html.

Uusi-Kämppä, J., Braskerud, B., Jansson, H., Syversen, N., & Uusitalo, R. (2000). Buffer zones and constructed wetlands as filters for agricultural phosphorus. *JEQ, 29*(1), 151-158. https://doi.org/10.2134/jeq2000.00472425002900010019x.

Uusitalo, R., & Aura, E. (2005). A rainfall simulation study on the relationships between soil test P versus dissolved and potentially bioavailable particulate phosphorus forms in runoff. *Agric. Food Sci., 14*(4), 335-345. https://doi.org/10.2137/1459606057758977l3.

Van Klaveren, R. W., & McCool, D. K. (1998). Erodibility and critical shear of a previously frozen soil. *Trans. ASAE, 41*(5), 1315-1321. https://doi.org/10.13031/2013.17304.

Wischmeier, W. H., & Smith, D. D. (1978). *Predicting rainfall erosion losses.* A quide to conservation planning. Agriculture Handbook No. 537. Supersedes Agriculture Handbook No. 282. Washington, DC: USDA. Retrieved from https://naldc.nal.usda.gov/download/CAT79706928/PDF.

Zhang, T., & Wang, X. (2006). Erosion and global change. In R. Lal (Ed.), *Encyclopedia of soil science* (2nd ed., Vol. 1, pp. 536-539). New York, NY: Taylor & Francis.

Zhang, X.-C. (2006). Erosion and sedimentation control: Amendment techniques. In R. Lal (Ed.), *Encyclopedia of soil science* (2nd ed., Vol. 1, pp. 544-552). New York, NY: Taylor & Francis.

Anaerobic Digestion of Agri-Food By-Products

Robert Bedoić
University of Zagreb, Faculty of Mechanical Engineering and Naval Architecture
Zagreb, Croatia

Tomislav Pukšec
University of Zagreb, Faculty of Mechanical Engineering and Naval Architecture
Zagreb, Croatia

Boris Ćosić
University of Zagreb, Faculty of Mechanical Engineering and Naval Architecture
Zagreb, Croatia

Neven Duić
University of Zagreb, Faculty of Mechanical Engineering and Naval Architecture
Zagreb, Croatia

KEY TERMS		
Substrates and feedstocks	Biogas chemistry	Biogas utilization
Pretreatment of feedstock	Inhibition parameters	Products
Operating modes	Biogas yield	Digestate management

Variables

$a, b, c,$ and d = number of atoms of carbon, hydrogen, oxygen, and nitrogen, respectively

$m_{substrate}$ = mass of substrate

Q = feedstock input in the digester

V = digester volume

$V_{N,biogas}$ = normalized cumulative volume of biogas

Introduction

Anaerobic digestion is a set of biochemical processes where complex organic matter is decomposed by the activity of bacteria in an oxygen-free atmosphere into biogas and digestate. Understanding the basic principles of anaerobic digestion (AD), and its role in the production of renewable energy sources, requires familiarity with the chemical composition of substrates, degradation stages in the process, and use of the products, both biogas and digestate. Agri-food by-products are recognized as a sustainable source of biomass for AD. Post-harvesting residues, food industry by-products and decomposed food can be utilized for AD to achieve environmental benefits (including a reduction of landfilling and greenhouse gas emissions) with added production of green energy. Biogas is a mixture of gaseous compounds, with the highest portion being methane (about 50–70% by volume), followed by carbon dioxide (30–50%). In a large-scale operation, biogas is usually utilized as a fuel to run a gas engine in the combined production of heat and electricity (CHP), or to produce biomethane (a gas similar in its characteristics to natural gas), through biogas upgrade processes. Digestate, another product of anaerobic digestion, is usually nutrient-rich, non-degraded organic material that can be used as a soil conditioner and replacement for conventional synthetic organic fertilizers.

Outcomes

After reading this chapter, you should be able to:

- Describe the chemistry of the AD (anaerobic digestion) process
- Calculate biogas production from different substrates based on their elemental composition
- Describe the factors influencing the AD process, including process inhibition
- Identify substrates suitable for direct use in the AD process and substrates that need pre-treatment before feeding the digester
- Describe methods of biogas and digestate utilization in the production of renewable energy

Concepts

Anaerobic Digestion Pathway

The concepts that underlie the AD process can be presented as a multi-step process (Lauwers et al., 2013), usually consisting of four main stages: *hydrolysis*, *acidogenesis*, *acetogenesis*, and *methanogenesis*.

Hydrolysis is the first stage of the AD process. In hydrolysis, large polymers (complex organic matter) are decomposed in the presence of hydrolytic enzymes into basic monomers: monosaccharides, amino acids, and long chain fatty acids. Hydrolysis can be represented using the following simplified chemical reaction:

$$C_6H_{10}O_4 + 2H_2O \rightarrow C_6H_{12}O_6 + H_2$$

The intensity of the hydrolysis process can be monitored through hydrogen production in the gas phase. Hydrolysis occurs at low rates because polymer molecules are not easily degradable into basic monomer compounds. Usually, this stage is also the rate-limiting stage of the overall AD process. The lower the rate of hydrolysis, the lower the production of biogas.

The second stage of AD, acidogenesis, includes conversion of monosaccharides, amino acids, and long chain fatty acids resulting from hydrolysis into carbon dioxide, alcohols, and volatile fatty acids (VFAs). The following two reactions illustrate the breakdown of monomers into ethanol and propionic acid:

$$C_6H_{12}O_6 \rightleftharpoons 2CH_3CH_2OH + 2CO_2$$

$$C_6H_{12}O_6 + 2H_2 \rightleftharpoons 2CH_3CH_2COOH + 2H_2O$$

Accumulation of VFAs caused by the acidogenesis process leads to a decrease in pH value. This phenomenon can contribute to significant problems in the operation of the AD process since it affects bacteria responsible for biogas production.

Acetogenesis is the third stage of the AD process, characterized by the production of hydrogen and acetic acid from basic monomers and VFAs. The reactions that describe the chemical processes occurring during acetogenesis are:

$$CH_3CH_2COO^- + 3H_2O \rightleftharpoons CH_3COO^- + HCO_3^- + H^+ + 3H_2$$

$$C_6H_{12}O_6 + 2H_2O \rightleftharpoons 2CH_3COOH + 4H_2 + 2CO_2$$

$$CH_3CH_2OH + 2H_2O \rightleftharpoons CH_3COO^- + 3H_2 + H^+$$

Acetogenesis and acidogenesis occur simultaneously; there is no time delay between the two processes. Hydrogen formed in acetogenesis could inhibit metabolic activity of acetogenic bacteria and decrease the reaction. On the other hand, hydrogen formed in acetogenesis could become the reactant for the last stage of AD.

Methanogenesis is the fourth stage of the AD process. In general, methanogenic bacteria can form methane from acetic acid, alcohols, hydrogen, and carbon dioxide, according to Bochmann and Montgomery (2013):

$$CH_3COOH \rightleftharpoons CH_4 + CO_2$$

$$CH_3OH + H_2 \rightleftharpoons CH_4 + H_2O$$

$$CO_2 + 4H_2 \rightleftharpoons CH_4 + 2H_2O$$

Biogas production is usually expressed in terms of a biogas yield—the amount of biogas produced by the mass of substrate. Al Seadi et al. (2008) and Frigon and Guiot (2010) have determined that each compound of biomass can be characterized by its theoretical biogas content and theoretical biogas yield, as presented in table 1.

Data in table 1 show that fats produce more biogas than proteins and carbohydrates, and that proteins and fats produce biogas with a higher methane

Table 1. Elemental formula, theoretical gas yields, and share of main compounds in biogas from different substrates (Al Seadi et al., 2008; Frigon and Guiot, 2010).

Polymers	Elemental Formula	Theoretical Biogas Yield (Nm³/kg TS)[a]	Biogas Composition CH$_4$ (%)	CO$_2$ (%)
Proteins	$C_{106}H_{168}O_{34}N_{28}S$	0.700	70 – 71	29 – 30
Fats	$C_8H_{15}O$	1.200 – 1.250	67 – 68	32 – 33
Carbohydrates	$(CH_2O)_n$	0.790 – 0.800	50	50

[a] Nm³ = normal cubic meter; TS = total solids or dry weight.

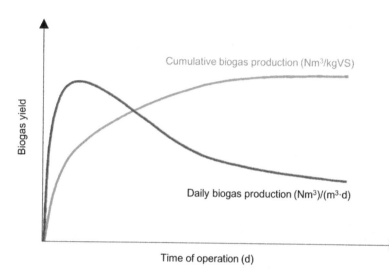

Figure 1. Theoretical biogas yield profiles for a batch test (Mähnert, 2006).

content than carbohydrates. The share of methane is relevant, since the efficiency of biogas utilization is based on the share of methane in the biogas. A gas analyzer is used to determine the composition of biogas—not only its methane content, but also its content of other components such as carbon dioxide, water, hydrogen sulphide, etc. The volume of biogas produced can be measured by several methods; the water displacement method is the most common one (Bedoić et al., 2019a). The recorded biogas volume is then usually adjusted to 0°C and 101,325 Pa pressure so that it can be compared to other reported values.

In batch AD tests the profile of biogas can also be presented over the daily production or cumulative production (figure 1). The biogas generation rate is highest at the start of the process and after a certain time, it decreases. When the profile of cumulative production of biogas remains constant over several days, it is an indication that the biodegradation of organic material has stopped.

The mass of the substrate can be expressed as total solids (TS) or volatile solids (VS). TS is the mass of solids in the substrate with water totally excluded (determined at 105°C), while VS is the mass of solids that is lost on ignition of TS at 550°C. TS and VS are used more commonly when anaerobic digestion is performed on a laboratory or pilot scale. For a large-scale operation the mass of fresh matter (FM; raw mass inserted into a digester, including water) is more commonly used.

Important Parameters for AD

Significant parameters for AD include constitution of the substrate that enters the reactor, pH, and temperature. The elements carbon (C), hydrogen (H), oxygen (O), nitrogen (N) and sulfur (S) are building blocks of organic matter that make up the polymeric carbohydrate, protein, and lipid molecules. Based on the elemental composition of a substrate used for anaerobic digestion it is possible to estimate its biodegradable properties. One of the most common ways to estimate the degradability of a substrate in AD is the carbon to nitrogen ratio (C:N). An optimum C:N of a substrate is between 25 and 30. Lower C:N values indicate a high nitrogen content, which could lead to ammonia generation and inhibition in the process. Higher C:N values indicate high levels of carbohydrates in a substrate, which makes it harder to disintegrate and produce biogas.

The pH range in the reactor depends on the feedstock used and its chemical properties. Liu et al. (2008) found that the optimum pH range inside the reactor is between 6.8 and 7.2, while the AD process can tolerate a range of 6.5 up to 8.0.

The anaerobic digestion process is highly sensitive to temperature changes. Van Lier et al. (1997) have studied the impact of temperature on the growth rate of bacteria in methanogenesis. In general, psychrophilic (2°C to 20°C) digestion is not used for commercial purposes, due to a high retention time of substrates. Feedstock co-digestion is usually performed in mesophilic (35°C to 38°C) or thermophilic (50°C to 70°C) conditions. Mesophilic anaerobic digestion is the most common system. It has a more stable operation than thermophilic anaerobic digestion, but a lower biogas production rate. Thermophilic anaerobic digestion shows advantages in terms of pathogen reduction during the process. Increasing the temperature of the process increases the organic acids inside the fermenter, but at the same time makes the process more unstable. Degradation of the feedstock under thermophilic conditions requires an additional heat supply to achieve and maintain such conditions in the digester.

Estimation of Biogas Yield of the Substrate Based on the Elemental Composition

The theoretical production of biogas can be estimated by knowing the elemental composition of the substrate. Gerike (1984) has determined an approach to find a molecular formula that represents the composition of the substrate in the form of $C_aH_bO_cN_d$ where a, b, c, and d represent the number of carbon, hydrogen, oxygen, and nitrogen atoms, respectively; this is estimated from the elemental composition of the substrate, on a dry basis (TS). To represent the entire AD process in one stage, the following chemical reaction is used:

$$C_aH_bO_cN_d + \frac{4a-b-2c+3d}{4}H_2O \rightarrow \frac{4a+b-2c-3d}{8}CH_4 + \frac{4a-b+2c+3d}{8}CO_2 + dNH_3$$

The reaction estimates that the entire organic compound, $C_aH_bO_cN_d$, is decomposed during the AD process into three products: methane, carbon dioxide and ammonia.

Substrates in AD are usually characterized through the value of oxygen demand. There are several types of oxygen demand; the ones usually used are the following.

- Biochemical oxygen demand (BOD) is the measure of the oxygen equivalent of the organic substrate that can be oxidized *biochemically* using aerobic biological organisms.
- Chemical oxygen demand (COD) is the measure of the oxygen equivalent of the organic substrate that can be oxidized *chemically* using dichromate

in an acid solution. The amount of methane produced in the anaerobic digestion per unit COD is around 0.40 Nm^3/kgO_2.
- Theoretical oxygen demand (ThOD) is the oxygen required to oxidize the organics to end products based on the elemental composition of the substrate. ThOD is used to estimate the oxygen demand of different substrates.

Koch et al. (2010) established a formula to estimate the ThOD of the substrate based on the molecular formula $C_aH_bO_cN_d$:

$$\text{ThOD} = \frac{16 \times (2a + 0.5(b-3d) - c)}{12a + b + 16c + 14d} \left(\frac{kg_{O_2}}{kgTS_{C_aH_bO_cN_d}} \right) \quad (1)$$

It does not reflect the fact that the organic substrate is not 100% degradable at any time.

Li et al. (2013) showed a way to estimate the theoretical biochemical methane potential (TBMP) of the organic substrate based on the elemental composition of the substrate, as:

$$\text{TBMP} = \frac{22.4 \times \left(\frac{a}{2} + \frac{b}{8} - \frac{c}{4} - \frac{3d}{8} \right)}{12a + b + 16c + 14d} \left(\frac{Nm^3_{CH_4}}{kgVS_{C_aH_bO_cN_d}} \right) \quad (2)$$

Since the organic matter cannot always be fully degraded during the AD process, it is valuable to conduct measurements (usually lab-scale) on the production of biogas, in order to investigate the actual degradability of the sample. For those purposes, biochemical gas potential (BGP) and biochemical methane potential (BMP) tests are used.

The BGP laboratory test is used in assessing the potential yield of a substrate in terms of biogas production and process stability (based upon pH and concentration of ammonia):

$$\text{BGP} = V_{N,\text{biogas}} / m_{\text{substrate}} \quad (3)$$

where $V_{N,\text{biogas}}$ = normalized cumulative volume of biogas (Nm^3)
$m_{\text{substrate}}$ = mass of substrate put in a reactor (kg FM or kg TS or kg VS).

The BMP test is usually used in assessing the feedstock potential, but in terms of biomethane production:

$$\text{BMP} = V_{N,CH_4} / m_{\text{substrate}} \quad (4)$$

where V_{N,CH_4} = normalized cumulative volume of biomethane (Nm^3)
$m_{\text{substrate}}$ = mass of substrate put in a reactor (kg FM or kg TS or kg VS).

Both tests contribute to the evaluation of the use of prepared feedstock in the AD process. The ratio of BMP and BGP represents the share of the most important compound of biogas, methane:

$$CH_4 \text{ in biogas} = BMP/BGP \qquad (5)$$

Degradation of a substrate is expressed as the ratio of the actual BMP to the TBMP, which depends on the chemical constitution of the substrate, as:

$$\text{Degradation (\%)} = \frac{BMP}{TBMP} \times 100 \qquad (6)$$

Large-Scale AD

Parameters that need to be considered during design, and controlled during the operation of the AD process in a large-scale biogas production at a satisfactory level, are *organic load rate* and *hydraulic retention time*.

Organic load rate (OLR) represents the quantity of organic material fed into the digester. It is usually higher for co-digestion compared to mono-digestion. If OLR is too high, foaming and instability of the process can occur due to higher levels of acidic components in the digester. OLR can be calculated based on the input volume of feedstock:

$$OLR = Q/V \qquad (7)$$

where Q = raw feedstock input in the digester per day (m^3 d^{-1})
V = digester volume (m^3)

A more common way to present the organic load rate is through the chemical oxygen demand:

$$OLR_{COD} = OLR \times COD \qquad (8)$$

where COD = chemical oxygen demand of feedstock per unit volume of feedstock (kg O$_2$ m^{-3}).

Interpretation of the input feedstock through COD includes taking into consideration the chemical properties of substrates in the feedstock.

Hydraulic retention time (HRT) represents the time (days) that a certain quantity of feedstock remains in the digester:

$$HRT = V/Q \qquad (9)$$

The required HRT depends on parameters such as feedstock composition, the operating conditions in the digester, and digester configuration. In order to avoid instability in the process usually caused by VFA accumulation, HRT should be >20 days.

Products of Anaerobic Digestion

Biogas is a mixture of gases, mainly composed of methane and carbon dioxide, while compounds like water, oxygen, ammonia, hydrogen, hydrogen sulfide, and nitrogen can be found in traces. Average data on the detailed composition of biogas is shown in table 2.

Table 2. Detailed composition of biogas.

Compound	Chemical symbol	Content (%)
Methane	CH_4	50 – 75
Carbon dioxide	CO_2	25 – 45
Water vapor	H_2O	2 (20°C) – 7 (40°C)
Oxygen	O_2	<2
Nitrogen	N_2	<2
Ammonia	NH_3	<1
Hydrogen	H_2	<1
Hydrogen sulfide	H_2S	<1

Methane is the most important compound of biogas since it is an energy-rich fuel. Therefore, the uses of biogas are primarily related to extracting benefits from methane recovery in terms of producing renewable energy.

Apart from biogas, a digestate is the second product of anaerobic digestion. It can be described as a macronutrients-rich indigestible material that can be used in improving the quality of the soil. Pognani et al. (2009) have studied the breakdown of macronutrients and total indigestible solids (TS g kg^{-1}), lignin, hemicellulose, and cellulose in digestate, based on different feedstock types as shown in table 3.

Table 3. Breakdown of macronutrients and indigestible compounds in digestate (Pognani et al., 2009).

		Macronutrients			Indigestible Compounds		
Feedstock	TS (g kg^{-1})	Total N (g kg^{-1} TS)	NH_4-N (g L^{-1})	Total P (g kg^{-1} TS)	Lignin (g kg^{-1} TS)	Hemicelluloses (g kg^{-1} TS)	Celluloses (g kg^{-1} TS)
Energy crops, cow manure slurry, and agro-industrial waste	35	105	2.499	10.92	280	42	68
Energy crops, cow manure slurry, agro-industrial waste, and OFMSW (organic fraction of municipal solid waste)	36	110	2.427	11.79	243	54	79

Digestate of agri-food by-products typically contains a very low quantity of total solids, around 3.5%; the remainder is water. Total nitrogen is usually the most abundant macronutrient, at about 11% of total solids. Phosphorus is present in much lower quantities in total solids, with about a 1% share. The main indigestible compound in digestate is lignin, at almost 30%.

Lignin is difficult to biodegrade during the AD process; hence, it can be found in the digestate. Mulat et al. (2018) have successfully applied pretreatment methods (steam explosion and enzymatic saccharification) to increase the biodegradability of lignin. The efficiency of pretreatment can be defined as the increase in BMP (or BGP) over the BMP (or BGP) of non-pretreated substrate:

> The word lignin comes from the Latin lignum, which means wood. It is a complex product of aromatic alcohols known as monolignols (Karak, 2016).

$$\text{Efficiency (\%)} = \frac{\text{BMP (after pretreatment)} - \text{BMP (without pretreatment)}}{\text{BMP (without pretreatment)}} \times 100 \quad (10)$$

Inhibition Parameters in AD

Many factors influence inhibition, i.e., reduced biogas production, of the AD process, but the most frequent is the use of inadequate substrates and their high loads. According to Xie et al. (2016), inhibition of the AD process is a result of the accumulation of several intermediates:

- free ammonia (FA), (NH_3)
- volatile fatty acids (VFAs)
- long-chain fatty acids (LCFAs)
- heavy metals (HMs).

Ammonia is a compound generated by the biological degradation of organic matter that contains nitrogen—primarily proteins. Typical protein-rich substrates for AD are slaughterhouse by-products (blood, rumen, stomach and intestinal content) and decomposed food (milk, whey, etc.). Process instability due to ammonia accumulation usually indicates accumulation of VFAs, which points to a decreasing of pH value (Sung and Liu, 2003). The critical ammonia toxicity range depends on the type of feedstock used in biogas production, but it goes from 3 to 5 $g(NH_4\text{-}N)\ L^{-1}$ and higher.

Inhibition limited by VFAs relates to the conversion of VFAs into acetic acid before methane is formed in the process of acetogenesis. Butyric acid is more likely to be converted into acetic acid compared to propionic acid. The ratio of concentrations of propionic and acetic acid in the digester is used as a valuable indicator of inhibition by VFAs; if the indicator is above 1.4, inhibition is present in the system.

Formation of LCFAs is more intense if the substrate contains more lipids. Examples of lipid-rich substrates are domestic sewage, oil-processing effluents, and slaughterhouse by-products. Ma et al. (2015) implied that a high concentration of LCFAs results in the accumulation of VFAs and lower methane yield. LCFAs can cause biochemical inhibition, increasing the degradation of microorganisms. Zonta et al. (2013) found that LCFAs could cause physical inhibition as a result of the adsorption of LCFAs on the surface of the microorganisms.

Heavy metals (HMs) are non-biodegradable inorganic compounds that can be found in the feedstock. Usually, municipal sewage and sludge are most dominant HM-rich substrates. During the AD process, HMs remain in the bulk volume in the digester. Therefore, the accumulation of HMs can reach a potentially dangerous concentration that can cause failure in the anaerobic digester operation. Some of the most notable HMs that could cause inhibition in the AD are Hg (mercury), Cd (cadmium), Cr (chromium), Cu (copper), As (arsenic), Zn (zinc), and Pb (lead).

Applications

Operating Modes of Anaerobic Digestion

Based on the number of substrates used in AD, there are two operating modes, *mono-digestion* and *co-digestion*. Mono-digestion is related to the use of only one substrate in AD, while co-digestion reflects the use of two or more substrates in preparing feedstock.

Usually, mono-digestion is an applicable method only on a farm level, where a single type of agricultural by-product is present, such as animal manure. Digestion of animal manure is usually performed in a small-scale digester that replaces inefficient storage of animal manure and contributes to the mitigation of greenhouse gas emissions (GHG).

Co-digestion involves mixing substrates in different ratios to keep properties of the mixture suitable for running the process with an optimum range. Co-digestion is a more advantageous method of energy recovery from organic material due to several benefits: better C:N ratio of the mixed feedstock, efficient pH and moisture content regulation, and higher biodegradability, thus a higher production of biogas (Das and Mondal, 2016). Patil and Deshmukh (2015) found that other variables important for the adequate running of AD could easily be adjusted through co-digestion performance, including moisture content and pH. Compared to mono-digestion, co-digestion has higher biogas yield, which is associated with the synergistic effects of the microorganisms present in the substrates.

AD processes are also classified by the moisture content of the feedstock into *wet AD* and *dry AD*. Wet AD is characterized by feedstock that can be mixed and pumped as liquid slurries, due to a low solid content (3% to 15%). Dry AD (sometimes called high solids AD) is performed in a pile, with the feedstock in stackable form. Tanks for large-scale AD are usually built of concrete with a corrosive-protective layer applied to the inner tank wall, in order to ensure longer durability in the gas/water interface zone.

AD processes can operate as large-scale *continuous* processes and as lab-scale *batch* processes, as well as intermediate fed-batch and semi-continuous processes. Large-scale biogas production processes (digester volume > 1,000 m^3) are performed in biogas plants. Energy production in biogas plants is directly linked to the efficiency of biological conversion of feedstock in a digester. OLR and feedstock properties are controlled to maintain a stable and efficient process. Biogas produced can be utilized in various ways, to produce heat, electricity, or natural-gas like biomethane.

Laboratory (lab-scale; digester volume <1 L) AD is usually performed in a batch mode with a goal to investigate the BMP of substrates for the purposes of AD. The basic principle of batch AD is to put feedstock in a small reactor, add inoculum (colony of bacteria), seal it well, deaerate it to remove oxygen from the digester atmosphere, and monitor the production and the quality of biogas over time by certain laboratory methods (generally water displacement or eudiometer with a pressure gauge). Bedoić et al. (2019a) studied co-digestion of residue grass and maize silage with animal manure in a 250 mL reactor with biogas measurement by a water displacement method. A heated bath was used to maintain the constant temperature in the reactor, since AD is a temperature-sensitive process. As the AD process ran, the generated biogas left the reactor through the outlet hose and entered the upside-down graduated measuring jug filled with water. The volume of the water ejected from the measuring jug represented the volume of the biogas generated in the AD.

In addition to the continuous process in large scale digesters and the batch process usually performed at a laboratory scale, Ruffino et al. (2015) described two additional operating modes for AD: fed-batch and semi-continuous. The fed-batch process is usually considered in a semi-pilot setup, where the digester is started as batch and after a certain period of time products are withdrawn from the reactor. These modes are in between large-scale continuous mode and a lab-scale batch mode. A certain portion of new substrates is added, and the process continues. If repeated several times, this operating mode is known as a repeated fed-batch. The semi-continuous process is considered as a pilot setup, where the process is driven in the continuous mode, but operates with a lower volume digester. Semi-continuous AD has shown many advantages compared to the batch operation when investigating AD, mainly due to a dynamic component in the process, which reflects the behavior of continuous-large scale operation.

AD of Agri-Food By-Products

This chapter stresses the use of agri-food material as feedstock for AD. Since agri-food material is represented through a variety of different substrates and compounds, it is important to concisely present basic issues regarding this topic. An attractive option to present agri-food material sources is by using the supply chain integrated into the concept of an *agricultural waste, co- and by-products (AWCB) value chain*, where the generation of biodegradable material is presented as three major steps (Bedoić et al., 2019b): cultivation/harvesting/farming, processing, and consumption. In each step of the AWCB value chain, there is a generation of organic matter that can be used for AD. Five commodities were selected to represent agri-food by-product sources in the AWCB value chain. More detailed information about agri-food sources in the AWCB value chain for selected commodities is shown in table 4.

Table 4. Agri-food residues suitable for AD, showing the sources in the AWCB value chain.

Commodity	Geographic Area	Cultivation/ Harvesting/ Farming	Processing	Consumption
Cattle, dairy cows	India, USA, China	manure	blood, fatty tissue, skin, feet, tail, brain, bones, whey	decayed beef, milk, butter, cheese
Rice	China, India, Indonesia	straw	bran, hull	decayed rice
Apple	China, E.U., USA	pruning residues and leaves	apple pomace (peel, core, seed, calyx, stems), sludge	decayed apples
Sugar beet	Russia, France, USA	sugar beet leaves	molasses, sugar beet pulp, wash water, factory lime, sugar beet tops and tails	wasted sugar
Olives	Spain, Italy, Greece, Northern Africa	twigs and leaves, woody branches	mill waste-water, olive pomace	wasted olive oil, decayed olives

During the first stage of the AWCB value chain (cultivation and harvesting), a certain amount of commodity is eaten or destroyed by animals (e.g., birds, rabbits, deer, wasps) or due to bad weather conditions and cannot be used as food (Bedoić et al., 2019b). By-products from this first stage of the AWCB value chain are mainly lignocellulosic matter, except for the case of manure. Since lignocellulosic matter contains an indigestible compound (lignin), intensive pretreatment methods are needed to enhance the degradation of this particular organic matter. On the other side, manure has a lower potential for biogas production compared to lignocellulosic matter, but it is important as a valuable source of nutrients.

The second stage of the AWCB value chain is the processing of commodities where additional residues are generated. Since there are many options to process a commodity, AWCB products in this stage require special consideration for AD. The most interesting, but at the same time the most challenging, AWCBs characterized by high oxygen demand are slaughterhouse remains. Slaughterhouse remains are characterized by an inappropriate (low) C:N of 6–14, which usually causes ammonia inhibition during AD (Moukazis et al., 2018). Co-digestion of olive pomace and apple co-products with cow slurry has demonstrated feasibility and economic attractiveness. Results of semi-continuous anaerobic co-digestion with different OLRs have shown that the mixture of this kind of substrates shows energy potential similar to mixtures of some energy crops and livestock combinations. Aboudi et al. (2016) studied mono-digestion of sugar beet cossettes and co-digestion with cow manure operating under mesophilic conditions in the semi-continuous anaerobic system. The results showed that co-digestion produced higher methane generation and no inhibition phenomena, compared to mono-digestion of sugar beet cossettes. Industrial crop by-products have shown potential to produce biogas through dry-AD with implemented technologies for pretreatment of substrates.

The third stage of the AWCB value chain is consumption, which includes materials such as food waste or spoiled food, mainly generated in households. It is quite difficult to estimate the composition of decayed food, due to the variety of different substrates present. However, some general facts about agri-food by-products as a feedstock for AD are that they are an ever present, everyday, nutrient rich, sustainable energy source. The nutrient-rich composition provides the potential for applying the digestate as a valuable soil conditioner. On the other side, some pretreatment techniques are required to increase relatively low biodegradability of food waste feedstock.

Pretreatment of Agri-Food By-Products to Enhance Biogas Production

Some organic compounds show low degradability if they enter the digester in their raw form. Ariunbaatar et al. (2014) presented several groups of pretreatment techniques that can be applied to increase the biodegradability of those substrates:

- mechanical — disintegration and grinding solid parts of the substrates, which result in releasing cell compounds and increasing the specific surface area for degradation
- thermal — used for pathogen removal, improving dewatering performance, and reducing the viscosity of the digestate; the most studied pretreatment method, applied at industrial scale
- chemical — used for destructing the organic compounds by means of strong acids, alkalis, or oxidants
- biological — includes both anaerobic and aerobic methods along with the addition of specific enzymes such as peptidase, carbohydrolase, and lipase

Pretreatments may be combined for further enhancement of biogas production and faster kinetics of AD. Usually, the applied combined pretreatment techniques are thermo-chemical and thermo-mechanical.

The influence of different pretreatment methods applied on substrates in terms of increased biogas production is shown in table 5. The effectiveness of the pretreatment method (increased biogas production) depends on the

Table 5. Influence of pretreatment techniques on biogas yield for different substrates.

Substrate	Pretreatment Technique	AD Operating Mode	Biogas and/or Biomethane Yield		Increased Production	Reference source
			Before Pretreatment	After Pretreatment		
OFMSW	rotary drum	thermophilic batch	346 mL CH_4/g VS	557 mL CH_4/g VS	61%	(Zhu et al., 2009)
	thermophilic pre-hydrolysis	thermophilic (continuous 2-stage)	223 mmol CH_4 /(L(reactor)·d))	441.6 mmol CH_4 /(L(reactor)·d))	98%	(Ueno et al., 2007)
Food waste	size reduction by beads mill	mesophilic batch	375 mL(biogas)/g COD	503 mL(biogas)/g COD	34%	(Izumi et al., 2010)
	thermal at 120°C (1 bar)	thermophilic batch	6.5 L(biogas)/L(reactor)	7.2 L(biogas)/L(reactor)	11%	(Ma et al., 2011)
	400 pulses with electroporation	mesophilic continuous	222 L CH_4/g TS	338 L CH_4/g TS	53%	(Carlsson et al., 2008)
Slaughterhouse waste	pasteurization (70°C, 1 h)	mesophilic fed-batch	0.31 L(biogas)/g VS	1.14 L(biogas)/g VS	268%	(Ware and Power, 2016)
	chemical pretreatment with NaOH	mesophilic batch	8.55 L(biogas)/kg FM	22.8 L(biogas)/kg FM	167%	(Flores-Juarez et al., 2014)
Lignocellulosic agro-industrial waste	enzymatic pretreatment of sugar beet residues	mesophilic fed-batch	163 mL(biogas)/d	183 mL(biogas)/d	12%	(Ziemiński et al., 2012)
	hydrothermal NaOH pretreated rice straw	mesophilic batch	140 L(biogas)/kg VS	185 L(biogas)/kg VS	32%	(Chandra et al., 2012)

applied pretreatment technique and substrate type. Significant effectiveness of pretreatment methods has been reported for slaughterhouse waste; since this material is not easily degradable, any process for biogas enhancement would be beneficial.

Biogas Utilization

Biogas generated from anaerobic digestion is an environmentally friendly, clean, renewable fuel. There are two basic end uses for biogas: production of heat and electricity (combined heat and power generation, or CHP), and replacement of natural gas in transportation and the gas grid. Raw biogas contains impurities such as water, hydrogen sulphide, ammonia, etc., which must be removed to make it usable in some applications.

CHP is usually done on-site in the biogas power plant. Internal combustion engines are most commonly used in CHP applications. A flow diagram of feedstock preparation, process operation, and the production of usable forms of energy in the CHP unit is shown in figure 2. Depending on the type of raw substrate used for the AD process, the application of pretreatment technologies is optional. Substrates that in general show lower biodegradability like lignocellulosic biomass, rotten food, etc. are ground and homogenized in a mixing tank. After a certain retention time in the mixing tank, the feedstock is pumped into a digester where the production of biogas happens. Generated biogas flows through a gasometer in order to monitor its production (and the quality, if available). For instance, if a reduced biogas production in the process occurs as a result of inhibition, it would be detected by the lower flow rate on the gasometer. Precautions in the operation of a biogas plant require the use of a gas flare, where biogas can be burned if not acceptable to be used as a fuel for an internal combustion engine. Some of the heat and electricity produced is used by the biogas plant itself to cover internal needs for energy supply: electromotors for pumps and mixers, temperature control in the digester, etc.; some heat and electricity is distributed to final consumers.

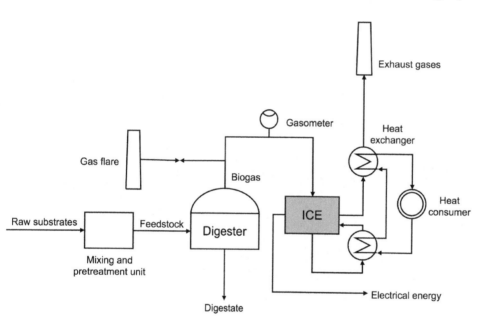

Figure 2. Flow diagram of biogas CHP cogeneration. ICE = internal combustion engine. (Adapted from Clarke Energy, 2020.)

Replacement of natural gas in transportation and the gas grid by biomethane is a relatively new approach in handling biogas from anaerobic digestion. The basic idea is to remove impurities in biogas, such as carbon dioxide, ammonia, and hydrogen sulphide, and produce biomethane, which further can be used as a replacement for natural gas in the gas grid or as a transportation fuel, either as CNG (compressed natural gas) or LNG (liquid natural gas). There are several technological solutions for removal of non-methane components from biogas.

- In pressure swing adsorption (PSA), carbon dioxide is removed from biogas by alternating pressure levels and its adsorption/desorption on zeolites or activated carbon.
- In chemical solvent scrubbing (CSS), carbon dioxide is trapped in dissolved compounds or liquid chemical, i.e., alkaline salt solutions and amine solutions.
- In pressurized water scrubbing (PWS), removal of carbon dioxide and hydrogen sulphide is based on their higher solubility in water compared to methane.
- In physical solvent scrubbing (PSS), instead of trapping carbon dioxide and hydrogen sulphide in water, some organic compounds can be used, i.e., glycols.
- Membrane separation is based on the different permeation rates of biogas compounds, when it undergoes high pressure across a nanoporous material (membrane) causing gas compound separation.
- Cryogenic distillation uses the condensing and freezing of carbon dioxide at low temperatures, at which methane is in the gas phase.
- Supersonic separation uses a specific nozzle to expand the saturated gas to supersonic velocities, which results in low temperature and pressure, which causes the change of aggregate state (condensation) and separation of compounds.
- The industrial (ecological) "lung" uses an enzyme, carbonic anhydrase, to pull carbon dioxide into an aqueous phase and absorbed.

Due to a low investment price, high removal efficiencies, high reliability, or a wide range of contaminants removal, the most commonly applied upgrading technologies are water scrubbing, PSA, and chemical scrubbing. A combination of technologies is often used to process larger quantities of biogas to biomethane. However, upgrade technologies are generally expensive to purchase and can be costly to operate and maintain.

Digestate Management

A digestate is composed of two fractions, liquid and solid. After separating digestate material into fractions, different utilization methods can be applied, as Drosg et al. (2015) studied. The liquid fraction of digestate usually contains high concentrations of nitrogen and, therefore, it can be applied directly as a soil liquid fertilizer, without any processing required. Also, the liquid fraction can be re-fed to the digester and recirculated in the AD process. The solid fraction generally consists of non-degraded material (primarily lignin) which can cause odor emissions. To prevent this outcome, the solid fraction of digestate can be used as a feedstock for a composting process. The resulting compost is a biofertilizer that slowly releases nutrients and improves soil characteristics. The other option for solid digestate fraction utilization is to remove remaining moisture by drying and produce solid state fuel (pellets); this approach is not satisfactory as valuable nutrients present in solid digestate are lost. So far, using digestate as a biofertilizer seems to be the most sustainable option.

Examples

Example 1: Theoretical oxygen demand and theoretical biochemical methane potential

Problem:

A lignocellulosic substrate was analyzed for its elemental composition (table 6). Calculate the (a) theoretical oxygen demand and (b) theoretical biochemical methane potential of this substrate.

Table 6. Elemental composition of the lignocellulosic substrate.

Elements	Based on Fresh Matter [%]	Based on Dry Matter [%]
Carbon	8.9	47.2
Hydrogen	1.1	5.8
Oxygen	8.5	44.2
Nitrogen	0.53	2.8

Solution:

(a) To calculate the theoretical oxygen demand, first estimate the elemental formula of the substrate ($C_a H_b O_c N_d$) based on the elements in the dry matter, since water (the remaining material) is not degradable during the AD process. Divide the share of elements by their relative atomic mass:

$$\frac{47.2}{12} : \frac{5.8}{1} : \frac{44.2}{16} : \frac{2.8}{14}$$

That results in the following values:

$$3.933 : 5.800 : 2.763 : 0.200$$

Then, it is necessary to divide all numbers by the lowest presented value, in this case 0.200:

$$(3.933 : 5.800 : 2.763 : 0.200)/0.200$$

The result of the applied action ($a : b : c : d$) is:

$$19.7 : 29 : 13.8 : 1$$

Which indicates the chemical formula of the lignocellulosic substrate as: $C_{19.7} H_{29} O_{13.8} N$.

(b) Estimate theoretical oxygen demand using equation 1:

$$ThOD = \frac{16 \times (2a + 0.5(b - 3d) - c)}{12a + b + 16c + 14d} \left(\frac{kg_{O_2}}{kgTS_{C_a H_b O_c N_d}} \right) \quad (1)$$

$$\text{ThOD} = \frac{16 \times (2 \times 19.7 + 0.5 \times (29 - 3 \times 1) - 13.8)}{12 \times 19.7 + 29 + 16 \times 13.8 + 14 \times 1} = \frac{1.235 \, kg_{O_2}}{kgTS_{C_{19.7}H_{29}O_{13.8}N}}$$

If the entire lignocellulosic substrate is degraded during the AD process, TBMP can be estimated using equation 2:

$$\text{TBMP} = \frac{22.4 \times \left(\frac{a}{2} + \frac{b}{8} - \frac{c}{4} - \frac{3d}{8}\right)}{12a + b + 16c + 14d} \left(\frac{Nm^3_{CH_4}}{kgVS_{C_aH_bO_cN_d}}\right) \quad (2)$$

$$\text{TBMP} = \frac{22.4 \times \left(\frac{19.7}{2} + \frac{29}{8} - \frac{13.8}{4} - \frac{3 \times 1}{8}\right)}{12 \times 19.7 + 29 + 16 \times 13.8 + 14 \times 1} = \frac{0.432 \, Nm^3}{kgVS_{C_{19.7}H_{29}O_{13.8}N}}$$

TBMP is 0.432 Nm³ of biomethane per kg of substrate VS.

Example 2: Degradation calculation

Problem:
The BMP tests of the lignocellulosic substrate in example 1 determined that the substrate has a BMP of 0.222 Nm³ kgVS⁻¹. Determine the degradation of the substrate.

Solution:
Calculate degradation using equation 6:

$$\text{Degradation (\%)} = \frac{\text{BMP}}{\text{TBMP}} \times 100 \quad (6)$$

$$\text{Degradation (\%)} = \frac{0.222 \, Nm^3 \, kgVS^{-1}}{0.432 \, Nm^3 \, kgVS^{-1}} \times 100 = 50.9\%$$

The result shows that during the AD tests performed on the lignocellulosic matter, 50.9% of the substrate was degraded and biomethane/biogas produced.

Example 3: Pretreatment efficiency determination

Problem:
Lignocellulosic substrate from example 1 has undergone a thermo-chemical pretreatment before entering the BMP test. Reported BMP of the pretreated substrate was 0.389 Nm³ kgVS⁻¹. Calculate the increase in biomethane production by applying the thermo-chemical pretreatment method, i.e., what is the efficiency of the pretreatment method?

Solution:

Calculate the efficiency of the pretreatment method (increase in biomethane production) using equation 10:

$$\text{Efficiency (\%)} = \frac{\text{BMP (after pretreatment)} - \text{BMP (without pretreatment)}}{\text{BMP (without pretreatment)}} \times 100 \quad (10)$$

$$\text{Efficiency (\%)} = \frac{0.389 \text{ Nm}^3 \text{ kgVS}^{-1} - 0.222 \text{ Nm}^3 \text{ kgVS}^{-1}}{0.222 \text{ Nm}^3 \text{ kgVS}^{-1}} \times 100 = 75\%$$

This case shows that the efficiency of the applied pretreatment technique is 75%.

Example 4: BGP test on anaerobic digestion of rotten food

Problem:

BGP tests have been conducted on the anaerobic digestion of a rotten food mixture with an average C:N ratio of 12. The working volume of the laboratory reactor is 250 mL. The mass of raw feedstock put in the reactor was 100 g, with an average dry matter content of 5%. Inoculum and feedstock were mixed in the ratio of 1:1 based on the total solids content. The reactor operated under mesophilic conditions, with a temperature of 38°C. Biogas production was measured by the water displacement method each day over a 40-day period. Table 7 presents the recorded volume of biogas during the AD operation (normalized to 0°C and 1 atm). Calculate and graph the daily and cumulative biogas production over the test period. If the average share of methane in biogas was recorded as 55%, calculate the BMP of the rotten food.

Table 7. Normalized biogas volume in the operation of batch AD of rotten food.

Time (day)	Produced Biogas per Day (NmL)	Time (day)	Produced Biogas per Day (NmL)
1	0	21	29
2	3	22	24
3	9	23	25
4	17	24	23
5	25	25	20
6	32	26	18
7	39	27	16
8	45	28	14
9	50	29	12
10	56	30	11
11	62	31	9
12	67	32	9
13	72	33	8
14	74	34	6
15	69	35	4
16	61	36	3
17	53	37	3
18	50	38	3
19	45	39	2
20	35	40	1

Solution:

Calculate the daily production of biogas in the studied example by dividing the volume of biogas produced each day by the reactor volume:

$$V(\text{digester}) = 250 \text{ mL}$$

$$\text{Daily production of biogas} = \frac{V_{N,\text{biogas}}}{V(\text{digester})}$$

The computed daily biogas production values are plotted in figure 3 in SI units $\text{Nm}^3/(\text{m}^3 \cdot \text{d})$.

Cumulative production of biogas is determined as the sequential sum of biogas volume produced each day, expressed over the mass of total solids of feedstock put in the reactor (figure 4).

The final value of cumulative biogas production (40th day), about 0.221 Nm³ kg⁻¹ TS, is the BGP of the rotten food sample. Determine the value of the BMP of the analyzed feedstock using equation 5 in the following form:

$$BMP = \text{share of methane} \times BGP$$

Insert reported values in the equation:

$$BMP = 0.55 \times 0.221 \text{ Nm}^3 \text{ kgTS}^{-1} = 0.121 \text{ Nm}^3 \text{ kgTS}^{-1}$$

BMP of the analyzed feedstock is calculated to be 0.121 Nm³ kgTS⁻¹.

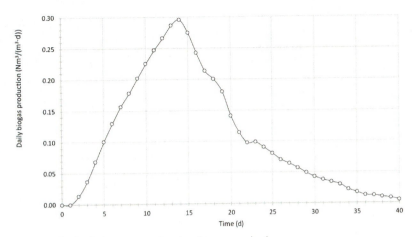

Figure 3. Daily biogas production for example 4.

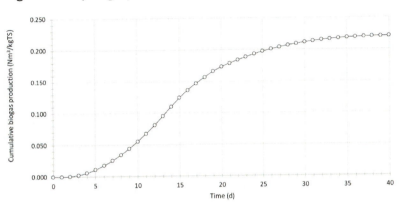

Figure 4. Cumulative biogas production for example 4.

Example 5: Biogas plant

Problem:

A biogas facility operates under mesophilic conditions (38°C) and produces biogas from food processing by-products. The digester volume is 3,750 m³ while the average hydraulic retention time is 50 days. Average COD of inlet stream processed at the biogas plant is 75 kgO₂ m⁻³.

(a) Determine the OLR expressed over the quantity of input stream and its chemical oxygen demand. Also, calculate the daily production of biogas in the digester, if the average methane share in biogas is about 65%.
(b) Assume the biogas facility started to operate with a different inlet stream, characterized with 40% higher COD value compared to inlet stream in (a). To keep the same organic load rate in terms of COD value, find the new HRT for the changed feedstock and the new production of methane.

Solution:

(a) To determine the required variables Q, Q_{biogas}, and OLR, first calculate the input volume of feedstock per day using equation 9:

$$HRT = V/Q \quad (9)$$

$$Q = V/HRT = 3{,}750 \text{ m}^3/50 \text{ d} = 75 \text{ m}^3 \text{ d}^{-1}$$

The calculated flow rate of feedstock is 75 m³ d⁻¹.

Input COD value of the feedstock per day is calculated as the product of volume flow rate (75 m³ d⁻¹) and COD (75 kgO₂ m⁻³)

$$COD_{input} = Q \times COD = 75 \text{ m}^3/\text{d} \times 75 \text{ kgO}_2/\text{m}^3 = 5{,}625 \text{ kgO}_2 \text{ d}^{-1}$$

This results in input COD of 5,625 kgO₂ d⁻¹.

As stated above, 1 kg of input COD in the AD can produce 0.40 Nm³ CH₄, so the flow rate of feedstock of 5,625 kgO₂ d⁻¹ can produce:

$$Q_{N,CH_4} = \frac{5{,}625 \text{ kgO}_2}{d} \times \frac{0.40 \text{ Nm}^3}{\text{kgO}_2}$$

This results in 2,250 Nm³ of CH₄ per day in the biogas production unit.

To find the production of methane in the digester during the process at a temperature of 38°C, it is necessary to apply the following relation:

$$Q_{38°C,CH_4} = Q_{N,CH_4} \times \left(\frac{273+38}{273}\right)$$

That resulted in the production of methane in the digester of 2,563 m³ at a temperature of 38°C.

Furthermore, to determine the production of biogas in the digester, divide the quantity of produced methane by the share of methane in biogas:

$$Q_{38°C,biogas} = \frac{Q_{38°C,CH_4}}{0.65}$$

This results in the daily biogas production rate of 3,943 m³ in a digester at 38°C.

To determine the organic load rate based on the input volume, OLR, use equation 7:

$$OLR = Q/V \tag{7}$$

$$OLR = \frac{75 \text{ m}^3/\text{d}}{3{,}750 \text{ m}^3} = 0.02 \text{ m}^3 \text{ d}^{-1} \text{ feedstock m}^{-3} \text{ digester.}$$

Then use equation 8 to express OLR in terms of COD value:

$$OLR_{COD} = OLR \times COD \tag{8}$$

$$OLR_{COD} = \frac{0.02 \text{ m}^3}{d \times m^3} \times \frac{75 \text{ kgO}_2}{m^3} = 1.5 \text{ kgO}_2 \text{ d}^{-1} \text{ m}^{-3} \text{ digester.}$$

(b) To determine the values of Q, HRT and $Q(CH_4)$ when new organic material (new feedstock) is entering the AD plant, first note that:

$$V = 3{,}750 \text{ m}^3$$
$$\text{COD}_{new} = \text{COD}_{old} + 0.40 \times \text{COD}_{old} = 1.4 \times \text{COD}_{old}$$

Therefore, the COD value of a new feedstock is assumed to be 105 kgO$_2$ m^{-3}.

Calculate the flow rate of the new feedstock using equation 8 in the following modified form:

$$\text{OLR}_{new} = \frac{\text{OLR}_{COD}}{\text{COD}_{new}}$$

$$\text{OLR}_{new} = \frac{1.5 \text{ kgO}_2 \text{ d}^{-1} \text{ m}^{-3} \text{ digester}}{105 \text{ kgO}_2 \text{ m}^{-3} \text{ feedstock}}$$

The new value of OLR is estimated to be 0.0143 m³ feedstock m^{-3} digester d^{-1}. Calculate the input flow rate of the new feedstock using equation 7:

$$Q_{new} = \text{OLR}_{new} \times V$$

$$Q_{new} = \frac{0.0143 \text{ m}^3}{\text{d} \times \text{m}^3} \times 3{,}750 \text{ m}^3$$

The flow rate of the new feedstock is 53.63 m³ d^{-1}. As expected, the input flow rate of the feedstock with higher COD is lower than the one in part (a) to maintain the same OLR$_{COD}$.

Since the input COD remains the same as in part (a), the production of methane is not changed, 2,563 m³ at 38°C.

HRT for the new feedstock is calculated with equation 9 and found to be 70 days. Since the new feedstock has a lower flow rate compared to the one in part (a), it is necessary to prolong the period of feedstock retention to achieve the same OLR$_{COD}$.

Image Credits

Figure 1. Mähnert, P. (2020). (CC by 4.0). Theoretical biogas yield profiles for a batch test. Retrieved from https://opus4.kobv.de/opus4-slbp/files/1301/biogas03.pdf

Figure 2. Clarke Energy. (2020). Flow diagram of biogas CHP cogeneration. ICE = internal combustion engine. Retrieved from https://www.clarke-energy.com/biogas/. [Fair Use].

Figure 3. Bedoić, R. (CC By 4.0). (2020). Daily biogas production for example 4.

Figure 4. Bedoić, R. (CC By 4.0). (2020). Cumulative biogas production for example, 4.

References

Aboudi, K., Álvarez-Gallego, C. J., & Romero-García, L. I. (2016). Biomethanization of sugar beet byproduct by semi-continuous single digestion and co-digestion with cow manure. *Bioresour. Technol.*, 200, 311-319. https://doi.org/10.1016/j.biortech.2015.10.051.

Al Seadi, T., Rutz, D., Prassl, H., Köttner, M., Finsterwalder, T., Volk, S., & Janssen, R. (2008). *Biogas handbook*. Esbjerg, Denmark: University of Southern Denmark Esbjerg.

Ariunbaatar, J., Panico, A., Esposito, G., Pirozzi, F., & Lens, P. N. L. (2014). Pretreatment methods to enhance anaerobic digestion of organic solid waste. *Appl. Energy, 123*, 143-156. https://doi.org/10.1016/j.apenergy.2014.02.035.

Bedoić, R., Čuček, L., Ćosić, B., Krajnc, D., Smoljanić, G., Kravanja, Z., ... Duić, N. (2019a). Green biomass to biogas—A study on anaerobic digestion of residue grass. *J. Cleaner Prod., 213*, 700-709. https://doi.org/https://doi.org/10.1016/j.jclepro.2018.12.224.

Bedoić, R., Ćosić, B., & Duić, N. (2019b). Technical potential and geographic distribution of agricultural residues, co-products and by-products in the European Union. *Sci. Total Environ., 686*, 568-579. https://doi.org/10.1016/j.scitotenv.2019.05.219.

Bochmann, G., & Montgomery, L. F. R. (2013). Storage and pre-treatment of substrates for biogas production. In A. Wellinger, J. Murphy, & D. Baxter (Eds.), *The biogas handbook: Science, production and applications* (pp. 85-103). Woodhead Publishing. https://doi.org/10.1533/9780857097415.1.85.

Carlsson, M., Lagerkvist, A., & Ecke, H. (2008). Electroporation for enhanced methane yield from municipal solid waste. *ORBIT 2008: Moving Organic Waste Recycling Towards Resource Management and Biobased Economy, 6*, 1-8.

Chandra, R., Takeuchi, H., & Hasegawa, T. (2012). Hydrothermal pretreatment of rice straw biomass: A potential and promising method for enhanced methane production. *Appl. Energy, 94*, 129-140. https://doi.org/10.1016/j.apenergy.2012.01.027.

Clarke Energy (2020). *Biogas*. Retrieved from https://www.clarke-energy.com/biogas/.

Das, A., & Mondal, C. (2016). Biogas production from co-digestion of substrates: A review. *Int. Res. J. Environ. Sci., 5*(1), 49-57.

Drosg, B., Fuchs, W., Al Seadi, T., Madsen, M., & Linke, B. (2015). *Nutrient recovery by biogas digestate processing*. IEA Bioenergy. Retrieved from http://www.iea-biogas.net.

Flores-Juarez, C. R., Rodríguez-García, A., Cárdenas-Mijangos, J., Montoya-Herrera, L., Godinez Mora-Tovar, L. A., Bustos-Bustos, E., ... Manríquez-Rocha, J. (2014). Chemically pretreating slaughterhouse solid waste to increase the efficiency of anaerobic digestion. *J. Biosci. Bioeng., 118*(4), 415-419. https://doi.org/10.1016/j.jbiosc.2014.03.013.

Frigon, J. C., & Guiot, S. R. (2010). Biomethane production from starch and lignocellulosic crops: A comparative review. *Biofuels, Bioprod. Biorefin., 4*(4), 447-458. https://doi.org/10.1002/bbb.229.

Gerike, P. (1984). The biodegradability testing of poorly water soluble compounds. *Chemosphere, 13*(1), 169-190. https://doi.org/10.1016/0045-6535(84)90018-3.

Izumi, K., Okishio, Y., Nagao, N., Niwa, C., Yamamoto, S., & Toda, T. (2010). Effects of particle size on anaerobic digestion of food waste. *Int. Biodeterioration and Biodegradation, 64*(7), 601-608. https://doi.org/10.1016/j.ibiod.2010.06.013.

Karak, N. (2016). Biopolymers for paints and surface coatings. In F. Pacheco-Torgal, V. Ivanov, & H. Jonkers (Eds.). *Biopolymers and biotech admixtures for eco-efficient construction materials* (pp. 333-368). Woodhead Publishing. https://doi.org/10.1016/B978-0-08-100214-8.00015-4.

Koch, K., Lübken, M., Gehring, T., Wichern, M., & Horn, H. (2010). Biogas from grass silage—Measurements and modeling with ADM1. *Bioresour. Technol., 101*(21), 8158-8165. https://doi.org/10.1016/j.biortech.2010.06.009.

Lauwers, J., Appels, L., Thompson, I. P., Degrève, J., Van Impe, J. F., & Dewil, R. (2013). Mathematical modelling of anaerobic digestion of biomass and waste: Power and limitations. *Prog. Energy Combust. Sci., 39*, 383-402. https://doi.org/10.1016/j.pecs.2013.03.003.

Li, Y., Zhang, R., Chen, C., Liu, G., He, Y., & Liu, X. (2013). Biogas production from co-digestion of corn stover and chicken manure under anaerobic wet, hemi-solid, and solid state conditions. *Bioresour. Technol., 149*, 406-412. https://doi.org/10.1016/j.biortech.2013.09.091.

Liu, C., Yuan, X., Zeng, G., Li, W., & Li, J. (2008). Prediction of methane yield at optimum pH for anaerobic digestion of organic fraction of municipal solid waste. *Bioresour. Technol., 99*, 882-888. https://doi.org/10.1016/j.biortech.2007.01.013.

Ma, J., Duong, T. H., Smits, M., Verstraete, W., & Carballa, M. (2011). Enhanced biomethanation of kitchen waste by different pre-treatments. *Bioresour. Technol., 102*(2), 592-599. https://doi.org/10.1016/j.biortech.2010.07.122.

Ma, J., Zhao, Q. B., Laurens, L. L. M., Jarvis, E. E., Nagle, N. J., Chen, S., & Frear, C. S. (2015). Mechanism, kinetics and microbiology of inhibition caused by long-chain fatty acids in anaerobic digestion of algal biomass. *Biotechnol. Biofuels, 8*(1). https://doi.org/10.1186/s13068-015-0322-z.

Mähnert, P. (2006). Grundlagen und verfahren der biogasgewinnung. *Leitfaden Biogas (FNR)*, 13-25. Retrieved from https://opus4.kobv.de/opus4-slbp/files/1301/biogas03.pdf.

Moukazis, I., Pellera, F. M., & Gidarakos, E. (2018). Slaughterhouse by-products treatment using anaerobic digestion. *Waste Manage., 71*, 652-662. https://doi.org/10.1016/j.wasman.2017.07.009.

Mulat, D. G., Dibdiakova, J., & Horn, S. J. (2018). Microbial biogas production from hydrolysis lignin: Insight into lignin structural changes. *Biotechnol. Biofuels, 11*(61). https://doi.org/10.1186/s13068-018-1054-7.

Patil, V. S., & Deshmukh, H. V. (2015). A review on co-digestion of vegetable waste with organic wastes for energy generation. *Int. J. Biol. Sci., 4*(6), 83-86.

Pognani, M., D'Imporzano, G., Scaglia, B., & Adani, F. (2009). Substituting energy crops with organic fraction of municipal solid waste for biogas production at farm level: A full-scale plant study. *Process Biochem., 44*(8), 817-821. https://doi.org/10.1016/j.procbio.2009.03.014.

Ruffino, B., Fiore, S., Roati, C., Campo, G., Novarino, D., & Zanetti, M. (2015). Scale effect of anaerobic digestion tests in fed-batch and semi-continuous mode for the technical and economic feasibility of a full scale digester. *Bioresour. Technol., 182*, 302-313. https://doi.org/10.1016/j.biortech.2015.02.021.

Sung, S., & Liu, T. (2003). Ammonia inhibition on thermophilic anaerobic digestion. *Chemosphere, 53*(1), 43-52. https://doi.org/10.1016/S0045-6535(03)00434-X.

Ueno, Y., Tatara, M., Fukui, H., Makiuchi, T., Goto, M., & Sode, K. (2007). Production of hydrogen and methane from organic solid wastes by phase-separation of anaerobic process. *Bioresour. Technol., 98*(9), 1861-1865. https://doi.org/10.1016/j.biortech.2006.06.017.

Van Lier, J. B., Rebac, S., & Lettinga, G. (1997). High-rate anaerobic wastewater treatment under psychrophilic and thermophilic conditions. *Water Sci. Technol., 35*, 199-206. https://doi.org/10.1016/S0273-1223(97)00202-3.

Ware, A., & Power, N. (2016). What is the effect of mandatory pasteurisation on the biogas transformation of solid slaughterhouse wastes? *Waste Manag., 48*, 503-512. https://doi.org/10.1016/j.wasman.2015.10.013.

Xie, S., Hai, F. I., Zhan, X., Guo, W., Ngo, H. H., Price, W. E., & Nghiem, L. D. (2016). Anaerobic co-digestion: A critical review of mathematical modelling for performance optimization. *Bioresour. Technol., 222*, 498-512. https://doi.org/10.1016/j.biortech.2016.10.015.

Zhu, B., Gikas, P., Zhang, R., Lord, J., Jenkins, B., & Li, X. (2009). Characteristics and biogas production potential of municipal solid wastes pretreated with a rotary drum reactor. *Bioresour. Technol., 100*(3), 1122-1129. https://doi.org/10.1016/j.biortech.2008.08.024.

Ziemiński, K., Romanowska, I., & Kowalska, M. (2012). Enzymatic pretreatment of lignocellulosic wastes to improve biogas production. *Waste Manag., 32*(6), 1131-1137. https://doi.org/10.1016/j.wasman.2012.01.016.

Zonta, Ž., Alves, M. M., Flotats, X., & Palatsi, J. (2013). Modelling inhibitory effects of long chain fatty acids in the anaerobic digestion process. *Water Res., 47*(3), 1369-1380. https://doi.org/10.1016/j.watres.2012.12.007.

Measurement of Gaseous Emissions from Animal Housing

Mélynda Hassouna
UMR Sol, Agro et hydrosystèmes et Spatialisation, INRAe, Agrocampus, France

Salvador Calvet
Institute of Animal Science and Technology, Universitat Politècnica de València, Valencia, Spain

Richard S. Gates
Agricultural & Biosystems Engineering, and Animal Sciences, Iowa State University, Ames, Iowa, USA

Enda Hayes
Air Quality Management Resource Centre, Department of Geography and Environmental Management, University of the West of England, Bristol, United Kingdom

Sabine Schrade
Ruminants Research Unit, Federal Department of Economic Affairs, Education and Research EAER, Agroscope, Ettenhausen, Switzerland

KEY TERMS		
Emission processes	Mass balance	Ammonia
Measurement techniques	Validation	Greenhouse gases
Sampling	Ventilation	

Variables

a = constant

A = cross-sectional area of the ventilation duct

c = CO_2 production

C_x = concentration of gas for position x

ER = emission rate

h = time at sampling

h_{min} = activity factor which relates to the time of day with minimum activity

m = animal mass

$m_{acid\ solution}$ = mass of the solution
$\dot{m}$ = mass flow rate
n = feed energy
s = mean airspeed
VR = ventilation rate
V_{sample} = volume of air

Introduction

Animal housing and manure storage facilities are two principal on-farm sources of gaseous emissions to the atmosphere. The most important pollutants emitted are ammonia (NH_3), methane (CH_4), and nitrous oxide (N_2O).

Ammonia (NH_3) is a colorless gas with a pungent smell that can have impacts on environmental and human health (figure 1). Ammonia is emitted by many agricultural activities, including crop production as well as animal production. Ammonia plays a key role in the formation of secondary particulate matter (PM) by reacting with acidic species such as sulfur dioxide (SO_2) and nitrogen oxides (NO_x) to form fine aerosols, and is thus called a particulate precursor. The PM created by the reaction of NH_3 and acidic species in the atmosphere contributes to poor air quality including regional haze. These particles have an aerodynamic diameter of less than 2.5 microns and are generally referred to as "PM-fine." They are readily inhaled and populations exposed to PM-fine have greater respiratory and cardiovascular health risks such as asthma, bronchitis, cardiac arrythmia and arrest, and premature death. Some emitted NH_3 is subsequently deposited on land and water downwind of facilities, and can acidify soils and freshwater. The addition of available nitrogen (N) to low-nutrient ecosystems disturbs their balance and can alter the relative growth and abundance of plant species.

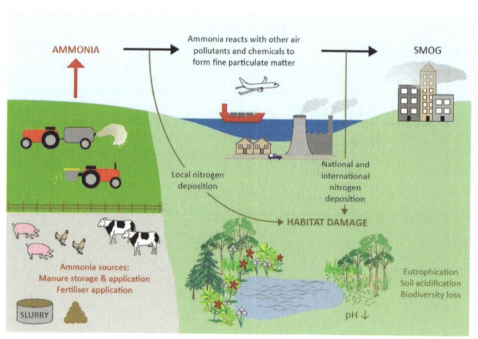

Figure 1. Environmental impacts of ammonia emissions.

Nitrous oxide and CH_4 are potent greenhouse gases (GHGs) which contribute to global warming. The global warming potential (GWP) is a factor specific to each GHG and allows comparisons of the relative global warming impacts between different GHGs. This factor indicates how much heat a given gas traps over a certain time horizon (usually 100 years), compared with an equal mass of carbon dioxide (CO_2). Nitrous oxide GWP for a 100-year time horizon is 265 with a lifetime in the atmosphere of 114 years. Methane GWP for a 100-year time horizon is 28 with a lifetime in the atmosphere of 12 years.

Nitrous oxide, also known as "laughing gas," is colorless and odorless, and contributes to the destruction of the atmospheric ozone layer. In agriculture, the main source of N_2O emissions is soil, from crop fertilizer use, soil cultivation, and spreading of urine and manure. Other sources include industrial processes, and natural processes involving soils and oceans.

Methane is a volatile organic compound, odorless and flammable. In agriculture, the main sources are enteric fermentation (fermentation that takes place in the digestive systems of animals) and the degradation of manure. Methane contributes to ozone formation in the lower atmosphere, and to ozone layer depletion in the upper atmosphere.

Researchers and engineers have developed different approaches to reliably measure and quantify emissions of NH_3, N_2O, and CH_4 from animal production facilities. The implementation of these methods helps to understand the production processes, to identify the influencing factors, and to develop mitigation techniques or practices. The specific characteristics of animal housing and the variability of the houses and animal production systems make the development and implementation of the different methods a real challenge.

Outcomes

After reading this chapter, you should be able to:

- Explain how gaseous emissions are formed and released from animal housing
- Outline the methods to measure concentrations of ammonia and greenhouse gases
- Estimate ventilation rates of selected animal housing
- Calculate emissions of ammonia and greenhouse gases from animal housing

Concepts

Animal Houses

Animal housing is designed to provide shelter and protection with control of feed consumption, diseases, parasites, and the interior thermal environment. An animal house is designed to take into account animal heat and moisture production, building characteristics (e.g., insulation and volume), and outdoor climate. Inside the house, animals produce the following critical components that affect emissions:

- sensible heat that is transferred to the interior air by means of convection, conduction, and radiation, and causes an increase of air temperature;

- latent heat that is generated through the evaporation of moisture from the lungs, skin, urine, and fecal material, and through increases of air humidity;
- mixtures of feces and urine, which become a source of gaseous emissions (NH_3, N_2O, CO_2, water vapor, CH_4), and heat; and
- CO_2 from animal respiration.

The control of temperature, moisture, gas concentrations and dust concentrations inside the house is essential to achieve optimal conditions for animal growth and production. The optimal conditions vary as a function of animal age, species, and breed, and rely on the implementation of ventilation systems. The ventilation system partly controls the rate and total emissions from the building. There are two types of ventilation systems: mechanical and natural. The RAMIRAN European network (RAMIRAN, 2011) defines the two systems as:

- mechanical ventilation, which is ventilation of a building, usually for pigs, poultry, or calves, through the use of electrically powered fans in the walls or roof that are normally controlled by the temperature in the building; and
- natural ventilation, which is ventilation of a building, e.g., for cattle, by openings or gaps designed into the roof and/or sides of the building.

NH_3, N_2O, and CH_4 Emissions Processes in Animal Housing and Influencing Factors

Ammonia is volatilized as a gas during manure management. It is mainly derived from the urea excreted in urine (or uric acid in the feces, in the case of poultry). The process of forming NH_3 from urea is relatively fast and is outlined in figure 2. Once excreted, urea is decomposed within a few hours to a few days into ammonium (NH_4) by means of the enzyme urease, which is widely present in feces and soils. Ammonium is in equilibrium with dissolved NH_3 enhanced at high pH values. A second physical equilibrium exists between dissolved and free NH_3 in the manure matrix. Finally, the free NH_3 can be released to the atmosphere. This is a mass transfer process affected by air velocity, diffusion from beneath the surface, and exposure to air on the surface of the manure.

This is a continuous process that starts in the animal housing itself and continues

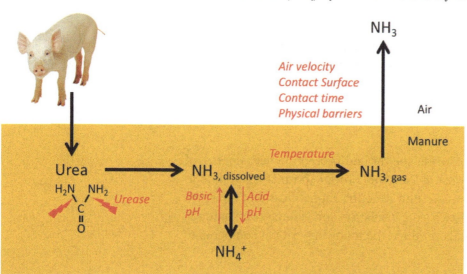

Figure 2. Process leading to ammonia emission and contributing factors (Snoek et al., 2014).

during manure management and land application. Several factors are involved in the amount of NH_3 emitted to the atmosphere:

- manure composition; the most relevant factors are the amount of urea excreted by the animals, the pH of the manure, and its moisture content;
- the environmental conditions, particularly temperature and wind speed above the emitting surface;
- the facilities for animal housing and manure management; and
- management practices, particularly those altering the contact of manure with air, and urine with feces, by reducing time of exposure or contact surface.

On farms, N_2O originates from the management of manure and its application to land as fertilizer. Emission of N_2O occurs from the successive nitrification and denitrification of NH_4. A first aerobic phase is required for the nitrification, while anoxic conditions are required for denitrification (figure 3). These conditions are characteristic of the following situations:

- composting with alternative wetting, mixing, and drying periods;
- aerobic treatment of slurry;
- air cleaners at the air exhaust of the animal house based on biological scrubbing of air; and
- application of manure to soil and subsequent drying-wetting events.

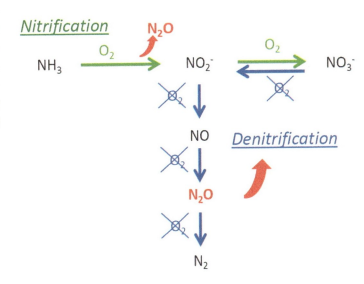

Figure 3. Reactions leading to N_2O emissions: nitrification (in green) and denitrification (in blue) (Wrage et al., 2001).

Methane is produced during the anaerobic decomposition of organic matter. This occurs mainly during digestion in ruminants and decomposition of manure. In this process, the microbial breakdown of organic matter occurs in different stages, from more complex molecules to the simplest. The main mechanism is presented in figure 4. Apart from the presence of organic matter and anoxic conditions, time (a few weeks) is needed to complete the process, and the process may be inhibited due to certain conditions, such as NH_3 accumulation. In contrast to NH_3, CH_4 has very low solubility in water and, once produced, is released to the atmosphere through a characteristic bubbling, in the case of slurries.

For enteric fermentation in ruminants, key factors are feed composition and genetics. More

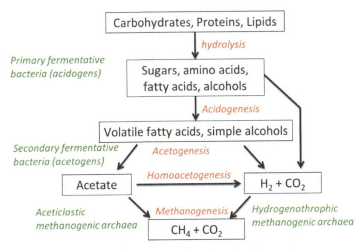

Figure 4. Process and microorganisms involved in methane formation.

digestible feeds reduce the amount of CH_4. Feed constituents such as lipids or essential oils may reduce CH_4 production through inhibition. Genetics also influence the amount of CH_4 produced and can be modified through animal genetic selection.

CH_4 emission during manure management is due to the presence of organic matter subjected to anaerobic conditions for sufficient time (about one month, at least) for methanogenic bacteria to develop. The amount and composition of organic matter determines the maximum potential for CH_4 formation. Manure management practices that interrupt anaerobic conditions, reduce the load of organic matter, or feature biogas capture are potentially effective for mitigating emissions.

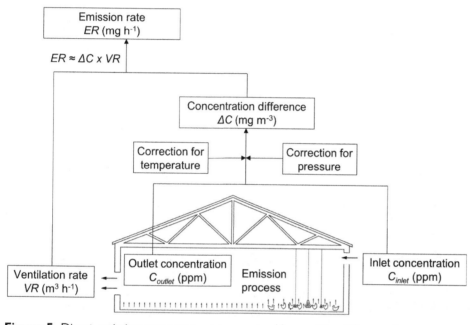

Figure 5. Direct emission measurement in an animal house. The difference in concentration between indoors and outdoors and the ventilation rate must be measured (adapted from Calvet et al., 2013).

Measuring Emissions from Animal Housing

The most common approach used to determine gas emission rates from animal houses is based on quantifying ventilation rates and inlet and outlet concentrations of the gas (figure 5). The mass flow rate of emitted gas (emission rate or ER) is proportional to the ventilation rate and the concentration difference between exhaust and outside air. Several different techniques are available to measure gas concentrations inside and outside the house and the ventilation rate.

Gas Concentration Measurement Techniques

The techniques used most often to measure NH_3, CH_4, and N_2O concentrations of animal houses are either physical (optical, gas chromatography, chemiluminescence) or chemical (acid traps, active colorimetric tubes).

Optical Techniques

Optical techniques are based on the Beer-Lambert absorption law, which indicates that the quantity of light of a given wavelength absorbed is related to the number of gas molecules in the light's path that are able to absorb it.

Optical techniques rely on the use of a light source, a chamber to contain the air sample during measurement, and a detector to quantify the target gas absorption. The main techniques used in animal houses are infrared (IR) spectroscopy (photoacoustic or Fourier transform), tunable diode laser absorption spectroscopy (TDLAS), off-axis integrated cavity output spectroscopy

(OA-ICOS), and cavity ring-down spectroscopy (CRDS). Differences between these techniques include the detection principle of absorption and the type and wavelength of light sources (quantum cascade laser, tunable diode laser, or IR source). Techniques using lasers (a monochromatic source with a narrow band of wavelengths) are more selective, accurate, and stable than techniques with a polychromatic IR source (i.e., large band of wavelengths) because selection of the absorption of a specific wavelength from a polychromatic IR source is difficult to achieve.

One main advantage of optical techniques is that they make monitoring of concentration dynamics in near real time possible, including monitoring several gases with different concentrations at the same time (Powers and Capelari, 2016). Advantages of optical instruments include linear responses over a wide range of concentrations and the ability to measure concentrations both inside (where there could be a high concentration level) and outside (low concentration level) the animal house with the same instruments. Most optical instruments have response times adapted to measurement in animal houses. They are portable and can be used on site. Nevertheless, they can be expensive, must be calibrated, and still require accurate estimation of ventilation rate.

Gas Chromatography

A gas chromatograph separates components in the sample and measures their concentrations. The equipment has four basic elements: an injector, a column, an oven surrounding the column, and a detector. The sample is vaporized in the injector and swept by the carrier gas through a heated column. The column separates each compound according to its polarity and boiling point. The detector identifies and quantifies the compounds separated. The detectors include a flame ionization detector for CH_4 and CO_2 and an electron capture detector for N_2O. This technique is accurate if the detector has been calibrated for the range of concentrations measured. It requires use of a carrier gas and regular calibration, which makes on-site implementation and continuous measurement difficult. It is often used to measure previously collected samples.

Chemiluminescence

Chemiluminescence is used to measure NH_3 concentration. NH_3 in the sample is first oxidized to N_2O by a catalytic converter, and then the N_2O is further oxidized to nitric oxide (NO) at high temperature and an elevated energy state. As the molecules return to a lower energy state, they release electromagnetic radiation at a specific wavelength, which is measured and quantified.

Acid Traps

An acid trap is a standard reference technique for measuring NH_3. A known volume of air is pumped through an acid solution and recorded (figure 6). The acid solution is later analyzed in the laboratory with a

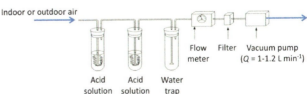

Figure 6. Acid trap configuration.

colorimetric or photometric method (Hassouna et al., 2016) to estimate the amount of NH_3 trapped in the solution, as:

$$NH_{3,trapped} = [N-NH_4^+]_{acid\ solution} \times m_{acid\ solution} \quad (1)$$

where $NH_{3,trapped}$ = amount of NH_3 trapped in the solution (kg)
$[N\text{-}NH_4^+]_{acid\ solution}$ = concentration of ammonium in the acid solution (kg kg^{-1})
$m_{acid\ solution}$ = mass of the solution (kg)

From this, NH_3 concentration in the air sample ($C_{N-NH_3,air}$ in kg m^{-3}) can be calculated as:

$$C_{N-NH_3,air} = \frac{NH_{3,trapped}}{V_{sample}} \quad (2)$$

where V_{sample} (m³) is the volume of air that passed through the solution.

Strong acid solutions are used, such as boric acid, orthophosphoric acid, nitric acid, and sulfuric acid. The trap can be used for a few hours or a few days depending on the NH_3 concentrations in the incoming air, the acid concentration, and volume of acid solution in the vials. Sampling time and concentration should be determined before the experiment as a function of the expected NH_3 concentrations. Two vials with acid solution are used sequentially to avoid saturating a single solution. This technique provides a mean NH_3 concentration over the sampling period and thus is not suitable for studies that require monitoring dynamics of NH_3 concentrations in a house. Nevertheless, as it is not expensive or too time-consuming, it can be used to check the consistency of measurements made, for instance, with optical techniques.

Active Colorimetric Tubes

Active colorimetric tubes are a manual technique that can be used to estimate NH_3, NO, volatile organic compounds (VOC), and CO_2 concentrations. Tubes are manufactured to react to a specific range of concentrations of a specific target gas. Before measurement, both ends of a sealed test tube are cut open. The tube is connected tightly to a hand pump, which draws air through the tube. If present in the air, the target gas reacts with reagents in the tube. The strength of the reaction is proportional to the concentration of the gas in the air. A graduated scale is used to read the degree of color change in the tube, which indicates the concentration of the target gas. Many of the reactions used are based on pH indicators, such as bromophenol blue to measure NH_3 concentrations. Gas concentration is expressed in ppm or mL m^{-3}. This technique is reliable and simple to use. It can be used to estimate concentrations, but not to measure them continuously or accurately.

Ventilation Rate

In mechanically ventilated houses with modern ventilation control systems, ventilation rate could be one of many data recorded continuously. In such situations, the data are thus easily available for emission calculations. For other houses, or if the time step of recording is not suitable or recorded data are not reliable, the ventilation rate must be measured or assessed. Different methods to estimate the ventilation rate have been evaluated and described in the literature (Ogink et al., 2013; Wang et al., 2016). The method chosen depends on the type of ventilation (natural or mechanical), the accessibility of the exhaust to make physical measurements, the level of ventilation rate, and the desired degree of accuracy. Some techniques are indirect (tracer gas, heat balance), while others are direct (fan wheel anemometer, specialized instruments).

Figure 7. Duct for dispersing an artificial tracer gas within an animal house.

Tracer Gas Techniques Using Artificial Tracer Gases

Tracer gas techniques are commonly used to quantify the ventilation rate in many kinds of houses, but mainly those with natural ventilation. An external tracer gas should be safe, inert, measurable, not produced in the house, and inexpensive (Phillips et al., 2001; Sherman, 1990). The most common tracer gas used in animal houses is sulfur hexafluoride (SF_6) (Mohn et al., 2018). A critical requirement is that of near-perfect air mixing inside the animal house to ensure that the tracer gas and the targeted gas (for emission calculations) being measured both disperse in a similar way. Air can be mixed artificially using a purpose-built ventilation duct (figure 7). Tracer gases can be dosed automatically using a mass flow controller and critical orifices (figure 8).

(a) (b)

Figure 8. (a) Tracer-gas dosing by steel tubes with critical orifices protected by steel elements next to the floors in a dairy housing; (b) gas bottles with mass-flow controller.

The basic principle for tracer gas techniques is conservation of mass (of both target gas and tracer gas). By monitoring the dosed mass flow and concentration at the sampling points of the tracer gas, the ventilation rate can be determined (figure 9). A tracer gas release

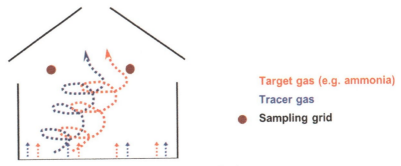

Figure 9. Principle of the tracer gas method.

technique is chosen based on the ventilation rate, the detection limit of the device used to monitor tracer gas concentration, and the ability to control and monitor the dosed mass flow accurately. According to Ogink et al. (2013), three tracer gas release techniques can be distinguished:

- constant injection method: tracer gas is injected at a constant rate, and its concentration is measured directly over a period of time and used to estimate the ventilation rate;
- decay method: tracer gas is injected until its concentration stabilizes, then injection is stopped and the decay in concentration is used to calculate the ventilation rate; and
- concentration method: tracer gas is distributed in the air of a house to a certain concentration to be constant.

Only the constant injection method and the decay method are common for measurements in animal houses.

To calculate the emission or mass flow of the target gas (e.g., NH_3, CH_4), a background correction of the concentration (C_x) must first be calculated for the target gases and the tracer gases:

$$C_x = C_{x,id} - C_{x,bgd} \quad (3)$$

where $x = T$ (tracer gas) or G (target gas)
 $C_{x,id}$ = indoor gas concentration ($\mu g\ m^{-3}$)
 $C_{x,bgd}$ = background gas concentration ($\mu g\ m^{-3}$)

The ratio of the background concentrations of emitted (target) gas, C_G, and tracer gas, C_T, then corresponds to the ratio of their mass flow rates ($\dot{m}$, g d^{-1}):

$$\frac{\dot{m}_G}{\dot{m}_T} = \frac{C_G}{C_T} \quad (4)$$

and thus

$$\dot{m}_G = \frac{\dot{m}_T C_G}{C_T} \quad (5)$$

Carbon Dioxide (CO_2) Mass Balance or Tracer Gas Methods Using an Internal Tracer

The CO_2 mass balance method is sometimes considered a tracer gas technique in which CO_2 is used as an internal tracer, that is, not dosed but produced by animal respiration and manure. It can be used in naturally or mechanically ventilated houses. It is based on the hypothesis that ventilation rate determines the relationship between CO_2 production in the house and the difference in CO_2 concentrations between the inside and outside of the house. This method has been widely described (Blanes and Pedersen, 2005; Estellés et al., 2011; Samer et al., 2012) and is more accurate in buildings with no litter and no gas heating system.

The ventilation rate for the house can be calculated as:

$$VR = \frac{\text{total heat per house} \times \text{ventilation flow per hpu}}{1000} \quad (6)$$

where the total heat produced for the entire house is expressed in heat production units (hpu; 1 hpu is 1 kW of total animal metabolic heat production at 20°C) and the ventilation flow per hpu is in $m^3\ h^{-1}\ hpu^{-1}$.

The International Commission of Agricultural Engineering provides a method to calculate total heat production (sensible plus latent) for different animal categories (Pedersen and Sällvik, 2002). For instance, for fattening pigs, the total heat produced for the entire house is calculated by multiplying the total heat per animal (in W animal^{-1}) by the number of animals and converting to heat production units as:

$$\text{total heat per house} = \text{total heat per animal} \times \text{number of animals} \quad (7)$$

$$\text{total heat per animal} =$$
$$(5.09 \times m^{0.75}) + \left[1 - (0.47 + 0.003 \times m)\right] \times \left[(n \times 5.09 \times m^{0.75}) - (5.09 \times m^{0.75})\right] \quad (8)$$

where m = animal mass (kg)
n = feed energy in relation to the heat dissipation due to maintenance (g d^{-1})

Ventilation flow per hpu varies as a function of animal activity at different times of the day and difference between indoor and outdoor CO_2 concentrations:

$$\text{ventilation flow per hpu} = \frac{c \times (\text{relative animal activity})}{\left(CO_{2,indoors} - CO_{2,outdoors}\right) \times 10^{-6}} \quad (9)$$

where c = CO_2 production (m^3 h^{-1} hpu^{-1}); varies as a function of animal type (Pedersen and Sällvik, 2002; Pedersen et al., 2008).
$CO_{2,indoors}$ and $CO_{2,outdoors}$ = measured indoor and outdoor CO_2 concentrations at time h (mL m^{-3})

Relative animal activity is calculated as:

$$\text{Relative animal activity} = 1 - a \sin\left[\left(\frac{2\pi}{24}\right) \times (h + 6 - h_{min})\right] \quad (10)$$

where a = constant expressing amplitude with respect to the constant value 1, which is a scaling factor based on empirical observation and which varies depending on the animal type (Pedersen et al., 2008)
h = time at sampling (this should be a decimal number $0 \leq h \leq 24$), e.g., (2:10 = 2.2)
h_{min} = activity factor that relates to the time of day with minimum activity (hours after midnight) (Pedersen and Sällvik, 2002)

Use of Sensors

Fan wheel or hot wire anemometers can be used to quantify ventilation rate in mechanically ventilated houses that draw outlet air through ducts or exhaust fans. One important requirement is having access to exhaust flow where the measurements are to be made, which is not possible in many animal houses.

The anemometer measures air velocity, and ventilation rate (VR) is calculated as follows:

$$VR = s\,A \qquad (11)$$

where s = mean airspeed (m h^{-1})
A = cross-sectional area of the ventilation duct or air stream (m^2)

Proper methods must be utilized to obtain representative mean air velocity over the flow area, for example by selecting a sufficient number of measurement points and applying either log-linear or log-Tchebycheff rules (ISO 3966, 2008) for measurement points spacing.

Use of anemometers is not recommended in naturally ventilated houses because of their rapid change in air fluxes and large size of the open area, which would require many sensors to obtain a representative estimate of the ventilation rate.

In mechanically ventilated houses, continuous monitoring of the static pressure differential and the operating status (on-off) of each fan can be used to estimate the fan's ventilation rate based on its theoretical or measured performance characteristics. Ideally, the in situ performance of each fan is determined first, and the house ventilation rate can be estimated by summing all operating fan flow rates. For example, Gates et al. (2004, 2005) developed and improved a fan assessment numeration system (FANS) to measure the in situ performance curve of ventilation fans operating in a negative pressure mechanically ventilated animal house (figure 10). This unit is placed either against a fan on the inside of the house, or at the fan exterior with appropriate flexible ducting (Morello et al., 2014) to direct all airflow through the unit. A series of anemometers traverse the entire flow area to obtain a single mean air velocity, which is multiplied by the calibrated unit cross sectional area. A series of these measurements taken at different building static pressures provides an empirical fan performance curve, obtained, for example, from the regression equation of measured flow on building static pressure. Then, measurements of fan run-time and concurrent static pressure can be used to determine reasonably accurate airflow rates for each fan, and their sum is the building ventilation rate. Previous work has clearly shown that neglecting to account for building ventilation by means of direct measurement results in substantial loss in accuracy of estimates for ER, due to the variation among fans.

Figure 10. Fan assessment numeration system (FANS) developed by Gates et al. (2005) to describe the performance curve of a ventilation fan in an animal house. The unit can be used on either the inlet or exhaust side of the ventilation fan.

The Mass Balance Approach for a Global Estimation of N and C Emissions and Emissions Measurement Validation

A mass balance approach estimates emissions based on changes in livestock over time, without the need to measure emissions directly (figure 11). The approach estimates total N or C emissions rather than emissions of specific gases (e.g., $N-NH_3$, $N-N_2O$, N_2, $C-CH_4$, $C-CO_2$) or emission dynamics. The accuracy of mass balance calculations depends on the technical and livestock management data available, characterization of the manure and feed, and, in certain cases, the length of the period considered. To test the validity of the data used to calculate an N or C mass

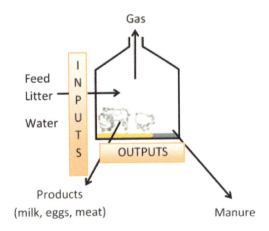

Figure 11. The mass balance approach in animal houses (from Hassouna et al., 2016)

balance, those of non-volatile elements such as phosphorus (P) and potassium (K) must be calculated. As the latter elements are non-volatile, their mass balance deficits (difference between inputs and outputs) should be zero, but the data used in calculations will have uncertainty, especially under commercial conditions. If the mass balance deficit for P and K is too high (e.g., > 15%), then the estimates of total N and C emissions must be reconsidered.

The estimation of N, C, or water emissions (X emissions, where X is for N, C, or water) over the production period can be calculated according to the following equation:

$$X_{inputs} - X_{outputs} = X_{emissions} \qquad (12)$$

X_{inputs} and $X_{outputs}$ are the quantity of X in all inputs and outputs. The estimation of these quantities requires careful data collection (quantities, chemical compositions) concerning animals, feed, eggs or milk (a function of animal production), litter, manure, and animal mortality. Models should be used to estimate the quantity of X in animals as a function of their weight.

Applications

Implementing the different emissions measurement methods in an animal house requires the development of a protocol based on the objectives of the project, the specifics of the animal house, the interior environment, and outdoor weather. Three important points that should be considered in a protocol concern the sampling, the data acquisition, and the validity check of the measurements.

Sampling and Sensor Locations

Evaluation of gas emissions requires measurement of inlet and outlet gas concentration. Sampling at air inlets or outlets is recommended if they can be identified, if their locations are fixed over the measurement period, and if they can be easily reached. When these conditions cannot be fulfilled, multiple locations inside the house are usually selected to provide a mean indoor concentration to accommodate spatial variability of indoor concentration. The same should be done for outdoor concentration.

The presence of animals inside the animal area makes installation of gas sensors or sampling tubes more complicated. Ideally, they should be installed when no animals are in the house, such as during an outdoor or vacancy period. They should be located where animals cannot bite, bump, or move them, and should be carefully protected from animals (figure 12). Successful sensor placement requires a trade-off between minimizing animal disturbance and maximizing the representativeness of measurements.

The environment inside animal housing is generally harsh for sensors and the air sampling system. Direct exposure to the combination of humidity, NH_3, and suspended particulate matter can damage the sensors (figure 13). Furthermore, indoor air is generally warmer and more humid than outdoor air because of animal heat production or the use of a heating or cooling system inside the house. These differences should be considered when sampling indoor air; for example, sampling tubes should be heated and insulated if air samples are analyzed in a cooler place and condensation within sampling lines might be expected.

Data Acquisition

During the production period, emissions can vary greatly during a 24-hour cycle and over longer time intervals. Variability is due to the same parameters that affect spatial variability, changes in animal behavior, their excretion patterns and quantity, and whether or not they have outdoor access. For instance, fattening pigs and poultry will excrete more total ammonia nitrogen (TAN) each day as they grow, yielding ever higher potential for NH_3 emissions. These two kinds of temporal dynamics (daily and production period) must be considered when measuring gas emissions.

Information concerning the production conditions (number of animals, feed and water consumption, animal mortality) and outdoor climate are also required for validation of measurements and comparison with emissions data already published. All operations made by the farmers or operators (for instance, feeding changes, litter supply, or cooling system implementation) and specific events (for instance, electric power shutdown) during the measurement period should be noted because they will be helpful for data analysis and interpretation.

Figure 12. Sensors must be installed beyond the reach of animals.

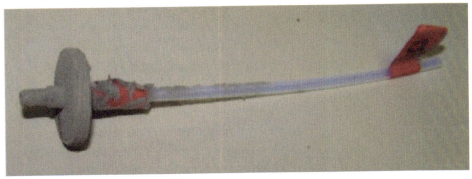

Figure 13. Clogging of a sampling tube and its dust filter after three months of measurement in a dairy cattle house.

Validation of Measurements

In order to achieve good quality measurements, data validation steps are necessary at several levels:

- Validation measurements for parts or the whole measuring setup should be carried out in advance, especially if the setup, single components, and/or the measurement objective (e.g., housing system) was not measured in this configuration before.
- Calibration of analytical devices and sensors has to be performed according to their specifications. For some analytical instruments, measurements with a reference method (e.g., acid traps for NH_3) are recommended.
- Frequent checks of operational mode and measurement values as well as housing and management conditions are necessary.
- Plausibility checks of raw data and emission values (e.g., comparison of courses of gas concentrations and wind speed, data check in view of a predetermined plausible range based upon user scientific and technological knowledge) help to find outliers, non-logical values, etc. These incorrect values have to be eliminated according to predefined criteria.
- Redundancy in measurements can enhance the reliability of the values. For instance, CO_2 concentrations could be measured with both a gas chromatograph and an optical gas analyzer during startup or periodically over the project.
- Comparison of the cumulative emissions for N and carbon (C) with N and C mass balance deficits over the measuring period.

Examples

Example 1: Calculate ammonia (NH_3) emission rate from a mechanically ventilated pig house using the carbon dioxide mass balance approach

Problem:

The following case study is based on a project undertaken on behalf of the Irish Environmental Protection Agency to test the suitability of existing NH_3 emission factors currently being used for different pig life stages and to explore the potential impact on Natura 2000 sites (i.e., Special Areas of Conservation and Special Protection Areas that may be sensitive to N). The monitoring equipment is described in detail by Kelleghan et al. (2016).

The house used mechanical ventilation, with air inlets along the side of the house and ceiling exhaust fans, but no access to the exhaust stream for direct measurement of ventilation rate. Gas concentrations were measured at 10 a.m. in a house with 406 fattening pigs of 81.25 kg animal^{-1} reared on a fully slatted floor. Indoor NH_3 concentrations in the house were measured using a Los Gatos Research ultraportable ammonia analyzer (UAA) in combination with an eight-inlet multiport unit allowing for multiple sampling points. Outdoor concentrations were not measured; for the purposes of this study and the calculations, they were assumed to be zero. Indoor and outdoor CO_2 was measured in the sample gas drawn by the UAA using a K30 CO_2 sensor (Senseair, Sweden). Measured gas concentrations are presented in table 1. Additional parameters adapted for finishing pigs are included in table 2. Calculate the building ventilation rate and the NH_3 emission rate.

Table 1. Concentrations measurement values.

	Measured
$CO_{2,indoors}$	915 ppm (mL m^{-3})
$CO_{2,outdoors}$	403 ppm (mL m^{-3})
$NH_{3,indoors}$	3.73 (mg m^3)
$NH_{3,outdoors}$	0 (mg m^3)

Table 2. Specific parameters useful for calculations.

Parameter	Value for This Experiment
a (equation 11)	0.53
h_{min} (equation 11)	1:40 AM = 1.7 in equation
n (equation 9)	3.38 (g day^{-1})
c (equation 10)	m^3 h^{-1} hpu^{-1}

Solution:

Calculate the building ventilation rate using the CO_2 mass balance approach expressed by equation 6:

$$VR = \frac{\text{total heat per house} \times \text{ventilation flow per hpu}}{1000} \quad (6)$$

To calculate the total heat per house, first, calculate the relative animal activity with equation 10 using the given values:

$$\text{Relative animal activity} = 1 - a \sin\left[\left(\frac{2\pi}{24}\right) \times (h + 6 - h_{min})\right] \quad (10)$$

$$\text{relative animal activity} = 1 - 0.53 \times \sin\left[\left(\frac{2\pi}{24}\right) \times (10 + 6 - 1.7)\right] = 1.2994$$

Next, calculate the total heat production per animal with equation 8:

$$\text{Total heat per animal} = (5.09 \times m^{0.75}) + \left[1 - (0.47 + 0.003 \times m)\right] \times \\ \left[(n \times 5.09 \times m^{0.75}) - (5.09 \times m^{0.75})\right] \quad (8)$$

$$\text{Total heat per animal} = (5.09 \times 81.25^{0.75}) + (1 - (0.47 + 0.003 \times 81.25)) \times \\ (0.00338 \times 5.09 \times 81.25^{0.75} - 5.09 \times 81.25^{0.75}) = 231.6 \text{ W animal}^{-1}$$

Thus, with 406 pigs, the total heat production for the house is:

$$406 \times 231.6 = 94026.5 \text{ W}$$

Next, calculate ventilation flow per heat producing unit (hpu) with equation 9:

$$\text{Ventilation flow per hpu} = \frac{c \times (\text{relative animal activity})}{(CO_{2,indoors} - CO_{2,outdoors}) \times 10^{-6}} \quad (9)$$

$$\text{Ventilation flow per hpu} = \frac{0.185 \times (1.2994)}{(915 - 403) \times 10^{-6}} = 469.5$$

Ventilation flow per hpu will equal 469.5 m³ h⁻¹ hpu⁻¹.

Finally, substitute the computed values into equation 6 to calculate the building ventilation rate:

$$VR = \frac{\text{total heat per house} \times \text{ventilation flow per hpu}}{1000} \quad (6)$$

$$VR = \frac{94026.5 \times 469.5}{1000} = 44{,}145.2$$

The building ventilation rate is 44145.2 m³ h⁻¹.

To calculate NH_3 emission rate, note that the NH_3 emission rate for the house is proportional to the difference between indoor and outdoor NH_3 concentrations multiplied by the ventilation rate:

$$\frac{(3.73 - 0) \times 44{,}145.2}{3600} = 45.7$$

The NH_3 emission rate for this example is 45.7 mL s⁻¹. This equates to 9.7 g day⁻¹ animal⁻¹.

Example 2: Calculate methane (CH_4) emissions from a naturally ventilated dairy cattle house using tracer gas (SF_6) measurement data

Problem:

This example is based on investigations carried out in experimental dairy housing for emission measurements (Mohn et al., 2018). The housing consists of two experimental compartments, each for 20 dairy cows, and a central section for milking, technical installations, an office, and analytics. The experimental compartments are naturally ventilated without thermal insulation and with flexible curtains as facades.

The diluted tracer gas, SF_6, was dosed continuously through steel tubes with critical capillaries (every third meter) next to the aisles to mimic the emission sources. Stainless steel tubes and critical orifices were protected with metal profiles from damage by animals and contamination with excrement. To adjust analyzed tracer gas concentration in an optimal range (> 0.05 µg m⁻³, < 1.5 µg m⁻³ SF_6), tracer gas flow was set according to meteorological and ventilation conditions (e.g., curtains open/closed) by mass flow controllers. Integrative air samples at a height of 2.5 m with a piping system consisting of Teflon tubes and critical glass orifices (every second meter) allow a representative sample. Teflon filters protected the critical orifices from dust and insects. Flow rates for individual orifices of the dosing and sampling systems were monitored before and after every measuring period using mass flow meters. The analytical instrumentation for CH_4 (cavity ring-down spectrometer, CRDS, Picarro Inc., Santa Clara, CA, USA) and SF_6 analysis (GC-ECD, Agilent, Santa Clara, CA, USA) was located in an air-conditioned trailer in the central section. The two compartments were sampled alternately for 10 min. each. Further, once per hour, the background (approximately 25 m from housing, unaffected by the housing) was sampled, so at least two 10-min samples per compartment were obtained every hour.

To describe the measurement situation, relevant accompanying parameters such as housing and outdoor climate, animal parameters (e.g., live weight, milk yield, milk composition, milk urea content, urine urea content), and feed (quality and quantity, amount of trough residue) were recorded.

Calculate the CH$_4$ emissions during one 10-min measurement taken on 23 September at 12:40 p.m. in a compartment with perforated floors holding 20 cows. The measured gas concentrations are presented in table 3.

Table 3. Concentrations of CH$_4$ and SF$_6$ at 12:40 p.m., 23 September.

Parameter	Value
SF$_6$ mass flow	2.879 g d^{-1}
SF$_6$ background	0.052 µg m^{-3}
SF$_6$ housing, sampling points	1.820 µg m^{-3}
CH$_4$ background	1384.2 µg m^{-3}
CH$_4$ housing, sampling points	9118.6 µg m^{-3}

Solution:

Calculate the background correction (C_x) according to equation 3 using measured concentrations of SF$_6$ (C_{SF6}) and CH$_4$ (C_{CH4}):

$$C_x = C_{x,sp} - C_{x,bgd} \qquad (3)$$

$$C_{CH4} = 9118.6 - 1384.2 = 7734.4 \text{ µg m}^{-3}$$

$$C_{SF6} = 1.820 - 0.052 = 1.768 \text{ µg m}^{-3}$$

Calculate emission or mass flow calculation of CH$_4$ using equation 5:

$$\dot{m}_G = \frac{\dot{m}_T \times C_G}{C_T} \qquad (5)$$

$$\dot{m}_{CH4} = \frac{2.879 \text{ g d}^{-1} \times 7734.4 \text{ µg m}^{-3}}{1.768 \text{ µg m}^{-3}} = 12{,}597.9 \text{ g d}^{-1}$$

ER per cow (20 cows per compartment): 629.7 g d^{-1}

Image Credits

Figure 1. Department for Environment Food & Rural Affairs. (CC By 4.0). (2018). Environmental impacts of ammonia emissions. Retrieved from https://www.gov.uk/government/publications/code-of-good-agricultural-practice-for-reducing-ammonia-emissions/code-of-good-agricultural-practice-cogap-for-reducing-ammonia-emissions

Figure 2. Calvet, S. (CC By 4.0). (2014). Process leading to ammonia emission and contributing factors. Adapted from Snoek et al.

Figure 3. Calvet, S. (CC By 4.0). (2001). Reactions leading to N2O emissions: nitrification (in green) and denitrification (in blue). Adapted from Wrage et al.

Figure 4. Calvet, S. (CC By 4.0). (2020). Process and microorganisms involved in methane formation.

Figure 5. Calvet, S. (CC By 4.0). (2020). Direct emission measurement in an animal house. The difference in concentration between indoors and outdoors and the ventilation rate must be measured (adapted from Calvet et al., 2013).

Figure 6. Hassouna, M. (CC By 4.0). (2020). Acid trap configuration.

Figure 7. Hassouna, M. (CC By 4.0). (2020). Duct for dispersing a tracer gas within an animal house.

Figure 8. Agroscope, Schrade, S. (CC By 4.0). (2020). (a) Tracer-gas dosing by steel tubes with critical orifices protected by steel elements next to the floors in a dairy housing; (b) gas bottles with mass-flow controller. Photos adapted from Agroscope. [Fair Use].

Figure 9. Agroscope, Schrade, S. (CC By 4.0). (2020). Principle of the tracer gas method.

Figure 10. Gates, R. (CC By 4.0). (2020). Fan Assessment Numeration System (FANS) developed by Gates et al. (2005) to describe the performance curve of a ventilation fan in an animal house. Note the unit can be used on either the inlet or exhaust side of the ventilation fan.

Figure 11. Hassouna, M. (CC By 4.0). (2016). The mass balance approach in animal houses.

Figure 12. Right. Agroscope, Schrade, S. (CC By 4.0). (2020). Sensors must be installed beyond the reach of animals.

Figure 12 Left. Hassouna, M. (CC By 4.0). (2020). Sensors must be installed beyond the reach of animals.

Figure 13. Hassouna, M. (CC By 4.0). (2020). Clogging of a sampling tube and its dust filter after three months of measurement in a dairy cattle house.

References

Blanes, V., & Pedersen, S. (2005). Ventilation flow in pig houses measured and calculated by carbon dioxide, moisture and heat balance equations. *Biosyst. Eng.*, 92(4), 483-493. https://doi.org/10.1016/j.biosystemseng.2005.09.002.

Calvet, S., Gates, R. S., Zhang, G., Estelles, F., Ogink, N. W. M., Pedersen, S., & Berckmans, D. (2013). Measuring gas emissions from livestock buildings: A review on uncertainty analysis and error sources. *Biosyst. Eng.*, 116(3), 221-231. https://doi.org/10.1016/j.biosystemseng.2012.11.004.

Estelles, F., Fernandez, N., Torres, A. G., & Calvet, S. (2011). Use of CO_2 balances to determine ventilation rates in a fattening rabbit house. *Spanish J. Agric. Res.*, 9(3), 8. https://doi.org/10.5424/sjar/20110903-368-10.

Gates, R. S., Casey, K. D., Xin, H., Wheeler, E. F., & Simmons, J. D. (2004). Fan assessment numeration system (FANS) design and calibration specifications. *Trans. ASAE*, 47(5), 1709-1715. https://doi.org/10.13031/2013.17613.

Gates, R. S., Xin, H., Casey, K. D., Liang, Y., & Wheeler, E. F. (2005). Method for measuring ammonia emissions from poultry houses. *J. Appl. Poultry Res.*, 14(3), 622-634. https://doi.org/10.1093/japr/14.3.622.

Hassouna, M., Eglin, T., Cellier, P., Colomb, V., Cohan, J.-P., Decuq, C., . . . Ponchant, P. (2016). Measuring emissions from livestock farming: Greenhouse gases, ammonia and nitrogen oxides. France: INRA-ADEME.

ISO 3966. (2008). Measurement of fluid flow in closed conduits. Velocity area method for regular flows using Pitot static tubes.

Kelleghan, D. B., Hayes, E. T., & Curran, T. P. (2016). Profile of ammonia and water vapor in an Irish broiler production house. ASABE Paper No. 162461252. St. Joseph, MI: ASABE. https://doi.org/10.13031/aim.20162461252.

Mohn, J., Zeyer, K., Keck, M., Keller, M., Zahner, M., Poteko, J., . . . Schrade, S. (2018). A dual tracer ratio method for comparative emission measurements in an experimental dairy housing. *Atmospheric Environ.*, 179, 12-22. https://doi.org/10.1016/j.atmosenv.2018.01.057.

Morello, G. M., Overhults, D. G., Day, G. B., Gates, R. S., Lopes, I. M., & Earnest Jr., J. (2014). Using the fan assessment numeration system (FANS) in situ: A procedure for minimizing errors during fan tests. *Trans. ASABE*, 57(1), 199-209. https://doi.org/10.13031/trans.57.10190.

Ogink, N. W. M., Mosquera, J., Calvet, S., & Zhang, G. (2013). Methods for measuring gas emissions from naturally ventilated livestock buildings: Developments over the last decade and perspectives for improvement. *Biosyst. Eng.*, *116*(3), 297-308. https://doi.org/10.1016/j.biosystemseng.2012.10.005.

Pedersen, S., & Sallvik, K. (2002). 4th Report of working group on climatization of animal houses. Heat and moisture production at animal and house levels. Research Centre Bygholm, Danish Institute of Agricultural Sciences.

Pedersen, S., Blanes-Vidal, V., Jorgensen, H., Chwalibog, A., Haeussermann, A., Heetkamp, M. J., & Aarnink, A. J. (2008). Carbon dioxide production in animal houses: A literature review. *Agric. Eng. Int.: CIGR J.*

Phillips, V. R., Lee, D. S., Scholtens, R., Garland, J. A., & Sneath, R. W. (2001). A review of methods for measuring emission rates of ammonia from livestock buildings and slurry or manure stores, Part 2: Monitoring flux rates, concentrations and airflow rates. *J. Agric. Eng. Res.*, *78*(1), 1-14. https://doi.org/10.1006/jaer.2000.0618.

Powers, W., & Capelari, M. (2016). Analytical methods for quantifying greenhouse gas flux in animal production systems. *J. Animal Sci.*, *94*(8), 3139-3146. https://doi.org/10.2527/jas.2015-0017.

RAMIRAN. (2011). Glossary of terms on livestock and manure management. Retrieved from http://ramiran.uvlf.sk/doc11/RAMIRAN%20Glossary_2011.pdf.

Samer, M., Ammon, C., Loebsin, C., Fiedler, M., Berg, W., Sanftleben, P., & Brunsch, R. (2012). Moisture balance and tracer gas technique for ventilation rates measurement and greenhouse gases and ammonia emissions quantification in naturally ventilated buildings. *Building Environ.*, *50*, 10-20. https://doi.org/10.1016/j.buildenv.2011.10.008.

Sherman, M. H. (1990). Tracer-gas techniques for measuring ventilation in a single zone. *Building Environ.*, *25*(4), 365-374. https://doi.org/10.1016/0360-1323(90)90010-O.

Snoek, D. J. W., Stigter, J. D., Ogink, N. W. M., & Groot Koerkamp, P. W. G. (2014). Sensitivity analysis of mechanistic models for estimating ammonia emission from dairy cow urine puddles. *Biosyst. Eng.*, *121*, 12-24. https://doi.org/10.1016/j.biosystemseng.2014.02.003.

Wang, X., Ndegwa, P. M., Joo, H., Neerackal, G. M., Stockle, C. O., Liu, H., & Harrison, J. H. (2016). Indirect method versus direct method for measuring ventilation rates in naturally ventilated dairy houses. *Biosyst. Eng.*, *144*, 13-25. https://doi.org/10.1016/j.biosystemseng.2016.01.010.

Wrage, N., Velthof, G. L., van Beusichem, M. L., & Oenema, O. (2001). Role of nitrifier denitrification in the production of nitrous oxide. *Soil Biol. Biochem.*, *33*(12), 1723-1732. https://doi.org/10.1016/S0038-0717(01)00096-7.

Plant Production in Controlled Environments

Timothy J. Shelford
Department of Biological and Environmental Engineering
Cornell University
Ithaca, NY, USA

A. J. Both
Department of Environmental Sciences
Rutgers University
New Brunswick, NJ, USA

KEY TERMS		
Psychrometric chart	Shading	Ventilation
Heating	Mechanical cooling	Installation cost
Lighting	Evaporative cooling	Operating cost

Variables

ρ_i = density of the greenhouse air

A_c = area of the greenhouse surface (walls and roof)

c_{pi} = specific heat of air in the greenhouse

h_{fg} = latent heat of vaporization of water at t_i

N = infiltration rate

q_{ccr} = heat loss by conduction, convection, and radiation

q_i = heat loss by infiltration

t_i = inside air temperature of the greenhouse

t_o = temperature of the ambient air (outside air)

T_d = dew point air temperature

T_{db} = dry-bulb air temperature

T_{wb} = wet-bulb air temperature

U = overall heat transfer coefficient

V = volume of the greenhouse

W_i = humidity ratio of the greenhouse air

W_o = humidity ratio of the outside air

Introduction

Controlled environment crop production involves the use of structures and technologies to minimize or eliminate the potentially negative impact of the weather on plant growth and development. Common structures include greenhouses (which can be equipped with a range of technologies depending on economics, crops grown and grower preferences), and indoor growing facilities (e.g., growth chambers, plant factories, shipping containers, and vertical farms in high-rise buildings). While each type of growing facility has unique challenges, many of the processes, principles, and technology solutions are similar. This chapter describes approaches to environmental control in plant production facilities with a focus on technologies used for crop production and light control.

Outcomes

After reading this chapter, you should be able to:

- List and explain the critical environmental control challenges for plant production in controlled environments
- Perform design calculations for systems used for plant production in controlled environments
- Calculate the installation and operating cost estimates of lighting systems for plant production in controlled environments

Concepts

Greenhouses were developed to extend the growing season in colder climates and to allow the production of perennial plants that would not naturally survive cold winter months. In providing an optimal environment for a crop, whether in a greenhouse or indoor growing facility, the air temperature is a critical factor that impacts plant growth and development. An equally important and related factor is the moisture content of the air (expressed as relative humidity). Plant growth depends on transpiration, a process by which water and nutrients from the roots are drawn up through the plant, culminating in evaporation of the water through the stomates located in the leaves. (Stomates are small openings that allow for gas exchange. They are actively controlled by the plant.) The transpiration of water through the stomates also results in cooling. Under high relative humidity conditions, the plant is unable to transpire effectively, resulting in reduced growth and, in some cases, physiological damage. Growers seek to create ideal growing environments in greenhouses and other indoor growing facilities by controlling heating, venting, and cooling (Both et al., 2015).

Psychrometric Chart

Knowledge of the relationship between temperature and relative humidity is critical in the design of heating, cooling, and venting systems to maintain the desired environmental conditions inside plant production facilities. The psychrometric chart (figure 1) is a convenient tool to help determine the properties of moist air. With values of only two parameters (e.g., dry-bulb temperature and relative humidity, or dry-bulb and wet-bulb temperatures), other air properties can be read from the chart (some interpolation may be necessary). The fundamental physical properties of air used in the psychrometric chart are described below.

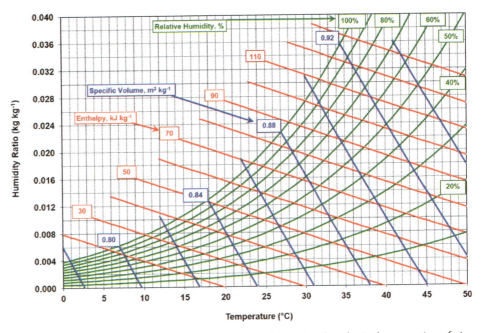

Figure 1. Example psychrometric chart used to determine the physical properties of air. Curved green lines: constant relative humidity; steep blue straight lines: constant specific volume; less steep red straight lines: constant enthalpy.

- Dry-bulb temperature (T_{db}, °C) is air temperature measured with a regular thermometer. In a psychrometric chart (figure 1), the dry-bulb temperature is read from the horizontal axis.
- Wet-bulb temperature (T_{wb}, °C) is air temperature measured when air is cooled to saturation (i.e., 100% relative humidity) by evaporating water into it. The energy (latent heat) required to evaporate the water comes from the air itself. The wet-bulb temperature can be measured by keeping the sensing tip of a thermometer moist (e.g., by surrounding it with a wick connected to a water reservoir) while the thermometer is moved through the air rapidly, or by blowing air through the moist (and stationary) sensing tip. In a psychrometric chart (figure 1), the wet-bulb temperature is read from the horizontal axis by following the line of constant enthalpy from the initial condition (e.g., the intersection of dry-bulb temperature and relative humidity combination) to the saturation line (100% relative humidity).
- Wet bulb depression is the difference between the dry- and wet-bulb temperature.
- Dewpoint temperature (T_d, °C) is the air temperature at which condensation occurs when moist air is cooled. In a psychrometric chart (figure 1), the dewpoint temperature is read from the horizontal axis after a horizontal line of constant humidity ratio is extended from the initial condition (e.g., the intersection of dry-bulb temperature and relative humidity combination) to the saturation line (100% relative humidity).
- Relative humidity (RH, %) is the level of air saturation (with water vapor). In a psychrometric chart (figure 1), curved lines are of constant relative humidity.

- The humidity ratio (kg kg⁻¹) is the mass of water vapor evaporated into a unit mass of dry air. In a psychrometric chart (figure 1), the humidity ratio is read from the vertical axis.
- Enthalpy (kJ kg⁻¹) is the energy content of a unit mass of dry air, including any contained water vapor. The psychrometric chart (figure 1) typically presents lines of constant enthalpy.
- Specific volume (m³ kg⁻¹) is the volume of a unit mass of dry air; it is the inverse of the air density. The psychrometric chart (figure 1) presents lines of constant specific volume.

Heating

A major expense of operating a greenhouse year-round in cold climates is the cost of heating. It is, therefore, important to understand the major modes of heat loss when designing or operating a greenhouse. Heat loss occurs from the structure directly through conduction, convection, and radiation. Depending on location, when estimating heat losses, it may be necessary to include heat loss around the outside perimeter, as well as the impact of high outside wind speeds and/or large temperature differences between the inside and outside of the greenhouse (Aldrich and Bartok, 1994).

Estimating Heat Needs

Estimating the heat losses due to conduction, convection, radiation, and infiltration, requires both the inside and outside air temperatures. The inside air temperature is usually based on the nighttime set point required by the crop. In the absence of specific crop requirements, typically 16°C can be used as a minimum. If the greenhouse is to be used year-round, typically the 99% winter design dry-bulb temperature is used for the outside temperature. The 99% winter design dry-bulb temperature is the outdoor temperature that is only exceeded 1% of the time (based on 30 years of data for the months December, January, and February collected at or near the greenhouse location). The term "exceeded" in the previous sentence means "colder than." Such values for many locations throughout the world are published by ASHRAE (2013).

Calculating the exchange of heat (by conduction, convection, and radiation) is a complex process that usually involves making many simplifying assumptions. Solutions often require iterative calculations that are tedious without the help of computing tools. Computing software such as EnergyPlus™ (Crawley, 2001) and Virtual Grower (USDA-ARS, 2019) are available for heat loss calculations. However, even software packages developed for heat loss calculations may not necessarily provide accurate results.

Other methods that greatly simplify performing heat loss calculations using heat transfer coefficients are available. Heat transfer coefficients combine the effects of conduction, convection, and radiation in a single coefficient. Since these processes depend on many factors other than the temperature differential, their accuracy is not high, especially when conditions are extreme, or outside of typical operating ranges. However, for quick estimates that are not computationally intensive, coefficient-based calculations may be useful to a designer or operator. Equation 1 provides a means to solve for the conductive, convective, and radiative heat losses:

$$q_{ccr} = U A_c (t_i - t_o) \quad (1)$$

where q_{ccr} = heat loss by conduction, convection, and radiation (W)
U = overall heat transfer coefficient (W m^{-2} °C^{-1})
A_c = area of the greenhouse surface (walls and roof) (m^2)
t_o = ambient (outside) air temperature (°C); the 99% design temperature is commonly used for this parameter (see text)

The overall heat transfer coefficients for typical greenhouse materials are listed in table 1.

Equation 2 is for solving the heat loss due to infiltration:

$$q_i = \rho_i N V \left[c_{pi} (t_i - t_o) + h_{fg} (W_i - W_o) \right] \quad (2)$$

where q_i = heat loss by infiltration (W)
ρ_i = density of the greenhouse air (kg m^{-3})
N = infiltration rate (s^{-1})
V = volume of the greenhouse (m^3)
c_{pi} = specific heat of the greenhouse air (J kg^{-1} °C^{-1})
t_i = greenhouse (inside) air temperature (°C)
t_o = outside air temperature (°C)
h_{fg} = latent heat of vaporization of water at t_i (J kg^{-1})
W_i = humidity ratio of the greenhouse air (kg$_{water}$ kg$_{air}^{-1}$)
W_o = humidity ratio of the outside air (kg$_{water}$ kg$_{air}^{-1}$)

Select heat transfer coefficients (U-values; table 1) and infiltration rates (table 2) with caution when performing heat loss calculations. Infiltration rates depend highly on the magnitude and direction of the wind, among other factors.

Cooling and Cooling Methods

During warmer periods of the year, the temperature inside the growing area of a plant production facility could be much higher than the outside temperature (as occurs inside a closed car on a sunny day). High temperatures inside greenhouses can depress plant growth and, in extreme cases, kill a crop. Cooling systems are

Table 1. Approximate overall heat transfer coefficients (U-values) for select greenhouse glazing methods and materials (ASAE Standards, 2003).

Greenhouse Covering	U Value (W m^{-2} °C^{-1})
Single glass, sealed	6.2
Single glass, low emissivity	5.4
Double glass, sealed	3.7
Single plastic	6.2
Single polycarbonate, corrugated	6.2–6.8
Single fiberglass, corrugated	5.7
Double polyethylene	4
Double polyethylene, IR inhibited	2.8
Rigid acrylic, double-wall	3.2
Rigid polycarbonate, double-wall[1]	3.2–3.6
Rigid acrylic, w/polystyrene pellets[2]	0.57
Double polyethylene over glass	2.8
Single glass and thermal curtain[3]	4
Double polyethylene and thermal curtain[3]	2.5

[1] Depending upon the spacing between walls.
[2] 32 mm rigid acrylic panels filled with polystyrene pellets.
[3] Only when the curtain is closed and well-sealed.

Table 2. Estimated infiltration rates for greenhouses by type and age of construction (ASAE Standards, 2003).

Type and Construction	Infiltration Rate (N)[1]	
	s^{-1}	h^{-1}
New construction:		
Double plastic film	2.13×10^{-4}–4.13×10^{-4}	0.75–1.5
Glass or fiberglass	1.43×10^{-4}–2.83×10^{-4}	0.50–1.0
Old construction:		
Glass, good maintenance	2.83×10^{-4}–5.63×10^{-4}	1.0–2.0
Glass, poor maintenance	5.63×10^{-4}–11.13×10^{-4}	2.0–4.0

[1] Internal air volume exchanges per unit time (s^{-1} or h^{-1}). High winds or direct exposure to wind will increase infiltration rates; conversely, low winds or protection from wind will reduce infiltration rates.

essential for plant production facilities that are used year-round.

Mechanical Cooling (Air Conditioning)

Although air conditioning of greenhouses is technically feasible, the installation and operating costs can be very high, particularly during the summer months. The most economical time to use air conditioners in greenhouses is during the spring and autumn when the heat load is relatively low and the crop may benefit from CO_2 enrichment. Air conditioning is an alternative to using ventilation to manage humidity and control temperature. By definition, air conditioning is a thermodynamic process that removes heat and moisture from an interior space (e.g., the interior of a controlled environment plant production facility) to improve its conditions. It involves a mechanical refrigeration cycle that forces a refrigerant through a circular process of expansion and contraction, resulting in evaporation and condensation, resulting in the extraction of heat (and moisture) from the plant growing area.

Mechanical cooling may be necessary for indoor growing facilities. Typically, indoor growing facilities operate with minimal exchange rates with the outside air, and so air conditioning becomes one of the ways to remove the humidity generated by plants during transpiration. It is essential to insulate and construct the building properly to minimize solar heat gain in indoor facilities that may add to the heat load. Additionally, it is crucial to know the heat load from electric lamps providing the energy needed for photosynthesis to size the air conditioner adequately.

Evaporative Cooling

Sometimes during the warm summer months, regular ventilation and shading (e.g., whitewash or movable curtains) are not able to keep the greenhouse temperature at the desired set point, thus, additional cooling is needed. Growers typically use evaporative cooling as a simple and relatively inexpensive cooling method. The process of evaporation requires heat. This heat (energy) comes from the surrounding air, thereby causing the air temperature to drop. Simultaneously, the humidity of the air increases as the evaporated water becomes part of the surrounding air mass. The maximum amount of cooling possible with evaporative cooling systems depends on the initial properties of the outside air, i.e., the relative humidity (the drier the air, the more water it can absorb, and the lower the final air temperature will be) and air temperature (warmer air can carry more water vapor compared to colder air). Two different evaporative cooling systems used to manage greenhouse indoor air temperatures during periods when using outside air for ventilation is not sufficient to maintain the set point temperatures are the pad-and-fan system and the fog system.

Pad-and-Fan System

Pad-and-fan systems include an evaporative cooling pad installed as a segment of the greenhouse wall, typically on the wall opposite the exhaust fans. Correctly installed pads allow all incoming ventilation air to pass through it before entering the greenhouse environment (figure 2). The pads are made from corrugated material (impregnated paper or plastic) glued together in a way that allows maximum contact with the air passing through the wet pad material. Water is introduced at the top of the pad and released through small holes along the entire length of the supply pipe. These holes are spaced uniformly along the whole length of the pad to provide even wetting. Excess water is collected at the bottom of the pad and returned to a sump tank for reuse. The sump tank is fitted with a float valve to manage make-up water that compensates for the portion of the recirculating water lost through evaporation and to dilute the salt concentration that may increase in the remaining water over time. It is common practice to continuously bleed off approximately 10% of the returning water to a designated drain to prevent excessive salt build-up (crystals) on the pad material that may reduce pad efficiency. During summer operation, it is common to "run the pads dry," i.e., to stop the flow of water while keeping the ventilation fans running at night to prevent algae build-up that can also reduce pad efficiency. The cooled (and humidified) air exits the pad and moves through the greenhouse picking up heat from the greenhouse interior. In general, pad-and-fan systems used in greenhouses experience a temperature gradient between the inlet (pad) and the outlet (exhaust fan). In properly designed systems, this temperature gradient is kept low (up to 4°–6°C is possible) to provide a uniform environment for all the plants.

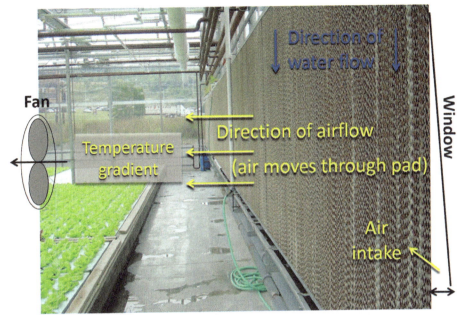

Figure 2. Main design features of a pad-and-fan evaporative cooling system. Water is supplied by a pump from a reservoir and is recirculated. (Photo by A. J. Both)

The required evaporative pad area depends on the pad thickness and can be calculated by:

$$A_{pad} = \frac{\text{total greenhouse ventilation fan capacity}}{\text{recommended air velocity through pad}} \qquad (3)$$

For example, for 10 cm thick pads, the fan capacity (in $m^3\ s^{-1}$) should be divided by the recommended air velocity through the pad, 1.3 $m\ s^{-1}$ (ASAE Standards, 2003). For 15 cm thick pads, the fan capacity should be divided by the recommended air velocity through the pad, 1.8 m s^{-1}. The recommended minimum pump capacity is 26 and 42 L s^{-1} per linear meter of the pad, and the minimum sump tank capacity is 33 and 41 L per m^2 of pad area for the 10 and 15 cm pads, respectively. For evaporative cooling pads, the estimated maximum water usage can be as high as 17–20 L h^{-1} per m^2 of pad area.

High-Pressure Fog System

The other evaporative cooling system commonly used is the fog system. This system is typically used in greenhouses with natural ventilation systems because natural ventilation does not have the force to overcome the additional resistance to airflow resulting from an evaporative cooling pad. The nozzles of a fog system are typically installed throughout the greenhouse to provide a more uniform cooling pattern compared to the pad-and-fan system. The recommended spacing is approximately one nozzle for every 5–10 m^2 of growing area. The water pressure used in greenhouse fog systems is relatively high (≥3,450 kPa) and enough to produce very fine droplets that evaporate before reaching plant surfaces. The water usage per nozzle is small, approximately 3.8–4.5 L h^{-1}. Water for fogging systems should be free of any impurities to prevent clogging of the nozzle openings. Therefore, fog systems require water treatment (filtration and purification) and a high-pressure pump. Thus, fog systems can be more expensive to install compared to pad-and-fan systems, but the resulting cooling is more uniform.

Ventilation

To maintain optimum growing conditions, warm and humid indoor air needs to be replaced with cooler and drier outside air. Plant production facilities use either mechanical or natural ventilation to accomplish this. Mechanical ventilation requires inlet openings, exhaust fans, and electric power to operate the fans. When appropriately designed, mechanical ventilation can provide adequate cooling and dehumidification under a wide range of weather conditions throughout many locations with temperate climates. The typical design specification for maximum mechanical ventilation capacity is 0.05 or 0.06 m^3 s^{-1} per m^2 of floor area for greenhouses with or without a shade curtain, respectively. When deliberate obstructions to the air intake are present (such as insect exclusion screens and an evaporative cooling pad), the inlet area should be carefully sized to overcome the increased resistance to airflow that would result in a reduction in the total air exchange rate relative to fully opened and unobstructed inlets. In that case, ventilation fans should be able to overcome the additional airflow resistance created by the screen or evaporative cooling pad. Multiple and staged fans can provide different ventilation rates based on environmental conditions. Variable-speed fan motors allow for more precise control of the ventilation rate and can reduce overall electricity consumption.

Natural ventilation works on two physical phenomena: thermal buoyancy (warm air is less dense and rises), and the wind effect (wind blowing outside a structure creates small pressure differences between the windward and leeward sides of the structure causing air to move towards the leeward side). All that is needed are carefully placed inlet and outlet openings, vent window motors, and electricity to operate the motors. In some naturally ventilated greenhouses, the vent window positions are managed manually (e.g., in a low-tech plastic tunnel production system), eliminating the need for motors and electricity, but this increases the amount of labor, especially where frequent adjustments are necessary. Electrically operated natural ventilation systems use much less power than mechanical (fan) ventilation systems. When using a natural ventilation system, additional cooling can be provided by a fog system, for example, provided the humidity of the air is not too high. Unfortunately, natural ventilation does not work very well on warm days when the wind velocity is low (less than 1 m s^{-1}) or when the facility uses a shade system that obstructs airflow. When using natural or forced ventilation alone, the indoor temperature cannot be lowered below the outdoor temperature without additional cooling capabilities (typically evaporative cooling).

For most freestanding greenhouses, mechanical ventilation systems usually move the air along the length of the greenhouse (i.e., the exhaust fans and inlet openings are installed in opposite end walls). To avoid excessive airspeed within the greenhouse, the inlet to fan distances are generally limited to 70 to 80 m, provided local climates are not too hot. Natural ventilation systems for freestanding greenhouses usually provide cross ventilation using sidewall windows and roof vents.

In gutter-connected greenhouses (figure 3), mechanical ventilation system inlets and outlets can be installed in the side or end walls, while natural ventilation systems usually consist of only roof vents. Sidewall vents have limited influence on the ventilation of interior sections in larger greenhouses. The ultimate natural ventilation system is the open-roof greenhouse design that allows for the indoor temperature to seldom exceed the outdoor temperature. This kind of effect is not attainable with mechanically ventilated greenhouses due to the substantial amounts of air that such systems would have to move through the greenhouse to accomplish the same results.

Whatever the ventilation system used, uniform air distribution inside the greenhouse is essential because uniformity in crop production is only possible when all plants experience the same

Figure 3. Gutter-connected greenhouses with mechanical ventilation. (Photo by A. J. Both)

environmental conditions. Therefore, the use of horizontal airflow fans is common to ensure proper air mixing. The recommended horizontal airflow fan capacity is approximately 0.015 m³ s⁻¹ per m² of the growing area.

Lighting and Shading

Since light is the driving force for photosynthesis and plant growth, managing the light environment of a growing facility is of prime importance. For many crops, plant growth is proportional to the amount of light the crop receives over the entire growing period. Both the instantaneous light intensity and the daily light integral are important parameters to growers. Plant scientists define light in the 400–700 nm waveband as photosynthetically active radiation (PAR). PAR represents the (instantaneous) light intensity and has the units µmol m⁻² s⁻¹ (ASABE Standards, 2017). When referring to the amount of light a crop receives over some time, such as an hour or a day, the sum of the instantaneous PAR intensities is calculated, and the resulting values are often called light integrals. Usually, growers measure light integrals over an entire day (sunrise to sunrise), resulting in the daily light integral (DLI), with the unit mol m⁻² d⁻¹. Instantaneous measures of PAR may be used to trigger control actions such as turning supplemental lighting on or off. Some growers deploy movable shade curtains to manage the light intensity. Daily light integrals (DLIs) can be used by growers to ensure a consistent level of crop growth by maintaining a consistent integral from day to day (whether from natural light, supplemental lighting, or a mix), or to track the accumulated radiation input that serves as the energy source for photosynthesis. The total DLI received by a plant canopy is the sum of the amount of sunlight received plus any contribution from the supplemental lighting system (for greenhouse production). Equation 4 determines the instantaneous PAR intensity (µmol m⁻² s⁻¹) necessary to meet a DLI target (mol m⁻² d⁻¹) over a specific number of hours:

$$\text{intensity}\left(\frac{\mu\text{mol}}{\text{m}^2\text{s}}\right) = \frac{\text{DLI}}{\text{h per day}} \times \frac{1\text{ h}}{3{,}600\text{ s}} \times \frac{1 \times 10^6\ \mu\text{mol}}{1\text{ mol}} \quad (4)$$

For example, using equation 4, an intensity of 197 µmol m⁻² s⁻¹ is needed to deliver a target DLI of 17 mol m⁻² d⁻¹ over 24 h (one day).

Plant Sensitivity to Light

Human eyes have a different sensitivity to (natural) light (or radiation) compared to how plants respond to light (figure 4). Human eyes are most sensitive to green wavelengths (peak at 555 nm), while most plants exhibit peak sensitivities in the blue (peaking at 430 nm) and orange-red part (peaking at 610 nm) of the visible light spectrum. This difference in sensitivity means the human eye is not a very useful "sensor" in terms of assessing whether a particular light environment is suitable for plant growth and development. While PAR is light across the 400–700 nm waveband, as shown in figure 4, plants are also sensitive to UV (280–400 nm) and far-red (700–800 nm) radiation. Therefore, it is best to use specially designed sensors (PAR sensors and spectroradiometers) to evaluate the light characteristics in environments used for plant production.

Natural and Electric Lighting

Natural light from the sun is an essential aspect of greenhouse production, both in terms of plant growth and development, but also in terms of energy balance (greenhouse heating and cooling). In indoor growing facilities, light is solely provided by electric lighting, though the amount of natural light striking the external surface of the building containing an indoor growing facility can also substantially affect the energy balance of the facility.

Direct and Diffuse Sunlight

The earth's atmosphere contains many particles (gas molecules, water vapor, and particulate matter) that can change the direction of the light from the sun. On a clear day, there are fewer particles in the atmosphere, and sunlight travels unimpeded before reaching the ground. This type of sunlight is called direct light or direct radiation. On cloudy days, the atmosphere contains more particles (mainly water vapor), and the interaction of sunlight with all those particles causes directional changes that are mostly random. As a result, on cloudy days, sunlight comes from many directions. This type of sunlight is called diffuse light or diffuse radiation. These frequent light-particle interactions will also result in a reduction in light intensity compared to direct radiation.

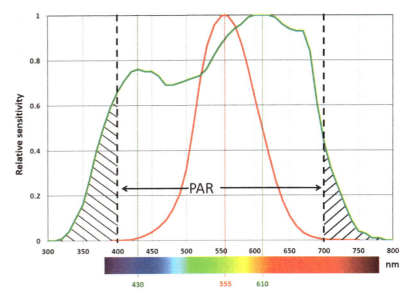

Figure 4. Differences in relative light sensitivity comparing the human eye (red line) to an average plant (green line). PAR = photosynthetically active radiation (400–700 nm). Horizontal axis: wavelength in nanometers. Sources: Commission Internationale de l'Eclairage (1931) and Sager et al. (1988).

Depending on the make-up of the atmosphere (cloudiness), sunlight will reach the surface as direct radiation, diffuse radiation, or a combination of the two. Direct radiation does not reach the lower canopy layers shaded by plant tissues (mostly leaves); however, because diffuse radiation is omnidirectional, it can penetrate deeper into a plant canopy (particularly in a multi-layered, taller canopy). Therefore, though the amount of diffuse radiation may appear small, it can boost plant production because it reaches more of the plant surfaces involved in photosynthesis. Some greenhouse glazing materials (e.g., polyethylene film) diffuse incoming solar radiation more than others (table 3), and while the overall light intensity is often lower in greenhouses covered with a diffusing glazing material, crop growth and development is not necessarily reduced proportionally because more of the canopy surfaces are receiving adequate light for photosynthesis.

The amounts of diffuse radiation are measured with a light sensor placed behind a disc that casts a precise shadow over the sensor, so it blocks all direct radiation. The amount of direct radiation is determined by using a second

Table 3. Characteristics of glazing materials.

Glazing Material	Direct PAR Transmittance (%)	Infrared (heat) Transmittance[a] (%)	Ultraviolet Transmittance[b] (%)	Life Expectancy (years)
Glass	90	0	60–70	30
Acrylic[c]	89	0	44	10–15
Polycarbonate3	80	0	18	10–15
Polyethylene[d]	90	45	80	3–4
PE, IR & AC[d][e]	90	30	80	3–4

[a] for wavelengths above 3,000 nm
[b] for wavelengths between 300 and 400 nm
[c] twin wall
[d] single layer
[e] polyethylene film with an infrared barrier and an anti-condensate surface treatment

sensor that measures total (direct plus diffuse) radiation (direct radiation = total radiation – diffuse radiation).

As sunlight reaches the external surfaces of the greenhouse structure, the light can be reflected, absorbed, or transmitted. Often these processes coincide. The quantities of reflected, absorbed, or transmitted light depend on the (glazing) materials involved, the time of day, the time of year, and whether the grower uses any control strategies (e.g., whitewash or shade curtains). Also, overhead equipment can block light and reduce the total amount of sunlight that reaches the plant canopy. It is not uncommon, even in modern greenhouses, for the plants to receive around 50–60% on average of the amount of sunlight available outside the greenhouse structure. Since every percent of additional light received by the plant canopy counts, it is essential to design greenhouses carefully with optimum light transmission in mind.

Effect of Greenhouse Orientation

Another consideration, particularly at higher latitudes, is the orientation of the greenhouse. At latitudes above 40 degrees, orienting the gutters of a greenhouse along an east-west direction can help capture the most amount of light during the winter months when the sun is low in the sky and the total amount of sunlight is also low. However, using such an orientation, shadow bands created by structural components and overhead equipment tend to move more slowly. This can be a particularly challenging issue when the crop is grown in the greenhouse for only a short amount of time (e.g., for leafy greens). In that case, it is preferable to orient the greenhouse north-south. Aside from any shadows, the intensity of sunlight is considered uniform throughout the growing area.

Shading

During bright sunny days, there is the risk of greenhouse crops being exposed to too much light, thus requiring the use of shade curtains to help reduce plant stress from high light intensities. On variably cloudy days, the light conditions inside a greenhouse can fluctuate rapidly from low light to high light conditions. Such swings in light conditions can negatively impact plant growth and development, so growers may have to deploy both the supplemental lighting system and the shade curtains to provide more stable growing conditions. Managing the supplemental lighting system often involves controlling the shade curtains.

Proper shading is essential for some crops. For example, lettuce grown in a greenhouse is subject to tipburn (figure 5) if light, temperature, and humidity conditions are not kept within specific ranges.

One strategy is to apply a whitewash treatment to the greenhouse during peak solar radiation months and to wash it off at the end of the natural growing season when light conditions diminish. Drawbacks include increased labor costs and additional requirements for supplemental lighting. Movable

> Tipburn is a physiological disorder caused by calcium deficiency in leaf tips. It renders the product unsalable.

Figure 5. Example of lettuce plants without (left, photo by A. J. Both) and with tipburn (right, photo courtesy of the Cornell University Controlled Environment Agriculture Program).

shade curtains are another effective strategy for managing tipburn, if properly designed and used. Deploying shade curtains too late during the day can cause tipburn in lettuce (too much light increases the growth rate beyond the point where the transport of calcium can keep up), and deploying them too early can result in extra hours of supplemental lighting. Movable shade curtains, depending on the design, can also reduce heat loss during the night, but this dual use is often a compromise between optimum shading capabilities and maximum energy retention. A more comprehensive solution is two different curtains, each optimized for its purpose, but such a dual curtain system doubles the installation cost.

Common Types of Artificial Lighting

The two most common types of greenhouse lighting are gas discharge and light-emitting diode (LED) lamps (figure 6). Gas discharge lamps, such as fluorescent (FL), metal halide (MH), and high-pressure sodium (HPS) lamps, produce light by passing a current through an ionized gas. The spectrum of light produced is a function of the gas used and the composition of the electrodes. MH lamps provide a more white-colored light, while HPS light is more yellowish orange (similar to traditional streetlamp light).

LED lamps use semiconductors that release energy in the form of photons when sufficient current is passed through them. The wavelength of light emitted is determined by the bandgap of the semiconductor and any phosphors used to convert the monochromatic LED light. Unlike gas discharge lamps, LEDs without phosphors produce light within a relatively narrow waveband. To get a broad-spectrum output, such as white light, manufacturers often use high-efficiency

Figure 6. High-pressure sodium fixtures mounted over a rose crop (left) and LED fixtures with a magenta color mounted within a tomato crop (right) (photos by A. J. Both).

blue LEDs and convert the output to white light using yellow phosphors. Some plants benefit from small amounts of UV radiation (280–400 nm), but people working in these environments should wear special eye and skin protection to minimize the harmful effects presented by UV radiation.

Lighting Efficacy

At the time of this writing (early 2020), the most energy-efficient lamps available for supplemental lighting are LED-based fixtures (Mitchell et al., 2012; Wallace and Both, 2016). However, not all LED fixtures are designed for plant growth applications. When comparing the efficiency of lights, the wall-plug energy use of the fixture must be considered. Some LED fixtures rely on active cooling using ventilation fans (in some cases water cooling) to prevent overheating that can shorten their lifespans. Active cooling installation requires additional energy, which must be considered, in addition to other losses, such as from transformers and drivers. Ideally, manufacturers publish an efficacy measurement, i.e., light output divided by energy input, or $\mu mol\ s^{-1}$ of PAR (light) output per W (electricity) input ($\mu mol\ J^{-1}$) for their fixtures (Both et al., 2017). Efficacies for lamps used in plant growth applications are shown in table 4. Fixture efficacies continue to increase with several LED fixtures now approaching 3 $\mu mol\ J^{-1}$. Higher efficacy fixtures use less electrical energy to produce the same amount of light.

A Note on Lighting Units

In the horticulture industry, it is still common to use light units of lux, lumens, or foot-candles. However, this is not particularly useful since lux, lumens, and foot candles are based on the sensitivity of the human eye, which is most sensitive to the green part of the visible light spectrum (figure 4). Ideally, the total light output of supplemental lighting devices should be reported in integrated PAR units ($\mu mol\ s^{-1}$). Note that this unit is not the same as the unit used for instantaneous PAR intensity ($\mu mol\ m^{-2}\ s^{-1}$). Users should be aware of this when purchasing lighting fixtures and make sure that the proper instruments were used to assess the light output.

Table 4. Selected fixture efficacies for several different lamp types used for horticultural applications (CMH = ceramic metal halide, HPS = high-pressure sodium, LED = light emitting diode).

Lamp Type	Power Consumption (W)	Efficacy ($\mu mol\ J^{-1}$)
Incandescent (Edison bulb)	102.4	0.32
Compact fluorescent (large bulb)	61.4	0.89
CMH (mogul base)	339	1.58
HPS (mogul base)	700	1.56
HPS (double ended)	1077	1.59
LED (bar, passively cooled)	214	2.39

Advantages and Disadvantages of Lighting Systems

HPS Lighting System

HPS lamps have long been the preferred lamp type for supplemental lighting applications (Both et al., 1997).

Advantages
- Both lamps and fixtures (including the ballasts and reflectors) are relatively inexpensive and easy to maintain (e.g., bulb replacement and reflector cleaning).
- Before LEDs became available, HPS lamps had the highest conversion efficiency (efficacy), and they produced a sufficiently broad spectrum that was acceptable for a wide range of plant species. The recent introduction of double-ended HPS lamps somewhat increased their efficacy.

Disadvantages
- A major drawback of HPS lamps is the production of a substantial amount of radiant energy, necessitating adequate distance between the lights and the plant surfaces exposed to this radiation.
- They require a warm-up cycle before they reach maximum output, and once turned off, need a cool-down period before using again.
- As with all lamps, the light output of HPS lamps depreciates over time, requiring bulb replacements every 10,000–15,000 hrs.

Since HPS lamps have been in use for several decades, researchers and growers have learned how best to produce their crops with this light source. For example, while the radiant heat production can be considered a drawback, it can also be used as a management tool to help maintain a desirable canopy temperature, and this radiant heat can help reduce the amount of heat energy (provided by the heating system) needed to keep the set point temperature.

LED Lighting Systems

LED lamps (often consisting of arrays of multiple individual LEDs) are a relatively new technology for horticultural applications, and their performance capabilities are still evolving (Mitchell et al., 2015).

Advantages

- The efficacy of carefully designed LED lamps has surpassed the efficacy of HPS lamps, and the heat they produce can be removed more easily by either natural or forced convection.
- The resulting convective heat (warm air emanating from the lamp/fixture) is easier to handle in controlled environment facilities than the radiant heat produced by HPS lamps because air handling is a common process while blocking radiant heat is not.
- LED lamps can be switched on and off rapidly and require a much shorter warm-up period than HPS lamps.
- It is possible to modulate the light intensity produced by LED lamps, either by adjusting the supply voltage or by a process called pulse width modulation (PWM; rapid on/off cycling during adjustable time intervals). By combining (and controlling) LEDs with different color outputs in a single fixture, growers have much more control over the spectrum that these lamps produce, opening up new strategies for growing their crops. This benefit in particular will require (a lot of) additional research to be fully understood or realized.
- LED lamps typically have a longer operating life (up to 50,000 h), but more testing is needed in plant production facilities to validate this estimate.

Disadvantages

- LED lamps (fixtures) are more expensive compared to HPS fixtures with similar output characteristics.
- LED lamps typically come as a packaged unit (including LEDs, housing, and electronic driver), making the replacement of failed components almost impossible.
- Because plants are most sensitive to blue and red light in terms of photosynthesis, growers often use LED fixtures that produce a combination of red + blue = magenta light. The magenta light (figure 6) makes it much more challenging to observe the actual color of plant tissue (which is essential for the observation of potential abnormalities resulting from pest and/or disease issues), and can make working in an environment with that spectrum more challenging (it has been reported to make some people uncomfortable).
- Some LED lamps have (unperceivable) flicker rates that can have health effects on humans with specific sensitivities (e.g., people with epilepsy).

Applications

Heating Systems in Greenhouses

Greenhouses can be heated using a variety of methods and equipment to manage heat losses during the cold season. Typically, fuel is combusted to heat either air or water (steam in older greenhouses) which is circulated through the greenhouse environment. Some greenhouses use infrared heating systems that radiate heat energy to exposed surfaces of the plant canopy. The use of electric (resistance) heating is minimal because of the high operating cost. However, as the costs of fossil fuels rise, electric heating could become competitive even for extensive greenhouse operations in various locations.

Unit Heaters and Furnaces

Typical air heating systems include unit heaters and furnaces (figure 7). Typically, the heat generated by the combustion process is transferred to the greenhouse air through a heat exchanger, or the air from the greenhouse used as the oxygen source for the combustion process and then released into the greenhouse. Using heat exchangers allows for the combustion air to remain separate from the greenhouse environment (separated combustion), thus minimizing the risk of releasing small amounts of potentially harmful gasses (e.g., ethylene, carbon monoxide) into the greenhouse environment. Also, it directly increases the air temperature without introducing additional moisture.

Using greenhouse air as a source of oxygen for combustion requires properly maintained combustion equipment and complete fuel combustion to ensure that only water vapor and carbon dioxide (CO_2) are released into the greenhouse environment. An intermediate approach is to use greenhouse air for combustion and vent the combustion gases outdoors.

Fans are usually incorporated in air heating systems to move and distribute the warm air to ensure even heating of the growing environment. Some greenhouses use inflatable polyethylene ducts (the poly-tube system) placed overhead or under the benches or crop rows to distribute the air. Some use strategically placed horizontal or vertical airflow fans. Air heating systems are relatively easy to install at a modest

Figure 7. Example of a unit heater delivering a jet of warm air to the greenhouse environment (Photo by A. J. Both).

cost, but have inadequate heat distribution compared to hot water heating systems.

Hot Water Heating Systems

Water-based heating systems consist of a boiler and a water circulation system (pumps, mixing valves, and plumbing) (figure 8). The boiler generates the heat to raise the temperature of the circulating water. The heated water is pumped to heat the greenhouse through a pipe network or tube distribution system. Usually, the heating pipes are installed on the support posts, around the perimeter, and overhead (sometimes near gutters to enhance snowmelt using the released heat, and spaced evenly between more widely spaced gutters to provide uniform heat delivery). Some greenhouses have floor or bench heating with additional heating tubes installed in the floor or on/near the benches for root-zone heating. These root-zone heating systems have the advantage of providing independent control of root-zone temperatures and delivering uniform heat very close to the plant canopy. However, root-zone heating systems are typically not able to provide sufficient heating capacity during the coldest times of the year, necessitating the use of additional heating in the form of perimeter and overhead heating pipes. A significant benefit of water-based heating systems is the ability to "store" heat in large insulated water tanks. Boilers can be used during the day to produce CO_2 for plant consumption, with any surplus heat stored for use during colder periods such as the night, when CO_2 supplementation is not required.

Figure 8. Hot water heating system including a boiler, mixing valves, pumps, and distribution plumbing (Photo by A. J. Both).

Infrared Heating Systems

Infrared heating systems have the advantage of immediate heat delivery once turned on, but only exposed (in terms of line-of-sight) plant canopy surfaces will receive the radiant heat. Infrared heating sometimes provides non-uniform heating, especially in crops with a multi-layered canopy. Also, infrared heating systems are typically designed as line sources and require some distance between the source and the radiated canopy surfaces to accomplish uniform distribution. Finally, like hot air systems, infrared heating systems accumulate little heat storage during operation, so that in case of an emergency shutdown, little residual heat is available to extend the heating time before the temperature drops below critical levels.

Alternative Energy Sources and Energy Conservation

The volatility in fossil fuel prices experienced during the last decades has put a greater emphasis on energy conservation and alternative energy sources. Energy conservation measures employed include relatively simple measures such as sealing unintended cracks and openings in the greenhouse glazing, improved insulation of structural components and heat transportation systems, and timely equipment maintenance, as well as more advanced measures such as movable insulation/shade curtains, new heating equipment with higher efficiencies (e.g., condensing boilers, heat pumps, combined heat and power systems), and novel control strategies (e.g., temperature integration, where growers are more concerned with the average temperature a crop receives, within set boundaries, rather than tightly maintaining a specific set point temperature). Some growers delay crop production to periods when the weather is warmer, while others use lower set point temperatures (often requiring more extended production periods and with potential impacts on plant physiology).

Alternative energy heating sources (i.e., non-fossil fuels) used for greenhouse applications include solar electric, solar thermal, wind, hydropower, biomass, and geothermal (co-generation and ground-source, shallow or deep). Many alternative energy installations are viable only under specific climatic conditions and may require significant investments that may require (local or national) financial incentives. Developing energy conservation and alternative energy strategies for greenhouse operations remains challenging because of the considerable differences in size, scope, and local circumstances. Selecting an alternative energy system includes considering economic viability for the greenhouse operation as well as protection of the environment.

Evaporative Cooling Systems

Growers or greenhouse managers often use evaporative cooling as the most affordable way of reducing the air temperature beyond what the ventilation system can achieve by air movement only. The maximum amount of cooling provided by evaporative cooling systems depends on the initial temperature and humidity of the ambient (i.e., outdoor) air. These parameters can be measured relatively easily with a standard thermometer and a relative humidity sensor. With these measurements, the psychrometric chart can be used to determine the corresponding properties of the air, such as the wet-bulb temperature, humidity ratio, enthalpy, etc. With the known wet-bulb temperature, the wet-bulb depression can be calculated to determine the theoretical temperature drop possible by evaporative cooling. Since few engineered systems are 100% efficient, the actual temperature drop achieved by the evaporative cooling system is likely to be 80–90% of the theoretical wet-bulb depression.

Lighting System Design

The concepts described earlier can be used to control the instantaneous intensity and integrated light intensities needed to assess the light conditions in plant growth facilities. The information can be used to determine the parameters needed to select fixtures to modify the light environment in plant growth facilities, e.g., switching the supplemental lighting system on or off, opening or closing a shade curtain (in greenhouses) and, when multi-spectral LEDs are used, can include changing the light spectrum and/or their overall intensity.

Light Requirements

In designing a lighting system for a greenhouse or indoor growing facility, the first step is to determine the light requirements of a particular crop. Research articles or grower handbooks for the crop of interest can provide information about the recommended light intensity and/or the optimum daily light integral (see, for example, Lopez and Runkle, 2017). For crops such as leafy greens grown in a greenhouse, the minimum daily target integral may be as low as 8 to 14 mol m^{-2} d^{-1}, or as high as 17 mol m^{-2} d^{-1} (the maximum daily integral for leaf lettuce before physiological damage occurs as a result of too much light). For vine crops, such as tomatoes, a minimum of 15 mol m^{-2} d^{-1} is typically tolerated, while the optimum target can exceed 30 mol m^{-2} d^{-1}. Generally, as a rule of thumb, for vegetable crops, a 1% increase in the DLI results in a 1% increase in growth (up to a point; Marcelis et al., 2006). Considering the high cost of providing the optimum growing environment, it usually makes economic sense to optimize plant growth whenever possible (Kubota et al., 2016).

Once the DLI for the crop has been determined, the next step is to determine how much supplemental lighting is required to make up any shortfall in natural light. In an indoor growing facility, all light must be supplied by electric lamps, while in a greenhouse, natural lighting typically provides the bulk of the DLI throughout the year. Even in relatively gloomy regions the sun can provide over 70% of the required light for a year-round greenhouse lettuce crop.

Supplemental lighting for greenhouse production is mostly used during the dark winter months when the sun is low and the days are short. Typically, greenhouse lighting systems are designed such that they can provide enough light during the darkest months of the year. To estimate the amount of light available for crop production at a particular location, ideally one would average several years of data so that an atypical year would have a minor impact on the overall trends. In the U.S., a useful resource is the National Solar Radiation Database maintained by the National Renewable Energy Laboratory (NREL) in Golden, Colorado (https://nsrdb.nrel.gov/).

The solar radiation data (i.e., shortwave radiation covering the waveband of approximately 300–3,000 nm) available from NREL is not specifically used for plant production and is usually expressed in units of J m^{-2} per unit of time (e.g., an hour or a day). To convert this to a form more useful for planning supplemental lighting systems, the following multiplier can be used (Ting and Giacomelli, 1987):

$$1\frac{\text{MJ}}{\text{m}^2\text{day}} \text{ short wave radiation} = 2.0804\frac{\text{mol}}{\text{m}^2\text{day}} \text{ PAR} \tag{5}$$

The NREL database covers several locations outside of the USA. For more specific location data, other weather databases maintained by national governments or local weather stations (e.g., radio or TV stations, airports) may have historic solar radiation data available from which average natural daily light integrals can be calculated.

For greenhouse production, the DLI does not have to be exactly the same each day to maximize production. During the seedling stage, many crops can tolerate DLIs much higher than during later stages of growth. For example, greenhouse lettuce typically is limited to 17 mol m^{-2} d^{-1} after the canopy has closed, to avoid damage from tipburn (Albright et al., 2000). However, seedlings can be provided with 20 mol m^{-2} d^{-1} and some varieties may even benefit from up to 30 mol m^{-2} d^{-1}. Generally, for hydroponic lettuce, deviating no more than 3 mol m^{-2} d^{-1} from the target DLI is acceptable, provided any surplus (or deficit) is compensated for over the following two days.

Once the amount of supplemental lighting necessary has been determined (whether 100% of the DLI for an indoor growing facility or some other fraction of the DLI for a greenhouse), the next step is to determine what intensity of light is required. For indoor facilities, determining the required crop light level is straightforward. For a crop such as lettuce where there is no requirement for a night break, 24 hours of light per day can be applied. For a greenhouse, the calculation is the same, however, a portion of the DLI will be supplied by natural light. It comes down to a judgement call by the designer with respect to how they want to size the lighting system, and if they want to over- or under-size the lighting capacity to consider extremely dark days when the supplemental lighting system would need to provide nearly all of the light in a greenhouse. Most commercial greenhouse supplemental lighting systems provide an instantaneous intensity between 50 and 200 µmol m^{-2} s^{-1} at crop level.

Figure 9 shows the increase in DLI that can be realized by adding supplemental lighting at three different intensities (50, 100, and 150 µmol m^{-2} s^{-1}), while operating the lamps for 18 hours per day during November, December, January, and February, for 12 hours per day during October, for 11 hours per day during March, and 2 hours per day during September and April for a total of 2,993 hours per year. As shown in figure 9, using this lighting schedule and an intensity of 150 µmol m^{-2} s^{-1} results in a more constant light integral over the course of a year.

A significant factor affecting the hours per day that supplemental lighting should be supplied is electricity pricing. Many utilities offer incentives to encourage off-peak usage of electricity, to even out the demand for electricity to all of their customers. It varies by utility providers, but savings as high as 40% on the supply charges of electricity are common for purchasing off-peak power. Typical off-peak periods correspond with nighttime and early morning, for example from 9:00 pm to 7:00 am (10 hours). In addition to saving on the supply price of electricity, it may be possible to

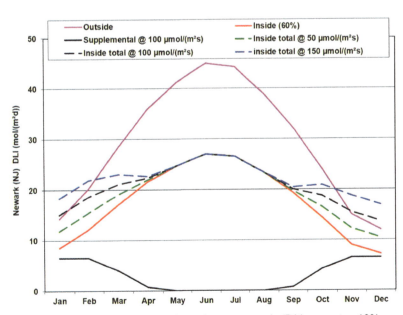

Figure 9. Outside and inside solar radiation integrals (DLI, assuming 60% transmission and averaged by month) for Newark, NJ, USA. The dashed lines indicate the inside radiation integrals after operating a supplemental lighting system at three different light intensities (50, 100, and 150 µmol m^{-2} s^{-1}) for different periods of time. See text for lighting system operating times.

avoid demand charges as well. In commercial operations that use a lot of power, electric utilities often collect demand charges based on the largest 15-minute on-peak consumption (kW) during a monthly billing cycle. The demand charge can easily add thousands of dollars to the monthly cost of a grower's electricity bill. During winter months, it may be unavoidable to light during peak use hours, but during the shoulder months when supplemental lighting is still necessary but is not used as much, it may be worthwhile to disable lighting during on-peak hours and make up any daily deficit the next day during off-peak hours.

Number of Fixtures to Achieve a Target Intensity

The number of fixtures needed to provide the desired intensity depends on the light output of each fixture and the mounting height. In addition, the characteristics of any reflector will affect both the uniformity and intensity of light delivered to the crop (Ciolkosz et al., 2001; Both et al., 2002). The mounting height is defined as the distance between the bottom of the lamp and the top of the plant canopy.

Ideally, the lighting manufacturer will have available an IES (Illuminating Engineering Society) file that contains data on the specific light output pattern of the fixture. Using the IES file and commercially available software, it is possible to design a layout to achieve a target light intensity at a specified mounting height. An additional consideration is the uniformity of the light. Ideally, the light should be as uniform as possible to produce consistent growth throughout the growing area. Keep in mind that, although the light intensity does not change much once the lamp density is determined (table 5), light uniformity significantly improves with increasing mounting height. For example, a 0.4 ha greenhouse (assuming an available mounting height of 2.44 m) would need approximately 383 HPS lamps (400 W each, not including the power drawn by the ballast) for a uniform light intensity of 49 $\mu mol\ m^{-2}\ s^{-1}$ and 786 lamps for the intensity of 100 $\mu mol\ m^{-2}\ s^{-1}$. Additional mounting patterns and resulting average light intensities are shown in table 5.

Table 5. Estimated average light intensities at the top of the plant canopy (in $\mu mol\ m^{-2}\ s^{-1}$) throughout a 0.4 ha greenhouse (10 gutter-connected bays of 7.3 m wide by 54.9 m long) for four different mounting heights and 400-watt HPS lamps. Note that these average light intensities are estimates without including edge effects (i.e., a drop in light intensity toward the outside walls) and these light intensities are estimates only; always consult with a trained lighting designer for an accurate calculation of expected light intensities in greenhouses.

Number of Lamps per Bay (lamps per row, with lamp placement staggered from row to row)	Floor Area per Lamp (m^2)	Light Intensity for a Mounting Height of 2.44 m ($\mu mol\ m^{-2}\ s^{-1}$)	Light Intensity for a Mounting Height of 2.13 m ($\mu mol\ m^{-2}\ s^{-1}$)	Light Intensity for a Mounting Height of 1.83 m ($\mu mol\ m^{-2}\ s^{-1}$)	Light Intensity for a Mounting Height of 1.52 m ($\mu mol\ m^{-2}\ s^{-1}$)
38 (13-12-13)	10.6	49	50	51	52
58 (15-14-15-14)	6.9	75	77	79	80
78 (16-15-16-15-16)	5.15	100	103	105	107
123 (21-20-21-20-21-20)	3.26	149	154	158	162
158 (23-22-23-22-23-22-23)	2.54	202	206	210	213

An additional consideration in greenhouses is that increasing the number of fixtures results in additional blockage of the natural light. Furthermore, power supply wires, ballasts, and reflectors can all block the transmission of natural light, and the greenhouse may require additional superstructure to provide a place to mount the fixtures and help support their weight.

Examples

Example 1: Greenhouse heating

Problem:
Determine the required heating system capacity for a greenhouse with the following characteristics:

- greenhouse dimensions: 330 by 330 m
- greenhouse surface area (roof plus sidewalls): 12,700 m²
- greenhouse volume: 50,110 m³
- outdoor humidity level: 45%
- nighttime temperature set point: 17°C
- indoor humidity level: 75%
- 99% design temperature (location specific): –15°C
- greenhouse U-value: 6.2 W m^{-2} °C^{-1}

Solution:
The required heating system capacity is a function of the structural heat loss (conduction, convection, and radiation), the infiltration heat loss, and the conversion efficiency of the fuel source for the heating system.

First, calculate the structural heat loss using equation 1:

$$q_{ccr} = U A_c (t_i - t_o) \qquad (1)$$

$$= 6.2 \times 12,700 \, [17 - (-15)] = 2,519,680 \text{ W} = 2,519.68 \text{ kW}$$

Next, determine the heat loss by infiltration using equation 2:

$$q_i = \rho_i NV \left[c_{pi} (t_i - t_o) + h_{fg} (W_i - W_o) \right] \qquad (2)$$

Some assumptions are required to solve equation 2. It is reasonable to assume that the air density of the greenhouse air is 1.2 kg m^{-3}. The infiltration rate N can be estimated from data included in table 2: a value of 0.0004 s^{-1} was selected (an older, glass-covered greenhouse with good maintenance). In order to determine the humidity ratios for the inside and outside air, we need to use the relative humidity of the inside and outside air. Using the psychrometric chart (figure 1), the humidity ratios for the inside and outside air are 0.0091 and 0.0005 kg kg^{-1}, respectively. The specific heat of greenhouse air at 17°C is 1.006 kJ kg^{-1} K^{-1} and the latent heat of vaporization of water at that temperature is approximately 2,460 kJ kg^{-1}. These values were determined from online calculators (Engineering ToolBox, 2004, 2010), but can also commonly be found in engineering textbooks regarding heat and mass transfer. Entering these values in equation 2:

$$q_i = \rho_i NV \left[c_{pi}(t_i - t_o) + h_{fg}(W_i - W_o) \right] \quad (2)$$

$$= 1.2 \times 0.0004 \times 50{,}110 \, \{1.006[17 - (-15)] + 2{,}460(0.0091 - 0.0005)\}$$

$$= 1{,}283{,}169 \text{ W} = 1{,}283.17 \text{ kW}$$

Thus, the resulting heat loss is the sum of the structural heat loss (conduction, convection, and radiation) and the infiltration heat loss: 2,519.68 + 1,283.17 = approximately 3803 kW.

The heating system capacity is the total heat loss divided by the conversion efficiency of the fuel source. For natural gas with a conversion efficiency of 85%, the required overall heating system capacity is 3803/0.85 = 4,474 kW.

Note that if these calculations are done in a spreadsheet, it is easy to adjust the assumptions that were made so that the sensitivity of the final answer to the magnitude of the assumptions can be assessed. Also, in colder climates, additional heat can be lost around the perimeter of a greenhouse where cold and wet soil is in direct contact with the perimeter walkway inside the greenhouse. To prevent this perimeter heat loss, vertically placed insulation boards can be installed extending from ground level to a depth of 50–60 cm.

Example 2: Evaporative Cooling Pad

Problem:
Find the expected temperature drop of the air passing through the evaporative cooling pad given the following information:

- the ambient (outside air) is at 25°C dry-bulb temperature and 50% relative humidity
- the evaporative cooling pad efficiency is 80%

Solution:
Using the psychrometric chart (figure 10) and the initial conditions of the outside air of 25°C dry-bulb temperature and 50% relative humidity, start at the intersection of the curved 50% RH line with the vertical line for a dry-bulb temperature of 25°C. At this intersection, determine the following environmental parameters:

- wet-bulb temperature = 17.8°C (from the starting point, follow the constant enthalpy line, 50.3 kJ kg^{-1} in this case, until it intersects with the 100% relative humidity curve)
- dew point temperature = 13.7°C
- humidity ratio = 0.0099 kg kg^{-1},

- enthalpy = 50.3 kJ kg^{-1}
- specific volume = 0.858 m^3 kg^{-1}

Thus, the wet-bulb depression for this example equals 25 − 17.8 = 7.2°C. Using an overall evaporative cooling system efficiency of 80% results in a practical temperature drop of approximately 5.8°C (7.2°C × 0.8). As the air continues to travel through the greenhouse on its way to the exhaust fans, the exiting air will be warmed, and moisture from crop transpiration will be added so the exiting air will have higher energy content and specific humidity than the air moving through the evaporative cooling pad.

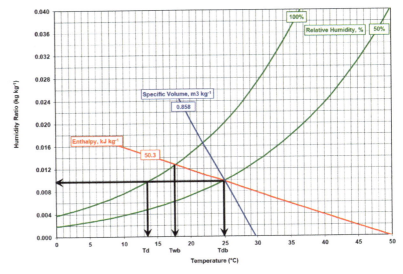

Figure 10. Simplified psychrometric chart used to visualize the evaporative cooling example described in the text. T_d = dew point temperature, T_{wb} = wet-bulb temperature, and T_{db} = dry-bulb temperature.

Example 3: Purchase and operating costs of crop lighting systems

Problem:

As mentioned previously, the performance of lamps in terms of their efficacy can vary significantly even when comparing the same type of lamp. For example, we measured HPS fixture efficacy values ranging from 0.94 to 1.7 µmol J^{-1}. Along with the efficacy, the unit cost of purchasing lamps is also an important consideration when deciding on a lighting system. In this example, we look at the cost of purchasing and operating two types of lighting systems, in both a greenhouse and an indoor growing facility.

Find the yearly cost savings of operating an LED vs. HPS system, and how long the systems should be operated to justify (payback) the higher purchase price of the LED lighting system, in both a greenhouse and indoor (no natural light) production system, given the following:

- HPS lighting system: 123 fixtures, each 400 W (plus 60 W for each ballast), cost of $300 per fixture (excluding installation cost), efficacy of 0.94 µmol J^{-1}
- LED lighting system: 55 fixtures, each 400 W (plus 20 W for each driver), cost of $1,200 per fixture (excluding installation cost), efficacy of 2.1 µmol J^{-1} (these LED fixtures are intended as a direct replacement for the HPS lighting system, meaning they deliver the same PAR intensity and distribution at crop level)
- Greenhouse: 2,200 hours of supplemental lighting per year (600 h during on-peak electricity rates and 1,600h during off-peak electricity rates)
- Indoor (no natural light) growing facility: 8,760 hours of lighting per year (5,100 h on-peak and 3,660 h off-peak)
- Electricity prices of $0.14 per kWh on-peak, and $0.09 per kWh off-peak.

Solution:

We can now compare the cost to purchase and operate the fixtures. The purchase price of the two systems is simply the unit cost multiplied by the number of units:

$$\text{HPS purchase cost} = \frac{\$300}{\text{fixture}} \times 123 \text{ fixtures} = \$36,900$$

$$\text{LED purchase cost} = \frac{\$1,200}{\text{fixture}} \times 55 \text{ fixtures} = \$66,000$$

For the greenhouse case, the electricity cost of the two lighting systems can be determined for both on-peak and off-peak use:

$$\text{HPS on-peak cost} = 123 \text{ fixtures} \times \frac{460 \text{ W}}{\text{fixture}} \times \frac{1 \text{ kW}}{1000 \text{ W}} \times \frac{600 \text{ h on-peak}}{\text{year}} \times \frac{\$0.14}{\text{kWh}} = \frac{\$4,753}{\text{year}}$$

$$\text{HPS off-peak cost} = 123 \text{ fixtures} \times \frac{460 \text{ W}}{\text{fixture}} \times \frac{1 \text{ kW}}{1000 \text{ W}} \times \frac{1,600 \text{ h off-peak}}{\text{year}} \times \frac{\$0.09}{\text{kWh}} = \frac{\$8,148}{\text{year}}$$

Adding these costs results in an annual electricity cost for HPS of $12,901 per year (excluding any potential demand charges).

$$\text{LED on-peak cost} = 55 \text{ fixtures} \times \frac{420 \text{ W}}{\text{fixture}} \times \frac{1 \text{ kW}}{1000 \text{ W}} \times \frac{600 \text{ h on-peak}}{\text{year}} \times \frac{\$0.14}{\text{kWh}} = \frac{\$1,940}{\text{year}}$$

$$\text{LED off-peak cost} = 55 \text{ fixtures} \times \frac{420 \text{ W}}{\text{fixture}} \times \frac{1 \text{ kW}}{1000 \text{ W}} \times \frac{1,600 \text{ h off-peak}}{\text{year}} \times \frac{\$0.09}{\text{kWh}} = \frac{\$3,326}{\text{year}}$$

Adding these costs results in an annual electricity cost for LED of $5,266 per year (excluding any potential demand charges).

The annual cost savings for electricity consumption by using the LED instead of the HPS fixtures amounts to $12,901 − $5,266 = $7,635.

The premium for purchasing LED instead of the HPS fixtures is $29,100 ($66,000 − $36,900). Therefore, it would take $\frac{\$29,100}{\$7,635} = 3.8$ years of operation to recover (pay back) the higher purchase price of the LED fixtures in the greenhouse situation.

For the case of an indoor growing facility, where all of the lighting had to be supplied by the lamp fixtures, and assuming the lights needed to operate

24 hours a day to meet the target light integral, the annual cost savings for electricity consumption by using the LED instead of the HPS fixtures amounts to \$34,933 (\$50,035 − \$24,102). Therefore, it would take $\frac{\$29{,}100}{\$34{,}933} = 0.83$ years of operation to recover (pay back) the higher purchase price of the LED fixtures.

Image Credits

Figure 1. Both, A. J. (CC By 4.0). (2020). Psychrometric chart.
Figure 2. Both, A. J. (CC By 4.0). (2020). Evaporative cooling system (pad and fan).
Figure 3. Both, A. J. (CC By 4.0). (2020). Gutter-connected greenhouses.
Figure 4. Both, A. J. (CC By 4.0). (2020). Sensitivity of human eye and plant.
Figure 5. Left. Both, A. J. (2020). Lettuce plants.
Figure 5. Right. Cornell University. (CC By 4.0). Lettuce Plants. Retrieved from https://urbanagnews.com/blog/prevent-tipburn-on-greenhouse-lettuce/. [Fair Use].
Figure 6. Both, A. J. (CC By 4.0). (2020). HPS and LED fixtures.
Figure 7. Both, A. J. (CC By 4.0). (2020). Unit heater.
Figure 8. Both, A. J. (CC By 4.0). (2020). Hot water heating system.
Figure 9. Both, A. J. (CC By 4.0). (2020). Solar radiation integrals.
Figure 10. Both, A. J. (CC By 4.0). (2020). Simplified psychrometric chart.

References

Albright, L. D., Both, A. J., & Chiu, A. J. (2000). Controlling greenhouse light to a consistent daily integral. *Trans. ASAE, 43*(2), 421-431. https://doi.org/10.13031/2013.2721.
Aldrich, R. A., & Bartok, J. W. (1994). Greenhouse engineering. NRAES Publ. No. 33. Retrieved from https://vdocuments.site/fair-use-of-this-pdf-file-of-greenhouse-engineering-nraes-33-by-.html.
ASABE Standards. (2017). ANSI/ASABE S640: Quantities and units of electromagnetic radiation for plants (photosynthetic organisms). St. Joseph, MI: ASABE.
ASAE Standards. (2003). ANSI/ASAE EP406.4: Heating, ventilating, and cooling greenhouses. Note: This is a withdrawn and archived standard. St. Joseph, MI: ASAE.
ASHRAE. (2013). ASHRAE Standard 169-2013: Weather data for building design standards. Atlanta, GA: ASHRAE.
Both, A. J., Albright, L. D., Langhans, R. W., Vinzant, B. G., & Walker, P. N. (1997). Electric energy consumption and PPF output of nine 400 Watt high pressure sodium luminaires and a greenhouse application of the results. *Acta Hortic., 418*, 195–202.
Both, A. J., Benjamin, L., Franklin, J., Holroyd, G., Incoll, L. D., Lefsrud, M. G., & Pitkin, G. (2015). Guidelines for measuring and reporting environmental parameters for experiments in greenhouses. *Plant Methods, 11*(43). https://doi.org/10.1186/s13007-015-0083-5.
Both, A. J., Bugbee, B., Kubota, C., Lopez, R. G., Mitchell, C., Runkle, E. S., & Wallace, C. (2017). Proposed product label for electric lamps used in the plant sciences. *HortTechnol., 27*(4), 544-549. https://doi.org/10.21273/horttech03648-16.
Both, A. J., Ciolkosz, D. E., & Albright, L. D. (2002). Evaluation of light uniformity underneath supplemental lighting systems. *Acta Hortic., 580*, 183–190.
Ciolkosz, D. E., Both, A. J., & Albright, L. D. (2001). Selection and placement of greenhouse luminaires for uniformity. *Appl. Eng. Agric., 17*(6), 875–882. https://doi.org/10.13031/2013.6842.
Commission Internationale de l'Eclairage. (1931). *Proc. Eighth Session*. Cambridge, UK: Cambridge University Press.

Crawley, D. B., Lawrie, L. K., Winkelmann, F. C., Buhl, W. F., Huang, Y. J., Pedersen, C.O., . . . Glazer, J. (2001). EnergyPlus: Creating a new-generation building energy simulation program. *Energy Build., 33*(4), 319-331. http://dx.doi.org/10.1016/S0378-7788(00)00114-6.

Engineering ToolBox. (2004). Air—Specific heat at constant pressure and varying temperature. Retrieved from https://www.engineeringtoolbox.com/air-specific-heat-capacity-d_705.html.

Engineering ToolBox. (2010). Water—Heat of vaporization. Retrieved from https://www.engineeringtoolbox.com/water-properties-d_1573.html.

Kubota, C., Kroggel, M., Both, A. J., Burr, J. F., & Whalen, M. (2016). Does supplemental lighting make sense for my crop? —Empirical evaluations. *Acta Hortic., 1134*, 403-411. http://dx.doi.org/10.17660/ActaHortic.2016.1134.52.

Lopez, R., & Runkle, E. S. (Eds.). (2017). Light management in controlled environments. Willoughby, OH: Meister Media Worldwide.

Marcelis, L. F. M., Broekhuijsen, A. G. M., Meinen, E., Nijs, E. M. F. M., & Raaphorst, M. G. M. (2006). Quantification of the growth response to light quantity of greenhouse grown crops. *Acta Hortic., 711*, 97-103. https://doi.org/10.17660/ActaHortic.2006.711.9.

Mitchell, C. A., Both, A. J., Bourget, C. M., Burr, J. F., Kubota, C., Lopez, R. G., . . . Runkle, E. S. (2012). LEDs: The future of greenhouse lighting! *Chron. Hortic., 52*(1), 6-12.

Mitchell, C. A., Dzakovich, M. P., Gomez, C., Lopez, R., Burr, J. F., Hernandez, R., . . . Both, A. J. (2015). Light-emitting diodes in horticulture. *Hortic. Rev., 43*, 1-87.

Sager, J. C., Smith, W. O., Edwards, J. L., & Cyr, K. L. (1988). Photosynthetic efficiency and phytochrome photoequilibria determination using spectral data. *Trans. ASAE, 31*(6), 1882-1889. https://doi.org/10.13031/2013.30952.

Ting, K. C., & Giacomelli, G. A. (1987). Availability of solar photosynthetically active radiation. *Trans. ASAE, 30*(5), 1453-1457. https://doi.org/10.13031/2013.30585.

USDA-ARS. (2019). Virtual grower 3.0. Washington, DC: U.S. Department of Agriculture. https://www.ars.usda.gov/research/software/download/?softwareid=309.

Wallace, C. & Both, A. J. (2016). Evaluating operating characteristics of light sources for horticultural applications. *Acta Hortic., 1134*, 435-443. https://doi.org/10.17660/ActaHortic.2016.1134.55.

Building Design for Energy Efficient Livestock Housing

Andrea Costantino
TEBE Research Group
Department of Energy
Politecnico di Torino
Torino, Italy
and
Institute of Animal Science and Technology
Universitat Politècnica de València
València, Spain

Enrico Fabrizio
TEBE Research Group
Department of Energy
Politecnico di Torino
Torino, Italy

KEY TERMS		
Energy balance	Supplemental heating	Climate control
Mass balance	Heat recovery	Ventilation
Energy management	Cooling systems	

Variables

α = solar absorption coefficient
γ = Boolean variable
ε = direct saturation effectiveness
ρ = volumetric mass density
ϕ = heat flow
b_{tr} = transmission correction factor
c = specific heat capacity
C = total heat capacity
E = energy
F_{sh} = shading correction factor
g_{gl} = total solar energy transmittance of transparent surface
I = solar irradiance

$\dot{m}$ = water vapor production
M = mass
n_a = number of animals
n_{hpu} = number of heat-producing units
R = surface heat resistance
t = time
T = temperature
U = thermal transmittance
V = volume
$\dot{V}$ = ventilation flow rate
w = live weight
x = air specific humidity
Y_{eggs} = egg production
Y_{feed} = coefficient related to the daily feed energy intake
Y_{milk} = milk production
$Y_{pregnancy}$ = number of days of pregnancy

Introduction

Energy usage on farms is considered direct when used to operate machinery and climate control systems or indirect when is used to manufacture feed and agro-chemicals. Direct on-farm energy consumption was estimated to be 6 EJ yr^{-1}, representing about 1.2% of total world energy consumption (OECD, 2008). If indirect energy is included, total farm energy consumption could be as much as 15 EJ yr^{-1}, representing about 3.1% of global energy consumption. Housed livestock require adequate indoor climate conditions to maximize both production and welfare, particularly avoiding thermal stress. The task of the engineer is to improve the energy use efficiency of livestock housing and to minimize energy consumption. This can be achieved by improving the energy performance of the equipment used for climate control and the design of the building.

The focus of this chapter is on building design for efficient energy management in livestock housing. Improving building design requires understanding the mass and energy balance of the system to specify materials, dimensions, and equipment needed to maintain safe operating conditions. The importance of understanding the energy needs of buildings is illustrated by the report of St-Pierre et al. (2003), who estimated the economic losses by the dairy industry in the U.S. at $1.69 to $2.36 billion annually due to heat stress. Understanding and being able to use fundamental concepts for animal housing design provides the foundation for desirable welfare and more efficient production-centric animal housing.

> **Outcomes**
>
> After reading this chapter, you should be able to:
>
> - Describe the energy needs for livestock housing
> - Explain the energy management requirements of a livestock house
> - Describe the main climate control systems used for livestock housing and the features that affect the energy management
> - Calculate energy balances for livestock houses

Concepts

Energy and Mass Balance of a Livestock House

Thermodynamically, a livestock house is an open system that exchanges energy and mass (such as air, moisture, and contaminants) between the indoor and outdoor environments and the animals that occupy the internal volume (the enclosure). The law of conservation of energy and mass is the basic principle for the mass balance. The building walls, floor, and roof represent the *control surfaces* and enclose the *control volume* of the thermodynamic system represented by the livestock house and its internal surfaces, such as animals, interior walls, and equipment. Energy and mass balance equations allow the analysis of the thermal behavior of a livestock house, but calculating these balances is challenging because many factors affect the thermal behavior of these buildings. It is essential to understand which terms to consider, and what to assume as negligible.

Energy Balance

Sensible heat is the amount of heat exchanged by a body and the surrounding thermodynamic system that involves a temperature change. *Latent heat* is the heat absorbed or released by a substance during a phase change without a change in temperature. These two forms of heat can be illustrated using an example of heating a pot of water on a stove. Initially, the water is at room temperature (say, 25°C), and as the water is heated, its temperature increases. The heat causing the temperature increase is sensible heat and for water is equal to 4,186 kJ kg^{-1} K^{-1}. When the water temperature reaches 100°C (boiling point of water at atmospheric pressure), the water changes phase from liquid to gas (steam). The heat provided during the phase change breaks the molecular bonds of the liquid water to transition to the gas phase, but the temperature does not change. The heat supplied to effect phase change is latent heat. The latent heat of vaporization for a unit of mass of water is 2,272 kJ kg^{-1} at 100°C and atmospheric pressure.

The energy balance of a livestock house, considering only the sensible heat, can be written as follows (Panagakis and Axaopoulos, 2008):

$$\phi_a + \phi_{tr} + \phi_{sol} + \phi_f + \phi_v + \phi_m + \left(\gamma_{fog} \cdot \phi_{fog}\right) + \left(\gamma_H \cdot \phi_H\right) = \sum_{k=1}^{n}(M_{el,k} \cdot C_{el,k}) \cdot \frac{dT_{air,i}}{dt} \quad (1)$$

where ϕ_a = sensible heat flow from the animals inside the enclosure (W)

ϕ_{tr} = sensible heat flow due to transmission through the control surfaces but excluding the floor (W)

ϕ_{sol} = sensible heat flow due to solar radiation through both opaque and glazed building elements (W)

ϕ_f = sensible heat flow due to transmission through the floor (W)

ϕ_v = sensible heat flow due to ventilation (W)

ϕ_m = sensible heat flow from internal sources, such as motors and lights (W)

γ_{fog} = Boolean variable for the presence ($\gamma_{fog}=1$) or not ($\gamma_{fog}=0$) of a fogging system inside the livestock house

ϕ_{fog} = sensible heat flow due to fogging system (W)

γ_H = Boolean variable for the presence ($\gamma_H=1$) or not ($\gamma_H=0$) of a supplemental heating system inside the livestock house

ϕ_H = sensible heat flow due to supplemental heating system (W)

$M_{el,k}$ = mass of the kth building element (kg)

$C_{el,k}$ = total heat capacity of kth building element (kJ kg^{-1} K^{-1})

$\dfrac{dT_{air,i}}{dt}$ = variation of the indoor air temperature $T_{air,i}$ with time t

When using equation 1 for calculations and sizing, pay attention to the heat flows because each term could be positive or negative depending on the physical context. Usually, heat flows coming into a control volume (the animal house) are positive, and the ones flowing out are negative. For example, in equation (1), the terms ϕ_a and ϕ_{sol} are always positive or zero, since they represent incoming heat flow from animals and solar radiation, respectively, while the values of ϕ_{tr} and ϕ_v could be positive or negative, depending on the difference in temperature inside and outside the animal house. The term ϕ_f depends on the floor construction. Although ϕ_{tr} and ϕ_f are both transmission heat flows through the control surface, they are always separated. Estimating the heat transfer through the ground is very challenging (Albright, 1990; Panagakis and Axaopoulos, 2008; Costantino et al., 2017), for example, in pig houses with ventilated pits for manure storage. To simplify the energy balance, the term ϕ_{tr} is often considered as the sum of ϕ_{tr} and ϕ_f and a corrective coefficient is used when ϕ_f is calculated.

The term ϕ_{fog} is always negative because it represents the sensible heat removed by water droplets of a fogging system. A fogging system provides cooling inside the animal house by putting a haze of tiny water droplets in the air to provide evaporative cooling for the animals. The term ϕ_H is always positive. The parameters γ_{fog} and γ_H should not have a value of 1 at the same time but can both be 0.

Sensible heat from the animals, ϕ_a, depends on species and body mass and ambient temperature. Sensible and latent heat values can be found in the literature, for example from ASABE (2012), Hellickson & Walker (1983), or Lindley & Whitaker (1996), and more detailed data are available in Pedersen and Sällvik (2002), who express sensible and latent heat from animals as a function of animal weight, indoor air temperature, and animal activity. In complex animal houses the sensible heat flow from internal sources, such as motors (fans and automatic feeding systems) and lights term (ϕ_m) can be included (Albright, 1990), but in many calculations it is excluded because is very small compared with ϕ_a (Midwest Plan Service, 1987). That exclusion is further justified when energy-efficient technologies such as LED/gas-discharge lamps and brushless motors are used.

The product $M_{el} \cdot C_{el}$ is the lumped effective heat capacity of a building element expressed in kJ K^{-1}. For each building element (walls and roof) the amount or mass of material must be known, and the amount of heat energy needed to raise the temperature of a unit mass of the material by one degree Celsius. The fraction $\frac{dT_{air,i}}{dt}$ represents the variation of the indoor air temperature through time. This side of the equation represents the change in temperature of the building itself.

It is possible to include additional terms to equation 1 (Albright, 1990; Esmay and Dixon, 1986) such as the sensible heat flow to evaporate the water inside the control volume from structures such as water troughs and a slurry store (ϕ_e). Some authors consider it important (Hamilton et al., 2016), while others do not (Midwest Plan Service, 1987). Liberati and Zappavigna (2005) consider sensible heat exchange between manure (especially when collected in pits) and the air inside the enclosure (ϕ_{man}) to be important in large-scale houses equipped with storage pits and manure when it is not removed frequently. A Boolean variable γ_{man} may generalize equation 1 further.

Equation 1 is a dynamic energy balance. If a large time step (perhaps a week or more) is assumed it can be written for steady-state conditions, meaning that the state variables that describe the system can be considered constant with time, and the terms of the balance represent the average values for the system. For large time steps or in steady-state conditions with constant indoor and outdoor air temperature, heat accumulation by the building itself can be considered to be zero, so equation 1 becomes:

$$\phi_a + \phi_{tr} + \phi_{sol} + \phi_v + \gamma_{fog} \cdot \phi_{fog} = 0 \qquad (2)$$

To obtain the energy balance of a livestock house in cold condition requiring supplemental heating, the energy balance becomes:

$$\phi_a + \phi_{tr} + \phi_H + \phi_{sol} + \phi_v = 0 \qquad (3)$$

Figure 1 presents an illustration of the sensible heat balance of equation 3 for simple dairy cow housing. Equation 3 can be used to design a basic livestock house. Undoubtedly, the presented formulation is a simplification, and in literature, other terms are introduced in the energy balance. The calculation of each term of the energy balance of equation 3 is provided in greater detail later in this chapter.

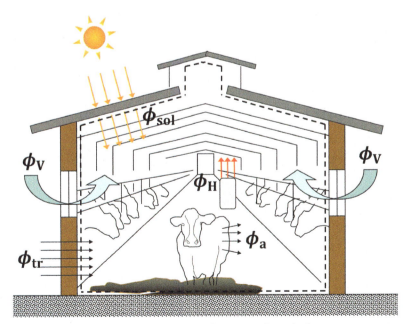

Figure 1. The sensible heat balance of equation 3 applied to a generic livestock house.

Mass Balance

Mass balances are necessary to plan the management of contaminants, such as carbon dioxide (CO_2), hydrogen sulfide (H_2S), and ammonia (NH_3), produced by the animals (Esmay and Dixon, 1986) and to regulate the indoor environment temperature, moisture content, and relative humidity. Along with temperature and relative humidity, the indoor air quality (IAQ) must be controlled by ventilation to avoid animal health problems. Calculating ventilation requirements for contaminant control is a mass balance problem. With low indoor air temperatures, a minimum ventilation flow rate (base ventilation) is used to dilute contaminants such as H_2S and NH_3. The minimum ventilation flow rate can be increased to reduce the moisture content. When the indoor air temperature is higher than the cooling setpoint temperature used to maintain animal comfort, the ventilation flow rate must be increased to cool the animals (Esmay and Dixon, 1986). The maximum ventilation flow rate must avoid high airspeeds that hurt animal welfare. If cooling cannot be achieved using mass flow, a fogging system can be used. The ventilation airflow can be expressed in $m^3 s^{-1}$, $m^3 h^{-1}$ or as ach (air changes per hour), which indicates how many times the volume of air inside the house is changed in one hour.

To estimate the ventilation flow rate for moisture control in a simple livestock house, equation 4 (Panagakis and Axaopoulos, 2008) can be used:

$$\dot{V}_{air} \cdot \rho_{air} \cdot \left(x_{air,i} - x_{air,o}\right) + \dot{m}_a + \left(\gamma_{fog} \cdot \dot{m}_{fog}\right) = \rho_{air,i} \cdot V_{air,i} \cdot \frac{dx_{air,i}}{dt} \quad (4)$$

where $\dot{V}_{air}$ = ventilation air flow rate ($m^3 s^{-1}$)
ρ_{air} = volumetric mass density ($kg\ m^3$)
$x_{air,o}$ = specific humidity of the outdoor air ($kg_{vapor}\ kg_{air}^{-1}$)
$x_{air,i}$ = specific humidity of the indoor air ($kg_{vapor}\ kg_{air}^{-1}$)
$\dot{m}_a$ = animal water vapor production ($kg_{vapor}\ s^{-1}$)
γ_{fog} = Boolean variable that indicates the presence of fogging system
$\dot{m}_{fog}$ = water added through fogging ($kg_{vapor}\ s^{-1}$)
$\rho_{air,i}$ = volumetric mass density of inside air ($kg\ m^{-3}$)
$V_{air,i}$ = volume of the inside air (m^3)
$\frac{dx_{air,i}}{dt}$ = variation of $x_{air,i}$ in time t

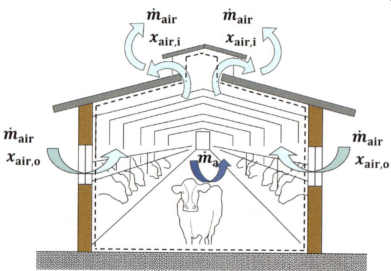

Figure 2. The vapor mass balance of equation 5 applied to a generic livestock house.

In steady-state conditions and not considering the presence of fogging systems, the mass balance (figure 2) can be simplified as:

$$\dot{m}_{air} \cdot x_{air,o} - \dot{m}_{air} \cdot x_{air,i} + \dot{m}_a = 0 \quad (5)$$

Equation 5 is the basic formulation of the moisture mass balance in steady-state conditions for livestock houses.

Energy Management Calculations

The basis for sensible energy and mass balance calculations for livestock housing are equations 3 and 5, respectively. In the following sections, the determination of each term of the energy balance (equation 3) is presented. A similar approach could be used for equation 5.

Heat Flow from the Reared Animals (ϕ_a)

The animals produce and contribute considerably to heat flow in their housing. In cool conditions, this heat flow can help warm the building and decrease the need for supplemental heat. In warm conditions, this heat flow should be removed to avoid overheating and causing animal heat stress. Animals need to emit heat (both sensible and latent heat) for regulating their body temperature and maintaining their body functions. As an animal grows (usually the desired outcome of a meat production system, but not for dairy and laying hens), the animal produces more heat. The amount of heat also depends on indoor air temperature, production targets (such as the mass of eggs, milk, or meat), and the energy concentration of the feedstuff. Estimating heat production is also essential to calculate ventilation requirements.

Standard values for heat production are available (CIGR, 1999; ASABE, 2012), but a specific calculation is possible (Pedersen and Sällvik, 2002). First the total heat produced, $\phi_{a,tot}$ (sum of the sensible and latent heat), for the animal house is calculated for an indoor air temperature of 20°C. The formulation of the equation depends on animal species and production:

Broilers:
$$\phi_{a,tot} = 10.62 \cdot w_a^{0.75} \cdot n_a \quad (6)$$

Laying hens:
$$\phi_{a,tot} = \left(6.28 \cdot w_a^{0.75} + 25 \cdot Y_{eggs}\right) \cdot n_a \quad (7)$$

Fattening pigs:
$$\phi_{a,tot} = \left\{5.09 + \left[1-\left(0.47+0.003 \cdot w_a\right)\right] \cdot \left[5.09 \cdot w_a^{0.75} \cdot \left(Y_{feed}-1\right)\right]\right\} \cdot n_a \quad (8)$$

Dairy cows:
$$\phi_{a,tot} = \left(5.6 \cdot w_a^{0.75} + 22 \cdot Y_{milk} + 1.6 \cdot 10^{-5} \cdot Y_{pregnancy}^3\right) \cdot n_a \quad (9)$$

where w_a = average animal live weight (kg)
n_a = number of animals inside the livestock house (animals)
Y_{eggs} = egg production (kg day^{-1}), usually between 0.04 (brooding production) and 0.05 kg day^{-1} (consumer eggs)
Y_{feed} = dimensionless coefficient related to the daily feed energy intake by the pigs (values of Y_{feed} are presented in table 1)
Y_{milk} = milk production (kg day^{-1})
$Y_{pregnancy}$ = number of days of pregnancy (days)

Next, the sensible heat produced (ϕ_a) at a given indoor air temperature is calculated. If the indoor air temperature is in the thermoneutral zone, that is, a temperature range where the animal heat dissipation is constant (Pedersen

Table 1. Values of Y_{feed} for fattening pigs (Pedersen and Sällvik, 2002).

Pig Body Mass (kg)	Y_{feed}		
	Rate of Gain: 700 g day^{-1}	Rate of Gain: 800 g day^{-1}	Rate of Gain: 900 g day^{-1}
20	3.03	3.39	3.39
30	2.79	3.25	3.25
40	2.60	3.22	3.43
50	2.73	3.16	3.41
60	2.78	3.16	3.40
70	2.84	3.12	3.40
80	2.83	3.04	3.38
90	2.74	2.79	3.18
100	2.64	2.57	2.98
110	2.52	2.40	2.78
120	2.36	2.25	2.60

and Sällvik, 2002) and the energy fraction used by animals for maintaining their homeothermy is at a minimum, at the house level ϕ_a can be calculated as:

Broiler house: $\phi_a = \left\{ 0.61 \cdot \left[1000 + 20 \cdot (20 - T_{air,i}) \right] - 0.228 \cdot T_{air,i}^2 \right\} \cdot n_{hpu}$ (10)

Laying hen house: $\phi_a = \left\{ 0.67 \cdot \left[1000 + 20 \cdot (20 - T_{air,i}) \right] - 9.8 \cdot 10^{-8} \cdot T_{air,i}^6 \right\} \cdot n_{hpu}$ (11)

Fattening pig house: $\phi_a = \left\{ 0.62 \cdot \left[1000 + 12 \cdot (20 - T_{air,i}) \right] - 1.15 \cdot 10^{-7} \cdot T_{air,i}^6 \right\} \cdot n_{hpu}$ (12)

Dairy cow house: $\phi_a = \left\{ 0.71 \cdot \left[1000 + 4 \cdot (20 - T_{air,i}) \right] - 0.408 \cdot T_{air,i}^2 \right\} \cdot n_{hpu}$ (13)

where $T_{air,i}$ = indoor air temperature (°C)
n_{hpu} = the number of heat-producing units (hpu) that are present inside the livestock house

One hpu is defined as the number of animals that produces 1000 W of total heat (sum of sensible and latent heat) at an indoor air temperature of 20°C and can be calculated as:

$$n_{hpu} = \frac{\phi_{a,tot}}{1000} \quad (14)$$

where $\phi_{a,tot}$ is calculated using equations 6, 7, 8, or 9 depending on species and production system. Out of the thermoneutral zone, no clear relationship can be found between indoor air temperature and total heat production, but values can be calculated using the formulations present in Pedersen and Sällvik (2002).

Transmission Heat Flow through the Building Envelope (ϕ_{tr})

The term ϕ_t is being taken to represent the heat flow through the walls, roof, windows, doors and floor. It is calculated as (European Committee for Standardization, 2007):

$$\phi_{tr} = \left[\sum_{j=1}^{n}\left(b_{tr,j} \cdot U_j \cdot A_j\right)\right] \cdot \left(T_{air,o} - T_{air,i}\right) \qquad (15)$$

where b_{tr} = dimensionless correction factor between 0 and 1
U_j = thermal transmittance of the j building element (W m^{-2} K^{-1})
A_j = total area of the j building element (m^2)
$T_{air,o}$ = outdoor air temperature (°C)

The factor b_{tr} is used to correct the heat flow when the forcing temperature difference is not the difference between the indoor and outdoor air, for example when the heat flow occurs toward unconditioned spaces (e.g. material storages and climate control rooms) or through the ground. In these cases, the air temperature difference between inside and outside can be used but the heat flow is decreased using b_{tr}. This coefficient can be computed in two cases: (1) if the adjacent space temperature is fixed and known, or (2) if all the heat transfer coefficients between the considered spaces can be numerically estimated. In most situations, b_{tr} (unitless) is obtained from standards, (e.g., table 2).

Table 2. Values of b_{tr} for different types of unconditioned spaces and floors (from EN 12831, European Committee for Standardisation, 2009).

Type of Unconditioned Space	b_{tr}
Space with 1 wall facing on the outdoor environment	0.40
Space with 2 walls facing on the outdoor environment (no doors)	0.50
Space with 2 walls facing on the outdoor environment (with doors)	0.60
Space with 3 walls facing on the outdoor environment (with doors)	0.80
Floor in direct contact with the ground	0.45
Ventilated floor (e.g. pits and under-floor cavity)	0.80

Heat Flow Due to a Supplemental Heating System (ϕ_H)

In most of the cases, ϕ_H is the unknown of the problem and the energy balance is solved with the aim of finding its value. A typical example is to solve the energy balance of equation 3 for finding ϕ_H and sizing the heating capacity of the supplemental heating system. In other cases, ϕ_H could be equal to zero and the unknown of the problem could be ϕ_V with the aim of finding the needed ventilation flow rate to maintain a certain indoor air temperature and to cool the reared animals. Rarely, ϕ_H has to be estimated. For example, ϕ_H has to be estimated when the energy balance is solved with the aim of evaluating the indoor air temperature in given specific boundary conditions. An easy way to estimate ϕ_H is to consider the heating capacity reported in the technical datasheet of the equipment for supplemental heating.

More details about the supplemental heating systems are described below, in the Application section.

Heat Flow from Solar Radiation (ϕ_{sol})

The heat flow due to solar radiation is dependent on the season, the farm location, and features of the building. In general terms, the solar heat flow can be split into two terms as follows (International Organization for Standardization, 2017):

$$\phi_{sol} = \sum_{n=1}^{q} \phi_{sol,op,q} + \sum_{n=1}^{k} \phi_{sol,gl,k} \qquad (16)$$

where $\phi_{sol,op,q}$ = heat flows on the q opaque (e.g. walls and roof) surfaces (W)
$\phi_{sol,gl,k}$ = heat flows on the k glazed (windows) surfaces (W)

For a generic opaque surface $\phi_{sol,op,q}$ is calculated as:

$$\phi_{sol,op,q} = A_q \cdot U_q \cdot \alpha_q \cdot R_{ex} \cdot I_{sol,q} \cdot F_{sh,q} \qquad (17)$$

where α_q = solar absorption coefficient of the considered surface depending on the surface color (0.3 for light colors, 0.9 for dark colors)
R_{ex} = external surface heat resistance (m² K⁻¹ W⁻¹), generally assumed equal to 0.04 m² K⁻¹ W⁻¹
$I_{sol,q}$ = solar irradiance incident on the considered surface (W m⁻²)
$F_{sh,q}$ = shading correction factor

For a generic glazed surface k, $\phi_{s,gl,k}$ is calculated as:

$$\phi_{s,gl,k} = A_k \cdot g_{gl} \cdot I_{sol,k} \cdot (1 - F_{fr}) \cdot F_{sh,k} \cdot F_{sh,gl,k} \qquad (18)$$

where g_{gl} = total solar energy transmittance of the transparent surface
F_{fr} = frame area fraction
$F_{sh,gl,k}$ = shading reduction factor for movable shading provisions

The shading factors for both opaque and glazed components can be excluded for most livestock housing because they increase the complexity of the calculation, but they do not greatly affect the results.

Heat Flow Due to the Ventilation System (ϕ_v)

The heat load due to the ventilation system can be expressed as

$$\phi_v = \rho_{air} \cdot c_{air} \cdot \dot{V} \cdot (T_{air,sup} - T_{air,i}) \qquad (19)$$

where ρ_{air} = air volumetric mass density (kg m⁻³)
c_{air} = air specific heat capacity (W h kg⁻¹ K⁻¹)
$\dot{V}$ = ventilation flow rate (m³ h⁻¹)
$T_{air,sup}$ = supply air temperature (°C)

In the cool season, $T_{air,sup}$ usually has the same value of $T_{air,o}$, since the ventilation uses outdoor air. In the warm season, $T_{air,sup}$ could have values lower than $T_{air,o}$, since outdoor air is cooled before entering inside the building. The value of $T_{air,sup}$ can be estimated using the direct saturation effectiveness ε (%) of an evaporative pad system, calculated as (ASHRAE, 2012):

$$\varepsilon = 100 \cdot \frac{T_{air,o,db} - T_{air,sup,db}}{T_{air,o,db} - T_{air,o,wb}} \qquad (20)$$

where $T_{air,o,db}$ = dry-bulb outdoor air temperature (°C)
$T_{air,sup,db}$ = dry-bulb temperature of the supply air leaving the cooling pad (°C)
$T_{air,o,wb}$ = wet-bulb temperature of the outdoor air entering in the pad (°C)

Equation 20 can be rearranged to estimate the air supply temperature ($T_{air,sup,db}$) in presence of evaporative pads for use in equation 19.

Applications

The concepts describe the basis for calculating the energy balance of a simple animal house. These are usually quite straightforward structures built to standard designs, which differ around the world but serve a similar function of making animal production more efficient for the farmer. The calculation for the design of the animal house (the control structure) necessarily assumes a typical or average environment. In reality, weather and production management mean that there have to be components of the system that are dynamic and respond to external conditions. In this section, some of the technology required to help maintain a safe and efficient living environment for the animals are discussed.

Heating Animal Houses

Supplemental Heating Systems

In cold weather, a supplemental heat source may be needed to reach the air setpoint temperature for guaranteeing adequate living conditions for the livestock. This is common at the beginning of the production cycle when animal heat production is small and in cold seasons of the year. This energy consumption represents a major fraction of the total direct energy consumption of the farm (table 3) and can be calculated using equation 3.

Supplemental heating systems can be classified into localized heating and space heating systems. Localized heating systems create temperature variations in the zones where animals are reared. This allows young animals to move to a zone for optimum thermal comfort. To design a localized heating system, the term ϕ_m (as used in equation 1) would have to be factored into the calculation to account for heat flow between the internal zones. Localized heating usually uses radiant heat, such as infrared lamps (for piglets) or infrared gas catalytic radiant heaters (for broilers). These systems emit 70% of their heat by radiation and the remaining 30% by convection; the radiation component directly heats the animals and floor while the convection component heats the air.

Space heating systems create a more uniform thermal environment. They are easier to design, manage, and control than localized heating systems, but they tend to have higher energy consumption. Space heating is usually based on a convection system using warm air. Heat is produced in boilers or furnaces and then is transferred into the building when needed.

An alternative is to use direct air heating in the house. Direct heating can be cheaper to install, but requires more maintenance to deal with contaminants, dust, and moisture (Lindley and Whitaker, 1996). Also, there is a need to vent

Table 3. List of energy uses and their percentages of the total energy consumption of different types of livestock houses in Italy (Rossi et al., 2013).

Livestock House	Operation	Percentage of Electrical Energy (of the total)	Percentage of Thermal Energy (of the total)
Broiler Houses	ventilation	39%	-
	supplemental heating	27%	96%
	lighting	9%	-
	feeding distribution	20%	-
	litter distribution and manure removal	-	3%
	manure transportation and disposal	-	1%
	product collecting and package	5%	-
Laying Hen Houses	ventilation	44%	-
	supplemental heating	-	-
	lighting	15%	-
	feeding distribution	5%	-
	litter distribution and manure removal	2%	33%
	manure treatment	27%	-
	manure transportation and disposal	-	67%
	product collecting and package	7%	-
Pig Houses	ventilation and supplemental heating	48%	69%
	lighting	2%	-
	feeding preparation	11%	-
	feeding distribution	19%	-
	litter care and manure removal	4%	1%
	manure treatment	4%	-
	manure transportation and disposal	12%	30%
Dairy Cow Houses	ventilation	20%	-
	lighting	8%	-
	feeding	17%	52%
	milking	16%	6%
	milk cooling	12%	-
	litter care	-	7%
	manure removal	8%	5%
	manure treatment	18%	4%
	manure transportation and disposal	1%	26%

exhaust fumes and CO_2 so ventilation flow rates have to be increased, requiring more energy consumption (Costantino et al., 2020). In other agricultural buildings, such as greenhouses, the warm air is recirculated to decrease energy consumption. In livestock houses this practice is strongly not recommended since the concentration of contaminants that are produced in the enclosure make the IAQ even worse.

Localized and space heating systems can be used together or coupled with floor heaters to improve the control of the indoor climate conditions. Floor heating is usually through hot water pipes or electric resistance cables buried directly in the floor, but this can cause greater evaporation and a rise in the air moisture content.

The most common energy sources for heating are electricity, natural gas, propane, and biomass. Solar energy represents an interesting solution for providing supplemental heating, but peak availability is during warm seasons and the daytime when heat demand is lowest.

Heat Recovery Systems

To maintain IAQ, indoor air is replaced by fresh outdoor air to dilute contaminants and decrease moisture content. During heating periods, every cubic meter of fresh air that is introduced inside the livestock house is heated to reach the indoor air set point temperature. The heat of the exhausted air is lost. When the outdoor air is cold, heating the fresh air requires considerable energy; ventilation accounts for 70% to 90% of the heat losses in typical livestock houses during the winter season (ASHRAE, 2011).

To improve energy performance especially in cold climates, heat recovery can be used. In livestock houses, air-to-air heat recovery systems are used to transfer sensible heat from an airstream at a high temperature (exhaust air) to an airstream at a low temperature (fresh supply air) (ASHRAE, 2012). The heat transfer happens through a heat exchange surface (a series of plates or tubes) that separates the two airstreams, avoiding the cross-contamination of fresh supply air with the contaminants in the exhaust air. The most common type of heat exchanger used in livestock houses is cross-flow (figure 3). The recovered heat directly increases the temperature of the fresh supply air, decreasing the supplemental heat that is needed to reach the indoor air set point temperature. Heat recovery systems mainly transfer sensible heat but, under certain psychrometric conditions, even part of the latent heat of the exhaust air can be recovered. For example, when the outdoor air is very cool, the water vapor contained in the exhaust air condenses and releases the latent heat of condensation increasing the temperature of the fresh air.

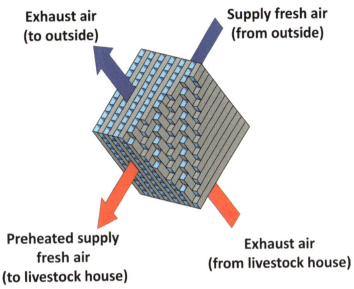

Figure 3. Diagram of the heat exchange surface in an air-to-air heat recovery system.

In practice, heat exchanger effectiveness is the ratio between the actual transfer of energy and the maximum possible transfer between the airstreams (ASHRAE, 1991). In livestock houses this is usually between 60% and 80%, because of freezing and dust accumulation on the heat-exchanging surfaces (ASHRAE, 2011). A buildup of dust reduces the heat transfer between the airstreams and reduces the flow rate. In addition, gases and moisture in exhaust air can damage the heat-exchanging surface. Filtration, automatic washing, insulation, and defrost controls can be used to avoid problems with heat exchange.

Cooling Animal Houses

Cooling Systems

In warmer conditions, cooling is required to reduce the indoor air temperature and to alleviate animal heat stress. Air flow driven by fans is used to remove the heat generated by animals and from solar radiation. With high indoor air temperature and in heat stress situations, greater air velocities around the animals are preferred because the skin temperature of the animals is reduced through the increasing convective heat exchange.

When the difference between outdoor air and indoor air temperatures is small, cooling ventilation is less effective because the needed air flow rates require air velocities too great for animal comfort. To overcome this problem, water cooling and evaporative cooling can be used (Lindley and Whitaker, 1996). Water cooling consists of sprinkling or dripping water directly on the animals to remove heat from their bodies through evaporation. Evaporative

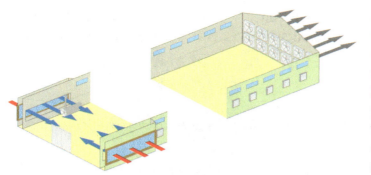

Figure 4. Diagram of a broiler house equipped with evaporative pads.

cooling uses heat from the indoor air to vaporize water and thus decrease indoor air temperature with either a fogging system or evaporative pads. Foggers release a mist of tiny water droplets directly inside the enclosure. Evaporative pads are used in livestock houses with exhaust ventilation systems (figure 4). In these systems, exhaust fans force out the indoor air creating a negative pressure difference between inside and outside the house. This pressure difference pulls the fresh outdoor air inside the house through the evaporative pads, decreasing its temperature by some degrees as a function of the direct saturation effectiveness, ε (ASHRAE, 2012) (equation 20). From a technical point of view, ε is the most exciting feature of an evaporative pad, and it ranges between 70% and 95% for commercially available evaporative pads. This value is directly proportional to the pad thickness (from 0.1 to 0.3 m) (ASHRAE, 2012) and inversely proportional to the air velocity through the pad. The highest efficiencies are with air velocity between 1.0 and 1.4 m s^{-1} (ASHRAE, 2011). The value of ε is also influenced by the age and the maintenance of the pad; ε can decrease to 30% in old and poorly maintained pads (Costantino et al., 2018).

Evaporative pads affect energy consumption in two ways. On the one hand, they decrease the temperature of the air that is used to ventilate the house, which means a reduction in the ventilation flow rate needed to maintain the indoor air setpoint temperature. On the other hand, they increase the pressure difference between the inside and outside the house, so for the same air flow rate, the fans in a livestock house equipped with evaporative pads require higher electricity consumption. Finally, the use of evaporative pads requires extra electrical energy due to the circulation pumps used to move the water from storage for wetting the top of the pads.

Ventilation Systems

The effectiveness of ventilation inside a livestock house depends on the selection, installation, and operation of the ventilation equipment, such as air inlets, outlets, control systems, and fans.

Fans are classified as centrifugal or axial, according to the direction of the airflow through the impeller (ASHRAE, 2012). Axial fans draw air parallel to the shaft axis (around which the blades rotate) and exhaust it in the same direction. Centrifugal fans exhaust air by deflection and centrifugal force. In centrifugal fans air enters next to the shaft due to the rotation of the impeller and then moves perpendicularly from the shaft to the opening where it is exhausted. Axial fans are usually used in livestock housing because the primary goal is to provide a high airflow rate and not to create a high-pressure difference across the fan. Fans cause considerable energy consumption in livestock houses (Costantino et al., 2016), as shown in table 3, but are typically bought based on purchase cost, not operating costs. When fans are installed in the livestock houses, a reduction in efficiency has to be expected due to the wear of the mechanical connections (ASHRAE, 2012).

Examples

Example 1: Heat flow through a building envelope

Problem:
Determine the total steady-state transmission heat flow through the building envelope of the gable roof broiler house presented in figure 5. The thermophysical properties of the envelope elements are shown in table 4. For the calculation, assume the indoor air temperature is 23°C and the outdoor air temperature is 20°C.

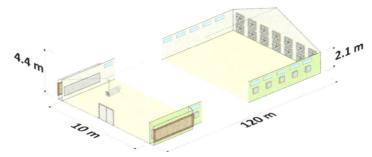

Figure 5. Diagram of the example broiler house with the main geometrical dimensions.

Solution:
The total transmission heat flow through the envelope should be calculated through equation 15. In the summation, all the envelope elements of the broiler house must be considered. In this broiler house, the various products $(b_{tr,j} \cdot U_j \cdot A_j)$ of the summation of equation 15 are:

$$\phi_{tr} = \left[\sum_{j=1}^{n} \left(b_{tr,j} \cdot U_j \cdot A_j \right) \right] \quad (15)$$

Table 4. Boundary conditions of the example broiler house.

Building Element	Area (m²)	U (W m⁻² K⁻¹)	b_{tr} (-)
North wall	195	0.81	1
South wall	195	0.81	1
East wall	18	0.81	1
West wall	33	0.81	1
Roof	1320	1.17	1
Floor	1200	0.94	0.45
Door (east)	15	1.51	1
North windows	57	3.60	1
South windows	57	3.60	1

$$b_{tr,walls} \cdot U_{walls} \cdot A_{walls} = 1 \cdot 0.81 \frac{W}{m^2 \cdot K} \cdot 441\ m^2 = 357.2 \frac{W}{K}$$

$$b_{tr,roof} \cdot U_{roof} \cdot A_{roof} = 1 \cdot 1.17 \frac{W}{m^2 \cdot K} \cdot 1320\ m^2 = 1544.4 \frac{W}{K}$$

$$b_{tr,doors} \cdot U_{doors} \cdot A_{doors} = 1 \cdot 1.51 \frac{W}{m^2 \cdot K} \cdot 15\ m^2 = 22.7 \frac{W}{K}$$

$$b_{tr,windows} \cdot U_{windows} \cdot A_{windows} = 1 \cdot 3.60 \frac{W}{m^2 \cdot K} \cdot 114\ m^2$$

$$= 410.4 \frac{W}{K}$$

The U-value of the floor of the broiler house is 0.94 W m⁻² K⁻¹. This value was calculated considering that the floor was made by a reinforced concrete screed and a waterproofing sheet directly in contact with the ground. In the transmission heat flow via ground, the b_{tr} coefficient has to be considered. Considering that the floor of the broiler house is in direct contact with the ground, $b_{tr,floor}$ can be assumed equal to 0.45 (value from table 2). The calculation is:

$$b_{tr,floor} \cdot U_{floor} \cdot A_{floor} = 0.45 \cdot 0.94 \frac{W}{m^2 \cdot K} \cdot 1200 \text{ m}^2$$

$$= 507.6 \frac{W}{K}$$

Considering the previously calculated values, the sum is:

$$\sum_{j=1}^{n}\left(b_{tr,j} \cdot U_j \cdot A_j\right) = 2842.3 \frac{W}{K}$$

Finally, the heat flow can be calculated considering the temperature difference between inside and outside as:

$$\phi_{tr} = \left(2842.3 \frac{W}{K}\right) \cdot (20°C - 23°C) = -8526.9 \text{ W}$$

Example 2: Sensible heat flow in a broiler house

Problem:
Determine the sensible heat flow produced at the house level by a flock of 14,000 broilers at an indoor air temperature of 23°C. The average weight of the broilers is 1.3 kg.

Solution:
The total heat production $\phi_{a,tot}$ from a broiler flock at an indoor air temperature of 20°C is defined by equation 6 that reads

$$\phi_{a,tot} = 10.62 \cdot w_a^{0.75} \cdot n_a \qquad (6)$$

Considering the given boundary conditions, equation 6 becomes:

$$\phi_{a,tot} = 10.62 \cdot 1.3^{0.75} \cdot 14,000 = 181,013.1 \text{ W}$$

Before calculating ϕ_a, n_{hpu} has to be calculated according to equation 14:

$$n_{hpu} = \frac{\phi_{a,tot}}{1000} \qquad (14)$$

$$n_{hpu} = \frac{181,013.1 \text{ W}}{1000 \frac{W}{hpu}} = 181.01 \text{ hpu}$$

Finally, ϕ_a calculated at 23°C of $T_{air,i}$ is (from equation 10):

$$\phi_a = \left\{0.61 \cdot \left[1000 + 20 \cdot (20 - T_{air,i})\right] - 0.228 \cdot T_{air,i}^2\right\} \cdot n_{hpu} \qquad (10)$$

$$\phi_a = \{0.61 \cdot [1000 + 20 \cdot (20-23°C)] - 0.228 \cdot (23°C)^2\} \cdot 181.01 = 81,959.2 \text{ W}$$

The broiler flock in this example produces around 82 kW of sensible heat.

Example 3: Solar based heat flow

Problem:
Determine the value of ϕ_{sol} considering the boundary conditions shown in table 5 and using the same broiler house of examples 1 and 2.

Solution:
The first step for determining ϕ_{sol} is to calculate $\phi_{sol,op}$ for each opaque building element according to equation 17, as:

Table 5. Boundary conditions of the example broiler house.

Building Element	Area (m²)	U (W m⁻² k⁻¹)	α (-)	g_{gl} (-)	I_{sol} (W m⁻²)
North wall	195	0.81	0.3	-	142
South wall	195	0.81	0.3	-	559
East wall	18	0.81	0.3	-	277
West wall	33	0.81	0.3	-	142
Roof	1320	1.17	0.9	-	721
Floor	1200	0.94	-	-	-
Door (east)	15	1.51	0.9	-	277
North windows	57	3.60	-	0.6	142
South windows	57	3.60	-	0.6	559

$$\phi_{sol,op,q} = A_q \cdot U_q \cdot \alpha_q \cdot R_{ex} \cdot I_{sol,q} \cdot F_{sh,q} \qquad (17)$$

$$\phi_{sol,op,wall,\,N} = 195 \text{ m}^2 \cdot 0.81 \frac{\text{W}}{\text{m}^2 \cdot \text{K}} \cdot 0.3 \cdot 0.04 \frac{\text{m}^2 \cdot \text{K}}{\text{W}} \cdot 142 \frac{\text{W}}{\text{m}^2} = 269.1 \text{ W}$$

$$\phi_{sol,op,wall,\,S} = 195 \text{ m}^2 \cdot 0.81 \frac{\text{W}}{\text{m}^2 \cdot \text{K}} \cdot 0.3 \cdot 0.04 \frac{\text{m}^2 \cdot \text{K}}{\text{W}} \cdot 559 \frac{\text{W}}{\text{m}^2} = 1059.5 \text{ W}$$

$$\phi_{sol,op,wall,\,E} = 18 \text{ m}^2 \cdot 0.81 \frac{\text{W}}{\text{m}^2 \cdot \text{K}} \cdot 0.3 \cdot 0.04 \frac{\text{m}^2 \cdot \text{K}}{\text{W}} \cdot 277 \frac{\text{W}}{\text{m}^2} = 48.5 \text{ W}$$

$$\phi_{sol,op,wall,\,W} = 33 \text{ m}^2 \cdot 0.81 \frac{\text{W}}{\text{m}^2 \cdot \text{K}} \cdot 0.3 \cdot 0.04 \frac{\text{m}^2 \cdot \text{K}}{\text{W}} \cdot 142 \frac{\text{W}}{\text{m}^2} = 45.5 \text{ W}$$

$$\phi_{sol,op,Roof} = 1320 \text{ m}^2 \cdot 1.17 \frac{W}{m^2 \cdot K} \cdot 0.9 \cdot 0.04 \frac{m^2 \cdot K}{W} \cdot 721 \frac{W}{m^2} = 40,086.4 \text{ W}$$

$$\phi_{sol,op,Door} = 15 \text{ m}^2 \cdot 1.51 \frac{W}{m^2 \cdot K} \cdot 0.9 \cdot 0.04 \frac{m^2 \cdot K}{W} \cdot 277 \frac{W}{m^2} = 225.9 \text{ W}$$

The sum of the calculated $\phi_{sol,op,q}$ values is:

$$\sum_{n=1}^{q} \phi_{sol,op,q} = 41,734.9 \text{ W}$$

The solar heat loads on glazed components can be estimated using equation 18:

$$\phi_{s,gl,k} = A_k \cdot g_{gl} \cdot I_{sol,k} \cdot (1 - F_{fr}) \cdot F_{sh,k} \cdot F_{sh,gl,k} \qquad (18)$$

Considering the given boundary conditions, $\phi_{sol,gl}$ for the glazed elements can be computed as:

$$\phi_{sol,gl,win,N} = 57 \text{ m}^2 \cdot 0.6 \cdot 142 \frac{W}{m^2} \cdot (1 - 0.2) = 3885.1 \text{ W}$$

$$\phi_{sol,gl,win,S} = 57 \text{ m}^2 \cdot 0.6 \cdot 559 \frac{W}{m^2} \cdot (1 - 0.2) = 15,294.2 \text{ W}$$

The sum of the calculated $\phi_{sol,gl,k}$ values is:

$$\sum_{n=1}^{k} \phi_{sol,gl,k} = 19,179.3 \text{ W}$$

Finally, the total solar heat load is:

$$\phi_{sol} = 41,734.9 \text{ W} + 19,179.3 \text{ W} = 60,914.2 \text{ W}$$

Example 4: Ventilation flow rate for temperature control

Problem:

Determine the volumetric ventilation flow rate (m³ h⁻¹) that has to be provided by the exhaust fans of the broiler house to maintain the indoor air temperature at 23°C. For the calculation, consider the absence of supplemental heating flow ($\phi_H = 0$ W) the heat flows calculated in example 1 (ϕ_{tr}), example 2 (ϕ_a) and example 4 (ϕ_{sol}). The supply air temperature is the same of the outdoor air (20°C, as in example 1).

Solution:

In the previous examples the following heat flows were calculated:

$$\phi_{tr} = -8{,}526.9 \text{ W} = -8.5 \text{ kW}$$

$$\phi_{a} = 81{,}959.2 \text{ W} = 82.0 \text{ kW}$$

$$\phi_{sol} = 60{,}914.2 \text{ W} = 60.9 \text{ kW}$$

The text of the problem states that no supplemental heating flow is present, therefore:

$$\phi_{H} = 0 \text{ kW}$$

Considering the given boundary conditions, the energy balance of equation 3 can be written as:

$$82.0 \text{ kW} - 8.5 \text{ kW} + 0 \text{ kW} + 60.9 \text{ kW} + \phi_{v} = 0$$

That becomes:

$$\phi_{v} = -134.4 \text{ kW}$$

Equation 19 can be expressed in $\dot{V}$ (the unknown of the problem, in kW) as:

$$\dot{V} = \frac{\phi_{v}}{\rho_{air} \cdot c_{air} \cdot (T_{air,sup} - T_{air,i})}$$

The value of ρ_{air} is assumed equal to 1.2 kg m^{-3} and c_{air} equal to 2.8×10^{-4} kWh kg^{-1} K^{-1} (1010 J kg^{-1} K^{-1}), even though for more detailed calculation ρ_{air} should be evaluated at the given indoor air temperature and atmospheric pressure. The ventilation air flow is provided with outdoor air, therefore, $T_{air,sup}$ is equal to $T_{air,o}$. Inputting the previously calculated value of ϕ_{v}, the previous equation reads:

$$\dot{V} = \frac{-134.4 \text{ kW}}{1.2 \frac{\text{kg}}{\text{m}^3} \cdot 2.8 \cdot 10^{-4} \frac{\text{kWh}}{\text{kg} \cdot \text{K}} \cdot (20°\text{C} - 23°\text{C})} = 133{,}333 \frac{\text{m}^3}{\text{h}}$$

To maintain the required indoor air temperature inside the livestock house, around 133,000 m³ h^{-1} of fresh outdoor air should be provided by the ventilation system.

Image Credits

Figure 1. Fabrizio, E. (CC By 4.0). (2020). The sensible heat balance of equation 3 applied to a generic livestock house.
Figure 2. Fabrizio, E. (CC By 4.0). (2020). The vapor mass balance of equation 5 applied to a generic livestock house.
Figure 3. Costantino, A. (CC By 4.0). (2020). Diagram of the heat exchange surface in an air-to-air heat recovery system.
Figure 4. Costantino, A. (CC By 4.0). (2020). Diagram of a broiler house equipped with evaporative pads.
Figure 5. Costantino, A. (CC By 4.0). (2020). Diagram of the example broiler house with the main geometrical dimensions.

References

Albright, L. (1990). *Environmental control for animals and plants*. St. Joseph, MI: ASAE.
ASABE Standards. (2012). ASAE EP270.5 DEC1986: Design of ventilation systems for poultry and livestock shelters. St. Joseph, MI: ASABE.
ASHRAE. (2012). *2012 ASHRAE handbook: HVAC systems and equipment*. Atlanta, GA: ASHRAE.
ASHRAE. (2011). *2011 ASHRAE handbook: HVAC applications*. Atlanta, GA: ASHRAE.
ASHRAE. (1991). ANSI/ASHRAE Standard 84-1991: Method of testing air-to-air heat exchangers. Atlanta, GA: ASHRAE.
CIGR. (1999). CIGR handbook of agricultural engineering (Vol. II). St. Joseph, MI: ASAE.
Costantino, A., Ballarini, I., & Fabrizio, E. (2017). Comparison between simplified and detailed methods for the calculation of heating and cooling energy needs of livestock housing: A case study. In *Building Simulation Applications* (pp. 193-200). Bolzano, Italy: Free University of Bozen-Bolzano.
Costantino, A., Fabrizio, E., Biglia, A., Cornale, P., & Battaglini, L. (2016). Energy use for climate control of animal houses: The state of the art in Europe. *Energy Proc. 101*, 184-191. https://doi.org/10.1016/j.egypro.2016.11.024
Costantino, A., Fabrizio, E., Ghiggini, A., & Bariani, M. (2018). Climate control in broiler houses: A thermal model for the calculation of the energy use and indoor environmental conditions. *Energy Build. 169*, 110-126. https://doi.org/10.1016/j.enbuild.2018.03.056
Costantino, A., Fabrizio, E., Villagrá, A., Estellés, F., & Calvet, S. (2020). The reduction of gas concentrations in broiler houses through ventilation: Assessment of the thermal and electrical energy consumption. *Biosyst. Eng.* https://doi.org/10.1016/j.biosystemseng.2020.01.002.
Esmay, M. E., & Dixon, J. E. (1986). *Environmental control for agricultural buildings*. Westport, CT: AVI.
European Committee for Standardisation. (2009). EN 12831: Heating systems in buildings—Method for calculation of the design heat load. Brussels, Belgium: CEN.
European Committee for Standardization. (2007). EN 13789: Thermal performance of buildings—Transmission and ventilation heat transfer coefficients—Calculation method. Brussels, Belgium: CEN.
Hamilton, J., Negnevitsky, M., & Wang, X. (2016). Thermal analysis of a single-storey livestock barn. *Adv. Mech. Eng. 8*(4). https://doi.org/10.1177/1687814016643456.
Hellickson, M. A., & Walker, J. N. (1983). *Ventilation of agricultural structures*. St. Joseph, MI: ASAE.
ISO. (2017). ISO 52016-1:2017: Energy performance of buildings—Energy needs for heating and cooling, internal temperatures and sensible and latent heat loads—Part 1: Calculation procedures. International Organization for Standardization.

Liberati, P., & Zappavigna, P. (2005). A computer model for optimisation of the internal climate in animal housing design. In *Livestock Environment VII, Proc. Int. Symp.* St. Joseph, MI: ASABE.

Lindley, J. A., & Whitaker, J. H. (1996). *Agricultural buildings and structures.* St. Joseph, MI: ASAE.

Midwest Plan Service. (1987). *Structures and environment handbook* (11th ed.). Ames, IA: Midwest Plan Service.

OECD. (2008). *Environmental performance of agriculture in OECD countries since 1990.* Paris, France: OECD.

Panagakis, P., & Axaopoulos, P. (2008). Comparing fogging strategies for pig rearing using simulations to determine apparent heat-stress indices. *Biosyst. Eng. 99*(1), 112-118. https://doi.org/10.1016/j.biosystemseng.2007.10.007.

Pedersen, S., & Sällvik, K. (2002). 4th report of working group on climatization of animal houses—Heat and moisture production at animal and house levels. Horsens, Denmark: Danish Institute of Agricultural Sciences.

Rossi, P., Gastaldo, A., Riva, G., & de Carolis, C. (2013). Progetto re sole—Linee guida per il risparmio energetico e per la produzione di energia da fonte solare negli allevamenti zootecnici (in Italian). Reggio Emilia, Italy: CRPA.

St-Pierre, N. R., Cobanov, B., & Schnitkey, G. (2003). Economic losses from heat stress by US livestock industries. *J. Dairy Sci. 86*, E52-E77. https://doi.org/10.3168/jds.s0022-0302(03)74040-5.

Freezing of Food

M. Elena Castell-Perez
Department of Biological and Agricultural Engineering
Texas A&M University, College Station, TX, USA

LIST OF KEY TERMS		
Sensible heat	Freezing point	Cooling load
Specific heat	Conduction	Freezing rate and time
Latent heat	Convection	Freeze drying

Variables for the Chapter

λ = latent heat of fusion
ρ = density
a = thickness
A = surface area
C_p = specific heat (also called specific heat capacity)
h = convective heat transfer coefficient
k = thermal conductivity
L = length
m = mass
m_A = mass of water in food, or moisture content
$\dot{m}_p$ = mass flow rate of product
M = mass or molecular weight
M_A = molecular weight of water (18 g/mol)
M_s = mass of solute in product
M_s = relative molecular mass of soluble solids in food
M_S = molecular weight of solute
M_{water} = amount of water in product or water content
P and R = parameters determined by the shape of the food being frozen
$Q_{conduction}$ = heat energy transferred through a solid by conduction

$Q_{convection}$ = heat energy transferred to a colder moving liquid from the warmer surface of a solid by convection

Q_L = heat energy removed to freeze the product at its freezing point; also known as latent heat energy

$\dot{Q}_p$ = rate of heat removed from the food, or cooling load

Q_p = heat energy removed to freeze the product to the target temperature

Q_S = sensible heat to change temperature of a food

R and P = parameters determined by the shape of the food being frozen

R = universal gas constant (8.314 kJ/kmol)

t_f = freezing time

T = temperature

T_a = freezing medium temperature; ambient temperature

T_f = freezing point temperature of the food

x = thickness of packaging material

Δx = thickness of the food

X_i = mass fraction of component i (example: for water, X_{water} or X_w)

Introduction

Freezing is one of the oldest and more common unit operations that apply heat and mass transfer principles to food. Engineers must know these principles to analyze and design a suitable freezing process and to select proper equipment by establishing system capacity requirements.

Freezing is a common process for long-term preservation of foods. The fundamental principle is the crystallization of most of the water—and some of the solutes—into ice by reducing the temperature of the food to −18°±3°C or lower (a standard commercial freezing target temperature) using the concepts of sensible and latent heat. These principles also apply to freezing of other types of materials that contain water.

If done properly, freezing is the best way to preserve foods without adding preservatives. Freezing aids preservation by reducing the rate of physical, chemical, biochemical, and microbiological reactions in the food. The liquid water-to-ice phase change reduces the availability of the water in the food to participate in any of these reactions. Therefore, a frozen food is more stable and can maintain its quality attributes throughout transportation and storage.

Freezing is commonly used to extend the shelf life of a wide variety of foods, such as fruits and vegetables, meats, fish, dairy, and prepared foods (e.g., ice cream, microwavable meals, pizzas) (George, 1993; James and James, 2014). The great demand for frozen food creates the need for proper knowledge of the mechanics of freezing and material thermophysical properties (Filip et al., 2010).

Outcomes

After reading this chapter, you will be able to:

- Describe the engineering principles of freezing of foods
- Describe how food product properties, such as freezing point temperature, size, shape, and composition, as well as packaging, affect the freezing process
- Describe how process factors, such as freezing medium temperature and convective heat transfer coefficient, affect the freezing process
- Calculate values of food properties and other factors required to design a freezing process
- Calculate freezing times
- Select a freezer for a specific application

Concepts

Process of Freezing

Freezing is a physical process by which the temperature of a material is reduced below its freezing point temperature. Two heat energy principles are involved: *sensible heat* and *latent heat*. When the material is at a temperature above its freezing point, first the sensible heat is removed until the material reaches its freezing point; second, the latent heat of crystallization (fusion) is removed, and finally, more sensible heat is removed until the material reaches the target temperature below its freezing point.

Sensible heat is the amount of heat energy that must be added or removed from a specific mass of material to change its temperature to a target value. It is referred to as "sensible" because one can usually sense the temperature surrounding the material during a heating or cooling process. *Latent heat* is the amount of energy that must be removed in order to change the phase of water in the material. During the phase change, there is no change in the temperature of the material because all the energy is used in the phase change. In the case of freezing, this is the *latent heat of fusion*. Heat is given off as the product crystallizes at constant temperature.

For pure water, the latent heat of fusion is a constant with a value of ~334 kJ per kg of water. For food products, the latent heat of fusion can be estimated as

$$\lambda = M_{water} \times \lambda_w \tag{1}$$

where λ = latent heat of fusion of food product (kJ/kg)
M_{water} = amount of water in product, or water content (decimal)
λ_w = latent heat of fusion of pure water (~334 kJ/kg)

The sensible and latent heat energy in the freezing of foods are quantified by equations 2-9. Table 1 presents values of the latent heat of several foods with specific moisture contents.

> The comparable term for boiling is the *latent heat of vaporization*.

Sensible Heat and Specific Heat

The sensible heat to change the temperature of a food is related to the specific heat of the food, its mass, and its temperature:

$$Q_S = mC_p(T_2 - T_1) \qquad (2)$$

where Q_S = sensible heat to change temperature of a food (kJ)
m = mass of the food (kg)
C_p = specific heat of the food (kJ/kg°C or kJ/kgK)
T_1 = initial temperature of food (°C)
T_2 = final temperature of food (°C)

The *specific heat* (also called *specific heat capacity*), C_p, of liquid water (above freezing) is 4.186 kJ/kg°C or 1 calorie/g°C. In foods, specific heat is a property that changes with the food's water (moisture) content. Usually, the higher the moisture or water content, the larger the value of C_p, and vice versa. As the water in the food reaches its freezing point temperature, the water begins to crystallize and turn into ice. When almost all of the water is frozen, the specific heat of the food decreases by about half (C_p of ice = 2.108 kJ/kg°C). Therefore, one must be careful when using equation 2 to use the correct value of C_p (above or below freezing; see example 1 and equations 5-7).

> The subscript "p" stands for "constant pressure," which is the method used to measure the specific heat of solids and liquids. For gases, C_v, or specific heat at constant volume, is used.

Values of C_p of a wide range of foods at a particular moisture content, above and below freezing, are available (Mohsening, 1980; Choi and Okos, 1986; ASHRAE, 2018; The Engineering Toolbox, 2019; see table 1 for some examples). When values of C_p or λ of the target foods are unavailable, they can be determined using several methods, ranging from standard calorimetry to differential scanning, ultrasound, and electrical methods (Mohsenin, 1980; Chen, 1985; Klinbun and Rattanadecho, 2017).

When these properties cannot be measured (e.g., because the sample is too small or heterogeneous, or equipment is unavailable), a wide range of models have been developed to predict the properties of food and agricultural materials as a function of time and composition. For instance, if detailed product composition data are not available, equation 3 can be used to approximate C_p for temperatures above freezing:

Table 1. Specific heat (C_p) and latent heat of fusion (λ) of selected foods estimated based on composition (ASHRAE, 2018).

Food	Moisture Content (%)	C_p Above Freezing (kJ/kg°C)	C_p Below Freezing (kJ/kg°C)	λ (kJ/kg)	T_f Initial Freezing Temperature (°C)[a]
Carrots	87.79	3.92	2.00	293	−1.39
Green peas	78.86	3.75	1.98	263	−0.61
Honeydew melon	89.66	3.92	1.86	299	−0.89
Strawberries	91.57	4.00	1.84	306	−0.78
Cod (whole)	81.22	3.78	2.14	271	−2.22
Chicken	65.99	4.34	3.62	220	−2.78

[a] Temperature at which water in food begins to freeze; freezing point temperature.

$$C_{p,\text{unfrozen}} = C_{pw}X_w + C_{ps}X_s \qquad (3)$$

where C_{pw} = specific heat of the water component (kJ/kg°C)
X_w = mass fraction of the water component (decimal)
C_{ps} = specific heat of the solids component (kJ/kg°C)
X_s = mass fraction of the solids component (decimal)

As a material balance, $X_w = 1 - X_s$. This method approximates the food as a binary system composed of only water and solids. When the main solids component is known, the C_p of the solids (C_{ps}) can be estimated from published data (e.g., table 2). For instance, if the food is mostly water and carbohydrates (e.g., a fruit), C_{ps} can be approximated as 1.5488 kJ/kgK (from table 2). If the target food is composed mostly of protein, then C_{ps} can be approximated as 2.0082 kJ/kg°C.

Table 2. Specific heat, C_p, of food components for −40°C to 150°C.

Food Component	Specific Heat (kJ/kg°C)
Protein	$C_p = 2.0082 + 1.2089 \times 10^{-3} T - 1.3129 \times 10^{-6} T^2$
Fat	$C_p = 1.9842 + 1.4733 \times 10^{-3} T - 4.8088 \times 10^{-6} T^2$
Carbohydrate	$C_p = 1.5488 + 19625 \times 10^{-3} T - 5.9399 \times 10^{-6} T^2$
Fiber	$C_p = 1.8459 + 1.8306 \times 10^{-3} T - 4.6509 \times 10^{-6} T^2$
Ash	$C_p = 1.0926 + 1.8896 \times 10^{-3} T - 3.6817 \times 10^{-6} T^2$

From Choi and Okos, 1986; ASHRAE, 2018. T in °C. An extensive database on food products composition is available in USDA (2019).

The specific heat of the food above its freezing point can be calculated based on its composition and the mass average specific heats of the different components as:

$$C_p = \sum_{i=1}^{n} X_i C_{pi} \qquad (4)$$

where X_i = mass fraction of component i (decimal, not percentage). For example, for water, $X_{\text{water}} = M_{\text{water}}/M$ where M = total mass of product
i = component (water, protein, fat, carbohydrate, fiber, ash)
C_{pi} = specific heat of component i estimated at a particular temperature value (kJ/kgK) (from table 2)

In the case of water, separate equations are available for liquid water at temperatures below (equation 5) and above (equation 6) freezing, while one equation applies for ice at temperatures below freezing (equation 7) (Choi and Okos, 1986):

For water −40°C to 0°C:

$$C_p = 4.1289 - 5.3062 \times 10^{-3} T + 9.9516 \times 10^{-4} T^2 \qquad (5)$$

For water 0°C to 150°C:

$$C_p = 4.1289 - 9.0864 \times 10^{-5} T + 5.4731 \times 10^{-6} T^2 \qquad (6)$$

For ice −40°C to 0°C:

$$C_p = 2.0623 + 6.0769 \times 10^{-3} T \tag{7}$$

Table 3. Examples of predictive models for calculation of specific heat of foods.

Model, Source	Equation (C_p in kJ/kgK)
Siebel (1892), above freezing[a]	$C_p = 0.837 + 3.348 X_w$
Siebel (1892), below freezing	$C_p = 0.837 + 1.256 X_w$
Chen (1985), above freezing[b]	$C_p = 4.19 - 2.30 X_s - 0.628 X_s^3$
Chen (1985), below freezing[c]	$C_p = 1.55 + 1.26 X_s + X_s [R\, T_0^2 / M_s T^2]$
Choi and Okos (1986)	$C_p = 4.180 X_w + 1.711 X_{protein} + 1.928 X_{fat} + 1.547 X_{carbohydrates} + 0.908 X_{ash}$

[a] X_w = moisture content, decimal; [b] X_s = mass fraction of solids, decimal; [c] R = universal gas constant, 8.314 kJ/kmol K; T_0 = freezing point temperature of water, K; M_s = relative molecular mass of soluble solids in food; T = temperature, K.

Many predictive models have been developed for the calculation of specific heat of various food products. Some examples are presented in table 3. An excellent description of these and other predictive models is presented in Mohsenin (1980).

Several models, such as a modified version of the model by Chen (1985), are available for simple calculation of the specific heat of a frozen food:

$$C_{p,\,frozen} = 1.55 + 1.26 X_s + \frac{(X_{w0} - X_b) \lambda_w T_f}{T^2} \tag{8}$$

where $C_{p,frozen}$ = apparent specific heat of frozen food (kJ/kgK)
X_s = mass fraction of solids (decimal)
X_{w0} = mass fraction of water in the unfrozen food (decimal)
X_b = bound water (decimal); this parameter can be approximated with great accuracy as $X_b \sim 0.4 X_p$ (Schwartzberg, 1976) with X_p = mass fraction of protein (decimal)
λ_w = latent heat of fusion of water (~334 kJ/kg)
T_f = freezing point of water = 0.01°C (can be approximated to 0.00°C)
T = food temperature (°C)

Latent Heat

The latent heat of a food product is:

$$Q_L = m\lambda \tag{9}$$

where Q_L = heat energy removed to freeze the product at its freezing point; also known as latent heat energy (kJ)
m = mass of product (kg)
λ = latent heat of fusion of product (kJ/kg)

For water, λ is approximated as 334 kJ/kg. Latent heat values for many food materials are also available (ASHRAE, 2018; table 1).

Freezing Point Temperature and Freezing Point Depression

The *freezing point temperature*, or *initial freezing point*, of a food product is defined as the temperature at which ice crystals begin to form. Knowledge of this property of foods is essential for proper design of frozen storage and freezing processes because it affects the amount of energy required to reduce the food's temperature to a specific value below freezing.

Although most foods contain water that turns into ice during freezing, the initial freezing point of most foods ranges from −0.5°C to −2.2°C (Pham, 1987; ASHRAE, 2018); values given in tables are usually average freezing temperatures. Foods freeze at temperatures lower than the freezing point of pure water (which is 0.01°C although most calculations assume 0.0°C) because the water in the foods is not pure water and, when removing heat energy from the food, the freezing point is depressed (lowered) due to the increase in solute concentration in the ice-water sections of the material. Therefore, the food will begin to freeze at temperatures lower than 0 to 0.01°C (table 1). This is called the *freezing point depression* (figure 1).

When designing cooling or freezing processes, the heat energy types (Q_S and Q_L) have a negative sign (heat energy is released from the system). When designing heating or thawing processes, these quantities are positive (heat energy is added to the system).

In general, 1 g-mol of soluble matter will decrease the freezing point of the product by approximately 1°C (Singh and Heldman, 2013). Consequently, the engineer should estimate the freezing point of the specific product and not assume that the food product will freeze at 0°C.

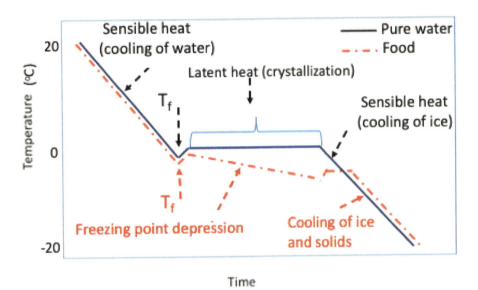

Figure 1. Freezing curves for pure water and a food product illustrating the concept of freezing point depression (latent heat is released over a range of temperatures when freezing foods versus a constant value for pure water).

Unfrozen or Bound Water

Water that is bound to the solids in food cannot be frozen. The percent of unfrozen (*bound*) water at −40°C, a temperature at which most of the water is frozen, ranges from 3% to 46%. This quantity is necessary to determine the heat content of a food (i.e., enthalpy) when exposed to temperatures that cause a phase change; in other words, its latent heat of fusion, λ.

A freezing point depression equation allows for prediction of the relationship between the unfrozen water fraction within the food (X_A) and temperature in a binary solution (i.e., water and solids mixture) over the range from −40°C to 40°C (Heldman, 1974; Chen, 1985; Pham, 1987):

Some foods with exceptionally low freezing points are dates (−15.7°C), salted yolk (−17.2°C), and cheeses (−16.3 to −1.2°C).

$$\ln X_A = \frac{\lambda}{R}\left(\frac{1}{T_0} - \frac{1}{T_f}\right) \tag{10}$$

where X_A = molar fraction of liquid (water) in product A (decimal). (The molar fraction is the number of moles of the liquid divided by the total number of moles of the mixture.)
λ = molar latent heat of fusion of water (6,003 J/mol)
R = universal gas constant (8.314 J/mol K)
T_0 = freezing point of pure water (K)
T_f = freezing point of food (K)

X_A is calculated as

$$X_A = \frac{\dfrac{m_A}{M_A}}{\dfrac{m_A}{M_A} + \dfrac{m_s}{M_s}} \tag{11}$$

where m_A = mass of water in food or moisture content (decimal)
M_A = molecular weight of water (18 g/mol)
M_s = mass of solute in product (decimal)
M_S = molecular weight of solute (g/mol)

Physics of Freezing: Heat Transfer Modes

During freezing of a material, heat is removed within the food by conduction and at its surface by convection, radiation, and evaporation. In practice, these four modes of heat transfer occur simultaneously but with different levels of significance (James and James, 2014). The contributions to heat transfer by radiation and evaporation are much smaller than for the other modes and, therefore, are assumed negligible (Cleland, 2003).

Heat transfer problems can be defined as steady- or unsteady-state situations. During a steady-state process, the temperature within a system (e.g., the food) only changes with location. Hence, temperature does not change with time at that particular location. This is the equilibrium state of a system. One example would be the temperature inside an oven once it has reached the target heating temperature after the food is placed inside the oven. On the other hand, an unsteady-state process (also known as a transient heat transfer problem) is one in which the temperature within the system (e.g., the food) changes with both time and location (the surface, the center, or any distance within the food). Freezing of a product until its center reaches the target frozen storage temperature is a typical unsteady-state problem while storage of a frozen product is a steady-state situation.

Heat Transfer by Conduction

In general, the rate of heat transfer within the food is dominated by conduction and calculated as

$$Q_{conduction} = kA \, \Delta T / \Delta x \tag{12}$$

where $Q_{conduction}$ = heat energy transferred through a solid by conduction (kJ)
k = thermal conductivity of the food (W/m°C)
A = surface area of the food (m²)
ΔT = temperature difference within the food (°C)
Δx = thickness of the food (m)

Equation 12 is valid for one-dimensional heat transfer through a rectangular object of thickness Δx under steady-state conditions (i.e., equilibrium). Variations of equation 12 have been developed for other geometries and are also available in heat transfer textbooks.

Heat Transfer by Convection

Convection controls the rate of heat transfer between the food and its surroundings and is expressed as

$$Q_{convection} = hA \, \Delta T \tag{13}$$

where $Q_{convection}$ = heat energy transferred to a colder moving liquid (air, water, etc.) from the warmer surface of a solid by convection (kJ) during cooling of a solid food
h = convective heat transfer coefficient (W/m²°C)
A = surface area of the solid food (m²)
ΔT = temperature difference between the surface of the solid food and the surrounding medium (air, water) = $T_{surface} - T_{medium}$ (°C)

The convective heat transfer coefficient, h, is a function of the type of freezing equipment and not of the type of material being frozen. The greater the value of h, the greater the transfer of heat energy from the food's surface to the cooling medium and the faster the cooling/freezing process at the surface of the food. Measurement and calculation of h values is a function of many factors (James and James, 2014; Pham, 2014). In the case of freezing, the convective heat transfer coefficient varies with selected air temperature and velocity. Table 4 shows values of h for different types of equipment commonly used in the food industry.

Table 4. Values of convective heat transfer coefficient, h, and operating temperature for different types of equipment used in food freezing operations.

Freezing Equipment	h (W/m²K)	Operating (ambient) Freezing Temperature T_a (°C)
Still air (batch)	5 to 20	−35 to −37
Air blast	10 to 200	−20 to −40
Impingement	50 to 200	−40
Spiral belt	25 to 50	−40
Fluidized bed	90 to 140	−40
Plate	100 to 500	−40
Immersion	100 to 500	−50 to −70
Cryogenic	1,500	−50 to −196

Design Parameters: Cooling Load, Freezing Rate, and Freezing Time

The engineer in charge of selecting a cooler or a freezer for a specific type of food needs to know two parameters: the cooling load and the freezing rate, which is related to freezing time.

Cooling Load

The *cooling load*, also called *refrigeration load requirement*, is the amount of heat energy that must be removed from the food or the frozen storage space. Here we assume that the rate of heat removed from the product (amount of heat energy per unit time) accounts for the majority of the refrigeration load requirement and that other refrigeration loads, such as those due to lights, machinery, and people in the refrigerated space can be neglected (James and James, 2014). Therefore, the rate of heat transfer between the food and the surrounding cooling medium at any time can be expressed as:

$$\dot{Q}_p = \dot{m}_p Q_p \tag{14}$$

where $\dot{Q}_p$ = rate of heat removed from the food, i.e., cooling load (kJ/s or kW)
$\dot{m}_p$ = mass flow rate of product (kg/s)
Q_p = heat energy in the product (kJ)

The computed cooling load is then used to select the proper motor size to carry out the freezing process.

Freezing Rate

The other critical design parameter is the product *freezing rate*, which relates to the freezing time. Basically, the freezing rate is the rate of change in temperature during the freezing process. A standard definition of the freezing rate of a food is the ratio between the minimal distance from the product surface to the *thermal center* of the food (basically the geometric center), d, and the time, t, elapsed between the surface reaching 0°C and the thermal center reaching 10°C colder than the initial freezing point temperature, T_f (IIR, 2006) (figure 2). The freezing rate is commonly given as °C/h or in terms of penetration depth measured as cm/h.

Freezing rate impacts the freezing operation in several ways: food quality, rate of throughput or the amount of food frozen, and equipment and refrigeration costs (Singh and

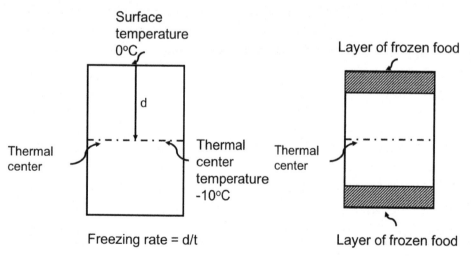

Figure 2. Schematic representation of freezing rate as defined by the International Institute of Refrigeration.

Heldman, 2013). The freezing rate affects the quality of the frozen food because it dictates the amount of water frozen into ice and the size of the ice crystals. Slower rates result in a larger amount of frozen water and larger ice crystals, which may result in undesirable product quality attributes such as a grainy texture in ice cream, ruptured muscle structure in meats and fish, and softer vegetables. Faster freezing produces a larger amount of smaller ice crystals, thus yielding products of superior quality. However, the engineer must take into account the economic viability of selecting a fast freezing process for certain applications (Barbosa-Canovas et al., 2005). Different freezing methods produce different freezing rates.

Freezing Time

Freezing rate and, therefore, freezing time, is the most critical information needed by an engineer to select and design a freezing process because freezing rate (or time) affects product quality, stability and safety, processing requirements, and economic aspects. In other words, the starting point in the design of any freezing system is the calculation of freezing time (Pham, 2014).

Freezing time is defined as the time required to reduce the initial product temperature to some established final temperature at the slowest cooling location, which is also called the thermal center (Singh and Heldman, 2013). The freezing time estimates the residence time of the product in the system and it helps calculate the process throughput (Pham, 2014).

Calculation of freezing time depends on the characteristics of the food being frozen (including composition, homogeneity, size, and shape), the temperature difference between the food and the freezing medium, the insulating effect of the boundary film of air surrounding the material (e.g., the package; this boundary is considered negligible in unpackaged foods), the convective heat transfer coefficient, h, of the system, and the distance that the heat must travel through the food (IRR, 2006).

While there are numerous methods to calculate freezing times, the method of Plank (1913) is presented here. Although this method was developed for freezing of water, its simplicity and applicability to foods make it well-liked by engineers. One modification is presented below.

In Plank's method, freezing time is calculated as:

$$t_f = \frac{\lambda \rho_f}{(T_f - T_a)} \left(\frac{P_a}{h} + \frac{R_a^2}{k_f} \right) \qquad (15)$$

where t_f = freezing time (sec)

λ = latent heat of fusion of the food (kJ/kg); if this value is unknown, it can be estimated using equation 1

ρ_f = density of the frozen food (kg/m³) (ASHRAE tables)

T_f = freezing point temperature (°C) (ASHRAE tables or equation 10)

T_a = freezing medium temperature (°C) (manufacturer specifications; table 4 for examples)

a = the thickness of an infinite slab, the diameter of a sphere or an infinite cylinder, or the smallest dimension of a rectangular brick or cube (m)

P and R = shape factor parameters determined by the shape of the food being frozen (table 5).

h = convective heat transfer coefficient (W/m²°C) (equation 13 or table 4)

Table 5. Shape factors for use in equations 15 and 16 (Lopez-Leiva and Hallstrom, 2003).

Shape	P	R
Infinite plate[a]	1/2	1/8
Infinite cylinder[b]	1/4	1/16
Cylinder[c]	1/6	1/24
Sphere	1/6	1/24
Cube	1/6	1/24

[a] A plate whose length and width are large compared with the thickness
[b] A cylinder with length much larger than the radius (i.e., a very long cylinder)
[c] A cylinder with length equal to its radius

k_f = thermal conductivity of the frozen food (W/m°C) (ASHRAE tables)

When the dimensions of the food are not infinite or spherical (for example, a brick-shaped product or a box), charts are available to determine the shape factors P and R (Cleland and Earle, 1982).

There are four common assumptions for using Plank's method to calculate freezing times of food products. Here freezing time is defined as the time to freeze the geometrical center of the product.

- The first assumption is that freezing starts with all water in the food unfrozen but at its freezing point, T_f, and loss of sensible heat is ignored. In other words, the initial freezing temperature is constant at T_f and the unfrozen center is also at T_f. The food product is not at temperatures above its initial freezing point and the temperature within the food is uniform.
- The second assumption is that heat transfer takes place sufficiently slowly for steady-state conditions to operate. This means that the food product is at equilibrium conditions and temperature is constant at a specified location (e.g., center or surface of the product). Furthermore, the heat given off is removed by conduction through the inside of the food product and convection at the outside surface, described by combining equations 11 and 12.
- The third assumption is that the food product is homogeneous and its thermal and physical properties are constant when unfrozen and then change to a different constant value when it is frozen.

This assumption addresses the fact that thermal conductivity, k, a thermal property of the product that determines its ability to conduct heat energy, is a function of temperature, more importantly below freezing. For instance, a piece of aluminum conducts heat very well and it has a large value of k. On the other hand, plastics are poor heat conductors and have low values of k. Relative to other liquids, water is a good conductor of heat, with a k value of 0.6 W/mK. In the case of foods, k depends on product composition, temperature, and pressure, with water content playing a significant role, similar to specific heat. One distinction is that k is affected by the porosity of the material and the direction of heat (this is called anisotropy). Thus, the higher the moisture content in the food, the closer the k value is to the one for water. Equations to calculate this thermal property as a function of temperature and composition are also provided by Choi and Okos (1986) and ASHRAE (2018). In the case of a frozen food, $k_{frozen\ food}$ is almost four times larger than the value of unfrozen food since k_{ice} is approximately four times the value of $k_{liquid\ water}$ (k_{ice} = 2.4 W/m°C, $k_{liquid\ water}$ = 0.6 W/m°C).

This third assumption also reminds us that the density, ρ, of food materials is affected by temperature (mostly below freezing), moisture content, and porosity. Equations to calculate density as a function of temperature and composition of foods are also provided by Choi and Okos (1986) and ASHRAE (2018). In the case of a frozen food, $\rho_{frozen\ food}$ is lower than the value of unfrozen food since ρ_{ice} is lower than $\rho_{liquid\ water}$ (e.g., ice floats in water).

- The fourth assumption is that the geometry of the food can be considered as one dimensional, i.e., heat transfers only in the direction of the radius of a cylinder or sphere or through the thickness of a plate and that heat transfer through other directions is negligible.

Despite its simplifying assumptions, Plank's method gives good results as long as the food's initial freezing temperature, thermal conductivity, and density of the frozen food are known. Modifications of equation 15 provide some improvement but still have limitations (Cleland and Earle, 1982; Pham, 1987). Nevertheless, Plank's method is widely used for a variety of foods.

One modified version of Plank's method (equation 15) that is commonly used was developed to calculate freezing times of packaged foods (Singh and Heldman, 2013):

$$t_f = \frac{\lambda \rho_f}{(T_f - T_a)} \left[PL\left(\frac{1}{h} + \frac{x}{k_2}\right) + \frac{R_a^2}{k_1} \right] \quad (16)$$

where L = length of the food (m)
  a = thickness of the food (m); assume the food fills the package
  x = thickness of packaging material (m)
  k_1 = thermal conductivity of packaging material (W/m°C)
  k_2 = thermal conductivity of the frozen food (W/m°C)

with other variables as defined in equation 15.

The term $\dfrac{1}{\left(\dfrac{1}{h} + \dfrac{x}{k_2}\right)}$ is known as the overall convective heat transfer coefficient. It includes both the convective ($1/h$) and the conductive (x/k_2) resistance to heat transfer through the packaging material.

Applications

Engineers use the concepts described in the previous section to analyze and design freezing processes and to select proper equipment by establishing system capacity requirements. Proper design of a freezing process requires knowledge of food properties including specific heat, thermal conductivity, density, latent heat of fusion, and initial freezing point, as well as the size and shape of the food, its packaging requirements, the cooling load, and the freezing rate and time (Heldman and Singh, 2013). All of these parameters can be calculated using the information described in this chapter.

When the freezing process is not properly designed, it might induce changes in texture and organoleptic (determined using the senses) properties of the foods and loss of shelf life (Singh and Helmand, 2013). Other disadvantages of freezing include the following:

- product weight losses often range between 4% and 10%;
- freezing injury of unpackaged foods in slow freezing processes causes cell-wall rupture due to the formation of large ice crystals;
- frozen products require frozen shipping and storage, which can be expensive;
- loss of nutrients such as vitamins B and C have been reported; and
- frozen foods should not be stored for longer than a year to avoid quality losses due to freezer burn (i.e., food surface gets dry and brown).

Because of the potential disadvantages of freezing, proper design of a freezing process also requires the following considerations:

- the parts of the equipment that will be in contact with the food (e.g., stainless steel) should not impart any flavor or odor to the food;
- the conditions in the processing plant should be sanitary and allow for easy cleaning;
- the equipment should be easy to operate;
- the packaging should be chosen to prevent freezer burn and other quality losses; and
- the properties of the food that is frozen rapidly may be different from when the food is being frozen slowly.

There is a wide variety of equipment available for freezing of food (table 6). The choice of freezing equipment depends upon the rate of freezing required as well as the size, shape, and packaging requirements of the food.

Table 6. Common types of freezers used in the food industry.

Type of Freezer	Freezing Rate Range
Slow (still-air, cold store)	1°C and 10°C/h (0.2 to 0.5 cm/h)
Quick (air-blast, plate, tunnel)	10°C and 50°C/h (0.5 to 3 cm/h)
Rapid (fluidized-bed, immersion)	Above 50°C/h (5 to 10 cm/h)

Sources: George (1993), Singh (2003), Sudheer and Indira (2007), Pham (2014).

Traditional Freezing Systems

Slow freezers are commonly used for the freezing and storage of frozen foods and are common practice in developing countries (Barbosa-Canovas et al., 2005). Examples of "still" freezers are ice boxes and chest freezers, a batch-type, stationary type of freezer that uses air between –20°C and –30°C. Air is usually circulated by fans (~1.8 m/s). This freezing method is low cost and requires little labor but product quality is low because it may take 3 to 72 h to freeze a 65-kg meat carcass (Pham, 2014).

Quick freezers are more common within the food industry because they are very flexible, easy to operate, and cost-effective for large-throughput operations (George, 1993). Air is forced over the food at 2 to 6 m/s, for an increased rate of heat transfer compared to slow freezers. Blast freezers are examples of this category and are available in batch or continuous mode (in the form of tunnels, spiral, and plate). These quick and blast freezers are relatively economical and provide flexibility to the food processor in terms of type and shape of foods. It takes 10 to 15 minutes to freeze products such as hamburger patties or ice cream (Sudheer and Indira, 2007). Throughput ranges from 350 to 5500 kg/hr.

Rapid freezers are well-suited for individual quick-frozen (IQF) products, such as peas and diced foods, because the very efficient transfer of heat through small-sized products induces the rapid formation of ice throughout the product and, consequently, greater product quality (George, 1993). Fluidized bed freezers are the most common type of freezer used for IQF processes. It usually takes three to four minutes to freeze unpacked peas (Sudheer and Indira, 2007). Throughput ranges from 250 to 3000 kg/hr.

Immersion freezers provide extremely rapid freezing of individual food portions by immersing the product into either a cryogen (a substance that produces very low temperatures, e.g., liquid nitrogen) or a fluid refrigerant with very low freezing temperatures (e.g., carbon dioxide). Immersion freezers also provide uniform temperature distribution throughout the product, which helps maintain product quality. It takes 10 to 15 minutes to freeze many food types (Singh, 2003).

Ultra-rapid freezers (e.g., cryogenic freezers) are suitable for high product throughput rates (over 1500 kg/h), require very little floor space, and are very flexible because they can be used with many types of food products, such as fish fillets, shellfish, pastries, burgers, meat slices, sausages, pizzas, and extruded products (George, 1993). It takes between one-half and one minute to freeze a variety of food items.

Freeze Drying

Freeze drying is a specific type of freezing process commonly used in the food industry (McHug, 2018). The process combines drying and freezing operations. In brief, the product is dried (i.e., moisture is removed) using the principle of sublimation of ice to water vapor. Hence, the product is dried at temperature and pressure below the triple point of water. (The triple point is the

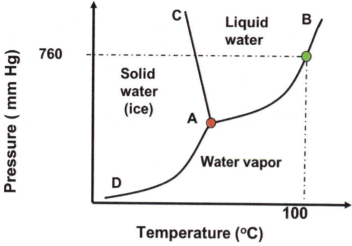

Figure 3. Phase diagram of water highlighting the different phases. A (red dot): Triple point of water, 0.01°C and 0.459 mm Hg. A-B line: Saturation (vaporization) line. A-C line: Solidification/fusion line. A-D line: Sublimation/deposition line. The green dot represents the T (100°C) at which water boils at atmospheric pressure (760 mm Hg).

temperature and pressure at which water exists in equilibrium with its three phases, gas, liquid and solid. At the triple point, $T = 0.01°C$ and $P = 611.2$ Pa; see figure 3.) The A-B line in figure 3 represents the saturation (vaporization) line when water transitions from liquid to gas or vice versa; the A-C line represents the line where water transitions from solid to liquid (fusion or melting) or from liquid to solid (solidification or freezing); and the A-D line represents the sublimation line when water transitions from solid to gas directly (as in freeze drying) or when it changes from gas to solid (deposition). Freeze drying is popular for manufacture of rehydrating foods, such as coffee, fruits and vegetables, meat, eggs, and dairy, due to the minimal changes to the products' physical and chemical properties (Luo and Shu, 2017). This phase change of water occurs at very low pressures.

New Freezing Processes

Alternatives to traditional freezing methods are evaluated to make freezing suitable for all types of foods, optimize the amount of energy used and reduce the impact in the environment. Processes such as impingement freezing and hydrofluidization (HF), an immersion type freezer that uses ice slurries, provide higher surface heat-transfer rates with increasing freezing rates, which has tremendous potential to improve the quality of products such as hamburgers or fish fillets (James and James, 2014). These methods use very high velocity air or refrigerant jets that enable very fast freezing of the product. Studies on their applications to foods and other biological materials are in progress. Operating conditions and feasibility of the techniques must be assessed before implementation.

Other promising technologies include high-pressure freezing (also called pressure-shifting freezing) (Otero and Sanz, 2012) and ultrasound-assisted freezing (Delgado and Sun, 2012), which facilitate formation of smaller ice crystals. Magnetic resonance and microwave-assisted freezing, cryofixation, and osmodehydrofreezing are other new freezing technologies. Another trend is "smart freezing" technology, which combines the mechanical aspects of freezing with sensor technologies to track food quality throughout the cold chain. Smart freezing uses computer vision and wireless sensor networks (WSN), real-time diagnosis tools to optimize the process, ultrasonic monitoring of the freezing process, gas sensors to predict crystal size in ice cream, and temperature-tracking sensors to predict freezing times and product quality (Xu et al., 2017).

Examples

Examples 1 through 6 show a few of the many options an engineer could consider when selecting the best type of freezing equipment and operational parameters to freeze a food product. Another critical aspect of design of freezing processes for foods is that many of the products are packaged and the packaging material offers resistance to the transfer of heat, thus increasing freezing time (Yanniotis, 2008). Example 7 illustrates this point.

Example 1: Calculation of refrigeration requirement to freeze a food product

Problem:
Calculate the refrigeration requirement when freezing 2,000 kg of strawberries (91.6% moisture) from an initial temperature of 20°C to −20°C. The initial freezing point of strawberries is −0.78°C (table 1).

Solution:
(1) identify the type of heat process(es) involved in this process and set up the energy balance; (2) calculate how much heat energy must be removed from the strawberries to carry out the freezing process; and (3) calculate the refrigeration requirement (in kW) for the freezing process.

The following assumptions are commonly made in this type of calculation:

- Conservation of mass during the freezing process. Thus, $m_{strawberries}$ = 2,000 kg remains constant because the fruits do not lose or gain moisture (or the changes in mass are negligible).
- The freezing point temperature is known.
- The temperature of the freezing medium (ambient temperature) and storage remains constant (i.e., a steady-state situation).

Step 1 *Identify the type of heat processes and set up the energy balance:*
- Sensible, to decrease the temperature of the strawberries from 20°C to just when they begin to crystallize at −0.78°C
- Latent, to change liquid water in strawberries to ice at −0.78°C
- Sensible, to further cool the strawberries to −20°C (using equation 2)

Thus, for the given freezing process, the heat energy balance is the sum of the three heat processes listed above.

Step 2 *Calculate the total amount of energy removed in the freezing process, Q:*

Sensible, from 20°C to −0.78°C, using equation 2, where $Q_s = Q_1$:

$$Q_1 = mC_{p,\text{unfrozen}}(T_2 - T_1) \qquad (2)$$

where $m = 2{,}000$ kg

$$T_1 = 20°C$$

$$T_2 = -0.78°C$$

C_p of unfrozen strawberries (at 91.6 % moisture) = 4.00 kJ/kg°C (table 1). See example 2 for calculation of the specific heat of a food product above freezing.

Thus,

$$Q_1 = (2{,}000 \text{ kg})(4.00 \text{ kJ/kg°C})(-0.78 - 20°C) = -166{,}240 \text{ kJ}$$

Note that this value is negative because heat is released from the product.

Latent, using equation 9:

$$Q_2 = m\lambda \text{ at } T = -0.78°C \qquad (9)$$

where $m = 2{,}000$ kg
λ = latent heat of fusion of strawberries at given moisture content = 306 kJ/kg (from table 1).

Thus,

$$Q_2 = (2{,}000 \text{ kg})(306 \text{ kJ/kg}) = -612{,}000 \text{ kJ}$$

Note that this value is negative because heat is being released from the product.

Sensible, to further cool to −20°C, again using equation 2:

$$Q_3 = mC_{p,\text{frozen}}(T_2 - T_1) \qquad (2)$$

where $m = 2{,}000$ kg

$$T_1 = -0.78°C$$

$$T_2 = -20°C$$

C_p of frozen strawberries (at 91.6 % moisture) = 1.84 kJ/kg°C (table 1). See example 3 for calculation of the specific heat of a frozen food product.

Thus,

$$Q_3 = (2{,}000 \text{ kg})(1.84 \text{ kJ/kg°C})(-20 + 0.78°C) = -70{,}729.6 \text{ kJ}$$

Adding all energy terms:

$$Q = Q_1 + Q_2 + Q_3 = -166{,}240 \text{ kJ} - 612{,}000 \text{ kJ} - 70{,}729.6 \text{ kJ}$$

$$= -848{,}969.6 \text{ kJ} = Q_{product}$$

The heat energy removed per kg of strawberries:

$$Q_{product} \text{ per kg of fruit} = -848{,}969.6 \text{ kJ}/2{,}000 \text{ kg} = -424.48 \text{ kJ/kg}$$

Thus, 848,969.6 kJ of heat must be removed from the 2,000 kg of strawberries (424.48 kJ/kg) initially held at 20°C to freeze them to the target storage temperature of –20°C.

Step 3 *Calculate the refrigeration requirement, or cooling load (in kW), for the freezing process. The cooling load $\dot{Q}_p$ (also called refrigeration requirement) to freeze 2,000 kg/h of strawberries from 20°C to –20°C is calculated with equation 14:*

$$\dot{Q}_p = \dot{m}_p Q_p \qquad (14)$$

where $\dot{m}_p$ = mass flow rate of product (kg/s)
Q_p = heat energy removed to freeze the product to the target temperature = –424.48 kJ/kg
$\dot{Q}_p$ = (2,000 kg/h × –424.48 kJ/kg) /3600 s = –235.82 kJ/s or kW

Note that 1 kJ/s = 1 kilowatt = 1 kW.

Example 2: Determine the initial freezing point (i.e., temperature at which water in food begins to freeze) and the latent heat of fusion of a food product

Problem:
Determine the initial freezing point and latent heat of fusion of green peas with 79% moisture.

Solution by use of tables:
From table 1, the T_f of green peas at the given moisture content is –0.61°C and the latent heat of fusion λ is 263 kJ/kg.

Solution by calculation:
If tabulated λ values are not available, λ of the product can be estimated using equation 1.

$$\lambda = M_{water} \times \lambda_w \tag{1}$$

$$\lambda = (0.79) \times (334 \text{ kJ/kg}) = 263.86 \text{ kJ/kg}$$

Example 3: Estimation of specific heat of a food product based on composition

Sometimes the engineer will not have access to measured or tabulated values of the specific heat of the food product and will have to estimate it in order to calculate cooling loads. This example provides some insight into how to estimate specific heat of a food product.

Problem:
Calculate the specific heat of honeydew melon at 20°C and at −20°C. Composition data for the melon is available as 89.66% water, 0.46% protein, 0.1% fat, 9.18% total carbohydrates (includes fiber), and 0.6% ash (USDA, 2019). Give answers in the SI units of kJ/kgK.

Solution:
Use equations 5-7 to account for the effect of product composition and temperature on specific heat. The specific heat of honeydew at $T = 20°C$ is calculated using equation 4:

$$C_p = \sum_{i=1}^{n} X_i C_{pi} \tag{4}$$

Thus,

$$C_{p,honeydew} = X_w C_{pw} + X_p C_{pp} + X_f C_{pf} + X_c C_{pc} + X_a C_{pa}$$

with subscripts $w, p, f, c,$ and a representing water, protein, fat, carbohydrates, and ash, respectively, and C_p in kJ/kg°C.

Step 1 *Calculate the C_p of water (C_{pw}) at 20°C using equation 6:*

$$C_p = 4.1289 - 9.0864 \times 10^{-5}\, T + 5.4731 \times 10^{-6} T^2 \tag{6}$$

Thus,

$$C_p = 4.1289 - (9.0864 \times 10^{-5})(20) + (5.4731 \times 10^{-6})(20)^2$$

$$C_{pw} = 4.127 \text{ kJ/kg°C}$$

Step 2 *Calculate the specific heat of the different components at T = 20°C using the equations given in table 2.*

Food Component	C_p of Honeydew at T = 20°C (kJ/kgK)
Water	4.127
Protein	2.032
Fat	2.012
Carbohydrate	1.586
Fiber	NA
Ash	1.129

Step 3 *Calculate the specific heat of honeydew at 20°C using equation 4:*

$$C_{p,honeydew} = (0.8966)(4.127) + (0.0046)(2.032) + (0.001)(2.012) \\ + (0.0918)(1.586) + (0.006)(1.129)$$

$$C_{p,honeydew} \text{ at } 20°C = 3.86 \text{ kJ/kg°C}$$

Step 4 *Calculate the specific heat of honeydew at −20°C using equation 8:*

$$C_{p,\text{ frozen}} = 1.55 + 1.26 X_s + \frac{(X_{w0} - X_b) L_0 T_f}{T^2} \qquad (8)$$

From the given composition of honeydew:

$$X_s = \text{mass fraction of solids} = 1 - 0.8966 = 0.1034$$

$$X_{w0} = \text{mass fraction of water in the unfrozen food} = 0.8966$$

$$X_b = \text{bound water} = 0.4 X_p = 0.4(0.0046) = 0.00184$$

$$T_f = \text{freezing point of food to be frozen} = -0.89°C \text{ (from table 1)}$$

$$T = \text{food target (or freezing process) temperature} = -20°C$$

Substituting the numbers into equation 8:

$$C_{p,\text{ frozen}} = 1.55 + 1.26 X_s + \frac{(X_{w0} - X_b) L_0 T_f}{T^2} \qquad (8)$$

$$C_{p,frozen} = 1.55 + 1.26(0.1034) + \frac{(0.8966 - 0.00184)\left(\frac{334 \text{ kJ}}{\text{kg}}\right)(-0.89°C)}{(-20°C)^2}$$

$$C_{p,frozen} = 2.397 \text{ kJ/kg°C}$$

The calculated C_p values can then be used to calculate cooling load as shown in example 1.

Observations:
- As expected, the specific heat of the frozen honeydew is lower than the value for the fruit above freezing.
- When values of the product's specific heat and initial freezing point are not available from tables, engineers should be able to estimate them using available prediction models and composition data.
- The C_p of the frozen product was calculated at −20°C, the freezing process temperature. Tabulated values are usually given when the food is fully frozen at a reference temperature of −40°C (ASHRAE, 2018). If we use −40°C in equation 8, then

$$C_{p,frozen} = 1.55 + 1.26(0.1034) + \frac{(0.8966 - 0.00184)\left(\frac{334 kJ}{kg}\right)(-0.89°C)}{(-40°C)^2}$$

$$C_{p,frozen} = 1.85 \text{ kJ/kg°C}.$$

This value is closer to the tabulated values. While the change in C_p as a function of temperature can be important in research studies, it does not influence the selection of freezing equipment.
- Many mathematical models are available for prediction of specific heat and other properties of foods (Mohsenin, 1980; Choi and Okos, 1986; ASHRAE, 2018). The engineer must choose the value that is more suitable for the specific application using available composition and temperature data.

Example 4: Calculation of initial freezing point temperature of a food product

Problem:
Consider the strawberries in example 1 and calculate the depression of the initial freezing point of the fruit assuming the main solid present in the strawberries is fructose (a sugar), with molecular weight of 108.16 g/mol.

Solution:
Calculation of the initial freezing point temperature requires a series of steps.

Step 1 *Collect all necessary data. From example 1, strawberries contain 91.6% water (m_A) and the rest is fructose (100 − 91.6 = 8.04% solids = m_s). Other information provided is $M_s = M_{fructose}$ = 108.16 g/mol, λ = 6,003 J/mol, M_A = 18 g/mol, R = 8.314 J/mol K, and T_0 = 273.15 K.*

Step 2 *Calculate X_A, the molar fraction of liquid (water) in the strawberries (decimal) using equation 11:*

$$X_A = \frac{\frac{0.916}{18}}{\frac{0.196}{18} + \frac{0.0804}{108.16}} = 0.9922$$

Step 3 *Calculate T_f of strawberries using equation 10:*

$$\ln X_A = \frac{\lambda}{R}\left(\frac{1}{T_0} - \frac{1}{T_f}\right) \tag{10}$$

Rearranged:

$$T_f = \left(\frac{R}{\lambda}\ln X_A - \frac{1}{T_0}\right)^{-1}$$

$$T_f = \left[\frac{8.314 \text{ J/mol K}}{6003 \text{ J/mol}}\ln(0.9922) - \frac{1}{273.15}\right]^{-1}$$

$$T_f = 272.34 \text{ K} = -0.81°C$$

Observation:

The presence of fructose in the strawberries results in an initial freezing point temperature lower than that for pure water.

Example 5: Calculation of freezing time of an unpackaged food product

An air-blast freezer is used to freeze cod fillets (81.22% moisture, freezing point temperature = −2.2°C, initial temperature = 5°C, mass of fish = 1 kg). Assume that each cod fillet is an infinite plate with thickness of 6 cm. Freezing process parameters for the air-blast freezer are: freezing medium temperature −20°C, convective heat transfer coefficient, h, of 50 W/m²°C (table 4), the density and thermal conductivity of the frozen fish are 992 kg/m³ and 1.9 W/m°C, respectively (ASHRAE, 2018). The target freezing time is less than 2 hours.

Problem:

Calculate the time required to freeze a fish fillet (freezing time, t_f), using Plank's method (equation 15):

$$t_f = \frac{\lambda \rho_f}{(T_F - T_a)}\left(\frac{P_a}{h} + \frac{R_a^2}{k_f}\right) \tag{15}$$

Solution:

Step 1 *Determine the required food and process parameters:*

λ = latent heat of fusion of cod fillet = 271.27 kJ/kg (from tables, ASHRAE, or calculated using equation 1, λ = (0.8122)(334 kJ/kg) = 271.27 kJ/kg = 271.27 × 10³ J/kg)

ρ_f = density of the frozen food, 992 kg/m³ (from ASHRAE, 2018)

T_f = freezing point temperature, −2.2°C (available in ASHRAE, 2018, or calculated using composition and equations 10 and 11)

T_a = freezing medium temperature, −20°C

a = thickness of the plate = 6 cm = 0.06 m

P and R = shape factor parameters, 1/2 and 1/8 (from table 6)

h = convective heat transfer coefficient, 50 W/m²°C (given)

k_f = thermal conductivity of the frozen food, 1.9 W/m°C (from ASHRAE)

Step 2 *Calculate the freezing time, t_f, from equation 15 as:*

$$t_f = \frac{\left(271.27 \times 10^3 \frac{J}{kg}\right)\left(992 \frac{kg}{m^3}\right)}{[(-2.2)-(20C)]}\left[\frac{(0.06\ m)}{2\left(\frac{50\ W}{m^2 C}\right)} + \frac{(0.06\ m)^2}{8\left(\frac{1.9\ W}{m\ C}\right)}\right]$$

t_f = 12,651.35 seconds/3600 = 3.5 h. The freezing time target would not be met.

Reminder:

Plank's method calculates the time required to remove the latent heat to freeze the fish. It does not take into account the time required to remove the sensible heat from the initial temperature of 5°C to the initial freezing point. This means that use of equation 15 might underestimate freezing times.

As shown in example 1, Q_S, the sensible heat removed to decrease the temperature of the fish from 5°C to just when it begins to crystallize at −2.2°C is calculated using equation 2:

$$Q_S = mC_p(T_2 - T_1) \qquad (2)$$

where m = mass of the food = 1 kg

C_p = specific heat of the unfrozen cod (at 81.22 % moisture) = 3.78 kJ/kg°C (from tables, ASHRAE, 2018).

$$T_1 = 5°C$$

$$T_2 = -2.2°C$$

Thus, Q_S = (1 kg)(3.78 kJ/kg°C)(−2.2 − 5°C) = −27.216 kJ = −27,216 J of heat energy removed per kg of fish. The quantity is negative because heat is released from the product when it is cooled. Also, although not negligible, this amount is much lower than the latent heat removal.

Example 6: Find ways to decrease freezing time of an unpackaged food product

Problem:
Find a way to decrease freezing time of the cod fillets in example 5 to less than 2 hours.

Solution:
Evaluate the effect (if any) of some process and product variables on the calculated freezing time, t_f, using Plank's method and determine which parameters decrease freezing time. Equation 15:

$$t_f = \frac{\lambda \rho_f}{(T_F - T_a)} \left(\frac{P_a}{h} + \frac{R_a^2}{k_f} \right) \tag{15}$$

Freezing process variables:

- Freezing time decreases when freezing medium temperature, T_a, decreases (colder medium):

$$t_f \propto \frac{1}{(T_F - T_a)}$$

- Freezing time decreases when the convective heat transfer coefficient, h, increases (faster removal of heat energy and thus faster freezing process):

$$t_f \propto \left(\frac{1}{h} \right)$$

Product variables:

- Freezing time decreases when the thickness, a, of the product decreases (smaller product):

$$t_f \propto \left(\frac{a}{h} + \frac{a^2}{k_f} \right)$$

- Freezing time decreases when the product shape changes from a plate to a cylinder or a sphere (greater surface area), i.e., P decreases from 1/2 to 1/6 and R decreases from 1/8 to 1/24:

$$t_f \propto \left(\frac{P}{h} + \frac{R}{k_f} \right)$$

- Freezing time decreases with a lower latent heat of fusion of the food, λ, a lower density of the frozen food, ρ_f, and a higher thermal conductivity of the frozen food, k_f:

$$t_f \propto \lambda \rho_f \left(\frac{1}{k_f} \right)$$

This highlights the need for accurate values for these variables when using this method to calculate freezing times.

- The effect of the initial freezing point is less significant due to the small range of variability among a wide variety of food products:

$$t_f \propto \frac{1}{(T_f - T_a)}$$

Changing freezing process variables. Do the calculation assuming a freezing medium temperature of −40°C (table 4) instead of the T_a = −20°C used in example 5, while holding everything else constant:

$$t_f = \frac{\left(271.27 \times 10^3 \frac{J}{kg}\right)\left(992 \frac{kg}{m^3}\right)}{[(-2.2)-(40C)]}\left[\frac{(0.06\ m)}{2\left(\frac{50\ W}{m^2 C}\right)} + \frac{(0.06\ m)^2}{8\left(\frac{1.9\ W}{m\ C}\right)}\right]$$

t_f = 5957.51 seconds ~1.66 h < 2 h. The freezing time target would be met.

This result makes sense because the lower the temperature of the freezing medium (air, in an air-blast freezer), the shorter the freezing time.

Next, consider increasing the convective heat transfer coefficient, h, for the air-blast freezer. Based on table 4, this variable can go as high as 200 W/m°C for this type of freezer. While holding everything else constant,

$$t_f = \frac{\left(271.27 \times 10^3 \frac{J}{kg}\right)\left(992 \frac{kg}{m^3}\right)}{[(-2.2)-(20C)]}\left[\frac{(0.06\ m)}{2\left(\frac{200\ W}{m^2 C}\right)} + \frac{(0.06\ m)^2}{8\left(\frac{1.9\ W}{m\ C}\right)}\right]$$

t_f = 5848.27 seconds ~ 1.63 h < 2 h. The freezing time target would be met.

This result also makes sense because the faster the freezing rate (due to higher h value), the shorter the freezing time.

Achieving the target freezing time of less than 2 hours would require a change in the freezing process parameters of the air-blast freezer, either the convective heat transfer coefficient h or the operating conditions (the freezing medium temperature, T_a).

Changing product variables. Try changing the thickness, a. Assume the fish is frozen as fillets that are 3 cm thick (half the thickness of the original design). Holding everything else constant except now a = 0.03 m,

$$t_f = \frac{\left(271.27 \times 10^3 \frac{J}{kg}\right)\left(992 \frac{kg}{m^3}\right)}{[(-2.2)-(20C)]}\left[\frac{(0.03\ m)}{2\left(\frac{50\ W}{m^2 C}\right)} + \frac{(0.06\ m)^2}{8\left(\frac{1.9\ W}{m\ C}\right)}\right]$$

t_f = 5430.53 seconds = 1.5 h < 2 h. The freezing time target would be met.

In this case, there is no need to change the operating conditions of the air-blast freezer.

Next, change the shape of the product. Fillets can be shaped as infinite (very long) cylinders (P and R = 1/4 and 1/16, respectively; table 5) with 6 cm diameter, instead of as long plates. Keeping everything else constant and using the original freezing process parameters:

$$t_f = \frac{\left(271.27 \times 10^3 \frac{J}{kg}\right)\left(992 \frac{kg}{m^3}\right)}{[(-2.2)-(20C)]}\left[\frac{(0.06\ m)}{4\left(\frac{50\ W}{m^2 C}\right)} + \frac{(0.06\ m)^2}{16\left(\frac{1.9\ W}{m\ C}\right)}\right]$$

t_f = 6325.68 seconds = 1.76 h < 2 h. The freezing time target would be met.

This result illustrates the significance of product shape on the rate of heat transfer and, consequently, freezing time. In general, a spherical product will freeze faster than one of similar size with the shape of a cylinder or a plate due to its greater surface area.

Example 7. Calculation of freezing time of a packaged food product

For this example, assume that the cod fish from example 5 is packed into a cardboard carton measuring 10 cm × 10 cm × 10 cm. The carton thickness is 1.5 mm and its thermal conductivity is 0.065 W/m°C.

Problem:
Calculate the freezing time using the original freezing process parameters (h = 50 W/m²°C, T_a = −20°C) and determine whether the product can be frozen in 2 to 3 hours. If not, provide recommendations to achieve the desired freezing time.

Solution:
Because the food is packaged, use the modified version of Plank (equation 16):

$$t_f = \frac{\lambda \rho_f}{(T_F - T_a)}\left[PL\left(\frac{1}{h} + \frac{x}{k_2}\right) + \frac{R_a^2}{k_1}\right] \quad (16)$$

Step 1 *Collect the information needed from example 5. Also,*

$$L = \text{length of the food} = 10 \text{ cm} = 0.1 \text{ m}$$

$$a = 10 \text{ cm} = 0.1 \text{ m}$$

$$x = \text{thickness of packaging material} = 1.5 \text{ mm} = 0.0015 \text{ m}$$

$$k_2 = \text{thermal conductivity of packaging material} = 0.065 \text{ W/m°C}$$

$$k_1 = \text{thermal conductivity of the frozen fish} = 1.9 \text{ W/m°C}$$

Step 2 *Calculate the freezing time:*

$$t_f = \frac{\left(271.27 \times 10^3 \frac{J}{kg}\right)\left(992 \frac{kg}{m^3}\right)}{17.8°C}\left[\frac{0.1 \text{ m}}{6}\left(\frac{1}{50} + \frac{0.0015}{0.065}\right) + \left(\frac{(0.1 \text{ m})^2}{24(1.9)}\right)\right]$$

$$t_f = 14{,}200.28 \text{ seconds} = 3.9 \text{ h} \ggg 2 \text{ to } 3 \text{ h.}$$

The freezing time target would not be met.

The freezing process must be modified. Note that freezing of the fish when packaged in cardboard takes longer than the unpackaged product.

Step 3 *Calculate some possible options to reduce the freezing time.*
- Shorten the freezing time by using a higher convective heat transfer coefficient, h, of 100 W/m²°C. Then,

$$t_f = \frac{\left(271.27 \times 10^3 \frac{J}{kg}\right)\left(992 \frac{kg}{m^3}\right)}{17.8°C}\left[\frac{0.1 \text{ m}}{6}\left(\frac{1}{100} + \frac{0.0015}{0.065}\right) + \left(\frac{(0.1 \text{ m})^2}{24(1.9)}\right)\right]$$

$$t_f = 11649.6 \text{ seconds} = 3.23 \text{ h.}$$

The freezing time target would not be met.
- Shorten the freezing time by using a yet higher h of 200 W/m²°C. Then,

$$t_f = \frac{\left(271.27 \times 10^3 \frac{J}{kg}\right)\left(992 \frac{kg}{m^3}\right)}{17.8°C}\left[\frac{0.1 \text{ m}}{6}\left(\frac{1}{200} + \frac{0.0015}{0.065}\right) + \left(\frac{(0.1 \text{ m})^2}{24(1.9)}\right)\right]$$

$$t_f = 10389.8 \text{ seconds} = 2.9 \text{ h.}$$

The freezing time target would be met.
- Shorten freezing time by using $h = 100$ W/m²°C and changing the temperature of the freezing medium, T_a, to −40°C:

$$t_f = \frac{\left(271.27 \times 10^3 \, \frac{J}{kg}\right)\left(992 \, \frac{kg}{m^3}\right)}{37.88°C} \left[\frac{0.1 \, m}{6}\left(\frac{1}{100} + \frac{0.0015}{0.065}\right) + \left(\frac{(0.1 \, m)^2}{24(1.9)}\right)\right]$$

$$t_f = 5485.8 \text{ seconds} = 1.5 \text{ h}.$$

This is closer to the target freezing time for the unpackaged fish.

Freezing time could also be reduced by using a different packaging material. For example, plastics have higher k_2 values than cardboard, decreasing product resistance to heat transfer.

Note that changing the shape of the packaging container to a cylinder would not have an effect on freezing time since P and R are the same as for a cube.

Example 8. Selection of freezer

The choice of freezer equipment depends on the cost and effect on product quality. Overall, the engineer will need to consider a faster freezing process when dealing with foods packaged in cardboard, compared to unpackaged products or food packaged in plastic.

Problem:

For this example, compare the freezing times for a typical plate freezer to that of a spiral belt freezer.

Solution:

Step 1 *Calculate the freezing time for the packaged product in example 7 using a plate freezer that produces $h = 300 \, W/m^2°C$ at $T_a = -40°C$:*

$$t_f = \frac{\left(271.27 \times 10^3 \, \frac{J}{kg}\right)\left(992 \, \frac{kg}{m^3}\right)}{37.8°C} \left[\frac{0.1 \, m}{6}\left(\frac{1}{300} + \frac{0.0015}{0.065}\right) + \left(\frac{(0.1 \, m)^2}{24(1.9)}\right)\right]$$

$$t_f = 4694.8 \text{ seconds} = 1.3 \text{ h}$$

Step 2 *Calculate the freezing time for the packaged product in example 7 using a spiral freezer that produces $h = 30 \, W/m^2°C$ at $T_a = -40°C$:*

$$t_f = \frac{\left(271.27 \times 10^3 \, \frac{J}{kg}\right)\left(992 \, \frac{kg}{m^3}\right)}{37.8°C} \left[\frac{0.1 \, m}{6}\left(\frac{1}{30} + \frac{0.0015}{0.065}\right) + \left(\frac{(0.1 \, m)^2}{24(1.9)}\right)\right]$$

$$t_f = 32893.2 \text{ seconds} = 9 \text{ h}.$$

This spiral freezer would not be suitable in terms of freezing time.

Image Credits

Figure 1. Castell-Perez, M. Elena. (CC By 4.0). (2020). Freezing curves for pure water and a food product illustrating the concept of freezing point depression (latent heat is released over a range of temperatures when freezing foods versus a constant value for pure water).

Figure 2. Castell-Perez, M. Elena. (CC By 4.0). (2020). Schematic representation of the International Institute of Refrigeration definition of freezing rate.

Figure 3. Castell-Perez, M. Elena. (CC By 4.0). (2020). Phase diagram of water highlighting the different phases. A (red dot): triple point of water, 0.00098°C and 0.459 mmHg. A-B line: saturation (vaporization) line. B-C line Solidification/fusion line. A-D line: Sublimation line. The green dot represents the T (100°C).

References

ASHRAE. (2018). Thermal properties of foods. In *ASHRAE Handbook—Refrigeration* (Chapter 19). American Society of Heating, Refrigeration and Air Conditioning. https://www.ashrae.org/.
 Note: This source should be available in libraries at universities and colleges with engineering programs. It consists of four volumes with the one on "Refrigeration" being the most relevant to this chapter.

Barbosa-Cánovas, G. V., Altukanar, B., & Mejia-Lorio, D. J. (2005). Freezing of fruits and vegetables. An agri-business alternative to rural and semi-rural areas. FAO Agricultural Services Bulletin 158. Food and Agriculture Organization of the United Nations. http://www.fao.org/docrep/008/y5979e/y5979e00.htm#Contents.

Chen, C. S. (1985) Thermodynamic analysis of the freezing and thawing of foods: Enthalpy and apparent specific heat. *J. Food Sci.* 50(4), 1158-1162. https://doi.org/10.1111/j.1365-2621.1985.tb13034.x.

Choi, Y., & Okos, M. R. (1986) Effect of temperature and composition on the thermal properties of foods. In M. Le Maguer and P. Jelen (Eds.), *Food Engineering and Process Applications* (Vol.1, pp.93-101). Elsevier.

Cleland, A. C., & Earle, R. L. (1982). Freezing time prediction for foods—A simplified procedure. *Int. J. Ref.* 5(3), 134-140. https://doi.org/10.1016/0140-7007(82)90092-5

Cleland, D. J. (2003). Freezing times calculation. In *Encyclopedia of Agricultural, Food, and Biological Engineering* (pp. 396-401). Marcel Dekker, Inc.

Delgado, A., & Sun, D. W. (2012) Ultrasound-accelerated freezing. In D. W. Sun (Ed.), *Handbook of Frozen Food Processing and Packaging* (2nd ed., Chapter 28, pp. 645-666). CRC Press.

Engineering Toolbox. 2019. Specific heat of food and foodstuff. https://www.engineeringtoolbox.com/specific-heat-capacity-food-d_295.html. Accessed 8 July 2019.

Filip, S., Fink, R., & Jevšnik, M. (2010). Influence of food composition on freezing time. *Int. J. Sanitary Eng. Res.* 4, 4-13.

George, R. M. (1993). Freezing processes used in the food industry. *Trends in Food Sci. Technol.* 4(5), 134-138. https://doi.org/10.1016/0924-2244(93)90032-6.

Heldman, D. R. (1974). Predicting the relationship between unfrozen water fraction and temperature during food freezing using freezing point depression. *Trans. ASAE* 17(1), 63-66. https://doi.org/10.13031/2013.36788.

IRR. (2006). *Recommendations for the Processing and Handling of Frozen Foods*. International Institute of Refrigeration.

James, S. J., & James, C. (2014). Chilling and freezing of foods. In S. Clark, S. Jung, & B. Lamsal (Eds.), *Food Processing: Principles and Applications* (2nd ed., Chapter 5). Wiley.

Klinbun, W., & Rattanadecho, P. (2017). An investigation of the dielectric and thermal properties of frozen foods over a temperature from −18 to 80°C. *Intl. J. Food Properties* 20(2), 455-464. https://doi.org/10.1080/10942912.2016.1166129.

Lopez-Leiva, M., & Hallstrom, B. (2003). The original Plank equation and its use in the development of food freezing rate predictions. *J. Food Eng. 58*(3), 267-275. https://doi.org/10.1016/S0260-8774(02)00385-0.

Luo, N., & Shu, H. (2017). Analysis of energy saving during food freeze drying. *Procedia Eng. 205*, 3763-3768. https://doi.org/10.1016/j.proeng.2017.10.330.

McHug, T. (2018). Freeze-drying fundamentals. *Food Technol. 72*(2), 72–74.

Mohsenin, N. N. (1980). *Thermal Properties of Plant and Animal Materials.* Gordon and Breach.

Otero, L., & Sanz, P. D. (2012) High-pressure shift freezing. In D. W. Sun (Ed.), *Handbook of Frozen Food Processing and Packaging* (2nd ed., Chapter 29, pp. 667-683). CRC Press.

Pham, Q. T. (1987). Calculation of bound water in frozen food. *J. Food Sci. 52*(1), 210-212. https://doi.org/10.1111/j.1365-2621.1987.tb14006.x.

Pham, Q. T. (2014). *Food Freezing and Thawing Calculations.* SpringerBriefs in Food, Health, and Nutrition. https://doi.org/10.1007/978-1-4939-0557-7.

Plank, R. Z. (1913). *Z. Gesamte Kalte-Ind. 20*, 109. The calculation of the freezing and thawing of foodstuffs. *Modern Refrig. 52*, 52.

Schwartzberg, H. G. 1976. Effective heat capacities for the freezing and thawing of food. *J. Food Sci. 41*(1), 152-156. https://doi.org/10.1111/j.1365-2621.1976.tb01123.x.

Siebel, E. (1892). Specific heats of various products. *Ice and Refrigeration, 2*, 256-257.

Singh, R. P. (2003). Food freezing. In *Encyclopedia of Life Support Systems* (Vol. III, pp. 53-68). EOLSS. https://www.eolss.net/.

Singh, R. P., & Heldman, D. R. (2013). *Introduction to Food Engineering* (5th ed.). Academic Press.

Sudheer, K. P., & Indira, V. (2007). *Post Harvest Technology of Horticultural Crops.* Horticulture Science Series Vol. 7. New India Publishing.

USDA. 2019. Food composition database. https://ndb.nal.usda.gov/. Accessed 7 March 2019.

Xu, J.-C., Zhang, M., Mujumdar, A. S., & Adhikari, V. (2017). Recent developments in smart freezing technology applied to fresh foods. *Critical Rev. Food Sci. Nutrition 57*(13), 2835-2843. https://doi.org/10.1080/10408398.2015.1074158.

Yanniotis, S. (2008). *Solving Problems in Food Engineering.* Springer Food Engineering Series.

Principles of Thermal Processing of Packaged Foods

Ricardo Simpson
Departamento de Ingeniería Química y Ambiental,
Universidad Técnica Federico Santa María, Valparaíso, Chile
Centro Regional de Estudios en Alimentos y Salud (CREAS)
Conicyt-Regional GORE Valparaíso Project R17A10001,
Curauma, Valparaíso, Chile

Cristian Ramírez
Departamento de Ingeniería Química y Ambiental,
Universidad Técnica Federico Santa María, Valparaíso, Chile
Centro Regional de Estudios en Alimentos y Salud (CREAS)
Conicyt-Regional GORE Valparaíso Project R17A10001,
Curauma, Valparaíso, Chile

Helena Nuñez
Departamento de Ingeniería Química y Ambiental,
Universidad Técnica Federico Santa Maria, Valparaíso, Chile

KEY TERMS		
Heat transfer	Bacterial inactivation	Food sterilization
Microorganism heat resistance	Decimal reduction time	Commercial sterilization

Variables

- α = thermal diffusivity
- ρ = density
- C_p = specific heat
- CUT = time required to come up to retort temperature
- D = decimal reduction time
- F_0 = cumulative lethality of the process from time 0 to the end of the process
- I = inactivation
- k = rate constant
- K_t = thermal conductivity
- N = number
- t = time

T = temperature
T_{ref} = reference temperature
TRT = retort temperature
z = temperature change

Introduction

Thermal processing of foods, like cooking, involves heat and food. However, thermal processing is applied to ensure food safety and not necessarily to cook the food. Thermal processing as a means of preservation of uncooked food was invented in France in 1795 by Nicholas Appert, a chef who was determined to win the prize of 12,000 francs offered by Napoleon for a way to prevent military food supplies from spoiling. Appert worked with Peter Durand to preserve meats and vegetables encased in jars or tin cans under vacuum and sealed with pitch and, by 1804, opened his first vacuum-packing plant. This French military secret soon leaked out, but it took more than 50 years for Louis Pasteur to provide the explanation for the effectiveness of Appert's method, when Pasteur was able to demonstrate that the growth of microorganisms was the cause of food spoilage.

The preservation for storage by thermal treatment and removal of atmosphere is known generically as *canning*, regardless of what container is used to store the food. The basic principles of canning have not changed dramatically since Appert and Durand developed the process: apply enough heat to food to destroy or inactivate microorganisms, then pack the food into sealed or "airtight" containers, ideally under vacuum. Canned foods have a shelf life of one to four years at ordinary temperatures, making them convenient, affordable, and easy to transport.

Outcomes

After reading this chapter, you should be able to:

- Identify the role of heat transfer concepts in thermal processing of packaged foods
- Describe the principles of commercial sterilization of foods
- Describe the inactivation conditions needed for some example microorganisms important for food safety
- Define some sterilization criteria for specific foods
- Apply, in simple form, the main thermal food processing evaluation techniques

Concepts

The main concepts used in thermal processing of foods include: (a) heat transfer; (b) heat resistance of microorganisms of concern; and (c) bacterial inactivation.

Heat Transfer

The main heat transfer mechanisms involved in the thermal processing of packaged foods are convection and conduction. Heat transfer by convection occurs due to the motion and mixing of flows. The term *natural convection* refers to the case when motion and mixing of flow is caused by density differences in different locations due to temperature gradients. The term *forced convection* refers to the case when motion and mixing of flow is produced by an outside force, e.g., a fan. Heat transfer by conduction occurs when atoms and molecules collide, transferring kinetic energy. Conceptually, atoms are bonded to their neighbors, and if energy is supplied to one part of the solid, atoms will vibrate and transfer their energy to their neighbors and so on.

The main heat transfer mechanisms involved in the thermal processing of packaged foods are shown in figure 1. Although the figure shows a cylindrical can (a cylinder of finite diameter and height), a similar situation will arise when processing other types of packaging such as glass containers, retortable pouches, and rigid and semi-rigid plastic containers. In general, independent of shape, food package sizes range from 0.1 L to 5 L (Holdsworth and Simpson, 2016).

The main mechanism of heat transfer from the heating medium (e.g., steam or hot water) to the container or packaging is convection. Then heat transfers by conduction through the wall of the container or package. Once inside the container, heat transfer through the covering liquid occurs by convection, and in solid foods mainly by conduction. In case of liquid foods, the main mechanism is convection.

The rate of heat transfer in packaged foods depends on process factors, product factors, and package types. *Process factors* include retort temperature profile, process time, heat transfer medium, and container agitation.

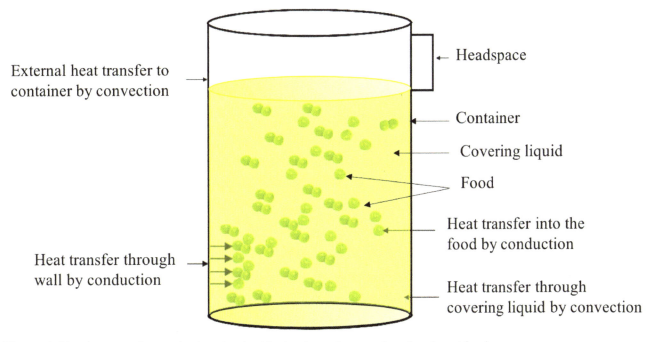

Figure 1. Main heat transfer mechanisms involved in the thermal processing of packaged foods.

Product factors include food composition, consistency, initial temperature, initial spore load, thermal diffusivity, and pH. Factors related to *package type* are container material, because the rate of heat transfer depends on thermal conductivity and thickness of the material, and container shape, because the surface area per unit volume plays a role in the heat penetration rate.

For liquid foods, the heating rate is determined not only by the thermal diffusivity α, but also by the viscosity. The thermal diffusivity is a material property that represents how fast the heat moves through the food and is determined as:

$$\alpha = K_t/(\rho\, C_p) \tag{1}$$

where α = thermal diffusivity (m²/s)
K_t = thermal conductivity (W/m-K)
ρ = density (kg/m³)
C_p = specific heat (W/s-kg-K)

It is extremely difficult to develop a theoretical model for the prediction of a time-temperature history within the packaging material. Therefore, from a practical point of view, a satisfactory thermal process (i.e., time-temperature relationship) is usually determined using the slowest heating point, the *cold spot*, inside the container.

Heat Resistance of Microorganisms of Concern

The main objective in the design of a sterilization process for foods is the inactivation of the microorganisms that cause food poisoning and spoilage. In order to design a safe sterilization process, the appropriate operating conditions (time and temperature) must be determined to meet the pre-established sterilization criterion. To establish this criterion, it is necessary to know the heat resistance of the microorganisms (some examples are given in table 1), the thermal properties of the food and packaging, and the shape and dimensions of the packaged food. From these, it is possible to determine the retort temperature and holding time (that is, the conditions for inactivation), how long it will take to reach that temperature (the come-up time), and how long it will take to cool to about 40°C (the cooling time) (Holdsworth and Simpson, 2016).

The pH of the food is extremely relevant to the selection of the sterilization process parameters, i.e., retort temperature and

Table 1. Some typical microorganisms heat resistance data (Holdsworth and Simpson, 2016).

Organism	Conditions for Inactivation
Vegetative cells	10 min at 80°C
Yeast ascospores	5 min at 60°C
Fungi	30–60 min at 88°C
Thermophilic organisms:	
Bacillus stearothermophilus	4 min at 121.1°C
Clostridium thermosaccharolyticum	3–4 min at 121.1°C
Mesophilic organisms:	
Clostridium botulinum spores	3 min at 121.1°C
Clostridium botulinum toxins Types A & B	0.1–1 min at 121.1°C
Clostridium sporogenes	1.5 min at 121.1°C
Bacillus subtilis	0.6 min at 121.1°C

holding time, because microorganisms grow better in a less acid environment. That is why the standard commercial sterilization process is based on the most resistant microorganism (Clostridium botulinum) at the worst-case scenario conditions (higher pH) (Teixeira et al., 2006). The microorganism heat resistance is greater in low-acid products (pH ≥ 4.5–4.6). On the other hand, medium-acid to acidic foods require a much gentler heat treatment (lower temperature) to meet the sterilization criterion. Based on that, foods are classified into three groups:

- low-acid products: *pH* > 4.5–4.6 (e.g., seafood, meat, vegetables, dairy products);
- medium-acid products: 3.7 < *pH* < 4.6 (e.g., tomato paste);
- acidic products: *pH* < 3.7 (e.g., most fruits).

Bacterial Inactivation

Abundant scientific literature supports the application of first-order kinetics to quantify bacterial (spores) inactivation as (Esty and Meyer, 1922; Ball and Olson, 1957; Stumbo, 1973, Holdsworth and Simpson, 2016):

$$\left(\frac{dN}{dt}\right)_I = -kN \tag{2}$$

where N = viable bacterial (microbial) concentration (microorganisms/g) after process time t

t = time

I = inactivation

k = bacterial inactivation rate constant (1/time)

Instead of k, food technologists have utilized the concept of *decimal reduction time*, D, defined as the time to reduce bacterial concentration by ten times. In other words, D is the required time at a specified temperature to inactivate 90% of the microorganism's population. A mathematical expression that relates the rate constant, k, from equation 2 to D is developed by separating variables and integrating the bacterial concentration from the initial concentration, N_0, to $N_0/10$ and from time 0 to D, therefore obtaining:

$$k = \frac{\ln 10}{D} = \frac{2.303}{D} \tag{3}$$

or

$$D = \frac{\ln 10}{k} = \frac{2.303}{k} \tag{4}$$

where k = reaction rate constant (1/min)

D = decimal reduction time (min)

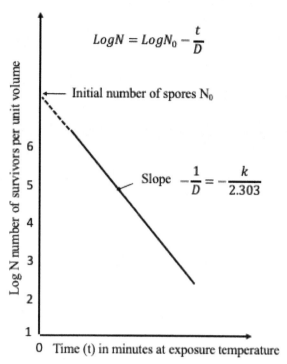

Figure 2. Semilogarithmic survivor curve.

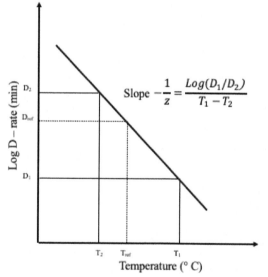

Figure 3. Thermal death time (TDT) curve.

A plot of the log of the survivors (log N) against D is called a *survivor curve* (figure 2). The slope of the line through one log cycle (decimal reduction) is $-1/D$ and

$$\log N = \log N_0 - \frac{t}{D} \tag{5}$$

where N = number of survivors
N_0 = N at time zero, the start of the process

Temperature Dependence of the Decimal Reduction Time, D

Every thermal process of a food product is a function of the thermal resistance of the microorganism in question. When the logarithm of the decimal reduction time, D, is plotted against temperature, a straight line results. This plot is called the *thermal death time (TDT) curve* (figure 3). From such a plot, the thermal sensitivity of a microorganism, z, can be determined as the temperature change necessary to vary TDT by one log cycle.

Bigelow and co-workers (Bigelow and Esty, 1920; Bigelow, 1921) were the first to coin the term *thermal death rate* to relate the temperature dependence of D. Mathematically, the following expression has been used:

$$\log D = \log D_{\text{ref}} - \frac{T - T_{\text{ref}}}{z} \tag{6}$$

or

$$D = D_{\text{ref}} 10^{\frac{T_{\text{ref}} - T}{z}} \tag{7}$$

where D = decimal reduction time at temperature T (min)
D_{ref} = decimal reduction time at reference temperature T_{ref} (min)
z = temperature change necessary to vary TDT by one log cycle (°C), e.g., normally $z = 10°C$ for *Clostridium botulinum*
T = temperature (°C)
T_{ref} = reference temperature (normally 121.1°C for sterilization)

The D value is directly related to the thermal resistance of a given microorganism. The more resistant the microorganism to the heat treatment, the higher the D value. On the other hand, the z value represents the temperature dependency but has no relation to the thermal resistance of the target microorganism. Then, the larger the z value the less sensitive the given microorganism is to temperature changes. D values are expressed as D_T. For example, D_{140} means the time required to reduce the microbial population by one log cycle when the food is heated at 140°C.

Food Sterilization Criterion and Calculation

Sterilization means the complete destruction or inactivation of microorganisms. The food science and engineering community has accepted the utilization of a first-order kinetic for *Clostridium botulinum* inactivation (equation 2). Again, this pathogen is the target microorganism in processes that use heat to sterilize foods. Theoretically, the inactivation time needed to fully inactivate *Clostridium botulinum* is infinite. According to equations 2 and 3 and assuming a constant process temperature and that k is constant, the following expression is obtained:

$$N_f = N_0 e^{-kt} = N_0 e^{-\frac{\ln 10}{D} t} \quad (8)$$

This equation shows that the final concentration of *Clostridium botulinum* (N_f) tends to zero when time (t) tends to infinity; therefore, it is not possible to reach a final concentration equal to zero for the target microorganism. Thus, it is necessary to define a *sterilization criterion* (or *commercial sterilization criterion*) to design a process that guarantees a safe product within a finite time.

The level of microbial inactivation, defined by the *microbial lethality value or cumulative lethality*, is the way in which the sterilization process is quantified. Specifically, the sterilizing value, denoted by F_0, is the required time at 121.1°C to achieve 12 decimal reductions (12D). In other words, F_0 is the time required to reduce the initial microorganism concentration from N_0 to $N_0/10^{12}$ at the process temperature of 121.1°C.

> F can be calculated for a process temperature other than 121.1°C.

The 12D sterilization criterion is an extreme process (i.e., overkill) designed to ensure no cells of *C. botulinum* remain in the food and, therefore, prevent illness or death. According to the FDA (1972), the minimum thermal treatment for a low-acid food should reach a minimum F_0 value of 3 min (that is larger than 12D; D for *C. botulinum* at 121.1°C is 0.21 min, then 12 × 0.21 = 2.52 min, which is lower than 3 min). Thus, a thermal process for commercial sterilization of a food product should have an F_0 value greater than 3 minutes.

The F_0 attained for a food can be calculated easily when the temperature at the center of the food during the thermal processing is known by:

$$F_0 = \int_0^t 10^{\frac{T - T_{ref}}{z}} dt \quad (9)$$

where F_0 = cumulative lethality of the process from time 0 to the end of the process (t)

T = temperature measured at the food cold spot, which is the place in the food that heats last

T_{ref} = temperature of microorganism reference; for sterilization of low-acid foods, T_{ref} = 121.1°C for *C. botulinum*

z = temperature change necessary to reduce D value by ten times; in the case of sterilization of low-acid foods, z = 10°C for *C. botulinum*

t = process time to reach F_0

Equation 9 can be calculated according to the general method proposed by Bigelow and co-workers 100 years ago (Bigelow et al., 1920; Simpson et al., 2003).

If the food is heated instantaneously to 121.1°C and maintained at this temperature for 3 min, then the F_0 value for this process will be 3 min. From equation 9,

$$F_0 = \int_0^t 10^{\frac{121.1-121.1}{10}} dt = \int_0^t 10^0 dt = \int_0^3 1\, dt$$

Since the time interval is between 0 to 3 min, then the integral solution is 3 min or $\int_0^3 1\, dt = 3 - 0 = 3$. However, in practice, due to the resistance of the food to the transfer of heat, the thermal sterilization process requires a longer time in order to get a $F_0 \geq 3$ min, because a significant part of the processing time is needed to raise the cold-spot temperature of the food and later to cool the food.

Applications

Commercial Sterilization Process

A general, simplified flow diagram for a typical commercial canning factory is presented in figure 4.

Stage 1: Selecting and preparing the food as cleanly, rapidly, and precisely as possible. Foods that maintain their desirable color, flavor, and texture through commercial sterilization include broccoli, corn, spinach, peas, green beans, peaches, cherries, berries, sauces, purees, jams and jellies, fruit and vegetable juices, and some meats (Featherstone, 2015). The preparation must be performed with great care and with the least amount of

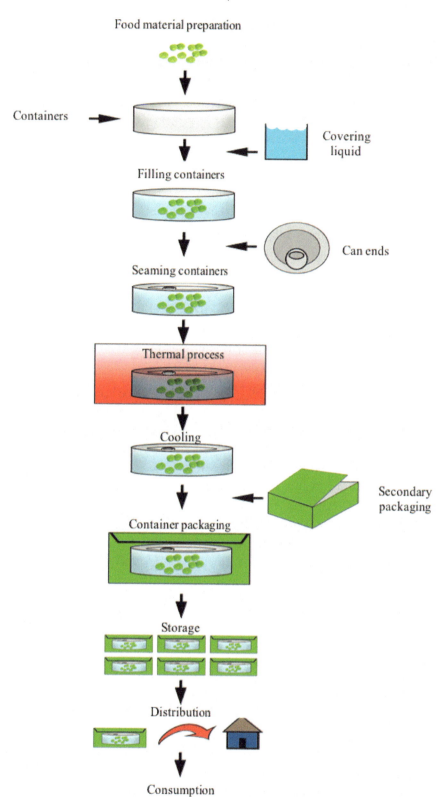

Figure 4. Stages of a typical commercial food canning process.

damage and loss to minimize the monetary cost of the operation. If foods are not properly handled, the effectiveness of the sterilization treatment is compromised.

Stage 2: Packing the product in hermetically sealable containers (jars, cans, or pouches) and sealing under a vacuum to eliminate residual air. A less common approach is to sterilize the food first and then aseptically package it (aseptic processing and packaging of foods).

Stage 3: Stabilizing the food by sterilizing through rigorous thermal processing (i.e., high temperature to achieve the correct degree of sterilization or the target destruction of the microorganisms present in the food), followed by cooling of the product to a low temperature (about 40°C), at which enzymatic and chemical reactions begin to slow down.

Stage 4: Storing at a temperature below 35°C, the temperature below which food-spoilage organisms cannot grow.

Stage 5: Labeling, secondary packaging, distribution, marketing, and consumption. Although not part of the thermal process per se, this stage addresses the steps required for commercialization of the treated foods.

Stage 3, thermal processing, is the focus of this chapter. The aim of the thermal process is to inactivate, by the effect of heat, spores and microorganisms present in the unprocessed product. The thermal process is performed in vessels known as retorts or autoclaves to achieve the required high temperatures (usually above 100°C).

As depicted in figure 5, a typical sterilization process has three main steps: come-up time, operator process time, and cooling. The first step, the come-up time (CUT), is the time required to reach the specified retort temperature (TRT), i.e., the target temperature in the retort. The second step is the holding time (P_t), also called operator process time, which is the amount of time that the retort temperature must be maintained to ensure the desired degree of lethality. This depends on the target microorganism or the expected microbiological contamination. The final step is the cooling, when the temperature of the product is decreased by introducing cold water into the retort. The purpose of cooling the food is to minimize the excessive (heat) processing of the food, and avoid the risk of thermophilic microorganism development. During the cooling cycle, it may be necessary to inject sterile air into the food packaging to avoid sudden internal pressure drops and prevent package deformation.

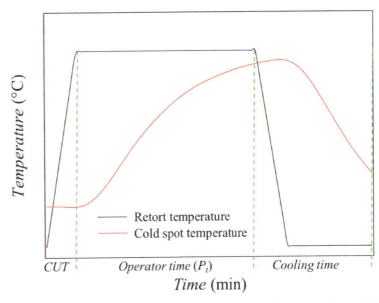

Figure 5. Temperature profiles for a typical thermal process, where *CUT* is come-up time and P_t is operator time.

Thermophilic organisms thrive on heat.

The concepts described in this chapter describe the key principles for applying a thermal process to packaged food to achieve the required lethality for food safety. These concepts can be used to design a thermal process to ensure adequate processing time and food safety while avoiding over processing the packaged food. This should ensure safe, tasty, and nutritious packaged foods.

Examples

Example 1: Calculation of microbial count after a given thermal process

Problem:
The $D_{120°C}$ value for a microorganism is 3 minutes. If the initial microbial contamination is 10^{12} cells per gram of product, how many microorganisms will remain in the sample after heat treatment at 120°C for 18 minutes?

Solution:
Calculate the number of remaining cells using equation 5 with $N_0 = 10^{12}$ cells/g, $t = 18$ minutes, and $D_{120°C} = 3$ minutes.

From equation 5,

$$\log N_{(t)} = \log N_0 - \frac{t}{D}$$

$$\log N_{(18)} = \log 10^{12} \frac{\text{cells}}{\text{g}} - \frac{18 \text{ min}}{3 \text{ min}}$$

Solving for $N_{(18)}$ yields:

$$N_{(18)} = 10^6 \text{ cells/g}$$

Discussion:
Starting with a known microbial concentration (N_0), the final concentration of a specific microorganism for a given thermal process at constant temperature can be calculated if the thermal resistance of the microorganism at a given temperature is known. In this case, $D_{120°C} = 3$ min.

Example 2: Calculation of z value for a particular microorganism

Problem:
D of a given bacterium in milk at 65°C is 15 minutes. When a food sample that has 10^{10} cells of the bacterium per gram of food is heated for 10 minutes at 75°C, the number of survivors is 2.15×10^3 cells. Calculate z for this bacterium.

Solution:

First, calculate D at the process temperature of 75°C, $D_{75°C}$, using equation 5. Then calculate z using equation 6 with $D_{65°C}$ = 15 minutes, $N_0 = 10^{10}$ cells/g, and t = 10 minutes at T = 75°C.

$$\log N_{(t)} = \log N_0 - \frac{t}{D} \qquad (5)$$

$$\log 2.15 \times 10^3 \text{ cells/g} = \log 10^{10} \text{ cells/g} - \frac{10 \text{ min}}{D_{75°C}}$$

and $D_{75°C}$ = 1.5 min.

To calculate z, recall equation 6:

$$\log D = \log D_{\text{ref}} - \frac{T - T_{\text{ref}}}{z}$$

Solving for z, equation 6 can be expressed as:

$$z = \frac{\Delta T}{\log\left(D_1 / D_2\right)}$$

with $\Delta T = (75 - 65)$°C, $D_1 = D_{65°C}$ and $D_2 = D_{75°C}$,

$$z = \frac{75 - 65}{\log \frac{15}{1.5}} = 10°C$$

Discussion:
As previously explained, the z value represents the change in process temperature required to reduce the D value of the target microorganism by ten times. In this case, the z value is 10°C and accordingly the D value was reduced 10 times, from 15 minutes to 1.5 minutes.

Example 3: Lethality of thermal processing of a can of tuna fish

Problem:
Table 2 presents the values of temperature measured in the retort (TRT) and the temperature measured at the cold spot of a can of tuna fish ($T_{\text{cold spot}}$) during a thermal process. The total process time was 63 min until the product was cold enough to be withdrawn from the retort.

(a) Determine CUT (the time required to come up to TRT), operator process time P_t, and cooling time.
(b) Determine the lethality value (F_0) attained for the can of tuna fish.

Table 2. Retort temperature (*TRT*) and cold spot ($T_{cold\ spot}$) during thermal processing of tuna fish in a can.

Time (min)	*TRT* (°C)	$T_{cold\ spot}$ (°C)
0.97	29.7	45.0
1.97	39.7	45.0
2.97	49.7	45.0
3.97	59.7	45.0
4.97	69.7	44.9
5.97	79.7	44.9
6.97	89.7	44.8
7.97	99.7	44.7
8.97	109.7	44.7
9.97	119.7	44.8
10.97	120.0	45.0
11.97	120.0	45.4
12.97	120.0	46.0
13.97	120.0	46.9
14.97	120.0	48.0
15.97	120.0	49.3
16.97	120.0	50.8
17.97	120.0	52.6
18.97	120.0	54.4
19.97	120.0	56.4
20.97	120.0	58.5
21.97	120.0	60.6
22.97	120.0	62.8
23.97	120.0	65.0
24.97	120.0	67.1
25.97	120.0	69.3
26.97	120.0	71.4
27.97	120.0	73.5
28.97	120.0	75.5
29.97	120.0	77.5
30.97	120.0	79.4
31.97	120.0	81.2
32.97	120.0	83.0
33.97	120.0	84.7
34.97	120.0	86.3
35.97	120.0	87.9
36.97	120.0	89.4
37.97	120.0	90.8

(continued)

Time (min)	TRT (°C)	$T_{cold\ spot}$ (°C)
38.97	120.0	92.2
39.97	120.0	93.6
40.97	120.0	94.8
41.97	120.0	96.0
42.97	120.0	97.2
43.97	120.0	98.3
44.97	120.0	99.3
45.97	120.0	100.3
46.97	120.0	101.3
47.97	120.0	102.2
48.97	120.0	103.0
49.97	120.0	103.9
50.97	120.0	104.7
51.97	120.0	105.4
52.97	120.0	106.1
53.97	120.0	106.8
54.97	120.0	107.4
55.97	120.0	108.0
56.97	120.0	108.6
57.97	120.0	109.2
58.97	120.0	109.7
59.97	120.0	110.2
60.97	120.0	110.7
61.97	120.0	111.1
62.97	120.0	111.6
63.97	120.0	112.0
64.97	120.0	112.4
65.97	120.0	112.8
66.97	120.0	113.1
67.97	120.0	113.4
68.97	120.0	113.8
69.97	120.0	114.1
70.97	120.0	114.4
71.97	120.0	114.6
72.97	120.0	114.9
73.97	120.0	115.2
74.97	120.0	115.4
76	25.0	115.6
77	25.0	115.8
78	25.0	116.0
79	25.0	116.2

(continued)

Table 2. Retort temperature (*TRT*) and cold spot ($T_{cold\ spot}$) during thermal processing of tuna fish in a can. (*continued*)

Time (min)	TRT (°C)	$T_{cold\ spot}$ (°C)
80	25.0	116.2
81	25.0	116.0
82	25.0	115.5
83	25.0	114.6
84	25.0	113.4
85	25.0	111.8
86	25.0	110.0
87	25.0	107.9
88	25.0	105.6
89	25.0	103.1
90	25.0	100.6
91	25.0	97.9
92	25.0	95.3
93	25.0	92.6
94	25.0	89.9
95	25.0	87.3
96	25.0	84.7
97	25.0	82.1
98	25.0	79.6
99	25.0	77.2
100	25.0	74.9
101	25.0	72.6
102	25.0	70.5
103	25.0	68.4
104	25.0	66.3
105	25.0	64.4
106	25.0	62.6
107	25.0	60.8
108	25	59.08
109	25	57.46
110	25	55.91
111	25	54.43
112	25	53.01
113	25	51.67
114	25	50.38
115	25	49.15
116	25	47.99
117	25	46.87
118	25	45.81

(*continued*)

Time (min)	TRT (°C)	$T_{cold\ spot}$ (°C)
119	25	44.8
120	25	43.84
121	25	42.92
122	25	42.05
123	25	41.22
124	25	40.43

Solution:

(a) To determine CUT and P_t, plot TRT and $T_{cold\ spot}$ against time, which produces the thermal profiles in figure 6.

Figure 6 shows that the CUT is approximately 10 min and P_t, during which process temperature is maintained constant at 120°C, is approximately 64 min.

(b) The lethality value, F_0, can be obtained through numerical integration of equation 9 using the trapezoidal rule (Patashnik, 1953). The calculations can be completed as follows or using software such as Excel.

As presented in table 3, for each time, we can evaluate equation 9:

$$F_0 = \int_0^t 10^{\frac{T-121.1}{10}} dt \quad (9)$$

where $T = T_{cold\ spot}$ and T_{ref} and z-value for *Clostridium botulinum* are 121.1°C and 10°C, respectively.

Given that F_0 corresponds to the integral of $10^{[(Tcold\ spot - Tref)/z]}$, this can be solved numerically by the trapezoidal rule method, i.e., by determining the area under the curve by dividing the area into trapezoids, computing the area of each trapezoid, and summing all trapezoidal areas to yield F_0. (More details about the trapezoidal rule are included in the appendix.) The calculations are summarized in table 3. In this particular case, F_0 was about 6.07 min. The change of F_0 along the thermal process is shown as the blue line in figure 7.

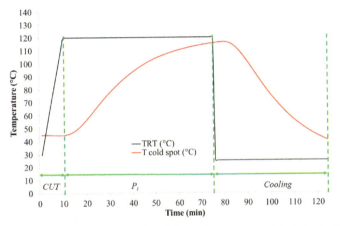

Figure 6. Temperature profile of thermal processing data in table 2.

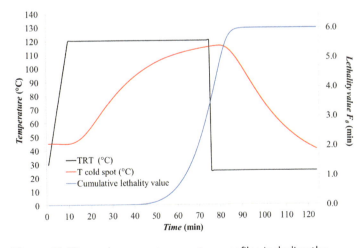

Figure 7. Thermal process temperature profiles including the cumulative lethality value (F at any time t).

Table 3. Numerical integration of equation 9 for the estimation of F_0.

Time (min)	TRT (°C)	$T_{cold\ spot}$ (°C)	$(T_{cold\ spot} - T_{ref})/z$	$10^{[(T_{cold\ spot} - T_{ref})/z]}$	Trapezoidal Area	Sum of Areas
0.97	29.67	45	−7.6	0.000	0.000	0.000
1.97	39.67	45	−7.6	0.000	0.000	0.000
2.97	49.67	44.99	−7.6	0.000	0.000	0.000
3.97	59.67	44.97	−7.6	0.000	0.000	0.000
4.97	69.67	44.93	−7.6	0.000	0.000	0.000
5.97	79.67	44.85	−7.6	0.000	0.000	0.000
6.97	89.67	44.76	−7.6	0.000	0.000	0.000
7.97	99.67	44.69	−7.6	0.000	0.000	0.000
8.97	109.67	44.68	−7.6	0.000	0.000	0.000
9.97	119.67	44.77	−7.6	0.000	0.000	0.000
10.97	120	45	−7.6	0.000	0.000	0.000
11.97	120	45.41	−7.6	0.000	0.000	0.000
12.97	120	46.03	−7.5	0.000	0.000	0.000
13.97	120	46.88	−7.4	0.000	0.000	0.000
14.97	120	47.97	−7.3	0.000	0.000	0.000
15.97	120	49.29	−7.2	0.000	0.000	0.000
16.97	120	50.83	−7.0	0.000	0.000	0.000
17.97	120	52.55	−6.9	0.000	0.000	0.000
18.97	120	54.42	−6.7	0.000	0.000	0.000
19.97	120	56.41	−6.5	0.000	0.000	0.000
20.97	120	58.49	−6.3	0.000	0.000	0.000
21.97	120	60.63	−6.0	0.000	0.000	0.000
22.97	120	62.79	−5.8	0.000	0.000	0.000
23.97	120	64.97	−5.6	0.000	0.000	0.000
24.97	120	67.14	−5.4	0.000	0.000	0.000
25.97	120	69.29	−5.2	0.000	0.000	0.000
26.97	120	71.41	−5.0	0.000	0.000	0.000
27.97	120	73.48	−4.8	0.000	0.000	0.000
28.97	120	75.5	−4.6	0.000	0.000	0.000
29.97	120	77.46	−4.4	0.000	0.000	0.000
30.97	120	79.36	−4.2	0.000	0.000	0.000
31.97	120	81.2	−4.0	0.000	0.000	0.000
32.97	120	82.97	−3.8	0.000	0.000	0.001
33.97	120	84.67	−3.6	0.000	0.000	0.001
34.97	120	86.31	−3.5	0.000	0.000	0.001
35.97	120	87.89	−3.3	0.000	0.001	0.002
36.97	120	89.4	−3.2	0.001	0.001	0.003
37.97	120	90.84	−3.0	0.001	0.001	0.004
38.97	120	92.22	−2.9	0.001	0.002	0.005
39.97	120	93.55	−2.8	0.002	0.002	0.007

(continued)

Time (min)	TRT (°C)	$T_{\text{cold spot}}$ (°C)	$(T_{\text{cold spot}} - T_{\text{ref}})/z$	$10^{[(T_{\text{cold spot}} - T_{\text{ref}})/z]}$	Trapezoidal Area	Sum of Areas
40.97	120	94.81	−2.6	0.002	0.003	0.010
41.97	120	96.02	−2.5	0.003	0.004	0.014
42.97	120	97.17	−2.4	0.004	0.005	0.018
43.97	120	98.27	−2.3	0.005	0.006	0.024
44.97	120	99.31	−2.2	0.007	0.007	0.032
45.97	120	100.31	−2.1	0.008	0.009	0.041
46.97	120	101.27	−2.0	0.010	0.012	0.053
47.97	120	102.17	−1.9	0.013	0.014	0.067
48.97	120	103.04	−1.8	0.016	0.017	0.084
49.97	120	103.86	−1.7	0.019	0.021	0.105
50.97	120	104.65	−1.6	0.023	0.025	0.130
51.97	120	105.39	−1.6	0.027	0.029	0.159
52.97	120	106.1	−1.5	0.032	0.034	0.193
53.97	120	106.78	−1.4	0.037	0.040	0.233
54.97	120	107.42	−1.4	0.043	0.046	0.279
55.97	120	108.04	−1.3	0.049	0.053	0.332
56.97	120	108.62	−1.2	0.056	0.060	0.393
57.97	120	109.18	−1.2	0.064	0.068	0.461
58.97	120	109.7	−1.1	0.072	0.077	0.538
59.97	120	110.21	−1.1	0.081	0.086	0.624
60.97	120	110.69	−1.0	0.091	0.096	0.720
61.97	120	111.14	−1.0	0.101	0.106	0.826
62.97	120	111.57	−1.0	0.111	0.117	0.943
63.97	120	111.99	−0.9	0.123	0.129	1.072
64.97	120	112.38	−0.9	0.134	0.140	1.212
65.97	120	112.75	−0.8	0.146	0.153	1.365
66.97	120	113.11	−0.8	0.159	0.165	1.530
67.97	120	113.44	−0.8	0.171	0.178	1.708
68.97	120	113.76	−0.7	0.185	0.191	1.899
69.97	120	114.07	−0.7	0.198	0.205	2.104
70.97	120	114.36	−0.7	0.212	0.219	2.323
71.97	120	114.63	−0.6	0.225	0.233	2.555
72.97	120	114.9	−0.6	0.240	0.247	2.802
73.97	120	115.15	−0.6	0.254	0.261	3.063
74.97	120	115.38	−0.6	0.268	0.275	3.338
76	25	115.61	−0.5	0.282	0.290	3.628
77	25	115.83	−0.5	0.297	0.304	3.932
78	25	116.02	−0.5	0.310	0.316	4.248
79	25	116.17	−0.5	0.321	0.322	4.570
80	25	116.19	−0.5	0.323	0.315	4.885
81	25	115.97	−0.5	0.307	0.290	5.175

(continued)

Table 3. Numerical integration of equation 9 for the estimation of F_0. (continued)

Time (min)	TRT (°C)	$T_{cold\ spot}$ (°C)	$(T_{cold\ spot} - T_{ref})/z$	$10^{[(T_{cold\ spot} - T_{ref})/z]}$	Trapezoidal Area	Sum of Areas
82	25	115.45	−0.6	0.272	0.248	5.422
83	25	114.58	−0.7	0.223	0.196	5.618
84	25	113.36	−0.8	0.168	0.143	5.761
85	25	111.81	−0.9	0.118	0.097	5.858
86	25	109.96	−1.1	0.077	0.062	5.920
87	25	107.87	−1.3	0.048	0.038	5.958
88	25	105.57	−1.6	0.028	0.022	5.980
89	25	103.12	−1.8	0.016	0.012	5.992
90	25	100.57	−2.1	0.009	0.007	5.999
91	25	97.94	−2.3	0.005	0.004	6.003
92	25	95.26	−2.6	0.003	0.002	6.005
93	25	92.58	−2.9	0.001	0.001	6.006
94	25	89.91	−3.1	0.001	0.001	6.007
95	25	87.26	−3.4	0.000	0.000	6.007
96	25	84.66	−3.6	0.000	0.000	6.007
97	25	82.11	−3.9	0.000	0.000	6.007
98	25	79.63	−4.1	0.000	0.000	6.007
99	25	77.22	−4.4	0.000	0.000	6.007
100	25	74.88	−4.6	0.000	0.000	6.007
101	25	72.62	−4.8	0.000	0.000	6.007
102	25	70.45	−5.1	0.000	0.000	6.007
103	25	68.35	−5.3	0.000	0.000	6.007
104	25	66.34	−5.5	0.000	0.000	6.007
105	25	64.41	−5.7	0.000	0.000	6.007
106	25	62.55	−5.9	0.000	0.000	6.007
107	25	60.78	−6.0	0.000	0.000	6.007
108	25	59.08	−6.2	0.000	0.000	6.007
109	25	57.46	−6.4	0.000	0.000	6.007
110	25	55.91	−6.5	0.000	0.000	6.007
111	25	54.43	−6.7	0.000	0.000	6.007
112	25	53.01	−6.8	0.000	0.000	6.007
113	25	51.67	−6.9	0.000	0.000	6.007
114	25	50.38	−7.1	0.000	0.000	6.007
115	25	49.15	−7.2	0.000	0.000	6.007
116	25	47.99	−7.3	0.000	0.000	6.007
117	25	46.87	−7.4	0.000	0.000	6.007
118	25	45.81	−7.5	0.000	0.000	6.007
119	25	44.8	−7.6	0.000	0.000	6.007
120	25	43.84	−7.7	0.000	0.000	6.007

(continued)

Time (min)	TRT (°C)	$T_{cold\ spot}$ (°C)	$(T_{cold\ spot} - T_{ref})/z$	$10^{[(Tcold\ spot - Tref)/z]}$	Trapezoidal Area	Sum of Areas
121	25	42.92	−7.8	0.000	0.000	6.007
122	25	42.05	−7.9	0.000	0.000	6.007
123	25	41.22	−8.0	0.000	0.000	6.007
124	25	40.43	−8.1	0.000	0.000	6.007

Discussion:

The cumulative lethality, F_0, was about 6.01 min, meaning that the process is safe according to FDA requirements, i.e., $F_0 \geq 3$ min (see the Food Sterilization Criterion and Calculation section above).

Example 4: Lethality of thermal processing of a can of mussels

Problem:

Temperatures measured in the retort and the temperature measured at the cold spot of a can of mussels during a thermal process performed at 120°C were recorded. The total process time was 113 min until the product was cold enough to be withdrawn from the retort. The measured thermal profiles (TRT and $T_{cold\ spot}$) were plotted, as was done in Example 3. The resulting plot (figure 8) shows that CUT was approximately 10 min and P_t was approximately 53 min. The lethality value, F_0, was obtained through numerical integration of equation 9. In this case, F_0 attained in the mussels can with a processing temperature of 120 °C was 2.508 min. The evolution of F_0 along the thermal process is shown in figure 9 as the blue line.

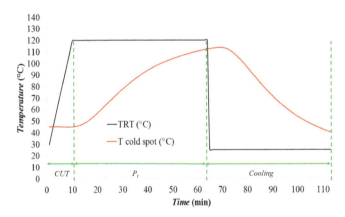

Figure 8. Temperature profile of thermal processing data for a can of mussels.

Discussion:

The cumulative lethality, F_0, attained along the thermal process was 2.5 min, meaning that the process is *not* safe according to FDA requirements ($F_0 \geq 3$ min). Thus, the thermal processing time of canning process of mussels must be extended in order to reach the safety value recommended by the FDA.

Example 5: Processing time at different retort temperatures

Problem:

Determine the required processing time to get a lethality of 6 min ($F_0 = 6$ min) when the retort

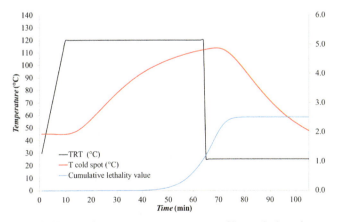

Figure 9. Thermal process temperature profiles including the cumulative lethality value (F at any time t).

Principles of Thermal Processing of Packaged Foods • 19

temperature is (a) 120°C and considered equal to the cold spot temperature, (b) 110°C, and (c) 130°C.

Solution:

The F_0 is typically set for the 12D value to give a 12 log reduction of heat-resistant species of mesophilic spores (typically taken as *C. botulinum*). The T_{ref} = 121.1°C and z = 10°C. Therefore, equation 9 can be used directly by replacing T by the retort temperature, given that cold spot temperature can be assumed equal to retort temperature:

$$F_0 = \int_0^t 10^{\frac{T-121.1}{z}} dt \tag{9}$$

$$6 = \int_0^t 10^{\frac{120-121.1}{10}} dt$$

$$t = \frac{6}{10^{\frac{120-121.1}{10}}}$$

Solving the integral yields a processing time, t, of 7.7 min.

(b) When the temperature of the retort is reduced to 110°C, the lethality must be maintained at 6 min. Solving equation 9:

$$6 = \int_0^t 10^{\frac{110-121.1}{10}} dt$$

gives the required processing time t of 77.2 min.

(c) When the temperature of the retort is increased to 130°C, and maintaining the F_0 = 6 min, the processing time is reduced to 0.77 min

$$6 = \int_0^t 10^{\frac{130-121.1}{10}} dt$$

Discussion:

The results showed that as the temperature in the food increased in 10°C increments, the processing time was reduced by one decimal reduction. This variation is due to a z value of 10°C.

Appendix: The Trapezoidal Rule

A trapezoid is a four-sided region with two opposite sides parallel (figure 10). The area of a trapezoid is the average length of the two parallel sides multiplied by the distance between the two sides. In figure 11, the area (A) under function $f(x)$ between points x_0 and x_n is given by:

$$A = \int_a^b f(x) dx \tag{10}$$

Figure 10. Example of a trapezoid.

An approximation of the area A is the sum of the areas of the individual trapezoids (T), where T can be calculated using equation 11:

$$T = \frac{1}{2}\Delta x_1 [f(x_0) + f(x_1)] + \frac{1}{2}\Delta x_2 [f(x_1) + f(x_2)] + \ldots + \frac{1}{2}\Delta x_n [f(x_{n-1}) + f(x_n)] \quad (11)$$

where $\Delta x_i = x_i - x_{i-1}$, for $i = 1, 2, 3, \ldots, n$

In the particular case where $\Delta x_1 = \Delta x_2 = \Delta x_3 = \ldots = \Delta x_n = \Delta x$, equation 11 can be expressed as:

$$T = \Delta x \left[\frac{f(x_0)}{2} + f(x_1) + f(x_2) + f(x_3) + \ldots + \frac{f(x_n)}{2} \right] \quad (12)$$

or, in the following reduced form:

$$T = \Delta x \left[\frac{f(x_0)}{2} + \sum_{i=1}^{n-1} f(x_i) + \frac{f(x_n)}{2} \right] \quad (13)$$

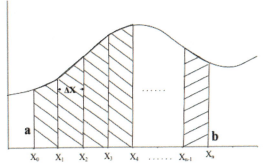

Figure 11. Curve divided into n equal parts each of length ΔX.

Finally, to estimate area A under the trapezoidal rule,

$$A = \int_{x_0}^{x_n} f(x)dx \cong \frac{1}{2}\Delta x_1 [f(x_0) + f(x_1)] + \frac{1}{2}\Delta x_2 [f(x_1) + f(x_2)] + \ldots + \frac{1}{2}\Delta x_n [f(x_{n-1}) + f(x_n)] \quad (14)$$

When all intervals are of the same size ($\Delta x_1 = \Delta x_2 = \Delta x_3 = \ldots = \Delta x_n = \Delta x$), the following expression can be applied:

$$A = \int_{x_0}^{x_n} f(x)dx \cong \Delta x \left(\frac{f(x_0)}{2} + \sum_{i=1}^{n-1} f(x_i) + \frac{f(x_n)}{2} \right) = \frac{1}{2}\Delta x (f(x_0) + 2\sum_{i=1}^{n-1} f(x_i) + f(x_n)) \quad (15)$$

Example

Problem:
Using the heat penetration data at the cold spot of a canned food in table 4, calculate the cumulative lethality, F_0, in the range of 23 to 27 min using the trapezoidal rule.

Solution:
From equation 9,

$$F_o = \int_{23}^{27} 10^{\frac{T-121.1}{10}} dt$$

Applying the trapezoidal rule and considering that all time steps are equal ($\Delta t = 1$ min), calculate F_0 using equation 15,

$$F_o = \int_{23}^{27} 10^{\frac{T-121.1}{10}} dt \cong \frac{1}{2}[f(23) + 2f(24) + 2f(25) + 2f(26) + f(27)]$$

where $\Delta t = 1$ (1 min interval), and:

$$f(23) = 10^{\frac{118.5 - 121.1}{10}} = 0.549541$$

Table 4. Heat penetration data at the slowest heating point.

Time (min)	Temperature (C)
...	...
23	118.5
24	118.7
25	118.9
26	119.1
27	119.3
...	...

$$f(24) = 10^{\frac{118.7-121.1}{10}} = 0.57544$$

$$f(25) = 10^{\frac{118.9-121.1}{10}} = 0.6025596$$

$$f(26) = 10^{\frac{119.1-121.1}{10}} = 0.63095734$$

$$f(27) = 10^{\frac{119.3-121.1}{10}} = 0.66069345$$

Replacing into equation (15):

$$F_o = \int_{23}^{27} 10^{\frac{T-121.1}{10}} dt \cong \frac{1}{2}(0.549541 + 2\times 0.57544 + 2\times 0.6025596 + 2\times 0.63095734 + 0.66069345)$$

Therefore, $F_0 \sim 2.41407394 \sim 2.41$ min.

Discussion:

The applied process to sterilize the target food is *not* safe since $F_0 < 3$ minutes.

Image Credits

Figure 1. Simpson, R. (CC By 4.0). (2020). Main heat transfer mechanisms involved in the thermal processing of packaged foods. Retrieved from https://onlinelibrary.wiley.com

Figure 2. Holdsworth, S. Donald-Simpson, R. (CC By 4.0). (2020). Semilogarithmic survivor curve. Retrieved from https://www.springer.com/la/book/9783319249025

Figure 3. Holdsworth, S. Donald-Simpson, R. (CC By 4.0). (2020). Thermal death time (TDT) curve. Retrieved from https://www.springer.com/la/book/9783319249025.

Figure 4. Simpson, R. (CC By 4.0). (2020). Stages of a typical food commercial canning factory.

Figure 5. Ramírez, C. (CC By 4.0). (2020). Temperature profiles for a typical thermal process, where CUT is come-up time and P_t is operator time.

Figure 6. Ramírez, C. (CC By 4.0). (2020). Temperature profile of thermal processing data in table 2.

Figure 7. Ramírez, C. (CC By 4.0). (2020). Thermal process temperature profiles including the cumulative lethality value (F at any time t).

Figure 8. Ramírez, C. (CC By 4.0). (2020). Temperature profile of thermal processing data (Table 4).

Figure 9. Ramírez, C. (CC By 4.0). (2020). Thermal process temperature profiles including the cumulative lethality value (F at any time t).

Figure 10. Simpson, R. (CC By 4.0). (2020). Example of a trapezoid.

Figure 11. Simpson, R. (CC By 4.0). (2020). Curve divided into n equal parts each of length ΔX.

References

Ball, C. O., & Olson, F. C. (1957). Sterilization in food technology—Theory, practice and calculations. New York, NY: McGraw-Hill.

Bigelow, W. D. (1921). The logarithmic nature of thermal death time curves. J. Infectious Dis., 29(5), 528-536. https://doi.org/10.1093/infdis/29.5.528.

Bigelow, W. D., & Esty, J. R. (1920). The thermal death point in relation to time of typical thermophilic organisms. J. Infectious Dis., 27(6), 602-617. https://doi.org/10.1093/infdis/27.6.602.

Bigelow, W. D., Bohart, G. S., Richardson, A. C., & Ball, C. O. (1920). Heat penetration in processing canned foods. Bull. No. 16. Washington, DC: Research Laboratory, National Canners Association.

Esty, J. R., & Meyer, K. F. (1922). The heat resistance of the spores of B. botulinus and allied anaerobes. J. Infectious Dis., 31(6), 650-663. https://doi.org/10.1093/infdis/31.6.650.

FDA. (1972). Sterilizing symbols. Low acid canned foods. Inspection technical guide. Ch. 7. ORO/ETSB (HFC-133). Washington, DC: FDA.

Featherstone, S. (2015). 7: Retortable flexible containers for food packaging. In A complete course in canning and related processes (14th ed.). Vol. 2: Microbiology, packaging, HACCP and ingredients (pp. 137-146). Sawston, Cambridge, U.K.: Woodhead Publ. https://doi.org/10.1016/B978-0-85709-678-4.00007-5.

Holdsworth, S. D., & Simpson, R. (2016). Thermal processing of packaged foods (3rd ed.). Springer. https://doi.org/10.1007/978-3-319-24904-9.

Patashnik, M. (1953). A simplified procedure for thermal process evaluation. Food Technol., 7(1), 1-6.

Simpson, R., Almonacid, S., & Teixeira, A. (2003). Bigelow's general method revisited: Development of a new calculation technique. J. Food Sci., 68(4), 1324-1333. https://doi.org/10.1111/j.1365-2621.2003.tb09646.x.

Stumbo, C. R. (1973). Thermobacteriology in food processing (2nd. ed.). New York, NY: Academic Press.

Teixeira, A., Almonacid, S., & Simpson, R. (2006). Keeping botulism out of canned foods. Food Technol., 60(2), Back page.

Deep Fat Frying of Food

Ram Yamsaengsung
Department of Chemical Engineering
Faculty of Engineering
Prince of Songkla University
Hat Yai, Thailand

Bandhita Saibandith
Department of Biotechnology
Faculty of Agro-Industry
Kasetsart University
Chatuchak, Bangkok, Thailand

KEY TERMS		
Frying chemistry	Mass transfer	Frying technology
Heat transfer	Mass and material balance	Industrial continuous frying systems
Heat balance	Product drying rate	Vacuum frying systems

Variables

- ρ = density of product
- A = surface area of product
- c = concentration of component (liquid, vapor, or oil)
- C_p = specific heat, also called specific heat capacity
- D = diffusion coefficient
- h = convective heat transfer coefficient
- ΔH = change in enthalpy
- k = thermal conductivity
- m = mass
- $\dot{m}$ = mass flow rate
- MC_t = moisture content at frying time t
- MC_e = equilibrium moisture content, i.e., at constant temperature and relative humidity
- MC_0 = moisture content at start of frying, time = 0
- MR = moisture ratio
- P = mass of potato chips
- q = heat flux
- $Q_{sensible}$ = sensible heat
- Q_{latent} = latent heat of evaporation

$Q_{total} = Q_{latent} + Q_{sensible}$
$Q_{heatsource}$ = heat from flame
R = mass of raw potato slices
t = time
T = product temperature
ΔT = difference in temperature
T_∞ = oil temperature
T_s = product surface temperature
$\dot{V}$ = rate of convective flow of liquid
W = mass of water
Δx = thickness of product

Introduction

This chapter introduces the basic principles of frying and its relevance to the food industry. To illustrate its importance, various fried products from around the world are described, and the mechanisms, equipment, and chemistry of the frying process are discussed. The pros and cons of frying food are presented in the context of texture, appearance, taste, and acceptability.

Frying is a highly popular method of cooking and has been used for thousands of years. A few examples from around the world include noodles, egg rolls, and crispy taros in China; tempuras (battered fried meats and vegetables) in Japan; fish and chips in the United Kingdom; and fried pork legs in Germany. In Latin countries and Tex-Mex restaurants, fried foods include tortilla-based products, such as tacos, nachos, and quesadillas. Examples of other popular fried foods include French fries, onion rings, and fried chicken, along with fried desserts such as doughnuts and battered, fried candy bars.

Traditionally, there are two major types of fried foods: (1) deep-fat fried (deep fried), such as potato chips, French fries, and battered fried chicken; and (2) pan fried, such as pancakes, eggs, and stir-fried dishes. This chapter focuses on atmospheric and vacuum deep-fat frying systems.

Outcomes

After reading this chapter, you should be able to:

- Describe various types of frying technology

- Describe basic frying chemistry and the heat and mass transfer mechanisms that are involved in the manufacture of different types of fried products

- Explain the advantages and disadvantages of the frying process

- Analyze the frying process using fundamental equations and calculate rate of water removal and amount of heat required during frying

Concepts

Frying Technology

Frying is defined as the process of cooking and drying through contact with hot oil. It involves simultaneous heat and mass transfer. Frying technology is important to many sectors of the food industry from suppliers of oils and ingredients; to fast-food outlets and restaurants; to industrial producers of fully fried, par-fried, and snack food products; and finally to manufacturers of frying equipment. The amount of fried food and oil used at both the industrial and commercial levels is massive.

Deep-Fat Frying (Deep Frying)

The process of immersing food partially or completely in oil during part or all of the cooking period at atmospheric pressure (760 mm Hg or 101.3 kPa absolute) is called deep-fat frying or deep frying. The food is completely surrounded by the oil, which is a very efficient heat-transfer medium. In addition to cooking the food, frying oil produces a crispy texture in food such as potato chips, French fries, and battered fried chicken (Moreira et al., 1999). The resulting product is usually golden brown in color with an oil content ranging from 8 to 25%.

A typical deep-fat fryer consists of a chamber into which heated oil and a food product are placed. The speed and efficiency of the frying process depend on the temperature and the overall quality of the oil, in terms of degradation of triglycerides and changes in thermal and physical properties such as color and viscosity (Moreira et al., 1999). The frying temperature is usually between 160° and 190°C. Cooking oil (such as sunflower oil, canola oil, soybean oil, corn oil, peanut oil, and olive oil) not only acts as the heat transfer medium, but it also enters into the product, providing flavor; table 1 lists the oil content of commonly deep-fat fried products.

In addition to frying at atmospheric pressure, food products can also be fried under a vacuum, where the pressure is reduced to about 60 mm Hg (8 kPa absolute). At this lower pressure, the boiling point of water is decreased to 41°C allowing for the frying oil temperature to be reduced to 90°–110°C. As a result, heat-sensitive products, such as fruits with a high sugar content (e.g., bananas, apples, jackfruits, durians, and pineapples) can be fried to a crisp. Furthermore, the fried products are able to maintain a fresh color and intense flavor, while the frying oil will have a longer life because of less contact with atmospheric oxygen.

Table 1. General oil content of products that are deep-fat fried using atmospheric frying (Moreira et al., 1999).

Product	Oil Content (%)
Potato chips	33–38
Tortilla chips	23–30
Expanded snack products	20–40
Roasted nuts	5–6
French fries	10–15
Doughnuts	20–25
Frozen foods	10–15

Chemistry of Frying

Sources of Oil Used in Frying

Oil seed crops are planted throughout the world to produce cooking oil. The seeds are washed and crushed before oil is removed using an extraction process. The oil is then refined to remove any unwanted taste, smell, color, or impurities. Some oils, such as virgin olive oil, walnut oil, and grapeseed oil, are pressed straight from the seed or fruit without further refining (EUFIC, 2014). Some other sources of frying oil include sunflower, canola, palm, and soybean.

Most vegetable oils are liquid at room temperature. When oils are heated, unsaturated fatty acids, which are the building blocks of triglycerides, are degraded. Monounsaturated-rich oils, such as olive oil or peanut oil, are more stable and can be re-used much more than polyunsaturated-rich oils like corn oil or soybean oil. For this reason, when deep-frying foods, it is important not to overheat the oil and to change it frequently.

Chemical Reactions

Many chemical reactions, including hydrolysis, isomerization, and pyrolysis, take place during frying and affect the quality and storage time of the oil. Several of these reactions lead to spoilage of the oil.

Hydrolysis is a chemical reaction in which a water molecule is inserted across a covalent bond and breaks the bond. Hydrolysis is the major chemical reaction that occurs during frying. As the food product is heated, water in the food evaporates and the water vapor diffuses into the oil. The water molecules cause hydrolysis in the oil, resulting in the formation of free fatty acids, reduction of the smoke point of the oil, and unpleasant flavors in both the oil and the food. The smoke point, or the burning point, of an oil or fat is the temperature at which it begins to produce a continuous bluish smoke that becomes clearly visible (AOCS, 2017). Baking powder also promotes hydrolysis of the oil (Moreira et al., 1999). Table 2 lists the smoke points of some common oils used in frying. For high temperature cooking (160–190°C), an oil with a low smoke point, such as unrefined sunflower oil and unrefined corn oil, may not be suitable.

Table 2. Smoke points of common oil used in frying (modified from Guillaume et al., 2018).

Cooking Oils and Fats	Smoke Point °C	Smoke Point °F	Cooking Oils and Fats	Smoke Point °C	Smoke Point °F
Unrefined sunflower oil	107°C	225°F	Grapeseed oil	216°C	420°F
Unrefined corn oil	160°C	320°F	Virgin olive oil	216°C	420°F
Butter	177°C	350°F	Sunflower oil	227°C	440°F
Coconut oil	177°C	350°F	Refined corn oil	232°C	450°F
Vegetable shortening	182°C	360°F	Palm oil	232°C	450°F
Lard	182°C	370°F	Extra light olive oil	242°C	468°F
Refined canola oil	204°C	400°F	Rice bran oil	254°C	490°F
Sesame oil	210°C	410°F	Avocado oil	271°C	520°F

Isomerization (polymerization) is the process by which one molecule is transformed into another molecule that has exactly the same atoms but arranged differently. Isomerization occurs rapidly during standby and frying periods. The bonds in the triglycerides are rearranged, making the oil more unstable and more sensitive to oxidation.

Pyrolysis results in the extensive breakdown of the chemical structure of the oil resulting in the formation of compounds of lower molecular weight.

Fried foods may absorb many oxidative products, such as hydro-peroxide and aldehydes, that are produced during frying (Sikorski & Kolakowska, 2002), thus affecting the quality of oil.

Repeated frying (using the same oil several times) increases the viscosity and darkens the color of the cooking oil. If the physico-chemical properties of cooking oil deteriorate, the oil must be discarded because it can prove to be harmful for human consumption (Goswami et al., 2015; Rani et al., 2010; Choe et al., 2007). Antioxidants, such as Vitamin E, added during frying are extremely effective in decreasing the rate of lipid oxidation, while enzymes such as superoxide dismutase, catalase, and peroxidase are also beneficial. Nonetheless, Vitamin E effectiveness decreases with increasing temperature (Goswami et al., 2015).

Heat and Mass Transfer Processes During Frying

The frying process, whether atmospheric or vacuum frying, is quite complicated involving coupled heat and mass transfer through a porous medium (the food), crust formation, and product shrinkage and expansion. These mechanisms all contribute to the difficulties in predicting physical and structural appearance of the final product. Thus, an understanding of the frying mechanism and the heat and mass transport phenomena is useful for food processors in order to produce and develop new fried and vacuum fried snack foods to meet the demands of consumers.

Heat Transfer

During the frying process, both heat and mass transfer take place, with water leaving and oil entering the product (figure 1). The heat transfer processes include radiation from the heat source to the fryer, conduction from the fryer outer wall to the inner surface, and from inner surface to oil. Once the oil is heated, heat energy is transferred by convection to the surface of the product. Due to the high temperature of frying (160–190°C), the convective heat transfer coefficient is much higher than air-drying processes. Finally, heat is conducted from the hotter surface to the colder center of the product, thus increasing its temperature.

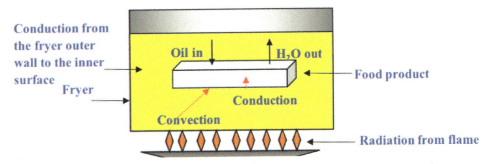

Figure 1. General schematic of the heat and mass transfer processes occurring during frying of a food product (Yamsaengsung, 2014).

Heat transfer during the frying process can be described using the three following simplifying assumptions (equations 1–3) relating to convection, conduction, and sensible heat.

The first assumption is that heat is transferred from the oil to the product surface via convection:

$$q = h\,\Delta T = hA(T_s - T_\infty) \tag{1}$$

where q = heat flux (J s^{-1}m^{-2} or W m^{-2}) (due to convection, in this case)
h = convective heat transfer coefficient (W m^{-2} °C^{-1})
A = surface area of product (m^2)
ΔT = difference in temperature (°C) between the product surface temperature and the oil temperature = $T_s - T_\infty$
T_s = product surface temperature (°C)
T_∞ = oil temperature (°C)

Table 3 lists ranges of values of the convective heat transfer coefficient (h) for several processes and media. Forced convection increases the heat transfer coefficient dramatically compared to free convection. At the same time, liquids have a much greater h value than gases, while a convection process with phase change can create a heat transfer coefficient as high as 2,500–100,000 W m^{-2}. Krokida et al. (2002) provide a good compilation of literature data on convective heat transfer coefficients in food processing operations and Alvis et al. (2009) is a good source for values of the coefficient in frying operations.

The second assumption is that heat is transferred from the product surface internally via conduction:

$$q = kA\frac{\Delta T}{\Delta x} = kA\frac{T_1 - T_2}{\Delta x} \tag{2}$$

where q = heat flux (J s^{-1}m^{-2}) or (W m^{-2}) (due to conduction in this case)
k = thermal conductivity (W m^{-1} °C^{-1})
A = surface area of product (m^2)
$\Delta T = T_1 - T_2$ = difference in temperature between the inner and outer surface of the product (°C)
Δx = thickness of product (m)

The third assumption is that the heat from the oil is also used as sensible heat (change in the temperature of the product without a change in phase) to increase the product temperature toward the oil temperature:

Table 3. Typical values of convective heat transfer coefficient (Engineering ToolBox, 2003).

Process	h (W m^{-2} K^{-1})
Free convection:	
Gases (e.g., air)	2–20
Liquids (e.g., water, oil)	50–1000
Forced convection:	
Gases (e.g., air)	25–300
Liquids (e.g., water, oil)	100–40,000
Convection with phase change:	
Boiling or condensation	2,500–100,000

$$Q = \Delta H = \dot{m} C_p \Delta T = \dot{m} C_p (T_1 - T_2) \tag{3}$$

where Q = sensible heat (J s^{-1})
ΔH = change in enthalpy (J)
$\dot{m}$ = mass flow rate (kg s^{-1})
C_p = specific heat (kJ kg^{-1} °C^{-1})
$\Delta T = T_1 - T_2$ = change in temperature of the material without undergoing a phase change (°C)

Table 4 gives the specific heat of water, vegetable oil, and common materials. As shown, the specific heat of vegetable oil is less than half that of liquid water, indicating that much less energy is needed to raise the temperature of the same amount of material by 1°C.

Sensible heat from the oil increases the water temperature to its boiling point. The release of heat energy at the boiling point is known as the latent heat of vaporization, or the heat required to evaporate the water or change its phase from liquid to gas. The latent heat of vaporization cools the product region during evaporation, keeping the product temperature near the boiling point (until most of the water has been removed).

Table 4. Specific heat of some common materials (Figura and Teixeira, 2007).

Material	Specific Heat (C_p) (kJ kg^{-1} °C^{-1})
Liquid water	4.18
Solid water (ice)	2.11
Water vapor	2.00
Vegetable oil	2.00
Dry air	1.01

Heat Balance

The simplified heat balance equation is:

$$\rho C_p \frac{dT}{dt} - div[k \nabla T] = Q_{heatsource} + h(T - T_\infty) \tag{4}$$

where ρ = density of product (kg m^{-3})
C_p = heat capacity of product (J kg^{-1} °C^{-1})
$div[k \nabla(T)]$ = conduction term = $\frac{\partial}{\partial x}\left(k\frac{\partial T}{\partial x}\right) + \frac{\partial}{\partial y}\left(k\frac{\partial T}{\partial y}\right) + \frac{\partial}{\partial z}\left(k\frac{\partial T}{\partial z}\right)$
x = direction x (m)
y = direction y (m)
z = direction z (m)
k = thermal conductivity (W m^{-1} °C^{-1})
$Q_{heatsource}$ = latent heat of evaporation term (J s^{-1}m^{-2})
h = convective heat transfer coefficient (W m^{-2} °C^{-1})
T_∞ = oil temperature at time t (°C)
T = product temperature at time t (°C)
t = time (s)

The simplified heat balance equation (equation 4), consists of the heat accumulation term $[\rho C_p(dT/dt)]$, the conduction term $div[k\nabla(T)]$, the heat source term ($Q_{heatsource}$) denoting the latent heat of vaporization, and the convection term, $h(T_{oil} - T)$, at the boundary surface, respectively. The heat accumulation term represents the change in the enthalpy of the system as a function of time. This change accounts for the heating of the product (change in enthalpy) and the transfer of the heat from the heated product toward evaporating the water vapor from the product. The conduction term accounts for the transfer of the heat from the product surface toward the center of the material, while the convection term accounts for the transfer of heat from the oil to the product surface and is dependent on the heat transfer coefficient of the cooking oil (Yamsaengsung et al., 2008).

Mass Transfer

The mass transfer processes during frying include (figure 1):

1. As the hot oil heats the product by conduction, the heat evaporates the water in the product when it reaches the water boiling temperature (Farkas et al., 1996).
2. As the water turns into vapor, it diffuses within the product and moves out of the product by convection.
3. Oil is driven into the product via capillary pressure (which is the pressure difference between two immiscible fluids in a thin tube), resulting from the interactions of forces between the fluids and the solid walls of the tube. Capillary pressure can serve as both an opposing force and a driving force for fluid transport (Moreira and Barrufet, 1998).
4. The final product is comprised of solids, water, air, and oil. In general, the product becomes more hygroscopic, i.e., readily attracts water from its surrounding, as frying proceeds. French fries are an excellent example of a product with a crispy surface or crust and a soft inner portion called crumb. In brief, after a specific time, the surface of the product becomes crispy, while the internal part of the product may retain a certain amount of moisture, leaving it with a softer texture.

Figure 2 depicts a typical non-hygroscopic material and a hygroscopic material (Figura and Teixeira, 2007). Each material consists of all three phases: gas, liquid water, and solid. One major difference is that in a hygroscopic material there is bound water. Bound water is defined as water that is bonded strongly to the inner surface of the pores of the materials (Yamsaengsung and Moreira, 2002) and very difficult to remove. In contrast, free water can be removed through capillary diffusion (Moreira et al., 1999) and convection flow from a pressure gradient. The bound water requires a longer drying and frying time to be removed. While more heat energy is required to remove this bound water, its removal leads to shrinkage of the material. Drying can also lead to shrinkage of the material, but frying can lead to additional puffing and expansion of the structure as the water vapor and gas expand during the later stages of the frying process (Yamsaengsung and Moreira, 2002).

> A hygroscopic material readily attracts water from its surroundings, while a non-hygroscopic material does not readily attract water from its surroundings.

> Capillary diffusion is the movement of fluids in unsaturated porous media due to surface tension and adhesive driving forces; capillarity.

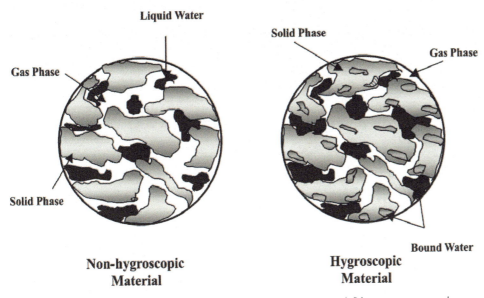

Figure 2. Schematic of non-hygroscopic and hygroscopic material (Yamsaengsung and Moreira, 2002).

Product Drying Rate

The percent moisture content of a food material can be expressed as wet basis (% w.b.) or dry basis (% d.b.) by mass. Percentage wet basis is commonly used in commercial applications while percentage dry basis is used in research reports.

The % w.b. is defined as:

$$\% \text{ w.b.} = \left(\frac{\text{water content(kg)}}{\text{total weight of product(kg)}} \right) \cdot 100 \quad (5)$$

The % d.b. is defined as:

$$\% \text{ d.b.} = \left(\frac{\text{kg of water}}{\text{kg of dried food}} \right) \cdot 100 \quad (6)$$

The drying rate of the product during the frying process is divided into constant and falling rate periods. During the constant rate period, water removal is fairly constant. During the falling rate period, the rate of water removal is drastically reduced. Each period is characterized by an averaged set of heat and mass transport parameters (Yamsaengsung, 2014). The moisture ratio (MR) is defined as:

$$MR = \frac{MC_t - MC_e}{MC_o - MC_e} \quad (7)$$

where MR = moisture ratio
MC_t = moisture content at frying time t (decimal d.b. or w.b.)
MC_e = equilibrium moisture content; the moisture content of the product under equilibrium conditions at constant temperature and relative humidity (d.b. or w.b.)
MC_0 = moisture content at start of frying, time = 0 (d.b. or w.b.)

Figure 3 illustrates the constant rate and the falling rate periods of the MR during the frying process.

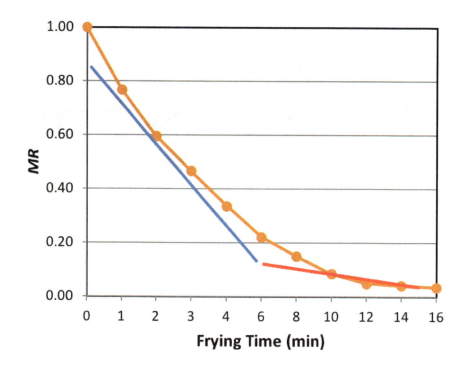

Figure 3. Typical drying curve (frying curve) showing the constant rate period (blue line) and the falling rate period (red line).

During the constant rate period, free water is removed as vapor via evaporation and diffusion from the product. Changes in the product surface as the crust forms are also taking place during this period. Typically, for crispy foods, the moisture content should be less than 5% w.b. During the falling rate period, a distinct crust region has been developed and bound water is being removed via vapor diffusion. The rate of oil absorption is proportional to the rate of moisture loss during the constant rate period, but is limited by the presence of the crust during the falling rate period. The development of crust and the increase in pressure inside the structure as the gas vapor expands with continuous heat absorption help to limit oil absorption, while causing the pores of the product to expand and the entire product to increase its thickness. This expansion is called puffing (Moreira et al., 1999).

In terms of mass transfer, the diffusion equation (equation 8) may be written to account for the convective flow of liquid and vapor as (Moreira et al., 1999):

$$\frac{dc}{dt} - div\left[D \cdot \nabla(c)\right] = \dot{V} \qquad (8)$$

where $\frac{dc}{dt}$ = change in concentration of component (liquid, vapor, or oil) (kg mol m^{-2} s^{-1})

$div[D \cdot \nabla(c)]$ = convective flow = $\frac{\partial}{\partial x}\left(D\frac{\partial c}{\partial x}\right) + \frac{\partial}{\partial y}\left(D\frac{\partial c}{\partial y}\right) + \frac{\partial}{\partial z}\left(D\frac{\partial c}{\partial z}\right)$

D = diffusion coefficient (m^{-2} s^{-1})

c = concentration of component (liquid, vapor, or oil) (kg mol m^{-3})

$\dot{V}$ = rate of convective flow of liquid (kg mol m^{-2} s^{-1})

When applied to each component, i.e., liquid, vapor, or oil, equation 8 is used to quantify the removal of liquid water and vapor from the product and the absorption of oil by the product during the frying process, i.e., as a function of time. The rate of water removal is estimated using the diffusion coefficient, while the change in the concentration of the component (liquid, vapor, or oil) is estimated using experimental data as a function of frying time (Yamsaengsung, et. al., 2008).

The heat and mass transfer equations allow calculation of the energy consumption, the heat required for heating of the cooking oil and removal of water from the product during the frying process, the amount of water that is being removed during frying, and, in many cases, the amount of product that would be obtained at the end of the frying period.

Material (Mass) Balance

Equations 4 and 8 describe heat and mass transfer during frying in three dimensions. Solving them requires advanced numerical methods, which are beyond the scope of this chapter. This section presents a model using simplified material balance (mass balance) equations, which accounts for the change in mass of each component during the process.

Equation 9 states the concept of the material, or mass, balance in words:

$$\left\{ \begin{array}{c} \text{accumulation} \\ \text{within} \\ \text{the system} \end{array} \right\} = \left\{ \begin{array}{c} \text{input} \\ \text{through} \\ \text{system} \\ \text{boundaries} \end{array} \right\} - \left\{ \begin{array}{c} \text{output} \\ \text{through} \\ \text{system} \\ \text{boundaries} \end{array} \right\}$$

$$+ \left\{ \begin{array}{c} \text{generation} \\ \text{within} \\ \text{the system} \end{array} \right\} - \left\{ \begin{array}{c} \text{consumption} \\ \text{within} \\ \text{the system} \end{array} \right\} \quad (9)$$

Accumulation refers to a change in mass (plus or minus) within the system with respect to time, whereas the input and output through the system boundaries refers to inputs and outputs of the system (Himmelblau, 1996). If considered over a time period for which the balance applies, equation 9 is a differential equation (consider, for example, the mass balance of water in figure 4). When formulated for an instant of time, equation 9 becomes a differential equation:

$$\frac{dm_{H_2O,\text{within system}}}{dt} = \dot{m}_{H_2O,\text{in}} - \dot{m}_{H_2O,\text{out1}} - \dot{m}_{H_2O,\text{out2}} \quad (10)$$

where m_{H_2O} is the mass of water and $\dot{m}_{H_2O}$ is the mass flow rate of water (mass/time, kg/s). When evaluating a process that is under equilibrium, or steady-state, condition, the values of the variables within the system do not change with time, and the accumulation (change in mass within the system with respect to time) term in equations 9 and 10 is zero.

To illustrate application of the mass balance, consider a frying operation to make potato chips (which are fried potato slices). For this example, 4 kg of peeled raw potato slices containing 83% water enter a fryer to make chips with

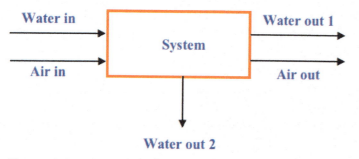

Figure 4. Process for a simple mass balance consisting of air and water entering and leaving a system.

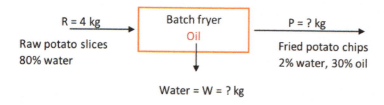

Figure 5. A schematic of the problem with the given data and the unknowns.

2% water and 30% oil. How many kg of water are evaporated from the product leaving the fryer, and how many kg of potato chips are produced in the process? The process is at steady state conditions.

The system is the fryer, and no accumulation, generation, or consumption occurs, since the process is steady state. Also, assume potatoes are made up of water and solids. The next step is to write the mass balance equations, total and for each component (% solids, % water), and solve for the unknowns.

The total mass balance of potato slices is:

in = out, as the time-dependent terms in equation 10 are zero

$$R + \text{oil (absorbed from fryer)} = W + P$$

The total material entering the fryer is given as 4 kg peeled raw potato slices. Substituting $R = 4$ kg in the total mass balance, yields one equation with two unknowns:

$$4 \text{ kg} + \text{oil (absorbed from fryer)} = W + P$$

Hence, a second equation is needed (basic material balances principle). The potatoes are composed of water and solids, hence, the terms in the total mass balance equation can be multiplied by the respective component percentages. The percent solids balance is:

$$4 \text{ kg } (1 - 0.83) + 0 = W(0) + P(1 - 0.32)$$

$$0.68 = 0.68\, P$$

$$P = 1 \text{ kg of potato chips produced}$$

Percent water balance is:

$$4 \text{ kg } (0.83) + 0 = W(1) + P(0.02)$$

$$3.32 = W + 0.02\, P$$

$$3.32 = W + (0.02)\, 1 \text{ kg} = W + 0.02$$

$$W = 3.3 \text{ kg of water removed}$$

Percent oil balance:

$$4 \text{ kg }(0) + \text{oil} = 3.3 \text{ kg}(0) + 1 \text{ kg}(0.3)$$

Thus, the mass of oil absorbed by the potatoes during frying is 0.3 kg.

The total material balance is 4 kg + 0.3 kg = 3.3 kg + 1 kg, or 4.3 = 4.3, which confirms the conservation of mass law.

Applications

In order to design and construct a productive and cost-effective frying system, product properties and characteristics, including size, shape, thickness, thermal conductivity, specific heat capacity, composition, and desired product attributes, such as color, texture (hardness, crispiness), smell, and flavor, all play a role and affect the frying time and temperature. Using equations for mass balance and heat transfer, the rate of water removal and a drying curve can be developed which can be used to predict the moisture loss and the product weight, which, in turn, affects the frying time and production capacity.

Industrial Continuous Deep-Fat Frying Systems

A typical continuous frying system consists of a fryer, a heat exchanger, an oil tank with a cooling system, a control panel, and a filter. Another common type of frying system consists of a combustor, an oil heat exchanger, and a fryer. In the combustor, a gas burner burns natural gas with fresh air and foul gas (vapors from the fryer) to produce combustion gases that flow through a heat exchanger to heat the frying oil that is re-circulated through the fryer. In many cases, exhaust gas recirculation is used to increase turbulence, provide combustor surface cooling, and reduce emissions. To reduce emissions and smells, vapors generated from the frying process are directed from the fryer to the combustor where they are incinerated (Wu et al., 2013).

Table 5 provides examples of mechanical specifications of some continuous frying systems and their throughputs, while table 6 gives pricing and frying time of some systems.

Table 5. Mechanical specifications of continuous frying systems and their throughputs (provided by Tsung Hsing Food Machinery).

Model	Dimensions (mm)			Effective Frying Space, length × width × height (mm)	Hp	Energy Consumption (kWh)	Edible Oil Capacity (L)	Production Capacity	
	length	width	height					Peanuts	Snacks
FRYIN-302-E	3450	2350	1950	2600 × 820 × 700	3	232.44	440	480 kg/hr	300 kg/hr
FRYIN-402-E	4950	2350	1950	4100 × 820 × 700	5	348.67	650	650 kg/hr	550 kg/hr
FRYIN-602-E	6450	2450	2070	5890 × 820 × 700	7.5	464.89	850		

Modified from https://www.tsunghsing.com.tw/en/product/oil_fryer-fryin_series.html

Table 6. Prices of typical automatic continuous frying systems and their throughputs (provided by Grace Food Machinery).

Machine Type	Products	Fuel Source/ Power Consumption	Capacity (kg/hr)	Frying Time (min.)	Cost, US$
Automatic continuous fried nuts processing line	nuts, almond, cashew, peanuts, etc.	diesel, LPG, gas, biofuel	N/A	N/A	$143,145
Automatic snacks frying machine	chips, meat, chicken, peanut	electric/10 kW	100–1,000	N/A	$14,314
Continuous banana chips fryer machine	chips, biscuit, donut, French fries, potato chips, banana chips, snacks	electric/25 kW	100–500	2–20	$25,766
Snack food fryer	snack foods	electric/25 kW	100–500	1–10	$143,145

Modified from https://www.gracefoodmachinery.com/continuous-fryer-systems.html

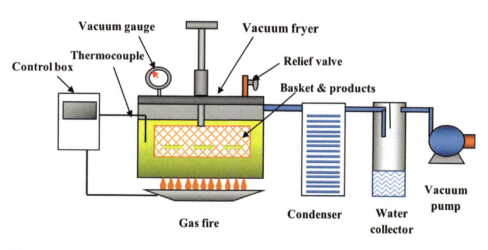

Figure 6. Schematic of a vacuum frying operation (Yamsaengsung, 2014).

Vacuum Frying Systems

The vacuum frying process, first developed in the 1960s and early 1970s, provides several benefits compared to the traditional (atmospheric) frying process. It is now widely used to process fruits in Asian countries. The principle behind vacuum frying is that using reduced pressure (below 101.3 kPa), the boiling point of water can be reduced from 100°C to as low as 45°C and the cooking oil temperature can also be reduced to less than 100°C (compared to atmospheric frying at 170°–190°C). As a result, products with high sugar content, such as ripened fruits, can be fried without burning and caramelization. Common methods to improve vacuum-frying of fruits include immersion in high sugar solutions and osmotic dehydration (Fito, 1994; Shyu and Hwang, 2001).

Figure 6 shows a schematic of a vacuum fryer (Yamsaengsung, 2014). In addition to the features shown in figure 6, a vacuum fryer must have a centrifuge to remove the oil content from the surface before the vacuum is broken. Table 7 provides a comparison of process operating conditions and applications for traditional and vacuum frying systems. The main components in the vacuum frying process are the vacuum fryer (8–10 mm thick wall and fryer cap), the condenser (for condensing water vapor), the water collector, and the vacuum pump (either rotary or liquid water ring type). However, the main drawbacks of vacuum frying are the high cost involved in purchasing the equipment and the more complicated process management. With the addition of a vacuum pump, a water condensing system, and much thicker fryer wall (8–10 mm vs. 1–2 mm), the cost of the vacuum fryer can be double the cost of an atmospheric fryer.

The benefits of vacuum frying include the ability to:

- fry high sugar content products such as fresh fruits;
- maintain original color, while adding intense flavor to the final product;
- reduce the amount of oil absorbed into the final product to as low as 1–3% depending on the machine (Garayo & Moreira, 2002); and
- extend the life of cooking oil by reducing its exposure to oxygen (lipid oxidation) and using a lower cooking oil temperature.

Moreover, even though Garayo and Moreira (2002) found that potato chips fried under vacuum conditions (3.115 kPa and 144°C) had more volume shrinkage, their texture was slightly softer and they were lighter in color than potato chips fried under atmospheric conditions (165°C). Yamsaengung and Rungsee (2003) also found that, compared to atmospheric frying, vacuum fried potato chips retained a lighter color and had a more intense flavor.

Table 7. Process settings and product characteristics for atmospheric vs. vacuum frying.[a]

Conditions/ Attributes	Atmospheric Frying	Vacuum Frying
Temperature	160°–190°C	90°–140°C
Pressure (absolute)	101.3 kPa	3.115 kPa
Convective heat transfer coefficient (h)	710-850 W m^{-2} K^{-1} (80-120°C)[b] 450-550 W m^{-2} K^{-1} (200-300°C)[b]	217–258 W m^{-2} K^{-1} (120°–140°C)[c] 700–1600 W m^{-2} K^{-1} (140°C)[d]
Oil absorption	25–40% w.b.	1–10% w.b.
Oil usage life	susceptible to lipid oxidation	minimal lipid oxidation longer usage life
High sugar content foods	not possible	possible
Major composition	high starch/ high protein	high starch high protein high sugar
Taste/texture	bland to salty/crispy	intense flavor/crispy
Color	intensity of color decreases	intensity of color is maintained
Investment cost	low	high

[a] Yamsaengsung (2014); [b] Farinu and Baik (2007) at 160°–190°C; [c] Pandey and Moreira (2012) at 120°–140°C; [d] Mir-Bel et al. (2012) at 140°C.

Examples

Example 1: Material balances

Problem:
Determine how many kg of raw potato slices containing 80% water must enter a batch fryer to make 500 kg of potato chips (fried potato slices) with 2% water and 30% oil. Also calculate how many kg of water are evaporated and leave the fryer. The process is at steady state conditions.

Solution:
Draw a schematic of the problem, enter the given data and identify the unknowns. Then, write down the material balance equations and solve for the unknowns.

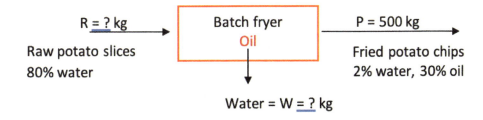

Deep Fat Frying of Food • 15

The system is the fryer, and no accumulation, generation, or consumption occurs, that is, it is at steady state. Also assume potatoes are made up of water and solids.

The total material balance in the fryer is:

$$\text{in} = \text{out}$$

$$R + \text{oil (in fryer)} = W + P$$

$$R + \text{oil} = W + 500 \text{ kg}$$

We have one equation and two unknowns, so we need another equation. Percent solids balance:

$$R(1 - 0.80) + 0 = W(0) + 500 \text{ kg}(1 - 0.32)$$

$$R = 1700 \text{ kg}$$

$R = 1700$ kg of raw potato slices containing 80% water are required

Percent water balance is:

$$1700 \text{ kg}(0.80) + 0 = W(1) + 500 \text{ kg}(0.02)$$

$$1360 = W + 10$$

$W = 1350$ kg of water removed from potato slices in the frying process

Finally, determine the amount of oil in the fried chips by conducting a percent oil balance:

$$1700 \text{ kg}(0) + \text{oil} = 1350 \text{ kg}(0) + 500 \text{ kg}(0.3)$$

$$\text{oil} = 150 \text{ kg}$$

Total material balance:

$$1700 \text{ kg} + 150 \text{ kg} = 1350 \text{ kg} + 500 \text{ kg}$$

This example illustrates the use of material balances using food initial and final composition data to calculate the amount of raw material entering the fryer to manufacture a product with specific composition characteristics.

Example 2: Moisture content of fried chips

Problem:
During a batch frying process, the weight of 50 kg of raw, fresh peeled potato slices decreases to 15 kg after frying. Each fried chip contains 30% oil content. If the initial moisture content of the fresh peeled potato is 80% (w.b), determine the final moisture content (% w.b.) of the fried chips.

Solution:
Draw a schematic of the problem, enter the given data and identify the unknowns. Then, write down the material balance equations and solve for the unknown moisture content of the fried chips, using the definition of percent wet basis.

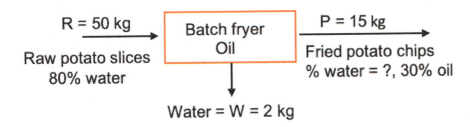

The system is the fryer, and no accumulation, generation, or consumption occurs (steady state). Also assume potatoes are made up of water and solids.
The total material balance in the fryer is:

$$\text{in} = \text{out}$$

$$R + \text{oil absorbed} = W + P$$

From the problem statement, mass of P = 15 kg with 30% oil.
Using percent oil content, calculate how much oil (in kg) the chips contain:

$$0.30 \times 15 \text{ kg} = 4.5 \text{ kg of oil in fried chips}$$

To figure the percent solids balance, note that in material balance applications in food engineering, the dry matter is constant. Hence, solids in = solids out.
From the raw materials with 80% water (and 20% solids), the dry matter (% solids) is:

$$50 \text{ kg} \times 0.2 = 10 \text{ kg}$$

Calculate how much water (in kg) is in the chips:

$$\text{chips} = \text{water} + \text{dry matter} + \text{oil}$$

$$15 \text{ kg} = \text{kg H}_2\text{O} + 10 \text{ kg} + 4.5 \text{ kg}$$

$$kg\ H_2O = 15\ kg - 10\ kg - 4.5\ kg = 0.05\ kg$$

Then, on a wet basis, the moisture content of the chips is (from equation 5):

$$\%w.b. = \frac{kg\ of\ water}{total\ of\ weight\ of\ product} \times 100 = \frac{0.05\ kg}{15\ kg} \times 100 = 0.333\%$$

This example shows how the final moisture content of the fried product can be calculated. Its importance lies in the effect of moisture on the crispiness of fried foods. Typically, for crispy snacks, the moisture content should be less than 5% w.b., so this fried product is considered crispy.

Example 3: Drying curve

Problem:

The following data represent the change in weight of vacuum fried bananas (70% w.b. moisture content) as a function of frying time. Also assume the moisture content in % d.b. at equilibrium (at the end of frying) is 0.02 kg water/kg dry matter. Neglect the weight of the oil absorbed (% oil content = 0.0%) and plot the drying curve as a function of the frying time (moisture ratio vs. time).

Time (min)	Weight (kg)	Solids (kg)	H$_2$O (kg)
0	10	3	7
1	8.4	3	5.4
2	7.2	3	4.2
3	6.3	3	3.3
4	5.4	3	2.4
6	4.6	3	1.6
8	4.1	3	1.1
10	3.65	3	0.65
12	3.4	3	0.4
14	3.35	3	0.35
16	3.3	3	0.3

Solution:

Calculate moisture ratio using equations 6 and 7:

$$\%\ d.b. = \left(\frac{kg\ of\ water}{kg\ of\ dried\ food}\right) \cdot 100 \quad (6)$$

$$MR = \frac{MC_t - MC_e}{MC_o - MC_e} \quad (7)$$

For the banana with 70% w.b. moisture, the percent solids content is 1 − 0.7 = 0.3 or 30%. For an initial weight of 10 kg, the solids content is 0.3 × 10 kg = 3 kg (a constant throughout the process).

Determine moisture content in % dry basis using equation 6 at each time *t*:

% d.b. = (total weight – weight of solids)/(weight of solids)
For example,

At t = 0 min $\quad MC_0\ (t=0) = (10\ kg - 3\ kg)/(3\ kg) = 2.33$

At t = 1 min $\quad MC_1\ (t=1) = (8.4\ kg - 3\ kg)/(3\ kg) = 1.80$

At t = 2 min $\quad MC_2\ (t=2) = (7.2\ kg - 3\ kg)/(3\ kg) = 1.40$

Repeat the procedure for all times.
Next, determine *MR* using equation 7. For example,

$$MR_t = (MC_t - MC_e)/(MC_0 - MC_e)$$

At t = 0 min $\quad MR_0 = (2.33 - 0.02)/(2.33 - 0.02) = 1.00$

At t = 1 min $\quad MR_1 = (1.80 - 0.02)/(2.33 - 0.02) = 0.77$

At t = 2 min $\quad MR_2 = (1.40 - 0.02)/(2.33 - 0.02) = 0.60$

Repeat the procedure for all times using table below.

Time (min)	Weight (kg)	Solids (kg)	H$_2$O (kg)	MC (d.b.)	MR
0	10	3	7	2.33	1.00
1	8.4	3	5.4	1.80	0.77
2	7.2	3	4.2	1.40	0.60
3	6.3	3	3.3	1.10	0.47
4	5.4	3	2.4	0.80	0.34
6	4.6	3	1.6	0.53	0.22
8	4.1	3	1.1	0.37	0.15
10	3.65	3	0.65	0.22	0.09
12	3.4	3	0.4	0.13	0.05
14	3.35	3	0.35	0.12	0.04
16	3.3	3	0.3	0.10	0.03

Plot *MR* vs. time (drying curve) as below. This shows that there is a constant rate of drying for about the first 10 min. followed by falling rate of drying period from 12-16 min.

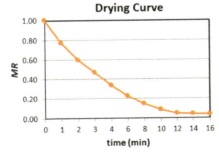

Example 4: Production throughput of a continuous fryer

Problem:
At a potato chip factory, 1,000 kg of potatoes are fed into a continuous vacuum fryer per hour.

(a) Assuming the initial moisture content of peeled potatoes is 80% (w.b.) and the final moisture content is 2% (w.b.), how much water is removed per hour?

(b) How much (in kg) of oil is added to the potato chips per hour? Potato chips have 2% w.b. moisture content and 30% oil.

(c) How many bags can be produced in one day if each bag contains 50 g of potato chips and the factory operates for 8 hours per day?

Solution:
Draw a schematic of the problem, enter the given data and identify the unknowns. Then, write down the material balance equations and solve for the unknowns. Neglect oil content in product. This is a continuous process (material/time).

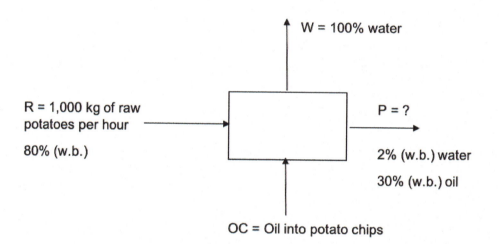

(a) Determine the amount of water removed from the raw potatoes, R, per hour.

Water in initial product:

$$1{,}000 \text{ kg/hr} \times 0.8 = 800 \text{ kg/hr of water}$$

Percent solids balance:

$$(1000 \text{ kg/hr})(1 - 0.8) = W(0) + (1 - 0.02 - 0.3)P$$

$$(1000 \text{ kg/hr})(0.2) = (0.68)P$$

$$200 \text{ kg/hr} = 0.68\,P$$

$$P = 294.12 \text{ kg/hr of potato chips}$$

Water in final product:

$$294.12 \text{ kg/hr} \times 0.02 = 5.88 \text{ kg/hr of water in potato chips}$$

Percent water balance:

$$(1000 \text{ kg/hr})(0.8) = W(1) + 5.88 \text{ kg/hr}$$

$$W = 800 \text{ kg/hr} - 5.88 \text{ kg/hr}$$

$$W = 794.12 \text{ kg of water removed from raw potatoes in one hour}$$

(b) Determine the amount of oil per hour added to the potato chips in the fryer.
   Chips have 30% oil. Thus,

$$0.30 = \frac{\text{oil in chips(kg/hr)}}{\text{total weight of chips(kg/hr)}}$$

$$0.30 = \frac{\text{oil in chips(kg/hr)}}{294.12 \text{ kg/hr}}$$

$$\text{oil in chips} = (0.30) \times 294.12 \text{ kg/hr}$$

$$\text{oil in chips} = 88.24 \text{ kg/hr}$$

(c) Determine number of bags per 8-hr day:

$$\text{amount of chips per day} = (294.12 \text{ kg/hr}) \times 8 \text{ hr/day}$$

$$= 2{,}352.96 \text{ kg chips/day} \times (1000 \text{ g/kg}) = 2{,}352{,}960 \text{ g/day}$$

$$\text{number of bags per day} = (2{,}352{,}960 \text{ g/day}) \times (1 \text{ bag/50 g})$$

$$= 47{,}059 \text{ bags per day}$$

This example illustrates how the engineer uses knowledge of material balances and food composition to determine production throughput of a continuous fryer.

Example 5: Energy requirement for an industrial fryer

Problem:
For an industrial fryer with a production capacity of 5,000 kg of corn chips per hour, how much energy is required to reduce the water content of the pre-baked masa (the product that will be fried to make the chips) from 50% w.b. to 4% w.b.? If the frying time takes 60 seconds at a frying temperature of 180°C, calculate:

(a) initial feed rate of the chips,
(b) total amount of water removed,
(c) amount of heat required to evaporate the water,
(d) total energy required for the frying process, and
(e) power required for the frying system.

Assume that the oil has already been pre-heated, the temperature of the oil does not drop during frying, but heat is needed to increase the corn chips feed temperature from 25°C to the frying temperature. The specific heat capacity (C_p) of the cooking oil is 2.0 kJ kg^{-1} °C^{-1}, the specific heats of the corn chips before and after frying are 2.9 kJ kg °C^{-1} and 1.2 kJ kg°C^{-1}, respectively, and the latent heat of evaporation of water at 100°C is 2,256 kJ/kg. (Hint: 1 kW = 1 kJ/s and water evaporates at 100°C.)

Solution:
Calculate the initial mass of the pre-baked masa using equation 5:

$$\% \text{ w.b.} = \left(\frac{\text{water content(kg)}}{\text{total weight of product(kg)}} \right) \cdot 100 \quad (5)$$

$$4\% = (\text{kg of water}/5{,}000 \text{ kg})$$

$$\text{weight of water} = 200 \text{ kg}$$

$$\text{weight of solid} = 5{,}000 \text{ kg} - 200 \text{ kg} = 4{,}800 \text{ kg}$$

$$MC = 50\% \text{ w.b.}$$

Find the mass of water using equation 5:

$$50\% = (\text{kg of water}/(\text{kg of water} + \text{kg of solid}))$$

$$50\% = W/(W + 4{,}800 \text{ kg})$$

$$0.5 \times (W + 4{,}800 \text{ kg}) = W$$

$$W = (0.5 \times 4{,}800 \text{ kg})/(1 - 0.5)$$

$$\text{weight of water} = 4{,}800 \text{ kg}$$

Initial feed rate of corn chips = weight of water + weight of solid = 9,600 kg/hr

Calculate the amount of water removed as initial – final:

$$\text{Initial weight of water} = 4{,}800 \text{ kg}$$

$$\text{Final weight of water} = 200 \text{ kg}$$

$$\text{Water removed} = 4{,}800 - 200 = 4{,}600 \text{ kg}$$

Calculate Q required to remove the water (to evaporate the water)

$$Q = \text{water removed} \times \text{latent heat of evaporation}$$

$$Q = (4{,}600 \text{ kg} \times 2{,}256 \text{ kJ/kg})$$

$$Q = 10{,}377{,}600 \text{ kJ}$$

Calculate sensible heat (25° – 100°C and 100° – 180°C) using equation 3:

$$Q = \Delta H = \dot{m} C_p \Delta T = \dot{m} C_p (T_1 - T_2) \quad (3)$$

$$Q = (9{,}600 \text{ kg}) \times (2.9 \text{ kJ/kg °C}) \times (100°C - 25°C)$$

$$Q = 2{,}088{,}000 \text{ kJ}$$

Likewise for 100° – 180°C,

$$Q = (5{,}000 \text{ kg}) \times (1.2 \text{ kJ/kg °C}) \times (180°C - 100°C)$$

$$Q = 4{,}800{,}000 \text{ kJ}$$

Calculate total Q as the sum of both sensible and latent heat:

$$Q_{total} = Q_{sensible} + Q_{latent}$$

$$Q_{total} = 2{,}088{,}000 \text{ kJ} + 4{,}800{,}000 \text{ kJ} + 10{,}277{,}600 \text{ kJ} = 17{,}265{,}600 \text{ kJ}$$

Calculate power as total heat per unit time:

$$\text{Power} = Q/t$$

$$t = 60 \text{ seconds}$$

$$Q = 17{,}265{,}600 \text{ kJ}$$

$$\text{Power} = 287{,}760 \text{ kW}$$

Example 6: Water removal rate during frying

Problem:
12 kilograms of fresh bananas were purchased at 0.50 US$/kg. After 4 kg of peel were removed, the bananas were sliced 2 mm thick and vacuum fried at 110°C for 45 minutes. This process reduced the moisture content from 75% w.b. to 2.5% w.b. Determine the rate of water removed from the fresh peeled bananas (kg/min) during the frying process. Assume bananas are composed of water and solids, and that the amount of oil absorbed is negligible (oil content = 0%).

Solution:
Draw a schematic of the problem, enter the given data, and identify the unknowns. Then, write down the material balance equations and solve for the unknowns.

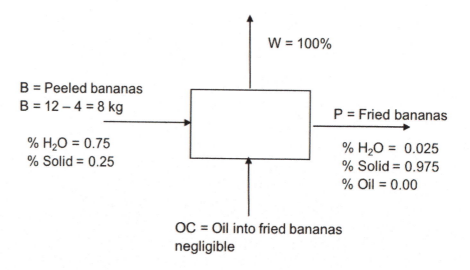

Total material balance: in = out
Remember that oil is zero in this example. Therefore, $B = W + P$

Solids balance: $(0.25)B = (0)W + (0.975)P$
Water balance: $(0.75)B = (1.00)W + (0.025)P$

From $B = 8$ kg, find P from the solids balance:

$$P = (0.25)(8 \text{ kg})/(0.975)$$

$$P = 2.05 \text{ kg of fried bananas}$$

From the total material balance, find the amount of water removed during the process:

$$W = B - P = 8 \text{ kg} - 2.05 \text{ kg}$$

$$W = 5.95 \text{ kg of water removed from the fresh peeled bananas}$$

The rate of water removal is the amount of water removed per unit time. Since the frying time was 45 minutes the rate of water removal in is:

$$\text{rate of water removal} = (5.95 \text{ kg})/(45 \text{ min})$$

$$\text{rate of water removal} = 0.132 \text{ kg water/min}$$

Why is this important? In a vacuum frying process, water is removed from the product during frying, trapped, and separated before it reaches the vacuum pump in order to maintain low pressure inside the frying system. The volume of the water trap and the capacity of the vacuum pump are needed in order to select the most efficient vacuum pump and heat exchanger for cooling the water vapor from the fryer. For example, if 20 kg of potato chips with an initial moisture content of 60% w.b. is fried, it can be assumed that almost 12 kg of water (approximately 12 L) must be removed and collected in a water trap. If the water is not condensed and collected, it will enter the pump and cause a decrease in vacuum pressure.

Image Credits

Figure 1. Yamsaengsung, R. (CC By 4.0). (2014). General schematic of the heat and mass transfer processes occurring during frying of a food product.
Figure 2. Yamsaengsung, R. (CC By 4.0). (2002). Schematic of non-hygroscopic and hygroscopic material.
Figure 3. Yamsaengsung, R. (CC By 4.0). (2020). Typical drying curve (frying curve) showing the constant rate period and the falling rate period.
Figure 4. Yamsaengsung, R. (CC By 4.0). (2020). Process for a simple mass balance consisting of air and water entering and leaving a system.
Figure 5. Yamsaengsung, R. (CC By 4.0). (2020). A schematic of the problem with the given data and the unknowns.
Figure 6. Yamsaengsung, R. (CC By 4.0). (2014). Schematic of a vacuum frying operation.
Example 1. Yamsaengsung, R. (CC By 4.0). (2020). Example 1: Material Balances.
Example 2. Yamsaengsung, R. (CC By 4.0). (2020). Example 2: Moisture content of fried chips.
Example 3. Yamsaengsung, R. (CC By 4.0). (2020). Example 3: Drying curve.
Example 4. Yamsaengsung, R. (CC By 4.0). (2020). Example 4: Production throughput of a continuous fryer.
Example 6. Yamsaengsung, R. (CC By 4.0). (2020). Example 6: Water removal rate during frying.

References

Alvis, A., Velez, C., Rada-Mendoza, M., Villamiel, M., & Villada, H.S. (2009). Heat transfer coefficient during deep-fat frying. *Food Control* 20(4), 321-325.
AOCS (2011). AOCS Official Method Cc 9a-48 Smoke, flash and fire points, Cleveland open cup method. In *Official methods and recommended practices of the AOCS*. Urbana, Ill.: American Oil Chemists' Society.
Choe, E., & Min, D.B. (2007). Chemistry of deep-fat frying oils. *J. Food Sci.* 72(5), R77-R86.
Engineering ToolBox. (2003). *Convective Heat Transfer*. Available at: https://www.engineeringtoolbox.com/convective-heat-transfer-d_430.html.

EUFIC. (2014). Facts on fats: The basics. The European Food Information Council. Retrieved from https://www.eufic.org/en/whats-in-food/article/facts-on-fats-the-basics.

Farkas, B. E., Singh, R. P., and Rumsey, T. R. (1996). Modeling heat and mass transfer in immersion frying, part I: Model development. *J. Food Eng.* 29(1996), 211-226.

Farinu, A., & Baik, O. D. (2007). Heat transfer coefficients during deep fat frying of sweetpotato: Effects of product size and oil temperature. *J. Food Res.* 40(8), 989-994.

Figura, L. O., & Teixeira, A. A. (2007). *Food physics: Physical properties—Measurement and applications.* Germany: Springer.

Fito, P. (1994). Modelling of vacuum osmotic dehydration of foods. *J. Food Eng.* 22(1-4), 313-328.

Garayo, J. and Moreira, R.G. (2002). Vacuum frying of potato chips. *J. Food Eng.* 55(2), 181-191. http://dx.doi.org/10.1016/S0260-8774(02)00062-6.

Goswami, G., Bora, R., Rathore, M.S. (2015). Oxidation of cooking oils due to repeated frying and human health. *Int. J. Sci. Technol. Manag.* 4(1), 495-501.

Guillaume C., De Alzaa, F., & Ravetti, L. (2018). Evaluation of chemical and physical changes in different commercial oils during heating. *Acta Sci. Nutri. Health* 26(2018), 2-11.

Himmelblau, D. M. (1996). *Basic principles in chemical engineering.* London: Prentice Hall Int.

Krokida, M. K., Zogzas, N. P., & Maroulis, Z. B. (2002) Heat transfer coefficient in food processing: Compilation of literature data. *Int. J. Food Prop.,* 5:2: 435-450. https://doi.org/10.1081/JFP-120005796.

Mir-Bel, J., Oria, R., & Salvador, M. L. (2012). Influence of temperature on heat transfer coefficient during moderate vacuum deep-fat frying. *J. Food Eng.* 113(2012), 167-176.

Moreira, R. G., & Barrufet, N. A. (1998). A new approach to describe oil absorption in fried foods: A simulation study. *J. Food Eng.* 35:1-22.

Moreira, R. G., Castell-Perez, M. E., & Barrufet, M. A. (1999). *Deep-fat frying: Fundamentals and applications.* Gaithersburg, MD: Aspen Publishers.

Pandey, A., & Moreira, R. G. (2012). Batch vacuum frying system analysis for potato chips. *J. Food Process. Eng.* 35(2012), 863-873.

Rani, A. K. S., Reddy, S. Y., & Chetana, R. (2010). Quality changes in trans and trans free fats/oils and products during frying. *Eur. Food Res. Technol.* 230(6), 803–811.

Shyu, S., & Hwang, L. S. (2001). Effect of processing conditions on the quality of vacuum fried apple chips. *Food Res. Int.* 34(2001), 133-142.

Sikorski, Z. E., & Kolakowska, A. (2002). *Chemical and functional properties of food lipids.* United Kingdom: CRC Press.

Wu, H., Tassou, S. A., Karayiammis, T. G., & Jouhara, H. (2013). Analysis and simulation of continuous frying processes. *Appl. Thermal Eng.* 53(2), 332-339. https://doi.org/10.1016/j.applthermaleng.2012.04.023.

Yamsaengsung, R., & Moreira, R. G. (2002). Modeling the transport phenomena and structural changes during deep fat frying Part I: model development. *J. Food Eng.* 53(2002), 1-10.

Yamsaengsung, R., Rungsee, C., & Prasertsit, K. 2008. Modeling the heat and mass transfer processes during the vacuum frying of potato chips. *Songklanakarin J. Sci. Technol.,* 31(1), 109-115.

Yamsaengsung, R., & Rungsee, C. 2003. Vacuum frying of fruits and vegetables. *Proc. 3th Ann. Conf. Thai Chem. Eng. Appl. Chem.,* Nakhon Nayok, Thailand, B-11.

Yamsaengsung, R. (2014). *Food product development: Fundamentals for innovations.* Hat Yai: Apple Art Printing House.

Irradiation of Food

Rosana G. Moreira
Department of Biological and Agricultural Engineering
Texas A&M University
College Station, Texas, USA

KEY TERMS		
Radiation sources	Depth-dose distribution	Food safety applications
Absorbed dose	Ionizing radiation effect	Kinetics of pathogen inactivation

Variables

- λ = wavelength
- η = throughput efficiency
- ρ = density of irradiated material
- A_c = cross-sectional area of food or package
- A_d = aerial density
- c = speed of light in a vacuum (3.0×10^8 m/s)
- C = rate of energy loss for e-beam treatment in water and water-like issues (2.33 MeV/cm)
- dE = energy in infinitesimal volume dv
- dm = mass in infinitesimal volume dv
- dm/dt = throughput or amount of product per time
- d_p = penetration depth of radiation energy per unit area
- d = thickness (or depth) of food
- d_{opt} = depth at which maximum throughput efficiency occurs for one-sided irradiation
- D = applied or absorbed dose or energy per unit mass
- D_{10} = radiation D value or kGy required to inactivate 90% of microbial population
- D_{max} = maximum dose
- D_{min} = minimum dose

D_{sf} = front surface dose, defined as the dose delivered at a depth d into the food, the target dose
DUR = dose uniformity ratio
E = maximum absorbed energy
$E_{50} = E_{mean}$ = absorbed energy at a depth of r_{50}
E_{ab} = energy deposited per incident electron
E_p = energy of a photon
f = radiation frequency
h = Planck's constant (6.626×10^{-34} J·s)
I_a = average current
I_A'' = current density
k = exponential rate constant
m = mass of food
N = microbial population at a particular dose
N_0 = initial microbial population
P = machine power
r_{max} = depth at which the maximum dose occurs
r_{opt} = depth at which the dose equals the entrance dose
r_{33} = depth at which the dose equals a third of the maximum dose
r_{50} = depth at which the dose equals half of the maximum dose
t = irradiation time
v = speed
w = scan width

Introduction

Food irradiation is a non-thermal technology often called "cold pasteurization" or "irradiation pasteurization" because it does not increase the temperature of the food during treatment (Cleland, 2005). The process is achieved by treating food products with ionizing radiation. Other common non-thermal processing technologies include high hydrostatic pressure, high-intensity pulsed electric fields, ultraviolet (UV) light, and cold plasma.

Irradiation technology has been in use for over 70 years. It offers several potential benefits, including inactivation of common foodborne bacteria and inhibition of enzymatic processes (such as those that cause sprouting and ripening); destruction of insects and parasites; sterilization of spices and herbs; and shelf life extension. The irradiation treatment does not introduce any toxicological, microbiological, sensory, or nutritional changes to the food products (packaged and unpackaged) beyond those brought about by conventional food processing techniques such as heating (vitamin degradation) and freezing (texture degradation) (Morehouse and Komolprasert, 2004). It is the only commercially available decontamination technology to treat fresh and fresh-cut fruits and vegetables, which do not undergo heat treatments such as pasteurization or sterilization. This is critical because many recent foodborne illness outbreaks and product recalls have been associated with fresh produce due to contamination with *Listeria*, *Salmonella*, and *Escherichia coli*. Approximately 76 million illnesses, 325,000 hospitalizations, and 5000 deaths occur in the United States annually and 1.6 million illnesses, 4000 hospitalizations, and 105 deaths in Canada (Health Canada, 2016). During 2018, these outbreaks caused 25,606 infections, 5,893 hospitalizations, and 120 deaths in the US (CDC, 2018).

Irradiation of foods has been approved by the World Health Organization (WHO) and the Food and Agriculture Organization (FAO) of the United Nations. At least 50 countries use this technology today for treatment of over 60 products, with spices and condiments being the largest application. In 2004, Australia became the first country to use irradiation for phytosanitary purposes, i.e., treatment of plants to control pests and plant diseases for export purposes (IAEA, 2015; Eustice, 2017). About ten countries have established bilateral agreements with the United States for trade in irradiated fresh fruits and vegetables. More than 18,000 tons of agricultural products are irradiated for this purpose around the world. The US has a strong commercial food irradiation program, with approximately 120,000 tons of food irradiated annually. Mexico, Brazil, and Canada are also big producers of irradiated products. China is the largest producer of irradiated foods in Asia, with more than 200,000 tons of food irradiated in 2010 (Eustice, 2017) followed by India, Thailand, Pakistan, Malaysia, the Philippines, and South Korea. Egypt and South Africa use irradiation technology to treat spices and dried foods. Russia, Costa Rica, and Uruguay have obtained approval for irradiation treatment of foods. Eleven European Union countries utilize food irradiation but the rest have been reluctant to adopt the technology due to consumers' misconceptions, such as thinking that irradiated foods are radioactive with damaged DNA or "dirty" (Maherani et al., 2016).

Food irradiation can be accomplished using different radiation sources, such as gamma rays, X-rays, and electron beams. Although the basic engineering principles apply to all the different sources of radiation energy, this chapter focuses on high-energy electron beams and X-rays to demonstrate the concepts because they are a more environmentally acceptable technology than the cobalt-60-based technology (gamma rays).

Outcomes

After reading this chapter, you should be able to:

- Explain the interaction of ionizing radiation with food products
- Quantify the effect of ionizing radiation on microorganisms and determine the dose required to inactivate pathogens in foods
- Select the best irradiation approach for different food product characteristics

Concepts

Food irradiation involves using controlled amounts of ionizing radiation with enough energy to ionize the atoms or molecules in the food to meet the desired processing goal. *Radiation* is the emission of energy that exists in the form of waves or photons as it travels through space or the food material (electromagnetic energy). In other words, it is a mode of energy transfer. The heat transfer equivalent would be the energy emitted by the Sun.

The type of radiation used in food processing is limited to high-energy gamma rays, X-rays, and accelerated electrons or electron beams (e-beams). Gamma and X-rays form part of the electromagnetic spectrum (like radio waves, microwaves, ultraviolet, and visible light rays), occurring in the short wavelength (10^{-8} to 10^{-15} m), higher frequency (10^{16} to 10^{23} Hz), high-energy (10^2 to 10^9 eV) region of the spectrum. High-energy electrons produced by electron accelerators in the form of e-beams can have as much as 10 MeV (megaelectronvolts = eV × 10^6) of energy (Browne, 2013).

The wavelength, or distance between peaks, λ, of the radiation energy is defined as the ratio of the speed of light in a vacuum, c, to the frequency, f, as follows:

$$\lambda = \frac{c}{f} \qquad (1)$$

where λ = wavelength (m)
$c = 3.0 \times 10^8$ (m/s)
f = radiation frequency (1/s)

From a quantum-mechanical perspective, electromagnetic radiation may be considered to be composed of photons (groups or packets of energy that are quantified). Therefore, each photon has a specific value of energy, E, that can be calculated as follows:

$$E_p = hf \qquad (2)$$

where E_p = energy of a photon (J)
h = Planck's constant (6.626×10^{-34} J·s)
f = radiation frequency (1/s)

The frequency, energy and wavelength of different types of electromagnetic radiation, calculated using equations 1 and 2, are given in table 1. The higher the frequency of the electromagnetic wave, the higher the energy, and the shorter the wavelength. Table 1 illustrates that X-rays and gamma rays are used in food irradiation processes because of their high energy. Table 1 also explains why exposure to UV light would only cause sunburn (lower energy electromagnetic radiation) while exposure to X-rays could be lethal (high-energy electromagnetic radiation).

Radiation Sources and Their Interactions with Matter

The properties and effects of gamma rays and X-rays on materials are the same, but their origins are different. X-rays are generated by machines while gamma rays come from the spontaneous disintegration of radionuclides, with cobalt-60 (^{60}Co) the most commonly used in food processing applications. X-ray machines with a maximum energy of 7.5 MeV and electron accelerators with a maximum energy of 10 MeV are approved by WHO worldwide because the energy from these radiation sources is too low to induce radioactivity in the food product (Attix, 1986). Likewise, although gamma rays are high energy radiation sources, the doses approved for irradiation of foods do not induce any radioactivity in products.

The difference in nature of the types of ionizing radiation results in different capabilities to penetrate matter (table 2). Gamma-ray and X-ray radiation can penetrate distances of a meter or more into the product, depending on the product density, whereas electron beams (e-beams), even with energy as high as 10 MeV, can penetrate only several centimeters. E-beam accelerators range from 1.35 MeV to 10 MeV (Miller, 2005). All types of radiation become less intense the further the distance from the radioactive material, as the particles or rays become more spread out (USNRC, 2018).

Absorbed Dose

Absorbed dose, or dose, D, is the quantity of ionizing radiation imparted to a unit mass of material. This quantity is used both to specify the irradiation process and to control it to ensure the product is not over- or under-exposed to the radiation energy. In food irradiation operations, dose values are average values because it is difficult to measure dose in small materials (IAEA, 2002).

Table 1. Frequency, energy level and wavelength of the different types of electromagnetic radiation calculated using equations 1 and 2.

Type of Electromagnetic Radiation	Frequency, f (Hz)	Energy, E (eV)	Wavelength, λ (cm)
Gamma rays	10^{20}	4.140×10^5	3.0×10^{-10}
X-rays	10^{18}	4.140×10^2	3.0×10^{-8}
UV light	10^{16}	4.140	3.0×10^{-6}
Infrared light	10^{14}	0.414	3.0×10^{-4}

Table 2. Different types of radiation sources and their characteristics (Attix, 1986; Lagunas-Solar, 1995; Miller, 2005).

Characteristics	Source		
	E-beams	X-rays	Cobalt-60 (gamma rays)
Energy (MeV)	10	5 or 7.5	1.17 and 1.33
Penetration depth (cm)	< 10	100	70
Irradiation on demand (machine can be turned off)	yes	yes	no
Relative throughput efficiency	high	medium	low
Dose uniformity ratio (D_{max}/D_{min})	low	high	medium
Administration process	authorization required[a]	authorization required[a]	authorization required[b]
Treatment time	seconds	minutes	hours
Average dose rate (kGy/s)	~3	0.00001	0.000061
Applications	low density products can be treated in cartons	low/medium density products can be treated in cartons or pellets	low/medium density products can be treated in cartons or pellets

[a] Standard registration required
[b] Complex and difficult process with extensive training

The SI unit of absorbed dose is the gray (Gy), where 1 Gy is equivalent to the absorption of 1 J per kg of material. Therefore, absorbed dose at any point in the target food is expressed as the mean energy, dE, imparted by ionizing radiation to the matter in an infinitesimal volume, dv, at that point divided by the infinitesimal mass, m, of dv:

$$D = \frac{dE}{dm} \qquad (3)$$

where D = dose (Gy)
dE = energy in infinitesimal volume dv (J)
dm = mass in infinitesimal volume dv (kg)

D represents the energy per unit mass which remains in the target material at a particular point to cause any effects due to the radiation energy (Attix, 1986).

In 1928, the roentgen was conceived as a unit of exposure, to characterize the radiation incident on an absorbing material without regard to the character of the absorber. It was defined as the amount of radiation that produces one electrostatic unit of ions, either positive or negative, per cubic centimeter of air at standard temperature and pressure (STP). In modern units, 1 roentgen equals 2.58×10^{-4} coulomb/kg air (Attix, 1986). In 1953, the International Commission on Radiation Units and Measurements (ICRU) recommended the "rad" as a new unit with 1 Gy equal to 100 rad. The term "rad" stands for "radiation absorbed dose." Absorbed dose requirements for various treatments involving food products range from 0.1 kGy to 30 kGy (table 3). Table 4 shows the maximum allowable dose for different products in the United States and worldwide.

The dose rate, or amount of energy emitted per unit time (dD/dt or $\frac{d}{dt}\left(\frac{dE}{dm}\right)$), determines the processing times and, hence, the throughput of the irradiator (i.e., the quantity of products treated per time unit). In those terms, 10 MeV electrons can

Table 3. Absorbed dose requirement for different food treatments (IAEA, 2002).

Treatment	Absorbed Dose (kGy)[a]
Sprout inhibition	0.1–0.2
Insect disinfestation	0.3–0.5
Parasite control	0.3–0.5
Delay of ripening	0.5–1
Fungi control	0.5–3
Pathogen inactivation	0.5–3
Pasteurization of spices	10–30
Sterilization (pathogen inactivation)	15–30

[a] 1 kGy = 10^3 Gy

Table 4. Maximum allowable dose for different foods in the United States and worldwide (WHO, 1981; ICGFI, 1999; Miller, 2005).

Purpose	Maximum Dose (kGy)	Product
Disinfestation	1.0	any food
Sprout inhibition	0.1–0.2	onions, potatoes, garlic
Insect disinfestation	0.3–0.5	fresh dried fruits, cereals and pulses, dried fish and meat
Parasite control	0.3–0.5	fresh pork
Delay of ripening	0.5–1.0	fruits and vegetables
Pathogen inactivation	3.0	poultry, shell eggs
Pathogen inactivation	1.0	fresh fruits and vegetables
Pathogen inactivation	4.5–7.0	fresh and frozen beef and pork
Pathogen inactivation	1.0–3.0	fresh and frozen seafood
Shelf life extension	1.0–3.0	fruits, mushrooms, leafy greens
Pasteurization	10–30	spices
Commercial sterilization	30–50	meat, poultry, seafood, prepared foods, hospital foods, pet foods

produce higher throughput (higher dose rate) compared to X-rays and gamma rays (table 2). Similar to absorbed dose, dose rates are average values.

Depth-Dose Distribution and Electron Energy

The energy deposition profile for a 10 MeV e-beam incident onto the surface of a water absorber has a characteristic shape (figure 1). The y-axis is the energy deposited per incident electron per unit area, E, also described as E_{ab}. This parameter is proportional to the absorbed dose, D. The x-axis is the penetration depth (also called mass thickness), d, in units of area density, g/cm^2, which is the thickness in cm multiplied by the volume density in g/cm^3:

$$d_p = d\rho \quad (4)$$

where d_p = penetration depth of radiation energy per unit area (g/cm^2)
d = thickness of irradiated material (cm)
ρ = density of irradiated material (g/cm^3)

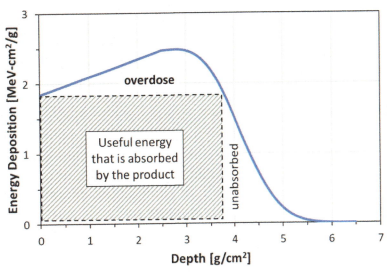

Figure 1. Energy deposition profile for 10 MeV electrons in a water absorber (adapted from Miller, 2005).

The penetration depth, d, of ionizing radiation is defined as the depth at which extrapolation of the tail of the dose-depth curve meets the x-axis (approximately 6 g/cm^2 in figure 1). Figure 1 also shows how the dose, D, tends to increase with increasing depth within the product to about the midpoint of the electron penetration range and then it quickly falls to low doses.

Because the electron energy deposition is not constant, there is a location in the product that will receive a minimum dose, D_{min}, and another position that will receive the maximum dose, D_{max}. A useful parameter for irradiator designers and engineers is the dose uniformity ratio (DUR), defined as the ratio of maximum to minimum absorbed dose:

$$DUR = \frac{D_{max}}{D_{min}} \quad (5)$$

A *DUR* close to 1.0 represents uniform dose distribution in the sample (Miller, 2005; Moreira et al., 2012). However, values greater than 1.0 are common in commercial applications and many food products can tolerate a higher *DUR*, of 2 or even 3 (IAEA, 2002).

The absorbed dose, D, at a particular depth, d, can be calculated as the product of the energy deposited times the current density times the irradiation time (Miller, 2005):

$$D(d) = E_{ab} I_A'' t \tag{6}$$

where D = dose (MeV/g) (1 Gy = 6.24 × 10^{12} MeV/kg)
  E_{ab} = energy deposited per incident electron (MeV-cm^2/g)
  I_A'' = current density (A/cm^2)
  t = irradiation time (s)

For a product with thickness, x, the energy represented by the dashed area in figure 1 is the useful energy absorbed in the product. The maximum efficiency will occur when the product depth is such that the back surface of the target product receives the same dose as the top surface. For instance, using figure 1 and assuming only energy penetration through the thickness of the material, the target with a minimum dose of 1.85 MeV/g (entrance dose) and the optimum depth of 3.8 g/cm^2 represents an effective absorbed energy of about 7 MeV (= 1.85 × 3.8). Therefore, using 10 MeV e-beams, the maximum utilization efficiency is 70% (Miller, 2005).

The depth in g/cm^2 at which the maximum throughput efficiency occurs for one-sided irradiation can be calculated as (Miller, 2005):

$$Depth_{optimum} = d_{opt} = 0.4 \times E - 0.2 \tag{7}$$

where E is the maximum absorbed energy (MeV).

Equation 7 provides a useful measure of the electron penetration power of the irradiator. The penetration of high-energy e-beams in irradiated materials increases linearly with the incident energy. The electron range (penetration) also depends on the atomic composition of the irradiated material. Materials with higher electron contents (electrons per unit mass) will have higher absorbed doses near the entrance surface, but lower electron ranges (penetration). For instance, because of its lack of neutrons, hydrogen has twice as many atomic electrons per unit mass as any other element. This means that materials with higher hydrogen contents, such as water (H_2O) and many food products, will have higher surface doses and shorter electron penetration than other materials (Becker et al., 1979).

In general, dose-penetration depth curves, such as the one represented by figure 1, show an initially marked increase (buildup) of energy deposition near the surface of the irradiated product. This buildup region is a phenomenon that happens in materials of low atomic number due to the progressive cascading of secondary electrons by collisional energy losses (IAEA, 2002). This is then followed by an exponential decay of dose to greater depths. The approximate value of the buildup depth for gamma rays (1.25 MeV) is 0.5 cm of water, while the depth for 10 MeV e-beams is 10.0 cm of water (IAEA, 2002).

Figure 2 shows the point of maximum dose (in kGy) and the absorption of energy for both electrons and photons (X-rays and gamma rays). The penetration depth of 10 MeV e-beams is limited as they deposit their energy over a short depth, with a maximum located after the entrance point. In the case of gamma rays, the energy is deposited over a longer distance, which results in a uniform dose distribution within the treated product. The penetration capabilities of both 7.5 MeV X-rays and gamma rays are comparable, but the higher energy of the X-rays results in a slightly more uniform distribution of the doses within the treated product. The configuration of the product strongly influences dose distribution within the product (IAEA, 2002).

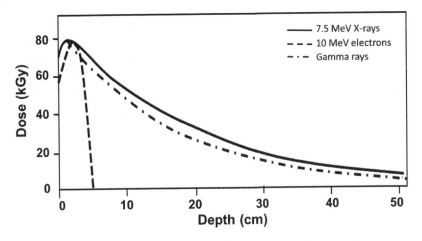

Figure 2. Dose-depth penetration for different radiation sources (X-rays, electron beams, and gamma rays) (adapted from IAEA, 2015).

Figure 3 shows the depth-dose distributions in water-equivalent products (such as fruits and vegetables) ranging from 1 to 10 MeV in terms of *relative dose* in percentage. For instance, for the 10 MeV curve, if the entrance (at the surface) dose of 1 kGy is 100%, the relative dose at a depth of 1 cm²/g is approximately 110% of the entrance dose or 1.1 kGy, and it is 0 and 1.40 kGy for 1 MeV and 5 MeV irradiation systems, respectively.

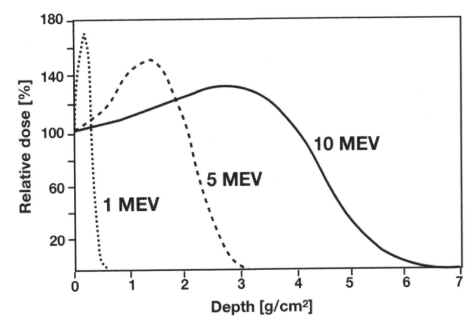

Figure 3. Typical depth-dose curves for electrons of various energies in the range applicable to food processing operations (adapted from IAEA, 2002). Here depth is penetration depth per unit area, d_p.

The shapes of the depth-dose curves shown in figure 3 can be better defined in terms of the penetration depth within the product (or product thickness) (figure 4). The parameters defined in figure 4, r_{max}, r_{opt}, r_{50}, and r_{33}, are useful to determine the maximum product thickness that can be irradiated using a particular type of electron beam (1, 5, or 10 MeV). Additionally, the deposited energy can be determined at a specific depth. For instance, E_{50} at a depth of r_{50} = 4.53 cm in water for a 10 MeV irradiation system is,

$$E_{mean} = E_{50} = Cr_{50} = 2.33(4.53\,\text{cm}) = 10.55\,\text{MeV} \tag{8}$$

where C is the rate of energy loss for e-beam treatment in water and water-like tissues = 2.33 MeV/cm (Strydom et al., 2005).

From figure 4 with r_{max} equal to 2.8 cm, the maximum dose is 130% or 1.3 kGy, and the entrance dose equals the exit dose at r_{opt} equal to 4 cm. This result means that if the irradiated product has a thickness between 2.8 and 4 cm, the DUR is constant with a value of 1.3 (DUR = 1.3 kGy/1.0 kGy). Such a DUR value suggests the irradiation process provides good uniformity in the dose distributed throughout the product thickness. If the process yields a DUR of 2 with a minimum dose of 0.67 kGy (DUR = 1.35 kGy/0.67 kGy), the maximum useful thickness of the irradiated product will be 4.5 cm or r_{50}, the depth at which the dose is half the maximum dose.

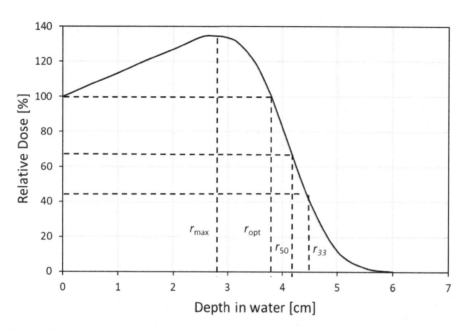

Figure 4. Depth-dose curve for 10 MeV electrons in water, where the entrance (surface) dose is 100% (adapted from IAEA, 2002). r_{max} is the depth in cm at which the maximum dose occurs, r_{opt} is the depth at which the dose equals the entrance dose (also described by equation 7), r_{50} is the depth at which the dose equals half of the maximum dose, and r_{33} is the depth at which the dose equals a third of the maximum dose.

Note that $r_{50} > r_{opt}$. Hence, if the product thickness exceeds r_{opt}, the DUR increases. As DUR approaches infinity at a depth of 6.5 cm for 10 MeV e-beam (figure 4), any part of the product beyond that depth will remain unexposed to the irradiation treatment. Therefore, the maximum processable product thickness for this irradiation system will be 6.5 cm. This result highlights a critical issue when using electron beam accelerators to pasteurize or sterilize food products, which need to be exposed in their entirety to the radiation energy.

The engineer has the option to apply the e-beams using the single e-beam configuration (which exposes the target food only on the top or bottom surface) or the double-beam configuration (which exposes the target food at both the top and bottom surfaces). Figure 5 illustrates the difference between one-sided and two-sided irradiation systems using 10 MeV electrons in water when DUR is 1.35.

Figure 5 shows that when irradiating from the top or bottom only, the maximum processable thickness will be close to 4 cm (shaded areas, figure 6), while the double-beam system increases the maximum processable thickness to about 8.3 cm (shaded area, figure 7). Therefore, to improve the penetration capability of a 10 MeV e-beam treatment, two 10 MeV accelerators, one irradiating from the top and the other from the bottom of a conveyor system, are frequently used in commercial applications (IAEA, 2002).

The depth at which the maximum throughput efficiency occurs for double-sided irradiation can be calculated as (Miller, 2005):

$$Depth_{optimum} = d_{opt} = 0.9 \times E - 0.4 \quad (9)$$

Measurement of Absorbed Dose

The effectiveness of ionizing radiation in food processing applications depends on proper delivery of the absorbed dose. To design the correct food irradiation process, the operator should be able to (1) measure the absorbed dose delivered to the food product using reliable dosimetry methods; (2) determine the dose distribution patterns in the product package; and (3) control the routine radiation process (through process control procedures). Dosimeters are used for quality and process control in radiation research and commercial processing.

Reliable techniques for measuring dose, called *dosimetry*, are crucial for ensuring the integrity of the irradiation process. Incorrect dosimetry can result in an ineffective food irradiation process. Dosimetry systems include physical or chemical dosimeters and measuring instrumentation, such as spectrophotometers and electron paramagnetic resonance (EPR) spectrometers. A *dosimeter* is a device capable of providing a reading that is a measure of the absorbed dose, D, deposited in its sensitive volume, V, by ionizing radiation. The measuring instrument must be well characterized so that it gives reproducible and accurate results (Attix, 1986).

There are four categories of dosimetry systems according to their intrinsic accuracy and usage (IAEA, 2002):

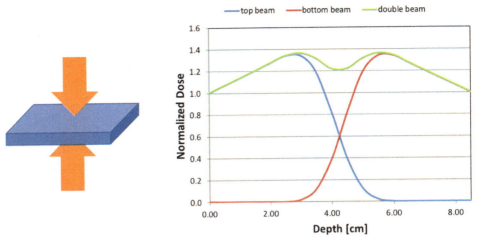

Figure 5. Depth-dose distributions for 10 MeV electrons in water for single-sided and double-sided configurations and *DUR* = 1.35. Normalized dose is the ratio of maximum to entrance dose (Miller, 2005).

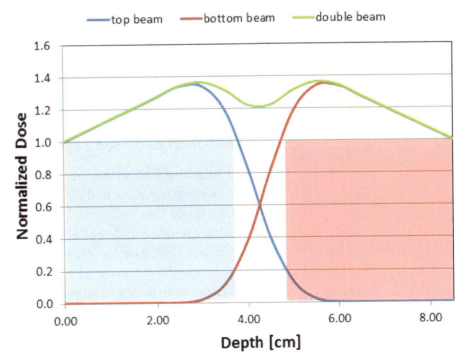

Figure 6. Maximum penetration thickness for top-only and bottom-only e-beam configurations using 10 MeV electrons in water and *DUR* = 1.35.

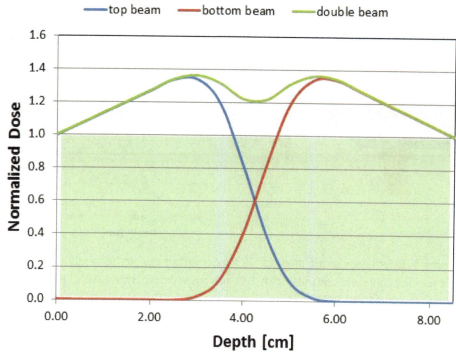

Figure 7. Maximum penetration thickness for double-sided e-beam irradiation using 10 MeV electrons in water (*DUR* = 1.35).

- *Primary standards* (ion chamber, calorimeters) measure the absolute (i.e., does not need to be calibrated) absorbed dose in SI units.
- *Reference standards* (alanine, Fricke, and other chemicals) have a high metrological quality that can be used as a reference standard to calibrate other dosimeters. They need to be calibrated against a primary standard, generally through the use of a transfer standard dosimeter.
- *Transfer standards* (thermoluminiscent dosimeter, TLD) are used for transferring dose information from a national standards laboratory to an irradiation facility to establish traceability to that standards laboratory. They should be used under conditions specified by the issuing laboratory. They need to be calibrated.
- *Routine dosimeters* (process monitoring, radiochromic films) are used in radiation processing facilities for dose mapping and for process monitoring for quality control. They must be calibrated frequently against reference or transfer dosimeters.

Food Irradiation and Food Safety Applications

Effect of Ionizing Radiation on Pathogens

Pathogen inactivation is the end effect of food irradiation. Exposure to ionizing radiation has two main effects on pathogenic microorganisms. First, the radiation energy can directly break strands (single or double) of the microorganism's DNA. The second effect occurs indirectly when the energy causes radiolysis of water to form very reactive hydrogen (H+) and hydroxyl ($\cdot$OH) radicals. These radicals can recombine to produce even more reactive radicals such as superoxide (HO_2), peroxide (H_2O_2), and ozone (O_3), which have an important role in inactivating pathogens in foods. Although DNA is the main target, other bioactive molecules, such as enzymes, can likewise undergo inactivation due to radiation damage, which enhances the efficacy of the irradiation treatment.

> The main pathogenic microorganisms of concern in food processing are *Salmonella* spp., *Escherichia coli* and *Listeria* spp. The abbreviation "spp." stands for more than one species in that genus, here meaning there are several types of bacteria in the same group.

Kinetics of Pathogen Inactivation

The traditional approach used in thermal processing calculations is to develop survival curves, which are semi-log plots of microorganism populations as a function of time at a given process temperature. This same approach can be used to develop radiation survival curves, i.e., plots of the log of the change in microbial populations as a function of applied dose. In this chapter, only first-order kinetics of microbial destruction are described.

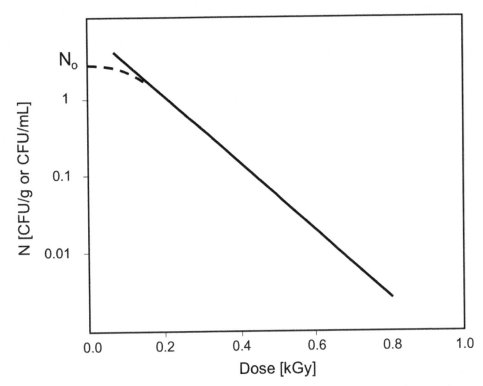

Figure 8. Typical survival curve showing first-order kinetics behavior with N_0 = initial microbial population and N = microbial population at a particular dose, both in CFU/g or CFU/mL. Dashed curve is initial nonlinear section of the curve.

Figure 8 is a survival curve obtained for inactivation of a pathogen in a food product due to exposure to radiation energy. Based on first-order kinetics (i.e., ignoring the initial non-linear section of the curve indicated by the arrow and the dashed line in figure 8), the microbial inactivation rate is described by:

$$\frac{dN}{dD} = -kD \tag{10}$$

where N = microbial population at a particular dose (CFU/g or CFU/mL; CFU stands for colony forming units)
D = the applied dose (kGy)
k = exponential rate constant (1/kGy)

The radiation resistance of the target microorganism is usually reported as the radiation D value, D_{10}, defined as the amount of radiation energy (kGy) required to inactivate 90% (or one log reduction) of the specific microorganism (Thayer et al., 1990). Using this definition and integrating equation 10 yields:

$$N = N_0 e^{-kD_{10}} \tag{11}$$

where N_0 = initial microbial population (CFU/g or CFU/mL)

Based on figure 8 and equation 11, the inverse of the slope of the line is the D_{10} value and is equivalent to the D-value used in thermal process calculations except that these have units of time as the slope of population change versus process time. The relationship between the D_{10} value and the rate constant is:

$$k = \frac{1}{D_{10}} \tag{12}$$

The D_{10} value varies with the target pathogen, type and condition of food (whole, shredded, peeled, cut, frozen, etc.), and the atmosphere in which it is packed (e.g., vacuum-packaged foods, pH, moisture, and temperature) (Niemira, 2007; Olaimat and Holley, 2012; Moreira et al., 2012). For instance, the D_{10}-values for *Salmonella* spp. and *Listeria spp.* in fresh produce can range between 0.16 to 0.54 kGy while *Escherichia coli* is slightly more resistant to irradiation treatment (sometimes up to 1 kGy) (Fan, 2012; Rajtowski et al., 2003). When tomatoes are irradiated, the D_{10}-values for *Escherichia coli* O157:H7, *Salmonella* spp., and *Listeria monocytogenes* are around 0.39, 0.56, and 0.66 kGy, respectively (Mahmoud et al., 2010). In commercial applications, the rule of thumb is to design an irradiation treatment for a five log or $5D_{10}$ reduction in the population of the target pathogen.

Applications

The goal of a food irradiation process is to deliver the minimum effective radiation dose to all portions of the product. Too high a dose (or energy) in any region of the target product could lead to wasted energy and deterioration of product quality.

To design a food irradiation process, the absorbed dose in the material of interest must be specified because different materials have different radiation absorption properties. In the case of food products, the material of interest is water because most foods behave essentially as water regardless of their water content. Dose requirements and maximum allowable doses should be used for specific applications (tables 3 and 4).

Cost estimates for food irradiation facilities include the capital cost of equipment, installation and shielding, material handling and engineering, and variable costs including electricity, maintenance, and labor. The approximate cost of an e-beam accelerator facility for a production rate of 2000 hours per year is between 2 and 5 million US dollars and has remained fairly steady (Morrison, 1989; Miller, 2005; University of Wisconsin, 2019).

Technology Selection

The selection of the right technology for a particular food irradiation application depends on many factors, including food product characteristics and processing requirements (Miller, 2005). Figure 9 shows the steps required to choose a food irradiation approach.

The first step is to define the product characteristics. What is the main goal of the process? What is the product state, i.e., frozen, unpackaged, etc.? What is the product's density, shape, and mass flow rate going through the accelerator?

The second step specifies the process requirements, including the product thickness and the acceptable DUR (equation 5). The final step is to select the appropriate radiation technology based on the product characteristics and process requirements. Selection includes determining the best technology (e-beams versus X-rays versus gamma rays), the size of the e-beam or X-rays accelerator(s), and, in the case of e-beams, whether single- or double-beam treatment will be more effective.

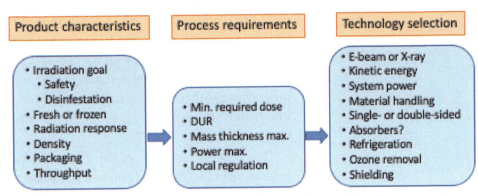

Figure 9. Steps needed to select the right irradiation technology for a food processing application (adapted from Miller, 2005).

A simplified flow diagram provides guidelines to follow in selecting the right technology for food irradiation (figure 10). The engineer must first determine if the product could be effectively irradiated at all based on maximum to minimum dose ratios and energy efficiency concepts. The penetration depth depends on the product mass thickness (g/cm^2), which is based on the product and/or package dimensions and density (equation 4). For food safety treatments, the DUR is based on the minimum dose requirement to reduce the population of a certain pathogen (i.e., the D_{10} value, equation 12) and the maximum dose allowed by local regulation or the dose a product can tolerate without degrading its quality. As indicated in figure 10, in general, the product will not be suitable for irradiation treatment when its mass thickness is greater than 50 g/cm^2 and DUR must be less than 3.

Finally, the engineer must select the product handling systems to transport the food product in and out of the e-beam and X-rays irradiators via conveyors. Orientation of the irradiators is an important consideration since e-beams are oriented vertically to the product while the higher-penetrating X-rays allow for horizontal irradiation of products. The dose rate is set by varying the speed of the conveyors. The engineer must also determine whether absorbers must be used to reduce the entrance dose; provide refrigeration of the facility, if needed, since many food products are perishable; include shielding of the facility (X-rays require thicker

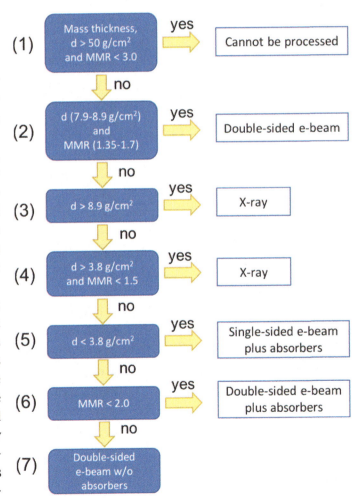

Figure 10. Decision flow diagram for selecting the correct irradiation approach (adapted from Miller, 2005). MMR is the acceptable range of maximum to minimum dose ratios (DUR).

Irradiation of Food • 15

walls than e-beam processing), and provide for ozone removal (a sub-product of irradiation from ionization of oxygen in the air) (Miller, 2005). Prior to entering the irradiation system, products are inspected in staging areas where products are palletized and loaded into containers to be transported on conveyers through the irradiators. Irradiated products are then loaded into transportation vehicles or stored in refrigerated chambers for distribution to retailers.

The speed, v, in cm/s, of the conveyor transporting the food through an e-beam scan facility is determined by (Miller, 2005):

$$v = \frac{1.85 \times 10^6 I_a}{w D_{sf}} \quad (13)$$

where I_a = average current (A), an e-beam accelerator configuration parameter

w = scan width (cm), an e-beam accelerator configuration parameter (see figure 11)

D_{sf} = the front surface dose (kGy), defined as the dose delivered at a depth d into the food (figure 11); the target dose

The conveyor speed is directly related to the throughput as:

$$v = \frac{dm/dt}{A_d w} \quad (14)$$

where dm/dt = throughput or amount of product per time (g/s)

A_d = aerial density (g/cm²) obtained from equation 15:

$$A_d = \rho d \quad (15)$$

where ρ = food density (g/cm³)

d = thickness (or depth) of food (cm)

Equations 13 and 14 show that for a system with fixed average current and scan width, the faster the speed of the conveyor, the more product is processed in the facility and the lower the dose it receives. Typical conveyor speeds range between 5 and 10 m/minute.

The total mass of product running through the conveyor belt is calculated as:

$$m = A_d A_c \quad (16)$$

where m = mass of food (kg)

A_d = aerial density from equation 15

A_c = cross-sectional area of food or package (m²)

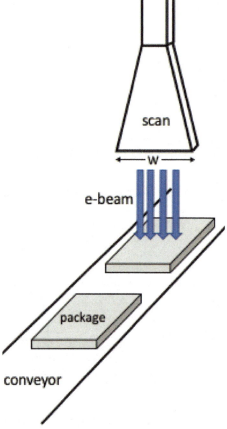

Figure 11. Typical electron beam irradiation configuration showing flat food packages placed onto the conveyor with the beam directed downward and scanned horizontally across the product. For double-beam irradiation a second beam is directed vertically upward (adapted from Miller, 2005).

The throughput requirements of electron beam facilities (dm/dt) are estimated based on the beam power, the minimum required dose, and irradiation mode (e.g., e-beam vs. X-rays) as follows (Miller, 2005):

$$\frac{dm}{dt} = \frac{\eta P}{D} \qquad (17)$$

where η = throughput efficiency, which is 0.025 to 0.035 at 5 MeV and 0.04 to 0.05 at 7.5 MeV for X-ray irradiation, and 0.4 to 0.5 for e-beam mode (Miller, 2005)
P = machine power (kW)
D = minimum dose requirement (kGy), which ranges from 250 Gy for disinfestation to 6–10 kGy for preservation of freshness for spices

Examples

Example 1: Interaction of ionizing radiation with matter

Problem:
If the incident current density at the surface is 10^{-6} A/cm² of the water absorber in figure 1 and the energy deposited per incident electron is 1.85 MeV-cm²/g, determine the absorbed dose in kGy after 1 second.

Solution:
Using equation 6:

$$D(d) = E_{ab} I_A^{''} t \qquad (6)$$

$$D(d) = \left(1.85 \text{ MeV} \frac{\text{cm}^2}{\text{g}}\right) \times \left(10^{-6} \frac{\text{A}}{\text{cm}^2}\right) \times 1 \text{ s}$$

with 1 MeV = 10^6 eV:

$$D(d) = \left(1.85 \times 10^6 \text{ eV} \frac{\text{cm}^2}{\text{g}}\right) \times \left(10^{-6} \frac{\text{A}}{\text{cm}^2}\right) \times 1 \text{ s}$$

In units of energy, 1 eV (electrovolt) equals 1.60218×10^{-19} Joules and 1 kJ = 1000 J

$$D(d) = \left(1.85 \times 10^6 \text{ eV} \frac{\text{cm}^2}{\text{g}}\right) \times \left(10^{-6} \frac{\text{A}}{\text{cm}^2}\right) \times 1 \text{ s} \left(\frac{1 \text{C}}{1 \text{ A} \times \text{s}}\right) \times$$

$$\left(\frac{1.6022 \times 10^{-19} \text{ J}}{1 \text{ eV}}\right) \times \left(\frac{1}{1.6022 \times 10^{-19} \text{C}}\right)$$

Finally, the dose in kGy is:

$$D(d) = 1.85 \frac{kJ}{kg} \text{ or kGy}$$

The absorbed dose after 1 second is 1.85 kGy.

Example 2: Calculation of dose uniformity ratio (DUR)

Problem:
Figure 1 shows that the absorbed dose increases at a depth of 2.75 g/cm² inside the irradiated water absorber. (a) Find the dose uniformity ratio (*DUR*). (b) Comment on the changes (if any) to this parameter as a function of depth in the irradiated target.

Solution:
(a) Based on figure 1 and using equation 5, calculate the *DUR*:

$$DUR = D_{max}/D_{min} = 2.5/1.85 = 1.35$$

The *DUR* value is within the acceptable range for dose uniformity in commercial irradiator systems (close to 1.0).

(b) Based on figure 1, the *DUR* remains constant (= 1.35) up to a depth of 3.8 g/cm². Beyond this depth, the minimum dose decreases which increases the *DUR*. This is clearly shown in figure 1 as the dose increases with increasing depth within the product and then it decreases.

Example 3: Product thickness for one sided e-beam irradiation

Problem:
Determine the maximum allowable product thickness for one-sided e-beam irradiation with 10 MeV electrons if a dose uniformity ratio of 3 is acceptable.

Solution:
From figure 4 and using equation 5, determine the depth in cm for *DUR* = 3

$$DUR = D_{max}/D_{min}$$

D_{max} = 130% or 1.3 kGy (figure 4) and D_{min} = 1.3/3 = 0.43 kGy or 43% relative dose

Again, from figure 4, the depth value is 4.8 cm = r_{33}.
Thus, the maximum allowable product thickness will be 4.8 cm and the exit dose equals a third of the maximum dose.

Example 4: Efficiency of single-sided vs. double-sided irradiation treatment

Problem:
Determine the depth at the maximum throughput efficiency for single-sided and double-sided 10 MeV irradiation of water (5 cm thick) when the energy absorbed is (a) 1.50 MeV-cm²/g, (b) 2.20 MeV-cm²/g, and (c) 2.40 MeV-cm²/g.

Solution:
Select the appropriate equation and calculate the depth in cm.
For single-sided irradiation use equation 7:

$$d_{opt} = 0.4 \times E - 0.2$$

(a) 1.50 MeV-cm²/g

$$d_{opt} = 0.4 \times (1.50) - 0.2 = 0.40 \text{ g/cm}^2$$

(b) 2.22 MeV-cm²/g

$$d_{opt} = 0.4 \times (2.22) - 0.2 = 0.68 \text{ g/cm}^2$$

(c) 2.40 MeV-cm²/g

$$d_{opt} = 0.4 \times (2.40) - 0.2 = 0.76 \text{ g/cm}^2$$

For double-sided irradiation use equation 9:

$$d_{opt} = 0.9 \times E - 0.4$$

(a) 1.50 MeV-cm²/g

$$d_{opt} = 0.9 \times (1.50) - 0.4 = 0.95 \text{ g/cm}^2$$

(b) 2.22 MeV-cm²/g

$$d_{opt} = 0.9 \times (2.22) - 0.4 = 1.60 \text{ g/cm}^2$$

(c) 2.40 MeV-cm²/g

$$d_{opt} = 0.9 \times (2.40) - 0.4 = 1.76 \text{ g/cm}^2$$

Energy Absorbed (MeV-cm²/g)	d_{opt} (g/cm²) Single-sided	d_{opt} (g/cm²) Double-sided
1.50	0.40	0.95
2.22	0.68	1.60
2.40	0.76	1.76

Results demonstrate that the double-beam configuration is more effective regarding penetration depth with minimum energy utilization, e.g., penetration of 0.95 g/cm² versus 0.40 g/cm² using electron beams with 1.5 MeV-cm²/g of energy.

Example 5: Interaction of ionizing radiation with food product and effect on dose penetration depth

Problem:

Comparisons of 10 MeV electron depth-dose distributions in a bag of vacuum-packed baby spinach leaves (mass thickness = 5.1 g/cm²) and ground beef patty (mass thickness = 5.1 g/cm²) are shown in figure 12. Determine the depth at which the maximum dose occurs for both food products and discuss your results.

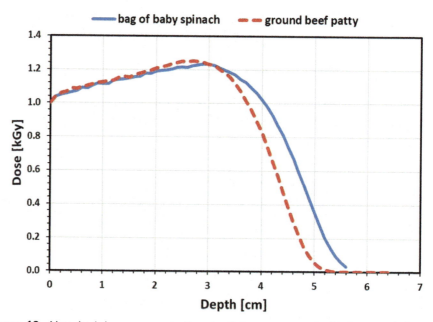

Figure 12. Absorbed dose vs. penetration depth in vacuum packed baby spinach leaves and ground beef patty at 10 MeV incident electron energy.

Solution:

Locate the depth (x-axis) at which dose (y-axis) is maximum. For the spinach, depth is 3.00 cm and for the ground beef patty, depth = 2.70 cm.

Both materials have very similar atomic composition and, therefore, absorb the incident energy very similarly.

Example 6: Calculation of radiation D_{10} value

Problem:
Romaine lettuce leaves were exposed to radiation doses up to 1.0 kGy using a 10 MeV e-beam irradiator to inactivate a pathogen. The population of survivors at each dose was measured right after irradiation (see table below).

Number of pathogens (CFU/g) in romaine lettuce leaves as a function of radiation dose:

Dose (kGy)	Population (log CFU/g)
0	6.70
0.25	5.50
0.50	4.30
0.75	3.30
1.00	2.00

(a) Calculate the D_{10} value of the pathogen in the fresh produce and determine the dose level required for a 5-log reduction in the population of the pathogen. The data point for a dose of 0 kGy represents the non-irradiated produce.

(b) If the maximum dose approved for irradiation of fresh vegetables is close to 1 kGy, is the irradiation treatment suitable?

Solution:
First, plot the logarithm of the population of survivors as a function of dose from the given data and determine the D_{10} value from the inverse of the slope of the line (figure 13).

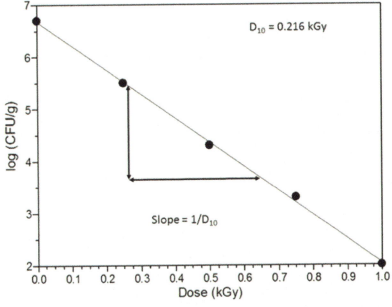

Figure 13. D_{10} value calculation assuming radiation inactivation as 1st order kinetics.

$$\text{Slope} = -\frac{\log N_1 - \log N_2}{D_1 - D_2} = -\frac{5-4}{0.375 - 0.591} = -\frac{1}{-0.216} = \frac{1}{0.216}$$

Then, determine the dose required for a 5-log reduction in microbial population, i.e., $5D_{10}$, and check if $5D_{10} < 1.0$ kGy. If yes, the process is suitable for treatment of the fresh produce. If $5D_{10} > 1.0$ kGy, another process should be considered.

$$5D_{10} = 5 \times 0.216 \text{ kGy} = 1.10 \text{ kGy}$$

This irradiation process would be suitable because the pathogen population in the romaine lettuce leaves will be reduced by 5 logs when exposed to a dose of approximately 1.0 kGy using 10 MeV electron beams.

Example 7: Selection of best irradiation technology

Problem:
A 10 MeV e-beam and a 5 MeV X-ray accelerator are available for irradiating the following products. Select the best irradiation technology to treat each of the products. Assume a minimum dose of 1 kGy.

(a) Ground beef patty contaminated with *Escherichia coli* O157:H7, $D_{max} = 1.25$ kGy (mass thickness = 8.5 g/cm²)
(b) Tomato contaminated with *Listeria monocytogenes*, $D_{max} = 1.4$ kGy (mass thickness = 3.2 g/cm²)
(c) Romaine lettuce contaminated with *Salmonella* Poona, $D_{max} = 1.37$ kGy (mass thickness = 4.1 g/cm²)

Solution:
Use the given information and the flow chart (figure 10) to determine whether e-beams or X-rays should be used for irradiation of the different products.

(a) *DUR* for beef patty (using equation 5, $DUR = D_{max}/D_{min}$) = 1.25 kGy/1 kGy = 1.25 = *MMR*
Following figure 10 with mass thickness $d = 8.5$ g/cm² and *MMR* = 1.25 leads to condition 4: $d > 3.8$ g/cm² and *MMR* < 1.5 and selection of X-ray as the appropriate technology for the beef patty.
(b) *DUR* for tomato sample (using equation 5, $DUR = D_{max}/D_{min}$): 1.4 kGy/1 kGy = 1.4 = *MMR*
Following figure 10 with mass thickness $d = 3.2$ g/cm² and *MMR* = 1.4 leads to condition 6 or 7: $d < 3.8$ g/cm² and selection of single or double-sided e-beam would be appropriate for the tomato sample.
(c) *DUR* for romaine lettuce (using equation 5, $DUR = D_{max}/D_{min}$): 1.37 kGY/1 kGy = 1.37 = *MMR*

Following figure 10 with mass thickness $d = 4.1$ g/cm² and $MMR = 1.37$ leads to condition 4: Mass thickness $d = 4.1$ g/cm². Since $d > 3.8$ g/cm² and $MMR < 1.5$, select X-ray as the appropriate technology for the romaine lettuce.

Product	Criteria	Choice of Radiation Technology
Beef patty	$d > 3.8$ g/cm², $MMR < 1.5$	X-rays
Tomato	$d < 3.8$ g/cm², $MMR < 1.5$	E-beams
Romaine lettuce	$d > 3.8$ g/cm², $MMR < 1.5$	X-rays

Example 8: Calculate the dose required for a 5-log reduction of pathogen population

Problem:
Calculate the dose required for a 5-log reduction of the pathogen for the three products from Example 7 using the following information. For each product, determine if the required dose is less than the maximum allowable dose for that product.

(a) Ground beef patty contaminated with *Escherichia coli* O157:H7 (D_{10} = 0.58 kGy)
(b) Tomato contaminated with *Listeria monocytogenes* (D_{10} = 0.22 kGy)
(c) Romaine lettuce contaminated with *Salmonella* Poona (D_{10} = 0.32 kGy)

Solution:
Given the D_{10} value for each pathogen, calculate 5D. The pathogen with the higher 5D value is the more resistant to irradiation and will require treatment at higher doses.

Product	Pathogen	5D (kGy)
Ground beef patty	*Escherichia coli* O157:H7	2.90
Tomato	*Listeria monocytogenes*	1.10
Romaine lettuce	*Salmonella* Poona	1.60

The *E. coli* in the beef patties will require higher doses to achieve a 5-log inactivation level than the doses required to treat the two fresh produces. The required treatment for the tomato samples falls within the acceptable dose level for fruits and vegetables (about 1 kGy). The *Salmonella* in the lettuce will require a slightly higher dose but the U.S. Food and Drug Administration (FDA, 2018) allows up to 4 kGy for treatment of leafy greens. The maximum allowable dose for pathogen inactivation in fresh and frozen beef ranges from 4.5–7.0 kGy in different countries (table 4).

Example 9: Calculation of conveyor speed in an e-beam system

Problem:
Calculate the conveyor speed required for a 1.5 kGy entrance dose (front surface dose) irradiation for a single-sided process using a 10 MeV, 1-mA beam with a scan width of 120 cm.

Solution:
Calculate the conveyor speed using equation 13:

$$v = \frac{1.85 \times 10^6 I_a}{w D_{sf}}$$

The conveyor speed, v, with the given values of D_{sf} = 1.5 kGy, I_a = 10^{-3} A and w = 120 cm is:

$$v = \frac{1.85 \times 10^6 I_a}{w D_{sf}} = \frac{1.85 \times 10^6 \times 10^{-3}}{120 \times 1.5} = 10.28 \text{ cm/s}$$

Conveyor speed varies according to product throughput. In this case, the conveyor must run at 10.28 cm/s (6 m/min) to ensure a 1.5 kGy entrance dose when treating the food with a 10 MeV e-beam accelerator in singled-sided mode and given current and scan width. The faster the conveyor speed, the lower the dose. For instance, if the required D_{sf} is 1 kGy, then the conveyor should run at 15.42 cm/s (9.25 m/min):

$$v = \frac{1.85 \times 10^6 I_a}{w D_{sf}} = \frac{1.85 \times 10^6 \times 10^{-3}}{120 \times 1} = 15.42 \text{ cm/s}$$

Example 10: Calculation of throughput rate for an e-beam system

Problem:
Calculate the throughput rate for e-beam disinfestation of papaya (minimum required dose of 0.26 kGy) with an e-beam irradiation (one-sided mode) with 12 kW of power and throughput efficiency of 0.5.

Solution:
(a) Calculate the throughput rate with P = 12 kW, D = 0.26 kGy, and η = 0.5.
From equation 17:

$$\frac{dm}{dt} = \frac{\eta P}{D}$$

Then:
$$\frac{dm}{dt}\left[\frac{kg}{s}\right] = \frac{0.5 \times 12 [kW]}{0.26 [kGy]} = 23.1 \text{ kg/s}$$

(b) Assuming an areal density of 7 g/cm² and a scan width of 120 cm, calculate the conveyor speed, v.

Find v using equation 14:

$$v = \frac{dm/dt}{A_d w}$$

with $A_d = 7$ g/cm², then:

$$v = \frac{dm/dt}{A_d \times w} = \frac{23.1 \left[\frac{kg}{s}\right] \times 1000 \left[\frac{kg}{g}\right]}{7 \left[\frac{g}{cm^2}\right] \times 120 [cm]} = 27.5 \text{ cm/s}$$

(c) If the product is arranged in cardboard boxes (figure 11), which have a cross sectional area of 7432 cm², calculate the total mass of food that should be placed in a box

Find m using equation 16:

$$m = A_d A_c$$

with $A_d = 7$ g/cm² and $A_c = 7432$ cm², then:

$$m = A_d \times A_c = \frac{7 \left[\frac{g}{cm^2}\right] \times 7432 \left[cm^2\right]}{1000 \left[\frac{g}{kg}\right]} = 52 \text{ kg}$$

Disinfestation treatment of papaya (dose of 0.26 kGy) using a one-sided e-beam can be achieved when 52 kg of the food is placed under the e-beam with the conveyor running at 27.5 cm/s.

Image Credits

Figure 1. Moreira, R. G. (CC By 4.0). (2020). Energy deposition profile for 10-MeV electrons in a water absorber (adapted from Miller, 2005).
Figure 2. Moreira, R. G. (CC By 4.0). (2020). Dose-depth penetration for different radiation sources (X-rays, electron beams and gamma rays) (adapted from IAEA, 2015).
Figure 3. Moreira, R. G. (CC By 4.0). (2020). Typical depth–dose curves for electrons of various energies in the range applicable to food processing operations (adapted from IAEA, 2002).
Figure 4. Moreira, R. G. (CC By 4.0). (2020). Depth–dose curve for 10 MeV electrons in water, where the entrance (surface) dose is 100% (adapted from IAEA, 2002).
Figure 5. Moreira, R. G. (CC By 4.0). (2020). Depth-dose distributions for 10 MeV electrons in water for single-sided and double-sided configurations ($DUR = 1.35$).
Figure 6. Moreira, R. G. (CC By 4.0). (2020). Maximum penetration thickness for top-only and bottom-only e-beam configurations using 10 MeV electrons in water ($DUR = 1.35$).
Figure 7. Moreira, R. G. (CC By 4.0). (2020). Maximum penetration thickness for double-sided e-beam irradiation using 10 MeV electrons in water ($DUR = 1.35$).
Figure 8. Moreira, R. G. (CC By 4.0). (2020). Typical survival curve showing first-order kinetics behavior.

Figure 9. Moreira, R. G. (CC By 4.0). (2020). Steps needed to select the right irradiation technology for a food processing application (adapted from Miller, 2005).

Figure 10. Moreira, R. G. (CC By 4.0). (2020). Decision flow diagram for selecting the correct irradiation approach (adapted from Miller, 2005). MMR is the acceptable range of maximum to minimum dose ratios (*DUR*).

Figure 11. Moreira, R. G. (CC By 4.0). (2020). Typical electron beam irradiation configuration.

Example 5. Moreira, R. G. (CC By 4.0). (2020). Example 5.

Example 6. Moreira, R. G. (CC By 4.0). (2020). Example 6.

References

Attix, F. H. (1986). Introduction to radiological physics and radiation dosimetry. New York, NY: Wiley Interscience Publ. https://doi.org/10.1002/9783527617135.

Becker, R. C., Bly, J. H., Cleland, M. R., & Farrell, J. P. (1979). Accelerator requirements for electron beam processing. Radiat. Phys. Chem., 14(3-6), 353-375. https://doi.org/10.1016/0146-5724(79)90075-X.

Browne, M. (2013). Physics for engineering and science (2nd. ed.). New York, NY: McGraw Hill/Schaum.

CDC. (2018). Preliminary incidence and trends of infections with pathogens transmitted commonly through food. Foodborne Diseases Active Surveillance Network, Morbidity and Mortality Weekly Report. 68(16), 369-373. CDC. Retrieved from https://www.cdc.gov/mmwr/volumes/68/wr/mm6816a2.htm?s_cid=mm6816a2_w.

Cleland, M. R. (2005). Course on small accelerators. 383-416 (CERN-2006-012). Zeegse, The Netherlands. https://doi.org/10.5170/CERN-2006-012.383.

Eustice, R. F. (2017). Global status and commercial applications of food irradiation. Ch. 20. In I. C. F. R. Ferreira, A. L. Antonio, & Cabo Verde, S. (Eds.), Food irradiation technologies. Concepts, applications and outcomes. Food chemistry, function and analysis No. 4. Cambridge, U. K.: Royal Society of Chemistry, Thomas Graham House.

Fan, X. (2012). Ionizing radiation. In V. M. Gomez-Lopez (Ed.), Decontamination of fresh and minimally processed produce. https://doi.org/10.1002/9781118229187.

FDA. (2018). Irradiation in the production, processing and handling of food. Code of Federal Regulations Title 21. 3, CFR179.26. Washington, DC: FDA. Retrieved from https://www.accessdata.fda.gov/scripts/cdrh/cfdocs/cfcfr/CFRSearch.cfm?fr=179.26.

Health Canada. (2016). Technical summary. Health Canada's safety evaluation of irradiation of fresh and frozen raw ground beef. Government of Canada, Health Canada. Retrieved from http://www.hc-sc.gc.ca/fn-an/securit/irridation/tech_sum_food_irradiation_aliment_som_tech-eng.php#toxicology.

IAEA. (2002). Dosimetry for food irradiation. Technical Reports Series No. 409. Vienna, Austria: International Atomic Energy Agency. Retrieved from https://www-pub.iaea.org/MTCD/Publications/PDF/TRS409_scr.pdf.

IAEA. (2015). Manual of good practice in food irradiation. Technical Reports Series No. 481. Vienna, Austria: International Atomic Energy Agency. Retrieved from https://www.iaea.org/publications/10801/manual-of-good-practice-in-food-irradiation.

ICGFI. (1999). Facts about food irradiation. Food and Environmental Protection Section. Joint FAO/IAEA Division of Nuclear techniques in Food and Agriculture. Vienna, Austria: International Consultative Group on Food Irradiation.

Lagunas-Solar, M. C. (1995). Radiation processing of foods: An overview of scientific principles and current status. J. Food Protection, 58(2), 186-192. https://doi.org/10.4315/0362-028x-58.2.186.

Maherani, B., Hossain, F., Criado, P., Ben-Fadhel, Y., Salmieri, S., & Lacroix, M. (2016). World market development and consumer acceptance of irradiation technology. Foods, 5(4): 2-21. https://doi.org/10.3390/foods5040079.

Mahmoud, B. S. M., Bachman, G., & Linton, R. H. (2010). Inactivation of Escherichia coli O157:H7, Listeria monocytogenes, Salmonella enterica and Shigella flexneri on spinach leaves by X-ray. Food Microbiol., 27(1), 24–28. https://doi.org/10.1016/j.fm.2009.07.004.

Miller, R. B. (2005). Electronic irradiation of foods: An introduction to the technology. Food Eng. Series. New York, NY: Springer.

Morehouse, K. M., & Komolprasert, V. (2004). Irradiation of food and packaging: An overview. Ch. 1. In K. M. Morehouse, & V. Komolprasert (Eds.), Irradiation of food and packaging. American Chemical Society. ACS Symposium Series. https://doi.org/10.1021/bk-2004-0875.

Moreira, R. G., Puerta-Gomez, A. F., Kim, J., & Castell-Perez, M. E. (2012). Factors affecting radiation d-values (D10) of an Escherichia coli cocktail and Salmonella typhimurium LT2 inoculated in fresh produce. J. Food Sci., 77(4), E104-E111. https://doi.org/10.1111/j.1750-3841.2011.02603.x.

Morrison, R. M. (1989). An economic analysis of electron accelerators and cobalt-60 for irradiating food. USDA Commodity Economics Division, Economic Research Service, Tech. Bull. No. 1762 (TB-1762). Rockville, MD.

Niemira, B. A. (2007). Relative efficacy of sodium hypochlorite wash versus irradiation to inactivate Escherichia coli O157:H7 Internalized in leaves of romaine lettuce and baby spinach. J. Food Protection, 70(11), 2526–2532. https://doi.org/10.4315/0362-028x-70.11.2526.

Olaimat, A. N., & Holley, R. A. (2012). Factors influencing the microbial safety of fresh produce: A review. Food Microbiol., 32(1), 1–19. https://doi.org/10.1016/j.fm.2012.04.016.

Rajtowski, K. T., Boyd, G., & Thayer, D. W. (2003). Irradiation D-values for Escherichia coli O157:H7 and Salmonella sp. on inoculated broccoli seeds and effects of irradiation on broccoli sprout keeping quality and seed viability. J. Food Protection, 66(5), 760–766. https://doi.org/10.4315/0362-028x-66.5.760.

Strydom, W., Parker, W., & Olivares, M. (2005). Electron beams: Physical and clinical aspects. In E. Podgorsakt (Ed.), Radiation oncology physics: A handbook for teachers and students. Vienna, Austria: IAEA.

Thayer, D. W., Boyd, G., Muller, W. S., Lipson, C. A., Hayne, W. C., & Baer, S. H. (1990). Radiation resistance of salmonella. J. Ind. Microbiol., 5(6), 383–390. doi: https://doi.org/10.1007/BF01578097.

University of Wisconsin. (2019). The food irradiation process. UW Food Irradiation Education Group. Retrieved from https://uw-food-irradiation.engr.wisc.edu/materials/FI_process_brochure.pdf.

USNRC. (2018). U.S. Nuclear Regulatory Commission. Retrieved from https://www.nrc.gov/about-nrc/radiation/health-effects/radiation-basics.html.

WHO. (1981). Wholesomeness of irradiated foods. Technical Report Series 659. Geneva, Switzerland: World Health Organization. Retrieved from https://apps.who.int/iris/bitstream/handle/10665/41508/WHO_TRS_659.pdf?sequence=1&isAllowed=y.

Packaging

Scott A. Morris
Departments of Agricultural & Biological Engineering and
Food Science & Human Nutrition
University of Illinois at Urbana-Champaign
Champaign, Illinois, USA

KEY TERMS		
Product protection	Permeation	Packaging cycle
Packaging design	Shelf life	Information cycle
Packaging materials	Packaging damage	

Variables

Δp = partial pressure gradient

σ = stress

A = reaction rate constant for Arrhenius equation

c = concentration

d = depth

D = diffusion coefficient

E_a = activation energy for the Arrhenius equation

J = diffusive flux

k = reaction rate constant

K = stress concentration factor

P = permeation rate

$\overline{P}$ = permeability

Q_{10} = quality-loss scaling factor

r = radius

R = universal gas constant

T = reaction temperature

TR = transmission rate

x = position

Introduction

Packaging is an engineering specialization that involves a systems-oriented means of preparing and distributing goods of all types. Packaging is responsible for several fundamental functions as well as having broad reach and wide impact beyond the consumer's immediate purchase. It is a much more complex system than most consumers (and many producers) realize and requires skills drawn from all facets of engineering. For that reason, integration of concepts is absolutely essential, and this chapter is best understood by considering the systems-cycle concepts laid out in the Applications section first, before pursuing isolated topics or calculations.

Packaging makes it possible to have a broad distribution of perishable items such as food and medicine. By considering the complete cycle of usage, conditions, handling, storage, and disposal, appropriate packaging can be designed for nearly any application, market, and regulatory structure. Thus, it is important for packaging to be included as early as possible in the product development cycle so that the proper packaging can be created and tested in time to meet production deadlines, and to highlight problems in the product that might make it susceptible to shipping damage or other harm.

Outcomes

After reading this chapter, you should be able to:

- Describe the large-scale packaging system, both physical and informational, beyond development of a simple container

- Apply basic materials data to calculate simple permeation (mass-transfer) problems for polymeric packaging applications

- Estimate shelf life of products and recognize some of the problems of relying solely on data-projection based estimation

- Describe how packaging designs and solutions vary depending on economics, available resources, and infrastructure, and how mimicking a solution from one market may be unproductive in another due to material availability or differing cost structure, particularly in different geographical regions

Concepts

Package Types

There are three package types: primary, secondary, and tertiary. The *primary package* material directly contacts the product, such as the plastic bottle containing water or a bag holding potato chips. For food, pharmaceuticals, cosmetics, and similar types of products, regulations require that the packaging material not transfer harmful material into the product (and the term primary package is usually used in relevant legislation) (Misko, 2019; USFDA, 2019). Current debates over bis-phenol A (BPA) content in packaging (for example, water bottles), and its health effects when consumed, is an example of this kind of material transfer that may cause material components to be banned in certain products or markets.

The *secondary package* usually surrounds the primary package. A box of cereal is a good example, with the product contained in the interior pouch (the primary package) and the exterior printed carton acting as the secondary package. The secondary package may act as advertising space on a store shelf, or to give a good first impression in e-commerce, and also carries information for point of sale (POS) operations.

Most often, the *tertiary package* is the shipping carton, carrier, or tray that carries unitized packages, i.e., packages that have been collected into groups for shipping, through the distribution system. In many cases, it is a corrugated shipping container, but for very strong types of packages such as glass jars and metal cans, it may be a simple overwrapped tray. This package must usually carry shipping information, and must frequently comply with relevant shipping regulations, rules, tariffs, and labeling requirements.

Material Types

Packaging is often described in terms of primary materials that make up the body or structure of the package. The most common primary packaging materials are plastics, metals (steel and aluminum), glass, and paper. Global use of material types is shown in figure 1. Other materials include traditional low-use materials such as structural wood in crates, as well as printing inks, adhesives, and other secondary materials. Secondary materials and components of the package are usually added to the primary structure and are often used for assembly, such as adhesives or a "closure"—the cap or lid on a container. Other components, such as inks used for printing, spray pumps, and other secondary features, may be included in the latter group.

Plastics

Plastics are most often created by the polymerization of petrochemical hydrocarbons, though there is substantial effort to create useful versions from naturally occurring carbohydrates, particularly from plant and algal sources as well as genetically engineered bacterial cultures. These polymers typically contain long carbon "backbone" chains of considerable length, and may or may not have bonds forming cross-links between the chains. A rule of thumb is that more cross-linking will create a stiffer, more brittle material. Additionally, plastics exhibit "crystallinity," which does not necessarily follow the strict definition of a crystal in the typical sense of a completely bound structure and very sharp melting point, but does exhibit a high degree of ordering: backbone chains arranged in regular patterns, usually emanating from a central nucleation site (figure 2). Polymers that have a low degree of ordering in their chain orientation are typically termed "amorphous," much like a bowl of cooked noodles. Additionally, melting would occur at a narrow range of temperatures depending on factors such as molecular weight distribution and additives rather than the broader, less well-defined softening and liquification range that an amorphous phase would exhibit.

For a given chain length, an ordered, crystalline polymer will have higher density, be more resistant to absorbing or permeating materials through the structure, and may be more brittle than amorphous materials, which will be tougher, more flexible, and more likely to absorb or transmit material through the molecular structure. For example, polyethylene is suitable for forming simple flexible structures such as milk cartons, but does not resist stress

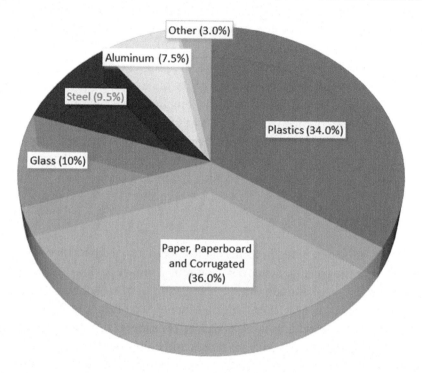

Figure 1. Global use of packaging materials by type (Packaging Distributors of America, 2016).

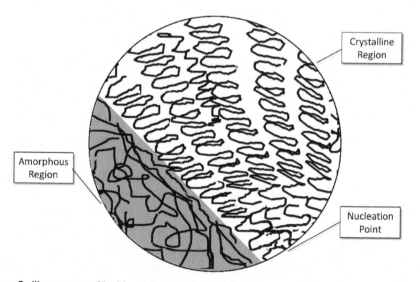

Figure 2. Illustration of highly ordered polymer chains in crystalline regions and disordered chains in amorphous regions.

cracking while flexing, so polypropylene is used for "living hinge" structures that are often seen as flip caps on containers.

Polymers may also have their structure altered by post-processing the sheet, film, or structure in a process called "orienting." This involves mechanically deforming the material so that the chains are pulled into alignment, creating a higher degree of crystallinity and better mechanical strength and barrier properties. This orientation may affect the density as well, since it will create order in the backbone chain. The relationship of chain length/molecular weight and crystallinity is illustrated in figure 3 (Morris, 2011).

For example, a polyethylene terephthalate (PET) soda bottle is first created as a molded "preform," roughly resembling a test-tube, with the threaded "finish" that the lid is attached to already formed. In the bottling plant, the preforms are heated to a very specific temperature and then rapidly inflated with compressed air inside a shaped mold. This "stretch-blow" process aligns the molecular structure of the body into a tight, two-way basket weave of polymer chains that is capable of resisting the tendency of the carbon dioxide (CO_2) to dissolve into the polymer and escape through the structure.

If a type of polymer is too brittle to be used properly in its intended function, it may also be modified by the addition of plasticizers that act as lubricants or spacers in the internal molecular structure and make the structure more ductile. This might be done for a squeeze dispenser or a structure that is too brittle at low temperatures.

Side groups bonded to the main carbon-carbon "backbone" chain usually define plastics that are commonly used in packaging. Since writing the entire structure of hundreds of thousands of units would be impractical, the structure is often represented by the repeat units that comprise the polymer backbone chain (figure 4). Several

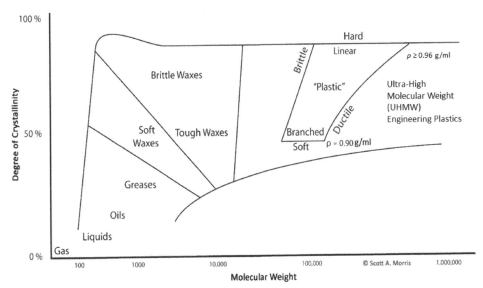

Figure 3. Relationship of crystallinity, molecular weight (which increases with chain length in this example), and physical properties for a typical linear polyolefin (polyethylene shown). ρ is density in g/ml. (Morris, 2011).

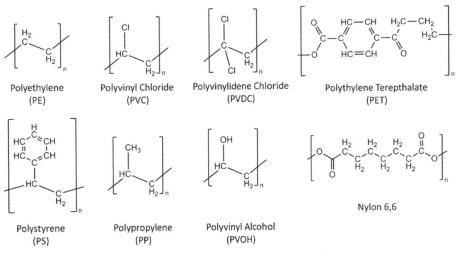

Figure 4. Repeat unit structures of common packaging polymers.

of these polymers may exhibit branching of the structures from the central backbone, but again, these are also composed of repeat units. For example, polyethylene has a simple side-structure of two hydrogen atoms, while polystyrene has a cyclic phenyl structure. The interaction of these rings with one another as long chains are produced results in the stiff, brittle behavior of unplasticized polystyrene.

Steel

Steel is used almost exclusively as cans for food, as well as larger drums for many types of products. When used as a food container for thermally processed foods, steel cans have an internal lining to reduce corrosion and reduce interaction with the product. Typically, the coating is of tin, which creates an anodic protection layer in the absence of oxygen, is non-toxic, and does not affect the flavor or texture of most products. There is usually an additional supplementary coating of some type of lacquer or synthetic polymer. Cans are formed either as two-piece or three-piece structures. The bottom and body of two-piece cans are formed from a single piece of material by progressive forcing through dies, with a seamed-on lid. The body of three-piece cans, which are increasingly uncommon since they are more costly to produce, is formed from a single piece of tinned sheet with a welded side-seam, a seamed-on bottom, and the lid seamed on after filling, as with the two-piece can. Steel cans have the advantage of resisting substantial loads both from stacking of many layers during storage and from the internal vacuum that is formed from condensing steam during the filling and lidding process that eliminates deteriorative oxygen in the headspace of the container.

Aluminum

Aluminum containers are used almost exclusively with beverages, since aluminum is quite ductile and relies on pressurization, either from the carbonation of a beverage or from the addition of a small quantity of pressurizing gas (typically nitrogen), to achieve sufficient strength. Aluminum cans are formed as two-piece cans and the interior of the can is coated with a sprayed-on resin to resist corrosion by the contents. For highly acidic products such as cola drinks, this critical step prevents the cans from corroding in a matter of days. The lid for aluminum cans has evolved as a masterpiece of production engineering since it attaches the tab with a formed "rivet" from a protrusion of the lid material rather than a third piece that would add prohibitive cost, and has a scored opening that reliably resists pressure until opened by the consumer.

Aluminum has two other substantial uses in packaging: foils and coatings. Since metal is inherently a very good barrier against gasses, light, and water, flexible foil layers are included in many types of paper/plastic laminates to provide protection for products. Similarly, an evaporated coating of aluminum is a common feature with flexible films, particularly with snack foods whose oily composition is susceptible to light, oxygen, and moisture.

Glass

Glass is formed by the fusion of sand, minerals, and recycled glass at temperatures above 1500°C. Forming a thick liquid, it is then dispensed in "gobs" that are carried to a mold that forms a preliminary structure called a "parison," then on to a final mold where the parison is inflated with air and takes its final form while still quite hot. Since glass has the combination of poor heat conduction and brittleness, the formed containers must then be cooled slowly in an annealing process, usually done in a slow conveyor-feed structure called a "lehr" that contains progressively lower temperature zones. This allows the molded container to cool slowly over a long period and prevents failure from residual thermal stress.

Once formed, glass containers are quite strong, although susceptible to brittle failure, particularly as the result of stress concentration in scratches or abrasion. For this reason, the containers have thicker areas called "shock bands" molded in and also are coated to reduce contact damage. Many glass-packaged products, particularly beverages, are shipped with an internal divider of inexpensive paperboard to separate the containers and prevent scratching. Glass is otherwise strong enough that it is often shipped with a simple tray and overwrap to unitize the containers until they are shelved at retail.

Glass is being replaced with plastic in many applications for several reasons, primarily fragility and weight. Since ingested glass shards represent an enormous hazard to the consumer, breakage during filling and handling operations requires stopping production and thorough cleaning for every occurrence and discarding nearby product whether contaminated or not (American Peanut Council, 2009). Additionally, weight savings can be significant: one peanut butter filling operation saved 84% of package weight by replacing glass containers with plastic containers (Palmer, 2014). Generally, the substitution of plastic for glass has resulted in both cost and liability reduction, although for products intended for thermal processing after filling, the designs can demand precise control of material properties and forming (Silvers et al., 2012).

Paper, Paperboard, and Corrugated Fiberboard

Paper materials are created from natural fibers, primarily from trees and recycled content. Other sources, such as rice straw, hemp, and bamboo, may be used. There is directionality in paper's preference for tearing, bending, and warping since the fibers will preferentially separate rather than break, causing paper to tear preferentially along the direction that the forming machine laid the fiber slurry (termed the "machine direction"). Since paper is a natural, fibrous material, there will be changes in the strength of the material because of moisture content. Since fibers typically swell in diameter (at right angles to the machine direction, termed the "cross machine direction") without significantly changing length, surface exposure to water or steam may cause the paper to curl around the machine direction axis.

While paper fibers can be processed in many ways, the basic approach is to separate the fibers into a slurry, then reform the slurry into long sheets (called "web") in one of two ways. The earliest process, the Fourdrinier process (named for the Fourdrinier brothers who developed it) mimics early hand-laid paper in that it pours the fiber slurry through a continuous wire-mesh belt (the "wire"). As the water drains, the web is eventually peeled from the wire, and put through rollers in several finishing and drying steps. This process is limited by drainage in its ability to create thick materials and only a few layers are possible.

A later development, the cylinder process, uses rotating cylinders to adhere fibers from the slurry to a continuous moving belt of absorbent material from underneath, circumventing the previous drainage limitation. This has the advantage of being able to form many layers for thick-section papers and paperboard. Paperboard, i.e., paper that is thicker than approximately 0.3 mm, is usually die-cut into cartons, dividers, or other more rigid structures. Paperboard is used in all types of consumer packaging from hanging cards to cartons, while paper is typically used in pouch structures and bags to add strength and good printing surfaces.

Corrugated fiberboard (colloquially called "cardboard") is a manufactured product that assembles paper into a rigid structural sheet, usually consisting of two outer "linerboard" layers and a crenulated internal "corrugating medium" layer. The linerboard may be pre-printed to match the product; this can allow much more sophisticated graphics to be used compared to printing after manufacture, which is limited by the irregular surface of the material. The medium is continuously formed using steam and a shaped roller and adhered between the linerboard sheets using starch-based glue. The sheets of corrugated board are then usually die cut into the necessary shapes for forming boxes, shipping containers, and other structures. Multiple layers are possible and are used for specialized applications such as household appliance shipping containers.

Product Protection

Packaging serves several functions. Protection of the product is of primary importance, particularly with products such as food. Fresh foods often require vastly different types of protection than processed and shelf stable foods that are meant to be stored for much longer periods of time. Proper packaging protects products from physical damage and reduces costs due to waste. Additional functions of packaging include utilization, communication, and integration with ordering, manufacturing, transportation, distribution, and retail systems as well as return logistics networks.

Definition of Food Damage or Quality Loss

Defining damage, spoilage, or unsuitability of food can be very difficult. While microbial contamination levels can be quantified, the effects of texture or color changes are often subtle and subjective. Far too often, a food product is considered spoiled based on a qualitative measure that is entirely subjective and may be motivated by other factors. It is important that the definition of unacceptable product be carefully considered (and perhaps contractually defined) to avoid subsequent conflict. Ingredients, components, and materials supplied to other manufacturing operations should always have quality criteria carefully and quantitatively defined to avoid arguments that may be motivated by an attempt at renegotiation of price or other commercial considerations (Bodenheimer, 2014; Pennella, 2006).

Since defining food product failure may be a difficult task, it can be useful to focus on the most easily degraded or damaged component that will cause the product to become unsafe or unacceptable if it fails—the *critical element* (Morris, 2011). This critical element may be defined as an easily degraded ingredient, a significant color change, mechanical failure, or an organoleptic quality (usually defined by a taste, texture, or odor, most often identified by human evaluators in a blind test) that fails an objective criterion for failure. The critical element to be used in a sampling plan must meet two criteria: its state must be determinable by objective analysis, and its failure conditions must be defined by objective criteria rather than subjective anecdote.

This approach can have several shortcomings. It is easy to focus on a particular aspect of quality to the more general detriment of the product, and it is tempting to choose the quality element because of its ease of assay rather than its impact or importance. Finally, this may be a moving target, as the critical element may become some other factor as circumstances change.

Transportation and Storage Damage

Damage resulting from static and dynamic effects during manufacture, storage, handling, and distribution may range from simple compression failure of a container to complex resonance effects in a vehicle-load of mixed product. An understanding of storage conditions and the transportation environment can help in the design of an efficient package capable of surviving distribution without over-packaging.

Light and Heat Damage

Damage to a food product may occur because of exposure to light or to temperature extremes, both high and low. Ultraviolet light may cause fading of the external printed copy and an unappealing appearance, but by itself does not penetrate into a transparent package. Certain products, however, are extremely sensitive to visible light. Skim milk exhibits a marked decrease in Vitamin A with exposure to fluorescent lights common in retail environments and beer's isohumulone flavoring will degrade into the compound 3-MBT (3-methyl-2-butene-1-thiol) causing a sulfurous "skunked" or "lightstruck" flavor to develop (deMan, 1981; Burns et al., 2001).

Thermal or heat damage may result from the long-term effects of both very high and very low temperature exposure, though low temperature exposure of a fragile product is more associated with the breakdown of texture and structure, usually from ice crystal growth or emulsion failure, than chemical changes. High temperatures will accelerate any thermally dependent degradation processes and may cause other problems, such as unexpectedly high permeation rates in packaging materials, because of transition from a glass to an amorphous state in polymers.

Gas and Vapor Transmission Damage

Gas and vapor transmission problems are very product-specific and may be situational. A carbonated drink may suffer from loss of carbonation, while another product may oxidize badly because of oxygen transmitted through the package. A confectionary product may have rapid flavor change because of loss of volatile flavor components that self-plasticize the packaging material. Volatile organic chemicals (VOCs) ranging from diesel fumes to flavor components may be transferred in or out through the package. Water vapor gain may cause spoilage of food or degradation of pharmaceuticals, while water vapor loss may cause staling of bread products. A good understanding of both the product properties and of the environment that it will face in distribution are important for proper design (Zweep, 2018).

Permeation in Permeable Polymeric Packaging Material

Permeation is the ability of one material (the permeant) to move through the structure of another. Many amorphous materials such as natural and artificial polymers are permeable because of substantial space between their molecular chains. Figure 5 shows this in schematic form, with permeation of vapor progressing from the high-concentration side to the low-concentration side via sorption into the high-concentration side, diffusion through the bulk matrix of the film membrane, and then desorption on the low-concentration side, all driven by the concentration differential across the material. Glass and metal packaging, on the other hand, are impermeable to everything except hydrogen because of their ordered structure or dense packing. Polymers in a highly ordered state also exhibit very low permeability relative to disordered structures.

The rate of permeation depends on the species of permeant, the type and state of the polymer, and any secondary factors such as coatings. The polymer may be glassy—essentially a low-order crystalline state (a good example of this is a brittle polystyrene drink cup)—or rubbery, which allows segmental motion of the polymer chains. With most polymers, this will have a measurable shift at a particular temperature, the glass transition temperature, with elasticity and permeation increasing when the polymer is above the glass transition temperature of the polymer.

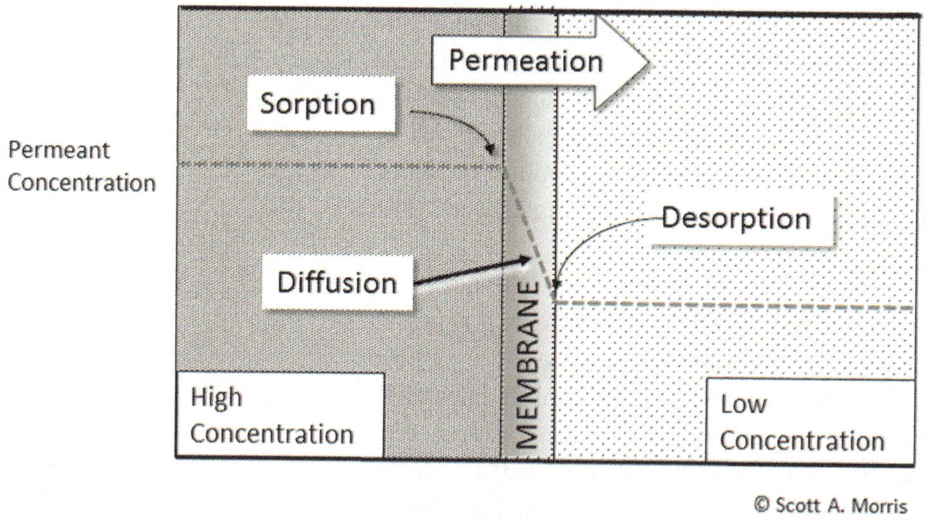

Permeability can be modeled as the concentration-gradient driven process (mass transfer process) of dissolving into the high-concentration surface, diffusing through the film membrane matrix materials, and then desorbing from the low-concentration

Figure 5. Permeation through packaging film membrane.

surface, much in the same way that heat is transmitted by conduction through the thickness of a wall (Suloff, 2002).

Mass-transfer equations can be constructed to create a simple model of the diffusive flux of the permeant (gas) based on the linear Fickian diffusion model (equation 1) (Fick, 1855). For movement of a permeant through a layer of material per surface area:

$$\left(\frac{\text{Quantity permeated per unit of time}}{\text{Area}}\right) = J = -D\frac{\delta c}{\delta x} \quad (1)$$

where J = diffusive flux (mol m^{-2} s^{-1})
D = diffusion coefficient (m^2 s^{-1})
c = concentration (mol m^{-3})
x = position (m); in figure 5, this would be the position within the cross section of the film membrane.

The transmission rate through composite structures, i.e., structures having several layers, can be calculated in a manner similar to thermal systems using equation 2:

For n layers of material,

$$TR_{total} = \frac{1}{TR_{layer\,1} + TR_{layer\,2} + ... + TR_{layer\,n}} \quad (2)$$

where TR_{total} = total transmission rate (mol s^{-1})
$TR_{layer\,n}$ = transmission rate of layer n

If the permeation of the material is known (or can be estimated), then estimating the permeation of a package design is a function of temperature, surface area, and partial pressure gradient, Δp. Partial pressure is defined as the pressure that would be exerted by a gas in a mixture if it occupied the same volume as the mix being considered. Usually Δp is defined by Dalton's law, i.e., in a mixture of non-reacting gases, the total pressure exerted is equal to the sum of the partial pressures of the individual gases, and, thus, the partial pressure is the product of both the partial pressure of the permeant species and the hydrostatic pressure (Dalton, 1802). Henry's law, which says that the amount of gas absorbed in a material is proportional to its pressure over the material, and the combination of hydrostatic pressure and permeant species prompts the selective nature of permeation by gasses that have differing partial pressures in a given polymer (Sanchez & Rogers, 1990).

Equations 1 and 2 are for idealized circumstances—a constant rate of permeation without chemical reaction between the polymer and the permeant at constant temperature and without physical distortion of the film—and are only valid for diffusion-based permeation. With holes, perforations, voids, or defects, the gas flow is explained by simple fluid-flow models. In real world applications, many conditions, such as temperature changes, fabrication methods, and handling stresses, will compromise this assumption. Diffusion in polymers is an ongoing field of research, and with the great array of volatile compounds in foods, the system may be complicated by several types of deviation from the ideal case. From a practitioner's standpoint, the permeation data provided by a supplier may be under idealized circumstances or for an initial production run, and will likely not accommodate variations that occur during manufacturing.

Permeability (frequently designated $\bar{P}$) has units that have been described as "... a nightmare of units" (Cooksey et al., 1999). The SI standard unit for this property of polymeric materials is mol/(m·s·Pa), though it is used inconsistently, even in academic literature and certainly in commercial data. Rates may be reported in any number of formats and improper mixes of US customary units, SI, cgs, or other measures, in results provided by various tests and manufacturers, so the practitioner will find it necessary to convert units in order to make use of the data. Most of these roughly conform to this format:

$$\bar{P} = \frac{\text{(quantity of permeating gas)(thickness)}}{\text{(time)(membrane area)(partial pressure difference across membrane)}} \tag{3}$$

Experimental Determination of Permeation Rate

Experimental determination of permeation rates and their derived constants is usually done using a test cell of known surface area with concentrated permeant (e.g., oxygen or CO_2) on one side of the package film (generally between 0.06 mm and 0.25 mm in thickness) and inert gas or air on the other side. As permeation progresses, the lag time (the time to achieve a steady rate of permeation) and the rate of concentration increase on the non-permeant side can be measured and used to calculate the solubility and diffusion coefficients (Mangaraj et al., 2015). Typical values of oxygen and water transmission rates and glass transition temperatures are shown in table 1.

For moisture permeation tests, a similar arrangement is used, except that a desiccant usually provides the partial pressure differential with a stream of humidified air circulating on the other side of the film membrane. The moisture gained by the desiccant, measured by weight change, is used to calculate the permeation rate (ISO, 2017). Additionally, there are dedicated test devices for oxygen and water permeability that rely on real-time determination of permeation rate using heated zircon and infrared-absorption detectors, respectively.

Permeation Modification in Packaging Films

Using the simple sorption-diffusion-desorption model of permeation shown in figure 5, one can find several ways to modify the barrier characteristics of packaging films, either by modifying the surface (sorption/desorption) characteristics or by affecting the diffusion characteristics of the overall film structure. Coatings and surface treatments can be used to modify sorption/desorption characteristics of polymer films. Foremost among these treatments is metallization, which is the evaporation of a thin layer of aluminum in a vacuum chamber. This may be done on either side of the film, but is most often done inside the package to avoid abrasion loss and may be laminated to prevent transfer of aluminum that would discolor the product. There are other surface chemistry modifications, such as fluorination, which, though challenging to implement in production, can convert the surface of simple polyolefins to a polyfluorinated compound with markedly better barrier characteristics. Other surface coatings and laminations are common. Printing, labeling, and other surface decorations may provide a degree of barrier properties over part of the product as well (Nakaya et al., 2015).

Table 1. Generalized properties of the common packaging polymers shown in figure 4 (Thermofisher Inc., 2019; Rogers, 1985; Sigma-Aldrich Inc., 2019). These properties are generalized from available literature, and unlike many engineering materials, there are no standard grades. The properties may vary widely between manufacturers, and will vary with density, crystallinity, orientation, and additives, among other factors. This table is provided for comparison only.

Polymer Type	Oxygen Transmission Rate[1]	Water Vapor Transmission Rate[2]	Glass Transition Temperature (°C)	Comments
Polyethylene (PE)	194	18	−25	Polyethylene properties vary significantly with density, branching, and orientation.
Polyvinyl chloride (PVC)	5	12	81	Both PVC and PVDC must be food grade (i.e., demonstrating no extractable vinyl chloride monomers) to be used with food products. Concerns over chlorinated films in the popular press reduced their use starting in the early 2000s.
Polyvinylidene chloride (PVDC)	5	30	−18	
Polyethylene terepthalate (PET)	5	18	72	PET will reduce its transmission rate drastically when it is oriented during fabrication.
Polystyrene (PS)	116–155	24	100	Polystyrene is very brittle, and must be plasticized to be useable in most applications. This increases transmission rates significantly.
Polypropylene (PP)	93	4	−8	Very impact resistant; used for snap caps and other multiple-use applications
Polyvinyl alcohol (PVOH)	0.8	8000+ (see note)	85	Water soluble; polyvinyl alcohol is a high oxygen barrier material, but must be kept dry, typically by layering between moisture barrier layers. Adsorption of moisture destroys barrier characteristics. PVOH film is also used by itself for water-soluble packets of household detergents and other consumer products.
Nylon 6,6	1.7	135	50	Hygroscopic; transmission rate varies with moisture content.

[1] In units of $\frac{cc \cdot \mu m}{m^2 \cdot 24h \cdot atm}$ tested at STP

[2] In units of $\frac{g \cdot \mu m}{m^2 \cdot 24h \cdot atm}$ tested at 37°C and 90% relative humidity

For a given polymeric material, modifying the internal structure of the polymer will change the diffusivity coefficient. Intentional modifications typically involve orienting the material by drawing it in one or more directions so that the polymer chains pack into a more orderly, denser structure (National Research Council, 1994). This produces better strength and barrier characteristics such as in the previously described stretch-blow molding of carbonated beverage bottles.

Polymers may also be modified with plasticizers—soluble polymer-chain lubricants—that reduce brittleness but allow chain mobility and create opportunities for permeants to penetrate the structure more readily. Plasticizers that contact food material must be approved for food use since they will likely migrate to the product in microscopic quantities. This has been the subject of several controversies as there is evidence of potential teratogenic (causing birth defects) activity in some plasticizers (EFSA, 2017). Food materials themselves, oils and fats most notably, may be plasticizers and may cause a package's material to change its barrier or physical characteristics.

Permeation Changes during Storage

Product ingredients or components dissolving into the package structure may result in decreased mechanical strength, reduced barrier properties and shelf life, or even the selective removal of flavor compounds (termed "flavor scalping"). This may create a mysterious reduction of shelf life because of synergistic effects. For example, a citrus flavoring compound rich with limonene may plasticize the packaging material and increase loss of both flavor and water, creating what appears to be a moisture loss problem (Sajilata et al., 2007). Similarly, volatile flavorings can increase oxygen permeation rates with harmful effects for the product, or may increase CO_2 loss rates in carbonated beverages.

Other Packaging Damage Occurring During Storage and Distribution

Corrosion of Tin-Plated Steel Cans

The electrochemistry of the tin-plated steel can is complex and depends on several factors in order to maintain the extraordinary shelf life that most consumers expect. Canning operations typically displace headspace air with live steam to both reduce oxygen in the can and provide vacuum once the steam condenses. After lidding, the can end is sealed by crimping the edge in a series of steps to provide a robust hermetic seal, and the environment in the package typically traverses three stages (Mannheim and Passy, 1982; Wu, 2016):

1. Initial oxidizing environment—Residual oxygen inside the freshly-sealed can and dissolved into the product is bound up in oxidation products in the product and can material. The tin layer is briefly cathodic, providing a positive charge during this stage and provides little protection until the oxygen is depleted. This typically takes a few days to conclude, depending on the composition of the product and processing conditions.

2. Reducing environment—In the absence of free oxygen, the electrochemistry then reverses, and the tin or chromium layer is anodic, slowly dissolving into the canned product to protect the steel of the can wall. This stage may last years, but may be affected by many factors, particularly product composition (e.g., pH level, acidifying agents, salts, and nitrogen sources). Each product must be considered unique, and product reformulation may cause significant changes in can corrosion properties.
3. Terminal corrosion—At the end of service life, the environment may still be anaerobic, keeping the electrochemistry anodic, but the protective coating of tin will have been depleted, allowing corrosion and pitting of the can. This can result in staining of the product or can surface, gas formation (hydrogen sulfide, producing so-called "stinkers") and, finally, pinholing of the can body and loss of hermeticity. Depending on the product, this may take from several months for highly acidic products, like pineapple juice and sauerkraut, to many decades.

Brittle Fracture and Glass Container Failure

Several failure modes are important to understand when working with glass packaging, particularly considering that there may be legal liabilities involved in their failure. Additionally, persistent glass failures in food production facilities can wreak havoc since dangerous glass shards are produced. As a brittle material, glass concentrates stress around thickness changes and scratches, since these provide a location for stress magnification as illustrated in equation 4 (Griffith, 1921):

$$\sigma_{max} = 2\sigma_{app}\left(\frac{d}{r}\right)^{1/2} \tag{4}$$

where σ_{max} = maximum stress at crack tip (N m^{-2})
σ_{app} = applied stress (N m^{-2})
d = depth of crack (m)
r = radius of crack tip (m)

A tiny scratch can create an enormous concentration of stress, and once the critical stress of the material is exceeded, a crack will form that will continue in the material until it fails or until it encounters a feature to re-distribute the stress. Stresses may occur as the result of thermal expansion or contraction since glass is not only brittle, but has poor thermal conductivity, so a section-thickness change may create a steep thermal gradient that causes a container to fail after fabrication or heat treatment. For carbonated beverages, the internal pressure combined with a surface scratch created during manufacture or handling may provide enough pressure and resultant stress in the package material to cause it to burst.

A stress concentration factor (K) can be developed from equation 4 as:

$$K = \frac{\sigma_{max}}{\sigma_{app}} = 2\left(\frac{d}{r}\right)^{1/2} \tag{5}$$

The stress concentration factor (*K*) becomes very large with scratches that have a very small crack tip, and even modest depth. The effects of scratches are avoided in design and manufacturing by providing "shock bands," which are thicker sections of material that are added to contact other bottles in manufacturing and handling, as well as by adding external surface coatings and putting dividers in shipping cartons.

Failure analysis on a broad scale is a specialty unto itself, but when determining the origin of the fracture, there are characteristic features that help identify the point of origin and direction of travel (figure 6). The point of origin in both ductile and brittle materials often has a different and distinct texture, usually mirror smooth, and as the failure progresses it will typically leave a distinctive pattern that radiates outward from the point of origin (Bradt, 2011).

When examining a failed glass container's reconstructed pieces, it is useful to consider the different failure modes that are common in glass structures. The most common failures are impact and pressure fractures, thermal failure, and hydrodynamic ("water hammer") failure (figure 7). Impact and pressure fractures often originate from a single point in the structure, with the fracture originating on the outer surface from impact, and from the inside from pressure, as determined by observation of magnified fracture edges at the point of initiation.

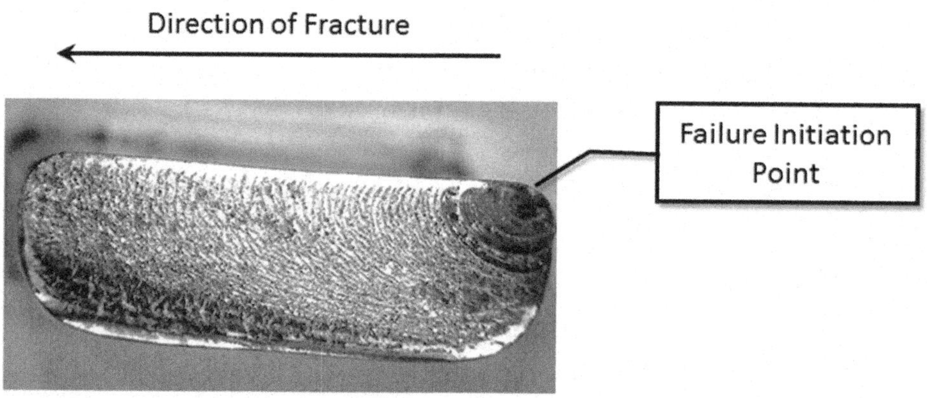

Thermal failure typically starts at a section thickness change (from thick to thin) as the container is heated or cooled abruptly and a large thermal differential generates shear stress in the material. This manifests itself most often in bottles and jars with a bottom that falls out of the rest of the container at the thickness change, perhaps with other cracks radiating up the sidewall.

Figure 6. Fracture failure in brittle material.

Water hammer failure is the result of hydraulic shock waves propagating through the product (usually from an impact that did not break the container directly) and causing localized formation of vapor bubbles that then collapse with enough force to break the container. This usually has the distinctive feature of a shattered ring completely around the container at a particular height (usually near the bottom) with obvious fragmentation

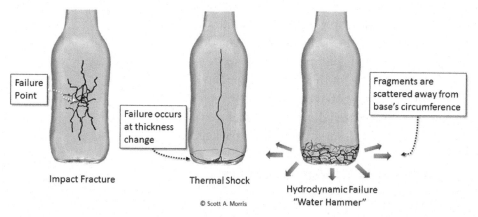

Figure 7. Illustration of glass failure types and significant indications of failure source.

outward from the pressure surge. Products with lower vapor pressures, particularly carbonated and alcoholic beverages, will fail with less energy input than liquid or gel products with high vapor pressures (Morris, 2011).

Shelf Life of Food Products and the Role of Packaging

Products have two shelf lives. The first is where the product becomes unusable or unsafe because of deterioration, contamination, or damage. The second shelf life is one of marketability; if the product's appearance degrades (such as color loss in food that can be seen while still on the shelf), then it will not appeal to consumers and will be difficult or impossible to sell.

The primary concern with packaged processed foods is usually microbial contamination, followed by the previously discussed gain or loss of food components. Since the food is not actively metabolizing, the usual problems apart from microbial growth result from oxidation, gain or loss of moisture or other components, and discoloration from light exposure. While barrier films and packaging can help with some of these problems, it may be useful to include active components such as sachets or other materials or devices that will bind up oxygen or moisture that infiltrates into the package. These are commonly seen on refrigerated-fresh products such as pasta, prepared meats, and others. Other types of active films or structures may incorporate an oxygen-absorbing barrier to extend shelf life. Light barriers may be a tough problem to contend with since many regulations prohibit packaging from hiding the product from view. Processed meat products such as sandwich meat, which is normally a pinkish color from the nitric oxide myoglobin formed during the curing process, will turn brown or grey under prolonged light exposure and will appear to be spoiled. Bacon has a substantial problem with light-promoted fat oxidation and in some countries is allowed to have a flip-up cover over the product window.

Unprocessed foods, such as fresh meat and vegetables, should be regarded as metabolically active. Fresh fruits and vegetables after harvest typically metabolize as they ripen, slowly consuming oxygen and stored carbohydrates and giving off CO_2, and may be ripening under the influence of ethylene gas self-production. It is possible to manipulate the oxygen level and strip ethylene from the products' environment—this is done on a large scale in commercial controlled-atmosphere (CA) storage facilities—but at the individual package level, the cost of specialized wrapping film and an ethylene adsorbent sachet may be prohibitive in markets with ready access to fresh fruits and vegetables. Other markets may find these expensively packaged fruits and vegetables appealing because of the ability to distribute fresh produce at great distance or in regions where it may be difficult to do directly. Since the early 2000s, the use of 1-methylcyclopropene (MCP), an ethylene antagonist, has allowed ripening prevention, but overexposure may permanently prevent ripening of some species (Chiriboga et al., 2011).

Freshly butchered meat will absorb oxygen, converting purple-colored reduced myoglobin to red oxymyoglobin and then to brown metmyoglobin. Most customers are not accustomed to seeing the purple color of very fresh meat, and expect it to be red in color, although the redness occurs through oxidation. This leads to the problem of extending the shelf life of meat products beyond a few days, since the packaging must allow oxygen in to provide the expected red coloration yet at the same time preventing the ongoing brownish discoloration as metmyoglobin is formed from oxygen. Work is ongoing with this. Many centralized meat packing facilities for large retailers may use carbon monoxide gas in the package to provide a near-fresh red color. This has created some controversy as it may disguise the age of the product and prevent some spoilage indication, but the practice is being widely adopted in order to take advantage of centralized processing facilities. Similar processes are being investigated for other meat and seafood products.

Shelf Life Testing and Estimation

In most practical applications, there is not enough time to actually wait for several iterations of the product's long intended shelf life in order to develop and refine a package. Once the initial design is laid out, it is often subjected to accelerated shelf life testing in order to allow an approximate assessment of protection over a shortened period. Shelf life modeling should be followed up with substantial quality-assessment data from actual distributed product over time, and attention should also be paid to errors in estimation methods, and their effect on longer-term predictions.

Q_{10} Accelerated Shelf Life Testing

For food and related products, shelf life testing may involve storing the test packages at high temperatures in order to accelerate the degradation that will occur over time. The core assumption with Q_{10} testing is that with an Arrhenius type reaction (equation 6), increasing the temperature by 10°C will cause the quality loss rate to increase by a scaling factor (k). The k value can be thought of as a magnification of effect over time by increasing the temperature of the test, within moderation. The general approach is commonly termed Q_{10} testing (Ragnarsson and Labuza, 1977):

$$k = Ae^{\frac{-E_a}{RT}} \qquad (6)$$

and

$$Q_{10} = \frac{\text{Time for product to spoil at temperature } T \text{ °C}}{\text{Time for product to spoil at temperature } (+10°C)} \qquad (7)$$

where k = reaction rate constant, in this context effectively the rate of quality deterioration
A = pre-exponential constant for the reaction
E_a = activation energy for the reaction (same units as RT)
R = universal gas constant
T = absolute temperature (kelvin)
Q_{10} = quality-loss scaling factor (dimensionless)

Typically Q_{10} values are in the range of 1.0 to 5.0 but must be verified by testing. Remember that shelf life is the result of many overlapping reactions, all of which may have very different kinetics, so the range of valid estimation is narrowly limited and the method and its results should be treated with great caution. There is a danger of attempting to do rapid testing at senselessly high temperatures, leading to grossly inaccurate estimations because of phase changes in the product, exceeding the packaging material's glass transition temperature, thawing, vaporization of compounds, and similar non-linear temperature effects that violate the simple Arrhenius kinetics assumed in many shelf life studies (Labuza, 2000).

Applications

The Packaging Cycle

Given the enormous variety of materials, structures, and components of packages (e.g., rigid vs. flexible, cans vs. pouches) for a global range of products, it is useful to consider packaging as a material-use cycle (figure 8) (Morris, 2011). This cycle originated with large-scale, industrialized types of packaging, but can be used to visualize the use of materials and design factors in other, smaller, or more specialized types of operations. When considering a new type of packaging material or new design, it provides a useful means for analyzing the resulting changes in sourcing and disposition beyond the immediate demands of the product.

Raw Materials
Raw materials of a full spectrum of major packaging materials and components consist of the resources needed to create the basic packaging materials. Raw materials are included in the packaging cycle because shifts in global resource production or supply may markedly affect package design and choice.

Conversion of Materials
Material conversion takes bulk, refined materials such as steel ingots or plastic resin pellets and converts them to an intermediate form, such as plastic film or metal foil, which is sent to manufacturers who create the finished package. Special processing may occur at this step, such as the plating of tin-plated steel for steel cans, or the aluminizing of plastic films for snack packaging. Because of the difficulties in molding molten glass, glass containers move directly from the refining furnace to finished containers in a single operation.

Finished Packages
Converted materials are made into ready-to-fill packages and necessary components such as jars, cans, bottles, boxes, and their lids or other closures. This step may occur in many places depending on the product involved. For example,

a dairy operation or soft drink bottler in a rural area may find it advantageous to be able to produce containers directly on-site. Other operations, such as canneries in crop producing areas, may have a nearby can or pouch fabricator serving several different companies to take advantage of local demand, or there may be a range of local producers working on contract to serve a large-scale local operation.

Package Filling Operation

The package filling operation brings together the package and the product to form a system intended to maintain and protect the product. In this step, the package is filled and sealed. Packaged products are then sent to any secondary treatment such as thermal sterilization, irradiation, or high-pressure treatment (omitted from figure 8). Once ready for shipment, the packages are usually unitized into multiples for greater handling and distribution efficiency.

This step also includes critical operations such as sealing, weight verification, label application, batch marking and "use by" date printing. Correct operation is imperative to deliver a consistent level of quality. Improvements in data management and control systems have offered efficiency improvements in this stage. For example, intra-system communication protocols, such as ISA-TR88.00.02 (often referred to as PackML, for packaging machine language), have been developed that define data used to monitor and control automated packaging and production systems and allow high levels of control and operation integration and increased production efficiency.

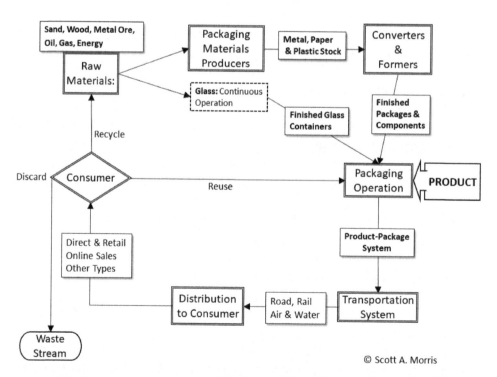

Figure 8. Packaging cycle showing the material use cycle from raw materials through package manufacturing, filling, distribution, and end-of-life (EOL) disposition.

Transportation System

The unitized product is sent out through a multitude of channels to distribution points, and is increasingly diverse with the rise of e-commerce. Typical modes of transportation are long-haul trucks, railcars, ships and barges, and aircraft. Each of these has a range of applicability and an economic envelope for efficient use. In areas with less developed infrastructure, distribution may operate very differently and high-value items, such as critical, perishable medication, may be flown in and then quickly distributed by foot, motorbike, or on pack animals. This "last mile" of distribution has become increasingly important in all markets. Even with e-commerce, distribution is left in the hands of delivery or postal services where products were previously handled by retail outlets and the customers themselves, and this introduces uncertainty and the possibility of different damage sources. Therefore, when designing a packaging system, the distribution chain must be considered to account for sources of damage. Additionally, each transportation type may have specific rules and regulations that must be followed to be considered acceptable for shipment and to limit liability.

Distribution to Consumer

Final distribution varies widely, and may have several modes in a single marketplace, such as direct-to-consumer (D2C), online retail, and traditional "shelved" retail outlets. All of these may vary in size and complexity depending on the culture, economy, market, and location. Rural markets in some countries have often responded well to small, sachet-sized manufactured products that are usually sold in larger containers elsewhere (Neuwirth, 2011), while large "club" stores may require large-volume packages, or unitized groups of product that are sold directly to consumers.

Consumer Decision about Disposal

When the product has been completely used, the final step for the packaging is disposal. The end user decides which form of disposal to use, with the decision being affected by economic incentives, cultural and popular habits, and available infrastructure. Discarded packaging is one of the most visible types of waste, since many people do not dispose or recycle it properly even when facilities are available, but is often a minority component of total municipal solid waste (MSW) relative to non-durable goods or other waste components. While collection and reuse of materials can be profitable when well-organized and when transportation and re-manufacturing infrastructure is available, many places do not have this in any functional sense. In addition, certain materials have fallen away from recyclability because of market changes. A good example is the recycling of EPS foam (expanded polystyrene, typically called Styrofoam™) in the United States. When fast food restaurants transitioned away from using EPS sandwich containers because of their environmentally unfriendly image, the ability to recycle any EPS was largely eliminated because of the loss of the largest stream of material, making most EPS recycling operations unprofitable.

Discarding into the Waste Stream
Packaging can be discarded via a collection system that collects MSW efficiently, either as landfill or as part of an energy conversion system, or it may be part of a less-centralized incineration or disposal effort. In the worst cases, there is no working infrastructure for collection, and packaging waste—particularly used food packaging—is simply left wherever is convenient and becomes a public health hazard. Recent concerns have emerged over the large-scale riverine dispersion of plastic waste into mid-oceanic gyres that create a Sargasso of waste that photodegrades very slowly, if at all, and may be a hazard to ocean ecosystems. Even in many locations with operating infrastructure, discarded materials are entombed in carefully constructed landfills that do not offer the possibility of degradation, while in others, MSW is used as an energy source for power generation. In some areas, organic material such as food and garden waste may be composted for use as fertilizer.

Reuse
Informal reuse schemes have been around as long as containers have existed. In more modern times, reuse of containers for various purposes is common, but the market for refilling in developed economies is somewhat limited to simple products such as filtered water. In some markets, the beverage industry requires that bottles be returned, with reused bottles recirculating for decades. Reuse has complications and liability concerns because of cleanliness issues and requires washing to remove secondary contaminants, such as fuels and pesticides, and inspection for contaminants that are not removed during the washing cycle.

Recycling
Recycling brings materials back into the cycle, and reuse of materials in some form is common in all cultures. The trajectory the materials take may vary widely, however. For example, the German Environment Ministry operates a "Green Dot" recycling system that requires manufacturers of packaged goods to pay into a system that collects and recycles used packaging. As of 2018, the city of Kamikatsu, Japan, which has taken on the mission of being the world's first "zero waste" community, had 45 different categories of recycling to be collected (Nippon.com, 2018). When properly conducted, recycling is the most efficient continued use of materials, but it depends on market demand and the ability to reprocess and reuse materials. For example, aluminum, which is intrinsically much cheaper to reuse from scrap than to reduce from bauxite ore, has had efficient recycling in place for more than half a century, whereas glass is often not recycled. Recycling is, in general, a function of economics, infrastructure, and regulations; in some markets, the waste disposal sites themselves are considered a resource for extracting materials such as steel and aluminum.

The Information Cycle

The information cycle (figure 9) is often as important as the actual material production cycle in that machine-readable coding allows the packages themselves to interface directly with point of sale (POS) systems, inventory and ordering software, and distribution infrastructure. Increasingly, this information is also used to create user profiles for product preferences, to optimize response to variations in demand, and to allow targeted marketing and distribution into niche markets.

Information continuously flows back from many points in the system to automatically create orders for store inventory, to track orders, and to forecast production levels for product manufacturers. Of course, this is not tightly integrated in all cases, but serves as an idealized representation. Other useful information is derived from the correlation of other data such as credit cards, loyalty programs, phone data, and in-store tracking. This is done to assist with marketing and demographic predictions, and to automate the creation of order lead-timing with the

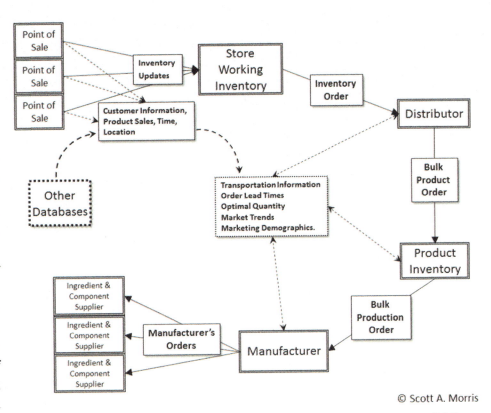

© Scott A. Morris

Figure 9. The information cycle, illustrating how information from point of sale (POS) as well as distribution and transportation sources use machine-readable information to create orders, manage inventory levels, and provide secondary information about customers, marketing trends, and distribution characteristics (Morris, 2011).

ultimate result of reducing store inventory to those items kept on the shelf, which is constantly replenished through various "just in time" systems to meet demand. This type of distribution system is appealing but can be brittle, breaking down in the event of large-scale disruption of the distribution chain unless large-scale contingencies are considered.

The current trend is to glean marketing information from combinations of this type of data and social media metrics. Extended use of informatics in distribution systems may also serve to locate diversion or counterfeiting of product, losses and theft, and other large-scale concerns in both commercial and aid distribution (GS1.org).

Examples

Example 1: Calculation of permeation failure in a package

Problem:
Consider a fried snack chip product that will fail a test for oxidative rancidity under STP when reacting with 1.0×10^{-4} mol of oxygen and working with a polymeric film material that has $\bar{P} = 23.7 \frac{cc \cdot \mu m}{m^2 \cdot atm \cdot day}$ and an exposed area of

0.1 m² at STP. It is assumed that there is no oxygen in the product or package headspace, and that the partial pressure of oxygen, from Dalton's law, is 0.21 atm. The maximum amount of permeant allowed (Q), as determined by product lab tests, is $Q = 1.0 \times 10^{-4}$ mol of oxygen = 2.24 cc at STP. Determine the film thickness necessary at STP to provide a shelf life of 180 days by keeping the oxygen uptake below Q.

Solution:

Solve equation 3 for quantity of permeating gas:

$$\bar{P} = \frac{(\text{quantity of permeating gas})(\text{thickness})}{(\text{membrane area})(\text{time})(\text{partial pressure difference across membrane})} \quad (3)$$

where area = 0.1 m²

$$\bar{P} = 23.7 \frac{\text{cc} \cdot \mu\text{m}}{\text{m}^2 \cdot \text{atm} \cdot \text{day}}$$

$$\Delta P = 0.21 \text{ atm}$$

$$\text{Quantity permeated} = \left(23.7 \frac{\text{cc} \cdot \mu\text{m}}{\text{m}^2 \cdot \text{atm} \cdot \text{day}}\right)(0.1 \text{ m}^2)(0.21 \text{ atm})$$

$$= 0.498 \frac{\text{cc} \cdot \mu\text{m}}{\text{day}}$$

$$\frac{2.24 \text{ cc}}{\left(0.498 \frac{\text{cc} \cdot \mu\text{m}}{\text{day}}\right)} = 4.501 \frac{\text{day}}{\mu\text{m}}$$

For a 6 month (180 day) shelf life,

$$\frac{180 \text{ days}}{\left(4.501 \frac{\text{day}}{\mu\text{m}}\right)} = 39.994 \ \mu\text{m or } 0.040 \text{ mm}$$

Example 2: Calculation of transmission rate (TR) of multi-layer film

Problem:

A composite plastic film with four layers is proposed as a packaging material. To determine its suitability, the overall transmission rate must be determined. The transmission rates, in units of (cc µm m⁻² atm⁻¹day⁻¹), of the individual layers are the following: Film A: 5.0, Film B: 20.0, Film C: 0.05, and Film D: 20.0. What is the overall transmission rate of the film?

Solution:

Calculate transmission rate (TR) using equation 2.

$$TR_{total} = \cfrac{1}{TR_{layer\,1} + TR_{layer\,2} + \ldots + TR_{layer\,n}} \qquad (2)$$

$$TR_T = \cfrac{1}{\cfrac{1}{5} + \cfrac{1}{20} + \cfrac{1}{0.05} + \cfrac{1}{20}} = 0.0493$$

All rates are in $\dfrac{cc \cdot \mu m}{m^2 \cdot atm \cdot day}$.

Example 3: Stress concentration in brittle materials (the case of a glass container)

Problem:
A packaging engineer knows that the stress concentration in a scratch can affect the initiation of a fracture in the materials of a container. In order to add enough additional material in the shock band to help prevent failure, the stress concentration factor must be determined. For a scratch in the sidewall of a glass container, with a depth of 0.01 mm and a crack tip radius of 0.001 mm, what is the stress concentration factor (K)?

Solution:
Calculate K using equation 5:

$$K = \frac{\sigma_{max}}{\sigma_{app}} = 2\left(\frac{d}{r}\right)^{1/2} \qquad (5)$$

or simply

$$K = 2\left(\frac{d}{r}\right)^{1/2}$$

where d = depth of crack = 1.0×10^{-5} m
r = radius of crack tip = 1.0×10^{-6} m

$$= 2\left(\frac{1.0 \times 10^{-6}\,m}{1.0 \times 10^{-5}\,m}\right)^{1/2}$$

$\cong$ 6.32 times the applied stress

Example 4: Identify the type of failure in glass

Problem:
Identify the type of failure experienced by the fractured glass in figure 10.

Solution:
The glass failed from thermal shock as evidenced by the crack traversing the region of transition from very thin sidewall to very thick base, the thickness change at the handle attachment point, and the lack of secondary fragmentation. The thick

Figure 10. Example of fractured glass (© Scott A. Morris).

sections change temperature much more slowly than the thin sidewall, creating shear stress from differential expansion and failure in the material.

Example 5: Q_{10} determination and shelf life estimation

Problem:

A new food product is being introduced, and a 180-day shelf life has been determined to be necessary. Because of the short timeframe for production, years of repeated long-term shelf-life tests are not practical. Spoilage of the food product is determined by testing for discoloration using a color analyzer. Shelf life estimations are conducted at temperatures of 25°C and 35°C for 15 days, and the time for the discoloration criteria to be exceeded is projected from the short-term data to be 180 days at 25°C and 60 days at 35°C. These values are useful for estimating the Q_{10} value for the new product. For a more accurate estimate of the 180-day shelf life when stored at 25°C, an accelerated test at a higher temperature is planned to determine if the product fails or not. Estimate the time required for the complete accelerated test of the 180-day shelf life at 25°C with testing conducted at 45°C.

Solution:

The first step is to calculate Q_{10} using equation 7:

$$Q_{10} = \left(\frac{\text{time for product to spoil at temperature } T_1}{\text{time for product to spoil at temperature } T_2} \right)^{\frac{10}{(T_2 - T_1)}} \quad (7)$$

$$Q_{10} = \left(\frac{180 \text{ days}}{60 \text{ days}} \right)^{\frac{10}{(35° - 25°)}} = 3.0$$

Under the simplest of linear-data circumstances (see cautionary note in text), the product shelf life will decrease by $1/Q_{10}$ for each Q_{10} interval (10°C in this case) increase in storage temperature. Thus, when stored at 45°C, which is two times the Q_{10} interval, the product would have a shelf life of 180 days × (1/3) × (1/3) = 20 days. The test time can also be calculated by using the Q_{10} value of 3.0 to solve equation 7 for the time for the product to spoil at 45°C:

$$3.0 = \left(\frac{180 \text{ days at } 25°C}{X \text{ days at } 45°C} \right)^{\frac{10}{(45 - 25)}}$$

$$X = \frac{180}{9} = 20 \text{ days}$$

This procedure allows the simple-case projected estimation of a 180-day shelf life using only 20 days of exposure at 45°C to estimate shelf life at 25°C. Such accelerated testing allows an approximate estimation of shelf life using increased temperatures and is useful for testing when product formulations or packaging change as well as contributing to ongoing quality control.

Note that errors in measurement or procedure at 45° will be amplified. A 5% error in measurement at 45°C will produce 5% × 180 days = ±9 days of error in the estimated shelf life. Results from accelerated testing are often very simplified, and may produce spurious results or failure from another condition not included in the model. Follow up tests with real-world products is an essential part of validating and correcting deficiencies in the model and is a common practice with many products.

Image Credits

Figure 1. Morris, S. A. (CC By 4.0). (2020). Global Use of Packaging Materials by Type. (Created with data from Packaging Distributors of America, 2016).

Figure 2. Morris, S. A. (CC By 4.0). (2020). Illustration of highly ordered polymer chains in crystalline regions and disordered chains in amorphous regions.

Figure 3. Morris, S. A. (CC By 4.0). (2020). Relationship of crystallinity, molecular weight (which increases with chain length in this example), and physical properties for a typical linear polyolefin (polyethylene shown). ρ is density in g/ml. Morris, S. A. (2011). Food and package engineering. New York, NY: Wiley & Son.

Figure 4. Morris, S. A. (CC By 4.0). (2020). Repeat unit structures of common packaging polymers.

Figure 5. Morris, S. A. (CC By 4.0). (2020). Permeation through packaging film membrane.

Figure 6. Morris, S. A. (CC By 4.0). (2020). Fracture failure in a brittle material.

Figure 7. Morris, S. A. (CC By 4.0). (2020). Illustration of glass failure types and significant indications of failure source.

Figure 8. Morris, S. A. (CC By 4.0). (2020). Packaging cycle showing the material use cycle from raw materials through package manufacturing, filling, distribution, and end-of-life (EOL) disposition. Morris, S. A. (2011). Food and package engineering. New York, NY: Wiley & Son.

Figure 9. Morris, S. A. (CC By 4.0). (2020). The information cycle illustrating how information from the point of sale (POS) as well as distribution and transportation sources use machine-readable information to create orders, manage inventory levels and provide secondary information about customers, marketing trends and distribution characteristics Morris, S. A. (2011). Food and package engineering. New York, NY: Wiley & Son.

Figure 10. Morris, S. A. (CC By 4.0). (2020). Fractured Glass Example.

References

American Peanut Council. (2009). Good manufacturing practices and industry best practices for peanut product manufacturers. Retrieved from https://www.peanutsusa.com/phocadownload/GMPs/2009%20APC%20GMP%20BP%20Chapter%207%20Peanut%20Product%20Manufacturers%2016%20Nov%2009%20Final%20Edit.pdf.

Bodenheimer, G. (2014). Mitigating packaging damage in the supply chain. *Packaging Digest*. Retrieved from https://www.packagingdigest.com/supply-chain/mitigating-packaging-damage-inthe-supply-chain140910.

Bradt, R. C. (2011). The fractography and crack patterns of broken glass. *J. Failure Analysis Prevention, 11*(2), 79–96. https://doi.org/10.1007/s11668-011-9432-5.

Burns, C. S., Heyerick, A., De Keukeleire, D., & Forbes, M. D. (2001). Mechanism for formation of the lightstruck flavor in beer revealed by time-resolved electron paramagnetic resonance.

Chiriboga, M. A., Schotsmans, W. C, Larrigaudière, C., Dupille, E., & Recasens, I. (2011). How to prevent ripening blockage in 1-MCP-treated 'Conference' pears. *J. Sci. Food Agric*, *91*(10), 1781-1788. https://doi.org/10.1002/jsfa.4382.

Cooksey, K., Marsh, K., & Doar, L. H. (1999). Predicting permeability & transmission rate for multilayer materials. *Food Technol.*, *5*(9), 60-63. https://www.ift.org/news-and-publications/food-technology-magazine/issues/1999/september/features/predicting-permeability-and-transmission-rate-for-multilayer-materials.

Dalton, J. (1802). Essay IV. On the expansion of elastic fluids by heat. *Memoirs of the Literary and Philosophical Society of Manchester*, *5*(2), 595-602.

deMan, J. M. (1981). Light-induced destruction of vitamin A in milk. *J. Dairy Sci.*, *64*(10), 2031-2032. https://doi.org/10.3168/jds.S0022-0302(81)82806-8.

EFSA. (2017). BisPhenol A. European Food Safety Authority. Retrieved from https://www.efsa.europa.eu/en/topics/topic/bisphenol.

Fick, A. (1855). Ueber diffusion. *Ann. Physik;* *9*(4), 59-86.

Griffith, A. A. (1921). VI. The phenomena of rupture and flow in solids. *Phil. Trans. Royal Soc. A*, *221*(1 January 1921), 582-593. http://dx.doi.org/10.1098/rsta.1921.0006.

ISO. (2017). ISO 2528:2017: Sheet materials — Determination of water vapour transmission rate (wvtr) — Gravimetric (dish) method. Geneva, Switzerland: International Organization for Standardization. Retrieved from https://www.iso.org/standard/72382.html.

Labuza, T. (2000). The search for shelf life. *Food Testing and Analysis*, *6*(3), 26-36.

Mangaraj, S., Goswami, T. K., & Panda, D. K. (2015). Modeling of gas transmission properties of polymeric films used for MA packaging of fruits. *J. Food Sci. Technol.*, *52*(9), 5456-5469. http://dx.doi.org/10.1007/s13197-014-1682-2.

Mannheim, C., & Passy, N. (1982). Internal corrosion and shelf life of food cans and methods of evaluation. *Crit. Rev. Food Sci. Nutr.*, *17*(4), 371-407. http://dx.doi.org/10.1080/10408398209527354.

Misko, G. G. (2019). The regulation of food packaging. *SciTech Lawyer*, *15*(2). Retrieved from https://www.packaginglaw.com/special-focus/regulation-food-packaging.

Morris, S. A. (2011). *Food and package engineering*. New York, NY: Wiley & Son.

Nakaya, M., Uedono, A., & Hotta, A. (2015). Recent progress in gas barrier thin film coatings on pet bottles in food and beverage applications. *Coatings*, *5*(4), 987-1001. https://doi.org/10.3390/coatings5040987.

National Research Council. (1994). 3. Manufacturing: Materials and processing. In *Polymer science and engineering: The shifting research frontiers* (pp. 65-115). Washington, DC: The National Academies Press. https://doi.org/10.17226/2307.

Neuwirth, B. (2011). Marketing channel strategies in rural emerging markets: Unlocking business potential. Chicago, IL: Kellogg School of Management. Retrieved from www.kellogg.northwestern.edu/research/crti/opportunities/~/media/Files/Research/CRTI/Marketing%20Channel%20Strategy%20in%20Rural%20Emerging%20Markets%20Ben%20Neuwirth.ashx.

Nippon.com. (2018). The Kamikatsu zero waste campaign: How a little town achieved a top recycling rate. Retrieved from https://www.nippon.com/en/guide-to-japan/gu900038/the-kamikatsu-zero-waste-campaign-how-a-little-town-achieved-a-top-recycling-rate.html.

Packaging Distributors of America. (2016). The numbers behind boxes and tape. Retrieved from http://www.pdachain.com/2016/11/30/packaging-statistics-that-might-surprise-you/.

Palmer, B. (2014). Why glass jars aren't necessarily better for the environment than plastic ones. *Washington Post*, June 23, 2014. Retrieved from https://www.washingtonpost.com/national/health-science/why-glass-jars-arent-necessarily-better-for-the-environment-than-plastic-ones/2014/06/23/2deecfd8-f56f-11e3-a606-946fd632f9f1_story.html?noredirect=on&utm_term=.30ac7c6f77.

Pennella, C.R. (2006). Managing contract quality requirements. Milwaukee, WI: American Society for Quality, Quality Press Publishing.

Ragnarsson, J. O., & Labuza, T. P. (1977). Accelerated shelf life testing for oxidative rancidity in foods—A review. *Food Chem., 2*(4), 291-308. https://doi.org/10.1016/0308-8146(77)90047-4.

Rogers, C. E. (1985) Permeation of gases and vapours in polymers. In J. Comyn (Ed.), *Polymer permeability* (pp. 11-73). London, UK: Chapman and Hall.

Sajilata, M. G., Savitha, K., Singhal, R. S., & Kanetkar, V. R. (2007). Scalping of flavors in packaged foods. *Comprehensive Rev. Food Sci. Food Saf., 6*(1), 17-35. https://doi.org/10.1111/j.1541-4337.2007.00014.x.

Sanchez, I. C., & Rogers, P. A (1990). Solubility of gases in polymers. *Pure & Appl. Chem., 62*(11), 2107-2114.

Sigma-Aldrich Inc. (2019). Thermal transitions of homopolymers: Glass transition & melting point. Retrieved from https://www.sigmaaldrich.com/technical-documents/articles/materials-science/polymer-science/thermal-transitions-of-homopolymers.html.

Silvers, K. W., Schneider, M. D., Bobrov, S. B., & Evins, S. E. (2012). PET containers with enhanced thermal properties and process for making same. U.S. Utility Patent No. US9023446B2. Retrieved from https://patents.google.com/patent/US9023446B2/en?oq=US9023446.

Suloff, E. C. (2002). Chapter 4. Permeability, diffusivity, and solubility of gas and solute through polymers. In *Sorption behavior of an aliphatic series of aldehydes in the presence of poly(ethylene terephthalate) blends containing aldehyde scavenging agents*. PhD diss. Blacksburg, VA: Virginia Polytechnic Institute and State University, Department of Food Science and Technology. http://hdl.handle.net/10919/29917.

Thermofisher Inc. (2019). Physical properties table. Brochure D20823. Retrieved from http://tools.thermofisher.com/content/sfs/brochures/D20826.pdf.

USFDA. (2019). Packaging & food contact substances (FCS). United States Food and Drug Administration. Washington, DC: USFDA. Retrieved from https://www.fda.gov/food/food-ingredients-packaging/packaging-food-contact-substances-fcs.

Wu, Y. W. (2016). Investigation of corrosion in canned chicken noodle soup using selected ion flow tube-mass spectrometry (SIFT-MS). MS thesis. Columbus, OH: The Ohio State University, Department of Food Science and Technology. Retrieved from https://etd.ohiolink.edu/apexprod/rws_olink/r/1501/6.

Zweep, C. (2018). Determining product shelf life. *Food Qual. Saf.*, October-November 2018. Retrieved from https://www.foodqualityandsafety.com/article/determining-product-shelf-life/.

Made in the USA
Columbia, SC
27 August 2021